THE MOLECULAR ORIGINS OF LIFE

The origin of life was an event probably unique in the Earth's history, and reconstructing this event is like assembling a puzzle made up of many pieces. These pieces are composed of information acquired from many different disciplines. The aim of this book is to integrate the most recent discoveries in astronomy, planetology, paleontology, biology, and chemistry, and to use this knowledge to present plausible scenarios that give us a better understanding of the likely origin of life on Earth. Twenty-two top experts contribute to chapters that discuss everything from the environment and atmosphere of the early Earth, through the appearance of organic molecules in the prebiotic environment, to primitive chiral systems capable of self-replication and evolution by mutation. The book also discusses various clues to the origin of life that can be obtained from the study of the past and present microbial world, as well as from the planets Titan and Mars.

Chemists, biologists, Earth scientists, and astronomers will find that this book presents a thought-provoking summary of our current state of knowledge of this extraordinary event.

Dr. André Brack, born in 1938 in Strasbourg, France, is Directeur de Recherche at the Centre National de la Recherche Scientifique, Centre de Biophysique Moléculaire, Orléans. He was trained in polymer chemistry at the University of Strasbourg. He received his M.S. in chemistry in 1962 and did his Ph.D. on polypeptide chemistry in 1970.

He started his career at the Centre de Recherche sur les Macromolécules in Strasbourg (1963–68) and moved to the Centre de Biophysique Moléculaire in Orléans in 1968. He spent one year (1975) at the Salk Institute for Biological Studies in San Diego. His field of expertise concerns the chemistry of the origin of life, as well as the search for life in the Solar System.

Dr. Brack is presently president of the International Society for the Study of the Origin of Life and is chairman of the Exobiology Science Team of the European Space Agency. He is also a member of the Comité des Programmes Scientifiques of the Centre National des Études Spatiales (France). He has published over 110 articles in international peer-reviewed scientific journals.

THE MOLECULAR ORIGINS OF LIFE

Assembling pieces of the puzzle

Edited by

ANDRÉ BRACK

PUBLISHED BY THE PRESS SYNDICATE OF THE UNIVERSITY OF CAMBRIDGE
The Pitt Building, Trumpington Street, Cambridge CB2 1RP, United Kingdom

CAMBRIDGE UNIVERSITY PRESS
The Edinburgh Building, Cambridge CB2 2RU, United Kingdom
40 West 20th Street, New York, NY 10011-4211, USA
10 Stamford Road, Oakleigh, Melbourne 3166, Australia

First published 1998

Printed in the United States of America

Typeset in Times Roman

Library of Congress Cataloging-in-Publication Data is available

A catalog record for this book is available from the British Library

ISBN 0 521 56412 3 hardback
ISBN 0 521 56475 1 paperback

Contents

Contributors

ANDRÉ BRACK *Centre de Biophysique Moléculaire–CNRS, Rue Charles Sadron, 45071 Orléans cédex 02, France*

JENS BURMEISTER *Institute for Organic Chemistry, Ruhr-University Bochum, Bochum, Universitätsstr. 150, 44801 Bochum, Germany*

JOHN R. CRONIN *Department of Chemistry and Biochemistry, Arizona State University, Tempe, AZ 85287-1604, U.S.A.*

DAVID W. DEAMER *Department of Chemistry and Biochemistry, University of California, 1156 High Street, Santa Cruz, CA 95064, U.S.A.*

CHRISTIAN DE DUVE *I.C.P., Avenue Hippocrate 75, 1200 Bruxelles, Belgium*

ARMAND H. DELSEMME *The University of Toledo, Toledo, OH 43606, U.S.A.*

JAMES P. FERRIS *Department of Chemistry, Rensselaer Polytechnic Institute, Troy, NY 12180, U.S.A.*

NILS G. HOLM and EVA M. ANDERSSON *Department of Geology and Geochemistry, Stockholm University, S-106 91 Stockholm, Sweden*

KENNETH D. JAMES and ANDREW D. ELLINGTON *Department of Chemistry, Indiana University, Bloomington, IN 47405, U.S.A.*

JAMES F. KASTING and LISA L. BROWN *Department of Geosciences, The Pennsylvania State University, University Park, PA 16802, U.S.A.*

CHRISTOPHER P. McKAY *Space Science Division, N.A.S.A. Ames Research Center, Moffett Field, CA 94035, U.S.A.*

MICHEL MAURETTE *Astrophysique du Solide, C.S.N.S.M., Bât 104, 91405 Orsay-Campus, France*

STANLEY L. MILLER *Department of Chemistry and Biochemistry, University of California, San Diego, La Jolla, CA 92093-0506, U.S.A.*

TOBIAS C. OWEN *University of Hawaii, Institute for Astronomy, 2680 Woodlawn Drive, Honolulu, HI 96822, U.S.A.*

FRANÇOIS RAULIN *L.I.S.A. Faculté des Sciences, Université Paris 7 et 12 – C.N.R.S., 61 Avenue du Général de Gaulle, 94010 Créteil, France*

J. WILLIAM SCHOPF *C.S.E.O.L. Geology Building, University of California, Los Angeles, CA 90095-1567, U.S.A.*

ALAN W. SCHWARTZ *Evolutionary Biology Research Group, Faculty of Science, University of Nijmegen, Toernooiveld, 6525 ED Nijmegen, The Netherlands*

KARL O. STETTER *Lehrstuhl für Mikrobiologie, Universität Regensburg, Universitätsstr. 31, D-93053 Regensburg, Germany*

GÜNTER WÄCHTERSHÄUSER *Tal 29, D-80331 München, Germany*

THE MOLECULAR ORIGINS OF LIFE

Introduction

ANDRÉ BRACK

Humans in every civilization have always been intrigued by their origin and by the question of the origin of life itself. During thousands of years, the comforting theory of spontaneous generation seemed to provide an answer to this enduring question. In ancient China, people thought that aphids were spontaneously generated from bamboos. Sacred documents from India mention the spontaneous formation of flies from dirt and sweat. Babylonian inscriptions indicate that mud from canals was able to generate worms.

For the Greek philosophers, life was inherent to matter; it was eternal and appeared spontaneously whenever the conditions were favorable. These ideas were clearly stated by Thales, Democritus, Epicurus, Lucretius, and even by Plato. Aristotle gathered the different claims into a real theory. This theory safely crossed the Middle Ages and the Renaissance. Famous thinkers like Newton, Descartes, and Bacon supported the idea of spontaneous generation.

The first experimental approach to the question was published in the middle of the 17th century, when the Flemish physician Van Helmont reported the generation of mice from wheat grains and a sweat-stained shirt. He was quite amazed to observe that the mice were identical to those obtained by procreation. A controversy arose in 1668 when Redi, a Tuscan physician, published a set of experiments demonstrating that maggots did not appear when putrefying meat was protected from flies by a thin muslin covering.

Six years after Redi's treatise, the Dutch scientist Anton Van Leeuwenhoek observed microorganisms for the first time through a microscope of his own making. From then on, microorganisms were found everywhere and the supporters of spontaneous generation took refuge in the microbial world. However, Van Leeuwenhoek was already convinced that the presence of microbes in his solutions was the result of contamination by ambient air. In 1718, his disciple Louis Joblot demonstrated that the microorganisms observed in solutions were, indeed, brought in from the ambient air, but he could not convince the naturalists.

Even Buffon, in the middle of the 18th century, thought that Nature was full of germs of life able to scatter during putrefaction and to gather again, later on, to reconstitute microbes. His Welsh friend John Needham did many experiments to support this view. He heated organic substances in water in a sealed flask in order to sterilize the solutions. After a while, all solutions showed a profusion of microbes. The Italian priest Lazzaro Spallanzani argued that sterilization was incomplete. He heated the solutions to a higher temperature and killed all the microbes, but could not kill the idea of microbial spontaneous generation.

The controversy reached its apotheosis one century later, when Felix Pouchet published his treatise in 1860. He documented the theory of spontaneous generation in the light of experiments that, in fact, were the results of contamination by ambient air. Pasteur gave the finishing blow to spontaneous generation when he designed a rigorous experimental setup for sterilization.

The beautiful demonstration of Pasteur raised the fascinating question of the historical origin of life. Since life can originate only from preexisting life, it has a history and therefore an origin, which must be understood and explained by chemists.

It is always difficult to define what is meant by the word *life*. This is the first difficulty encountered by scientists who try to reconstruct the birth of life on Earth. The second difficulty is related to the factor time. Because of time flow and evolution, primitive life was different from life today and only hypothetical descriptions of primitive life can be proposed. Because of the limitations of time, prebiotic chemistry can never be repeated in the laboratory. Therefore, simulations may only represent possible supports for plausible hypotheses from the point of view of historical legitimacy. A way around these difficulties is to collect clues from different disciplines. Today, these clues are like pieces of a puzzle that we can begin to assemble. The purpose of this book is to integrate the most recent discoveries in astronomy, planetology, geology, paleontology, biology, and chemistry into plausible scenarios for a better understanding of the origins of life.

The minimal requirements for primitive life can be tentatively deduced from the following definition: Primitive life is an aqueous chemical system able to transfer its molecular information and to evolve. The concept of evolution implies that the system transfers its molecular information fairly faithfully but makes occasional accidental errors.

Since most of Earth's early geological history has been erased by later events, we remain ignorant of the true historical facts on Earth from the time when life started (Chapters 1 and 2). We must, therefore, imagine simple informative molecules as well as the machinery for information transfer. Both functions must then be tested in the laboratory.

Water molecules are widespread in the universe as grains of solid ice or as very dilute water vapor. Liquid water is a fleeting substance that can persist only above 0 °C and under an atmospheric pressure higher than 6 mbars. Therefore, the size of a planet and its distance from the star are two basic characteristics that will determine the presence of liquid water. If a body is too small, like Mercury or the Moon, it will not be able to retain any atmosphere or, therefore, liquid water. If the planet is too close to the star, the mean temperature rises due to starlight intensity. Any seawater present will evaporate, delivering large amounts of water vapor to the atmosphere and thus contributing to the greenhouse effect, which causes a further temperature rise. Such a positive feedback loop could lead to a runaway greenhouse: All of the surface water would be transferred to the upper atmosphere, where photodissociation by ultraviolet light would break the molecules into hydrogen and oxygen. Loss of the atmosphere would result from the escape of hydrogen to space and the combination of oxygen with the crust. A planet that is far from the star may permit the existence of liquid water, provided that the planet can maintain a constant greenhouse atmosphere. However, water could provoke its own disappearance. The atmospheric greenhouse gas CO_2, for instance, would be dissolved in the oceans and finally trapped as insoluble carbonates by rock weathering. This negative feedback could lower the surface pressure and, consequently, the temperature to such an extent that water would be largely frozen. The size of Earth and its distance from the Sun are such that the planet never experienced either a runaway greenhouse or a divergent glaciation.

Present-day life is based on organic molecules made of carbon, hydrogen, oxygen, nitrogen, sulphur, and phosphorus atoms. However, primitive chemical information may have been stored in mineral crystals. This idea has been developed by Cairns-Smith at the University of Glasgow. According to this theory, some clays offered a potential repository for genetic information in the intrinsic pattern of crystal defects. These clays proliferated and their replicating defects became more abundant. Certain lines developed photochemical machinery producing nonclay species such as polyphosphates and small organic molecules. Natural selection favored these lines because the newly formed organic molecules modified the clay's properties in favorable ways, thus catalyzing synthesis of such lines. Organic polymers of specified monomer composition appeared but, at first, served only structural roles. Based-paired polynucleotides replicated, giving rise to a minor genetic material. This material proved to be useful in the alignment of amino acids for polymerization. Finally, the clay machinery was dispensed with in favor of a polynucleotide-based replication; a "genetic takeover" whereby nucleic acids became the genetic libraries. Unfortunately, since 1966 when Cairns-Smith began to advocate his interesting ideas, practically no experimental evidence has been provided to support such a primitive mineral genetic material.

It is generally believed, through analogy with contemporary life, that primitive life originated from the processing of organic molecules by liquid water.

Alexander Oparin, in 1924, suggested that the small, reduced organic molecules needed for primitive life were formed in a primitive atmosphere dominated by methane. The idea was tested in the laboratory by Stanley Miller (Chapter 3). He exposed a mixture of methane, ammonia, hydrogen, and water to electric discharges. Among the compounds formed, he identified 4 of the 20 naturally occurring amino acids, the building blocks of proteins. Since this historical experiment, 17 natural amino acids have been obtained via the intermediary formation of simple precursors such as hydrogen cyanide and formaldehyde. Spark discharge synthesis of amino acids occurs efficiently when a reducing gas mixture containing significant amounts of hydrogen is used. However, the true composition of primitive Earth's atmosphere remains unknown. Today, geochemists favor a nonreducing atmosphere dominated by carbon dioxide. Under such conditions, the production of amino acids appears to be very limited.

More recently, Günter Wächtershäuser suggested that life started from carbon dioxide. The energy source required to reduce carbon dioxide was provided by the oxidative formation of pyrite (FeS_2) from iron sulfide (FeS) and hydrogen sulfide. Pyrite has positive surface charges and binds the products of carbon dioxide reduction, giving rise to a surface chemistry (developed in Chapter 9). Experiments are presently being run to test this new hypothesis.

Deep-sea hydrothermal systems may also represent likely environments for the synthesis of prebiotic organic molecules (Chapter 4). Experiments have been carried out to test whether amino acids can be formed under conditions simulating hydrothermally altered oceanic crust.

Organic chemistry is universal. So far, more than 50 different organic molecules have been identified by radioastronomy in dense clouds of the interstellar medium. Synthesis of these molecules is thought to be initiated by collisions of high-energy cosmic ray particles with hydrogen and helium or by photochemical processes. The most important compounds for prebiotic chemistry that have been identified are probably hydrogen cyanide, formaldehyde, acetylene, and acetonitrile.

Comets also show substantial amounts of organic material. According to Delsemme's analysis, Comet Halley contains 14% organic carbon by mass. About 30% of cometary grains are dominated by the light elements C, H, O, and N, and 35% are close in composition to carbonaceous chondrites. Among the molecules identified in comets are hydrogen cyanide and formaldehyde.

Comets, therefore, may have been an important source of organics delivered to the primitive Earth (Chapter 5).

The study of meteorites, particularly the carbonaceous chondrites that contain up to 5% by weight of organic matter, has allowed close examination of extraterrestrial organic material (Chapter 6). Eight proteinaceous amino acids have been identified in the Murchison meteorite among more than 70 amino acids. Engel reported that L-alanine was more abundant that D-alanine in the Murchison meteorite (see Chapter 6). This rather surprising result has been recently confirmed by Cronin. The latter found a racemic composition (equal mixture of L and D enantiomers) for norvaline and α-amino-n-butyric acid, which can racemize by abstraction of the C_α hydrogen atom. More interestingly, Cronin found enantiomeric excesses of about 10% for isovaline, α-methyl norvaline, and α-methyl isoleucine, which cannot racemize by proton abstraction. The enantiomeric excesses found in the Murchison meteorite may help us understand the emergence of a primitive homochiral life.

Homochirality of present-day life is believed by many researchers to be not just a consequence of life but also a prerequisite for life. Present terrestrial life is dominated by proteins, which catalyze biochemical reactions; nucleic acids, which carry genetic information; and a lipidic micellar system, which forms the cellular protecting membranes. Most of the constituents, that is amino acids, sugars, and lipids, contain at least one asymmetric carbon atom.

The 19 chiral proteinaceous amino acids belong without exception to the L-configuration class, whereas the two sugars found in nucleic acids are related to the D-series. The biopolymers themselves form asymmetric helical structures and superstructures, the α- and β-conformations of polypeptides, the A-, B-, and Z-forms of nucleic acids, and the helical conformations of polysaccharides.

Any chemical reaction producing chiral molecules in statistically large numbers that is run in a symmetrical environment yields a racemic mixture, that is, a mixture of equal quantities of right- and left-handed enantiomers. However, in view of the importance of optical purity in present-day life, it is difficult to believe that, at the beginning, a completely racemic life form arose using biomolecules of both configurations simultaneously in the same protocell.

Theoretical models of the origin of chirality on Earth can be divided into two classes, those which call for a chance mechanism and those which call for a determinate mechanism resulting from an asymmetrical environment originating from the universe or from the Earth.

The proponents of the chance mechanism argue that the notion of equimolarity of a racemic mixture is only relative. For a relatively small set of molecules, random fluctuations may favor one enantiomer over the other. Both theoretical and experimental models are available. In a rather simple kinetic model initially proposed by Franck, an open flow reactor, run in far-from-equilibrium conditions, is fed by achiral compounds and forms two enantiomers reversibly and autocatalytically. If the two enantiomers can react to form an irreversible combination flowing out of the reactor, by precipitation for instance, and if certain conditions of fluxes and concentrations are reached, the racemic production may become metastable and the system may switch permanently toward the production of either one or the other enantiomer, depending on a small excess in one enantiomer.

Spontaneous resolution of enantiomers via crystallization represents the most effective means of breaking chiral symmetry since optically pure enantiomers can be isolated on scales ranging from grams to tons with the help of homochiral seeds. Did terrestrial quartz, known to be an asymmetric adsorbent and catalyst, undergo such a spontaneous resolution during crystallization? Close examination of over 27,000 natural quartz crystals gave 49.83% L and 50.17% D.

Parity nonconservation has raised many hopes among those who call for a determinate mechanism. This fundamental asymmetry of matter has been examined from various aspects. For example, circularly polarized photons emitted by the slowing down of longitudinally polarized electrons might be capable of inducing degradation reactions or stereoselective crystallization of racemic mixtures. No experiment has convincingly supported these theoretical considerations for the origin of a dominant enantiomer on Earth. Either the results were shown to be artifacts or to be so weak that they are doubtful.

Parity is also violated in the weak interactions mediated by neutral bosons Z^0. All electrons are intrinsically left-handed (their momentum and spin are more likely to be antiparallel). The antimatter counterpart, the positrons, are intrinsically right-handed. Therefore, L-serine, made of left-handed electrons, and D-serine, also made of left-handed electrons, are not true enantiomers but diastereoisomers. There is a very tiny, parity-violating energy difference in favor of L-amino acids in their preferred conformations in water and in favor of D-sugars. The energy difference is about 3.10^{-19} eV, corresponding to one part in 10^{17}, for the excess of L-molecules in a racemic mixture at thermodynamic equilibrium at ambient temperature.

Other chiral force fields that could have been acting on the Earth's surface have been researched: Asymmetric synthesis and degradation have been achieved with circularly polarized light, and an original approach to enantiomer resolution – using Earth gravity and a macroscopic vortex – has been tested.

Unfortunately, the classical electromagnetic interactions, such as circularly polarized light or other fields that can be imagined acting on Earth, would probably never result in a very high yield of optically pure compounds. Such asymmetric force fields would also probably cancel their effects on a time and space average. The chiral (one handed) amino acids found in the Murchison meteorite push the problem of the origin of biological chirality out into the cosmos. Among the possible extraterrestrial sources of circularly polarized light, sunlight is generally discounted as probably being too weak and not of a consistent handedness for sufficiently long periods. According to Bonner and Rubenstein (1987), synchrotron radiation from the neutron star remnants of supernova events is a better candidate. Interaction of neutron star circularly polarized light with interstellar grains in dense clouds could produce chiral molecules in the organic mantles by partial asymmetric photolysis of mirror-image molecules.

Micrometeorites (Chapter 7), also referred to as cosmic dust or interplanetary dust particles, have been extracted both from black sediments collected from the melt zone of the Greenland ice cap and directly from Antarctic old blue ice. In the 50–100-μm size range, a constant high percentage of 80% of unmelted chondritic micrometeorites has been observed, indicating that many particles cross the terrestrial atmosphere without drastic thermal alteration. In this size range, the carbonaceous micrometeorites represent 80% of the samples and contain 2% carbon; they might have brought about 10^{20} g of carbon over a period of 300 million years, corresponding to the late terrestrial bombardment phase, assuming that the flux was 1,000 times more intense than today. This delivery represents more carbon than that now present in the biosphere (about 10^{18} g).

It is generally believed, based on analogy with contemporary living systems, that primitive life emerged as a cell, thus requiring, at the least, boundary molecules able to isolate the system from the aqueous environment (membrane), catalytic molecules providing the basic chemical work of the cell (enzymes), and informative molecules allowing the storage and the transfer of the information needed for replication (RNA).

Fatty acids are known to form vesicles when the hydrocarbon chains contain more than 10 carbon atoms. Such vesicle-forming fatty acids have been identified in the Murchison meteorite. However, the membranes obtained with these simple amphiphiles are not stable over a broad range of conditions. Stable neutral lipids can be obtained by condensing fatty acids with glycerol or with glycerol phosphate, thus mimicking the stable contemporary phospholipid. Primitive membranes could initially have also been formed by simple terpenoids (Chapter 8).

Wächtershäuser (Chapter 9) suggests that the carbon source for life was carbon dioxide. He proposes that the energy source required to reduce carbon dioxide was provided by the oxidative formation of pyrite from iron sulfide and hydrogen sulfide. An attractive point in this hypothesis is that pyrite has positive surface charges and binds the products of carbon dioxide reduction, giving rise to a two-dimensional reaction system, a surface metabolism that would be more efficient than a system in which the products can freely diffuse away.

Most of the chemical reactions in a living cell are catalyzed by proteinaceous enzymes. Proteins are built up from 20 different amino acids. Each amino acid, with the exception of glycine, exists in two enantiomeric forms, L and D, although only L-amino acids occur in proteins. Proteins adopt asymmetrical rigid geometries, α-helices and β-sheets, which play a key role in the catalytic activity. According to de Duve, the first peptides appeared via thioesters and began to develop their catalytic properties in a thioester world (Chapter 10).

In contemporary living systems, the hereditary memory is stored in nucleic acids built up with purine and pyrimidine bases, sugars, and phosphate groups. Nonenzymatic replication has been demonstrated by Orgel and his coworkers at the Salk Institute in California. The preformed chains align the nucleotides by base-pairing to form helical structures that bring the reacting groups into close proximity (Chapter 11). However, the prebiotic synthesis of nucleotides remains an unsolved challenge. Although purines and pyrimidines have been obtained in model syntheses, the formation of nucleosides is a difficult problem. Condensation of formaldehyde leads to ribose, among a large number of other sugars. The synthesis of purine nucleosides, the covalent combination of purine and ribose, has been achieved by heating the two components in the solid state, but the yields are very low. No successful preparation of a pyrimidine nucleoside has been reported. Nucleoside phosphorylation is possible by thermal activation but without any regioselectivity. Chemists are now considering the possibility that early living systems used simpler informative molecules. For example, ribose has been replaced by glycerol in the backbone of one such candidate. In the light of the first experiments, however, the chemistry of these simplified informative molecules does not appear to offer any advantages. The straightforward and selective formation of ribose diphosphates from glycolaldehyde phosphate suggests that pyranosyl-RNA may have been involved in the early forms of life. Intense experimental work quoted in Chapter 11 is presently being conducted in Eschenmoser's laboratory along this new avenue.

Even if clays did not participate in the early stages of life as informative molecules, they probably played a key role as catalysts. Efficient clay-catalyzed condensation of nucleotides into oligomers has been recently reported (Chapter 12). On the other hand, Cech at the University of Colorado found that self-splicing and maturation of some introns do not require the help of any peptidic enzymes. Fragments of introns markedly increase the rate of hydrolysis of oligoribonucleotides. They also can act as polymerization templates since chains up to the 30-mer can be obtained starting from a pentanucleotide. The catalytic spectrum of these ribozymes has been considerably enlarged by directed molecular evolution (Chapter 13). RNAs have been shown to be able to act simultaneously as informative and catalytic molecules. They are often viewed as the first living systems on the primitive Earth (RNA world).

However, since the accumulation of nucleotides under prebiotic conditions seems unlikely, many chemists are now tempted to consider that primitive life was supported by simpler informative molecules, and great efforts are devoted to autocatalytic systems (Chapter 14) including simple organic molecules and micelles.

To what extent can the nature of primitive life be illuminated by back-extrapolation of present-day life and particularly by biopolymer phylogeny? Firstly, a good picture of the common microbial ancestor would help to understand the different steps of evolution of early life. Recent advances in molecular phylogeny suggest that the first microorganisms were hyperthermophilic prokaryotes (Chapter 15), but arguments have also been provided suggesting that prokaryotes and eukaryotes emerged simultaneously from the same mesophilic common ancestor.

The geological record obviously provides important information. The isotopic signatures of the organic carbon of the Greenland metasediments bring indirect evidence that life may be 3.8 billion years old. The isotopic signatures are fully consistent with a biological origin and the remarkable diversity of the microflora discovered by Schopf (Chapter 16). Eleven species of cellularly preserved filamentous microbes, comprising the oldest diverse microbial assemblage now known in the geologic record, have been discovered in shallow water cherts interbedded with lava flows of the Early Archean Apex Basalt of northwestern Western Australia. This prokaryotic assemblage establishes that trichomic cyanobacterium-like microorganisms were extant and were both morphologically and taxonomically diverse at least as early as ~3.465 million years ago, thus suggesting that oxygen-producing photoautotrophy may have already evolved by this early stage in biotic history. The existence of the Apex microfossils demonstrates that the paleobiologically neglected Archean rock record is a fruitful source of direct evidence regarding the earliest history of life.

Unfortunately, the direct clues that may help chemists to understand the emergence of life on Earth about 4 billion years ago have been erased by the Earth's turbulent geological history, the permanent presence of liquid water, and by life itself when it conquered the whole planet. We remain ignorant of the true historical facts on Earth from the time when life started. Titan, the largest satellite of Saturn, offers a nice control of a planetary laboratory probably not modified by "living" systems. Titan's organic chemistry is believed to have remained almost unchanged over geological periods. Its active atmospheric chemistry will be studied by the Huygens probe of the Cassini mission (Chapter 17).

The early histories of Mars and Earth clearly show similarities. Geological observations collected from Martian orbiters suggest that liquid water was once stable on the surface of Mars, attesting to the presence of an atmosphere capable of decelerating C-rich micrometeorites. Therefore, primitive life may have developed on Mars as well. Liquid water seems to have disappeared from the surface of Mars about 3.8 Ga ago. The Viking missions did not find any organic molecules or clear-cut evidence for microbial activities at the surface of the Martian soil. These experiments do not exclude the existence of organic molecules and fossils of microorganisms that may have developed on early Mars before liquid water disappeared. The Martian subsurface perhaps keeps a frozen record of the early evolution of life (Chapter 18).

References

Frank, F. C. 1953. On spontaneous asymmetric synthesis. *Biochim. Biophys. Acta* 11:459–463.

Bonner, W. A., and Rubenstein, E. 1987. Supernovae, neutron stars and biomolecular chirality. *BioSystems* 20:99–111.

Part I

Setting the stage

1

The origin of the atmosphere

TOBIAS C. OWEN
Institute for Astronomy
University of Hawaii, Honolulu

1. Introduction

The origin and early evolution of our atmosphere is a subject that has been repeatedly explored during the last four decades, without a definitive conclusion having yet been reached. Recent reviews have been published by Hunten (1993), who included the other terrestrial planets, and Kasting (1993), who focused on the Earth. The volume edited by Atreya, Pollack, and Matthews (1989) contains a number of relevant chapters by specialists in several fields, with a comprehensive bibliography.

The fundamental problem can be divided into a search for **sources** of the volatiles that ultimately became the oceans and the atmosphere, and an evaluation of the **processes** that could have changed the initial composition to produce the mixture we observe today. This chapter will concentrate on the sources. We will adopt a comparative approach, using new information about Mars and comets to put the terrestrial atmosphere in a larger context. The publication of possible evidence for ancient life on Mars (McKay et al. 1996) makes the inclusion of this planet particularly relevant. The following chapter by Kasting will examine some of the processes involved in subsequent atmospheric evolution.

Ever since Brown (1952) emphasized the depletion and fractionation of atmospheric noble gases, it has been clear that our atmosphere is not composed simply of captured solar nebula gas. Ideas about the origin of the oceans and the atmosphere have therefore focused on two principal sources: the rocks that formed the bulk of the planet's mass (the internal reservoir), plus a late-accreting veneer of material that originated well outside the Earth's orbit (the external reservoir).

The formation of the internal reservoir is a natural consequence of the accretion of the planet itself. This process has been studied extensively by Wetherill (1990), who concluded that collisions with planetary embryos on eccentric orbits should have thoroughly homogenized the rocky composition of Venus, Earth, and Mars. The late-accreting veneer contributed by some

Table1.1. *Atmospheres of inner planets**

PLANET	TOTAL (bars)	CO_2 (%)	N_2 (%)	Ne (ppm)	Ar	^{36}Ar	Kr	Xe	H_2O
VENUS	92	96.5	3.5	7	70	35	0.05	<0.04	30 to 200
EARTH	1.013	0.033	78	18.2	9,340	31	1.14	0.087	≤ 3%
⊕**	70	98	1.5	0.4	190	0.6			
MARS	0.006	95.3	2.7	2.5	16,000	5	0.3	0.08	≤100

*After Hunten (1993).

⊕** represents the total surficial volatile inventory on Earth, from Anders and Owen (1977).

combination of volatile-rich meteorites and comets should also have been uniform from planet to planet. Thus we might expect that we can use knowledge about the volatiles on one planet to help us understand the atmospheres of the others. As we shall see, however, in practice this expectation is not as rigorously satisfied as one might like. For example, the oxygen isotope abundances in the SNC meteorites (Shergottites, Nakhlites, Chassignites, all of which originated on Mars) are distinctly different from those in terrestrial rocks (Clayton and Mayeda 1983), so these two planets do not in fact have identical compositions.

The need for a volatile-rich veneer is not universally accepted (e.g., Lewis and Prinn 1984). However, the widely held hypothesis that the Earth's moon was formed by a giant impact between the Earth and a Mars-size planetary embryo (e.g., Stevenson 1987) appears to require such a veneer to replace any atmosphere produced by the planet's early accretion (Ahrens, O'Keefe and Lange 1989). Furthermore, the possibility that the early atmosphere underwent massive hydrodynamic loss would also require the subsequent replacement of at least some of the missing volatiles (Hunten, Pepin and Walker 1987; Zahnle, Kasting and Pollack 1990; Pepin 1991).

The same terminal bombardment that caused the impact craters on the surface of every solid body in the Solar System from 4.5 to 3.8 AE provides an obvious delivery system. It is then a question of establishing the mass and composition of the objects that carried out this bombardment to see if they could have supplied the proper proportions of the elements we need.

For the purposes of this book, the volatiles we are most interested in are water and compounds of nitrogen and carbon (Table 1.1). We can safely assume that if we bring in those molecules, then the phosphorous, calcium, and other elements essential to life will be delivered as well. For obvious reasons, it is convenient to use the chemically inert, heavy noble gases – neon, argon, krypton, and xenon – and their isotopes as tracers. As we will see, these elements place illuminating constraints on models for volatile delivery.

This review takes the position that meteorites alone or in combination with planetary rocks could not have produced the Earth's entire volatile inventory. A significant contribution from icy planetesimals is required. Lacking direct data on the composition of such planetesimals, A. Bar-Nun and I have relied on laboratory experiments that demonstrate the ability of amorphous ice forming at temperatures below 100 K to trap ambient gases. We have combined these data with new information on the composition of the interstellar medium, comets, and planetary atmospheres to develop an "icy impact" model for the formation of inner planet atmospheres (Owen and Bar-Nun 1995a, b). This approach leads to the suggestion that icy planetesimals (comets) must have played a major role in forming the atmospheres of all the planets in the Solar System. As ice is the commonest solid in the universe, we could then argue that at least the essential biogenic elements should have been delivered by icy planetesimals to planets everywhere.

We begin with a discussion of the shortcomings of a meteoritic source of volatiles and then explain why comets may offer a better alternative. In Section 4, we use the Martian atmosphere to test this idea, because the SNC meteorites from Mars provide a rich source of data on both atmospheric gases and atmosphere–surface interactions. We then discuss the origin of water on Mars and the Earth, as water is a sine qua non for life as we know it. Section 6 focuses on neon; this gas appears to give us a record of events on Earth prior to delivery of the veneer from which most of the volatiles are derived. The final section includes a description of tests of the hypothesis that icy impacts played such a key role in delivering the volatiles. Many of these tests should be carried out during the next few years.

2. Problems with meteorites

It has been customary to consider the late-accreting veneer as composed of materials similar to the volatile-rich meteorites classified as carbonaceous chondrites. This is the heterogeneous accretion model for the formation of the Earth (Anders 1968; Turekian and Clark 1969). The latter authors also used this model to make a comparative study of the atmospheres of Earth and Venus (Turekian and Clark 1975). It was further developed by Anders and Owen (1977) for application to the newly obtained data on the atmosphere of Mars. The most recent elaboration of this idea has been the comprehensive geochemical study of Dreibus and Wänke (1987, 1989).

Meteorites have dominated people's thinking about the volatile-rich veneer because these rocks have been so well studied in the laboratory and because we have so many contemporary examples of their impacts with the Earth. It seems hard to avoid a contribution to the Earth from meteorites. On the other hand, there have always been some concerns about this approach. For example, Wetherill and Chapman (1988) have pointed out that the meteorites we see today are almost certainly fragments produced by relatively recent collisions within the asteroid belt. Perhaps the rocks that make up the bulk of the Earth are not represented in our meteorite collections or even in the asteroid belt – they are all in the planet! Perhaps the carbonaceous chondrites were not impacting the Earth in significant numbers during the late heavy bombardment.

This argument is difficult to prove, but there is other, empirical, evidence in the form of the composition of our atmosphere that strongly indicates the inadequacy of an exclusively meteoritic source. This is the issue of atmos-

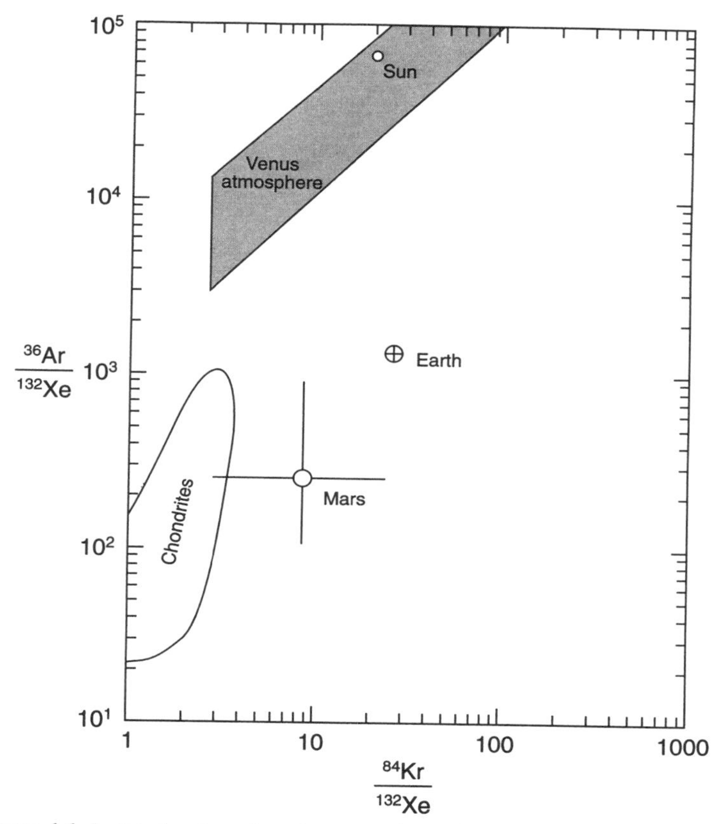

Figure 1.1. In this log–log plot, the proportions of noble gases found in ordinary and carbonaceous chondrites cluster at the lower left, whereas the proportions found in planetary atmospheres are dramatically different. This difference results from the excess xenon in the meteorites. As this gas has not yet been measured on Venus, only the limits corresponding to the shaded trapezoid at the top of the diagram can be set.

pheric xenon. Neither the relative abundance nor the isotopic makeup of this gas matches what we find in any meteorites except those that come from Mars. The ordinary and carbonaceous chondrites exhibit a ratio of $^{132}Xe/^{84}Kr$ that is typically over 10 times the value found in our atmosphere. It has been customary to assume that this "missing xenon" was hidden somewhere on the crust of the planet in shales or in ice. However, careful searches have failed to find it (Wacker and Anders 1984; Bernatowicz, Kennedy and Podosek 1985). Evidently it simply isn't here. This same discrepancy exists on Mars (Figure 1.1).

Similarly, the pattern of isotopic abundances in atmospheric xenon is distinctly different from that in the chondrites or in the solar wind. Once again, there is a pronounced similarity with the xenon found in the atmosphere of

Mars (except for ^{129}Xe) as exemplified by gas trapped in the glassy inclusions of the Shergottite EETA 79001 (Pepin 1989).

The xenon problem doesn't rule out *any* contribution from meteorites; it simply says that they could not be the *only* source of the Earth's volatiles. A firm upper limit on the size of the meteoritic contribution can be estimated by assuming that meteorites gave us all of the atmospheric xenon we see today. (This cannot be true because of the difference in isotopes previously mentioned, hence our certainty that it is an upper limit.) Using Heymann's (1971) value of 5×10^{-9} cc/gm for ^{132}Xe in type 1 carbonaceous meteorites, we find an upper limit of 2×10^{25} gm for the meteoritic contribution. This may be compared to the range of $1–4 \times 10^{25}$ gm calculated by Chyba, Owen and Ip (1994) for the mass delivered to the Earth in the late heavy bombardment, derived from an analysis of the lunar cratering record. This is a reassuring consistency, but it still leaves us a factor 10 short of the krypton and argon that we need. Any source that brings in these gases will also bring in xenon, which further lowers the upper limit on the meteoritic contribution (Owen and Bar-Nun 1995a).

3. A cometary solution

Although some meteorites inevitably struck the Earth, our planet must also have been hit by a large number of icy planetesimals. Oró (1961) suggested that comets may have been an important source of organic matter on the early Earth. Sill and Wilkening (1978) proposed that comets could have delivered substantial amounts of carbon and nitrogen to the Earth if these elements formed gases that were trapped in clathrate hydrates in the icy comet nuclei. Sill and Wilkening (1978) used Wetherill's (1975) early study of the possible cometary contribution to the cratering record on the Moon to buttress their argument for cometary impact. Ip and Fernandez (1988) studied cometary bombardment of inner planets in more detail and concluded that $6 \times 10^{24}–6 \times 10^{25}$ g of cometary material could have been delivered to the Earth by scattering of comets out of the Uranus–Neptune region at the time of the formation of the great Oort Cloud of comets that surrounds the Solar System. This is equivalent to 4–40 times the present mass of the oceans, assuming ~50% of the cometary mass is ice.

Thus the potential supply of cometary material was more than adequate. What about the xenon problem? Unfortunately, no noble gases have been detected in comets yet. We must therefore turn to laboratory experiments on the ability of ice condensing at low temperatures to trap gases in order to see what proportion of xenon/krypton comets might deliver. An extensive series of such experiments has been carried out by Akiva Bar-Nun and his colleagues (Bar-Nun et al. 1985; Laufer, Kochavi and Bar-Nun 1987; Bar-Nun, Kleinfeld and Kochavi 1988; Owen, Bar-Nun and Kleinfeld 1991). These experiments demonstrate that ice forming at 30 K in the presence of a solar mixture traps argon, krypton, and xenon in their original proportions. As the formation temperature increases, however, the trapped gases exhibit an increasing degree of fractionation, dominated by a steady decrease in the relative amount of argon that is sequestered in the ice. There is an accompanying decrease in the total amount of gas the ice can trap (Owen, Bar-Nun and Kleinfeld 1991). However, the ratio of Xe/Kr in the gas mixture trapped in the ice remains close to the value found in the terrestrial atmosphere for a range of ice condensation temperatures from 30 to 75 K. It never exhibits the low value found in the chondritic meteorites (Figure 1.2). Hence, if the laboratory experiments duplicate the conditions under which ice formed in the outer solar nebula, it appears that early bombardment of the Earth by icy planetesimals could indeed solve the missing xenon problem.

How realistic is the laboratory setting? Lunine et al. (1991) have investigated the formation of ice when the original fragment of an interstellar cloud collapsed to form the solar nebula. These authors concluded that the interstellar ice particles would first sublime and then recondense on refractory grains that had assumed the local equilibrium temperature. This process is very close to the laboratory experiments, in which ice is condensed on a cold plate at a controlled temperature in the presence of a known mixture of gases. Thus we may have some confidence that the results obtained in the laboratory should apply to the natural setting.

The relative abundances of CO, Ar, N_2, and CH_4 that are trapped in ice forming at different temperatures vary dramatically (Bar-Nun, Kleinfeld and Kochavi 1988). These experiments should therefore allow us to predict the volatile content of cometary ices formed at different distances from the sun. We have shown that observed abundances of ions of N_2 and CO in comets are consistent with our model's predictions for the incorporation of these gases in ice forming in the Uranus–Neptune region, suggesting that the prediction for the incorporation of noble gases will also be valid (Owen and Bar-Nun 1995a; Notesco and Bar-Nun 1996). Notesco and Bar-Nun (1997) have performed a similar analysis for CH_4 and C_2H_6, successfully accounting for the proportions of these two gases found in Comet Hyakutake by Mumma et al. (1996).

The detection of both HDO and DCN in Comet Hale–Bopp (Meier, Owen, Matthews et al. 1998; Meier, Owen, Jewitt et al. 1998) supports our central hypothesis that interstellar abundances are well preserved through the process of comet formation in the solar nebula. The ratios of D/H in H_2O and HCN are distinctly different from each other with an ~7 times higher value in HCN, just what is observed in interstellar clouds. If the comets had formed from

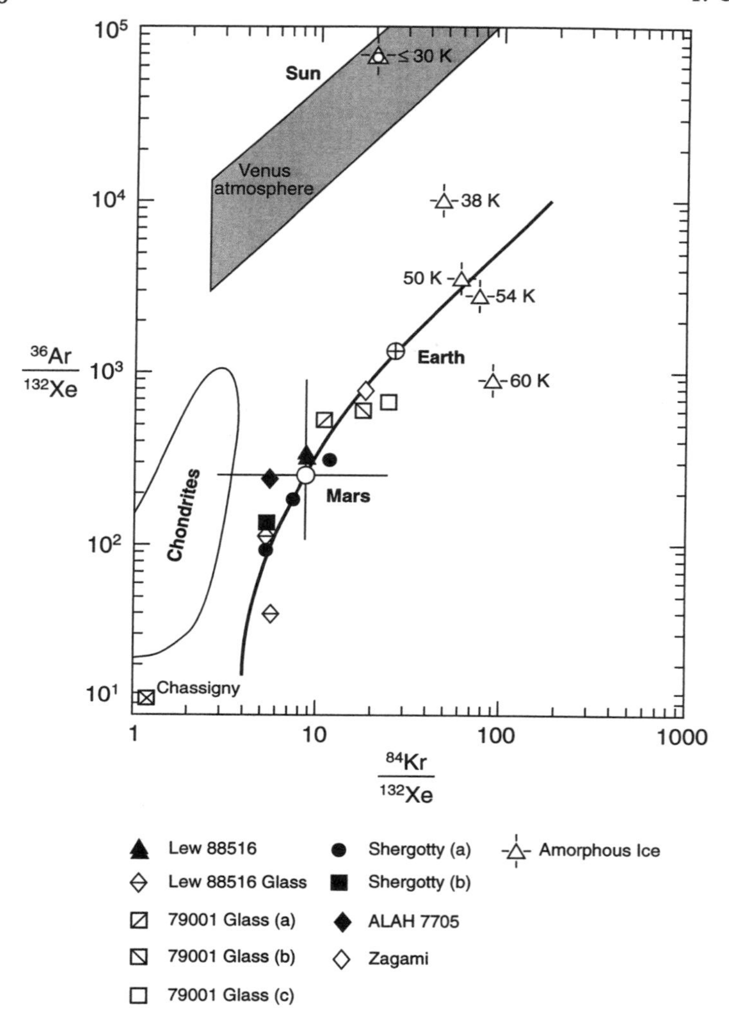

Figure 1.2. A simple mixing line connecting the atmospheres of Mars and Earth appears curved on this log–log plot. The Shergottites exhibit proportions of noble gases that extend along this line, whereas the gases trapped in ice formed at low temperatures in the laboratory intersect an extrapolation of the mixing line beyond the terrestrial point. The "external reservoir" is represented by this intersection; the "internal reservoir" occupies the lower left corner of the diagram.

solar nebula gas that was warmed and homogenized by radial mixing, the values of D/H in both molecules would be much lower and identical.

Davies et al. (1997) have presented near-IR spectra of Hale–Bopp at 7 AU from the sun that show the absorption features of water ice for the first time in any comet. These authors argue that the absence of absorption at

1.65 μm indicates that the ice is amorphous, consistent with the laboratory experiments.

Taken together, these various results suggest that the laboratory work provides a good model for the ice in cometary nuclei. However, the basic test of this model will be provided only by direct detection of noble gases in a comet. Lacking this, we can study the atmospheres of other planets to look for evidence indicating that comets were indeed important in delivering volatiles. The atmosphere of Mars is especially useful because it is so thin that even geologically recent cometary impacts can be expected to have had measurable effects on its composition.

4. Lessons from Mars

The discovery that the present atmosphere of Mars has an average surface pressure of only 7.5 mbar has to rank as one of the greatest disappointments of the space age. Intuition suggested that a planet forming farther from the sun than Earth should be richer in volatiles, reflecting the lower temperature at which its materials condensed from the nebula. Indeed, there is extensive evidence of ancient water erosion on the Martian surface, indicating a thicker, warmer atmosphere in the past (e.g., Carr 1981; Baker 1982; Squyres and Kasting 1994). What happened to this early atmosphere? One could argue that atmospheric CO_2 dissolved in water and formed carbonates, but a proportional amount of atmospheric nitrogen must also have been removed, which is much more difficult (Owen and Bar-Nun 1995a). At the present time, the best explanation for the surprising thinness of the Martian atmosphere appears to be the phenomenon of impact erosion (Cameron 1983; Walker 1986; Ahrens, O'Keefe and Lange 1989; Melosh and Vickery 1989). Impacts during the first 800 million years of the planet's history not only brought volatiles in; because of the low mass of Mars, large impacts could blow off the fraction of the atmosphere above the plane tangent to the impact point. Melosh and Vickery (1989) estimate that at least 100 times the present mass of the atmosphere could have been lost in this way.

Impact erosion is a nonfractionating removal process. The relative abundances and isotopic ratios of the residual constituents will not be changed. This provides an explanation for the otherwise mysterious enhancement of $^{129}Xe/^{132}Xe$ on Mars, which is close to 2.5 times the solar or telluric value (Owen et al. 1977; Marti et al. 1995). Evidently, early blow-off of "normal" xenon (along with everything else) left a thin atmosphere that was subsequently enriched in ^{129}Xe provided by the decay of ^{129}I. Thus the small mass of the present Martian atmosphere does not mean that Mars had a dramatically different source of volatiles than the Earth.

On the contrary, we have argued that the two atmospheres both obtained most of their volatiles from the same two sources: the inner and outer reservoirs described in Section 1 (Owen and Bar-Nun 1995a, b). It has been obvious for some time that the atmospheric abundances of CO_2 and N_2 on Venus are nearly the same as the abundances of these compounds in the Earth's near-surface inventory of volatiles (Table 1.1). Although the Martian atmosphere is 150 times thinner than ours, we find it also exhibits approximately the same value of CO_2/N_2. The danger in such comparisons is that one cannot account accurately for C and N hidden in surface deposits on other planets. Thus we have concentrated on comparing the noble gases, which will remain in the atmosphere. If we accept the idea that Mars and Earth had a common source of volatiles, we can then draw a simple mixing line through the Mars and Earth points on our diagram (Figure 1.2).

Such a line implies that both planets obtained their volatiles from essentially identical reservoirs, one lying at either end of the line. The external reservoir is assumed to be the gases trapped in icy planetesimals. The laboratory experiments by Bar-Nun and colleagues (Owen, Bar-Nun and Kleinfeld 1991) demonstrated that ice formed at about 50 K would trap the heavy noble gases in proportions that fall right on the upward extrapolation of this mixing line (Figure 1.2). This temperature corresponds to the region in the solar nebula where Uranus and Neptune formed. Calculations of the formation process have demonstrated that icy planetesimals that were not incorporated in these giant planets were scattered out of this region. Some went into orbits that eventually populated the great Oort cloud of comets at 50,000 AU from the sun; others must have ended their lives in the inner Solar System, crashing into the sun or the inner planets (Ip and Fernandez 1988; Weissman 1991; Chyba, Owen and Ip 1994). The laboratory experiments demonstrate that these icy planetesimals (comets) could have carried the noble gases to Mars and Earth in the correct proportions.

In the case of Venus, we must invoke a different source. The laboratory data show that ice forming at temperatures near 30 K will trap all the noble gases in their original proportions in the solar nebula (the point labeled "sun" in Figures 1.1 and 1.2). Hence, comets formed at this low temperature would be capable of delivering the mixture found on Venus. Such comets would be members of the so-called Kuiper Belt, a disk of icy planetesimals found to occupy the space beyond the orbit of Neptune, with Pluto being the largest representative (Jewitt and Luu 1993, 1997).

The idea that the planetary atmospheres represent mixing between two reservoirs finds additional support from the data on noble gases trapped in the SNC meteorites, which fall along the same mixing line in Figure 1.2. These are the "meteorites from Mars" referred to in the introduction. It is the

Shergottites (the S in SNC) that fit this particular mixing line. The position of Chassigny is also shown, but the Nakhlites cannot be depicted on this plot as the argon they contain is contaminated (Ott 1988). That the fit of the Shergottites is not simply a solubility effect, as suggested by Ozima and Wada (1993), is demonstrated by the fact that ^{129}Xe is steadily increasing as one moves up the line. Isotope ratios will not be changed in solution, but this is exactly the effect that would occur as more atmospheric gas is added to a nonatmospheric component (Owen and Bar-Nun 1993, 1995b).

We have thus established a relationship between the atmosphere of Mars and the volatile inventory of the Earth through the relative abundances of the noble gases and their isotopes. We can then use this relationship to predict abundances in the Martian volatile inventory, based on our knowledge of volatiles on our own planet. By indexing to ^{84}Kr, we can avoid potential problems associated with possible depletion of ^{36}Ar by sputtering from O^+ ions (Luhmann, Johnson and Zhang 1992; Jakosky et al. 1994). We then compare the Martian C/ ^{84}Kr ratio with values found on Earth and Venus and find that 10 times the present atmospheric CO_2 abundance must be missing on Mars. Multiplying this value by Melosh and Vickery's (1989) lower limit to the magnitude of atmospheric erosion, we find that Mars once had the equivalent of at least 7.5 bars of CO_2, with an associated water abundance equivalent to a global layer at least 750 m deep (Owen and Bar-Nun 1995a). This entire reservoir of volatiles may not have been present at any given time, but it is sufficient to have permitted several epochs of clement conditions on early Mars, during which the famous fluvial features could have been carved into the surface.

Comets contain approximately solar abundances of the elements, except for hydrogen, nitrogen, and the noble gases (Geiss 1988; Owen and Bar-Nun 1995a). That means that during the early epochs on Mars when running water was available on the surface, all the biogenic elements were also present. Thus there must have been opportunities for the origin of life on Mars during the same period when life developed on the Earth. Much of the water that was originally delivered to and outgassed on Mars should still be on the planet. It would have been at least partially protected from impact erosion by being stored on and under the surface, including frozen forms such as permafrost and polar ice caps. Liquid water should still exist at depth today, in places where the thermal gradient reaches temperatures above the freezing point and subsurface ice is present. These locations can provide a subterranean habitat for microorganisms similar to those found at great depths on Earth (Pedersen 1993; Shock 1997). The search for fossils, and even highly adapted living microbes, remains an attractive, viable enterprise (Brack 1996; Klein 1996; McKay et al. 1996; Nealson 1997).

5. Water on Mars and Earth

Although differing mixtures of Oort Cloud and Kuiper Belt comets can explain the abundances of volatiles on Mars, Earth, and Venus, the model is obviously becoming increasingly ad hoc, which is not a pleasant situation. Nevertheless, as we learn more about the importance of impacts in shaping the early Solar System, we can appreciate the likelihood of random effects. We obviously need some direct measurements of noble gases in comets to test the laboratory-based predictions. Meanwhile we can ask if there is any other evidence for comets impacting the inner planets.

There may be some cometary gas trapped in the Shergottites, if it was a cometary impact that was responsible for the expulsion of these rocks from Mars (Owen and Bar-Nun 1995b). We have suggested that deviations from the mixing line exhibited by a plot of $^{129}Xe/^{132}Xe$ vs. $^{84}Kr/^{132}Xe$ for Shergottites and the separate mixing line for Nakhlites on such a graph could both be due to this effect, but other alternatives are possible (Drake et al. 1994; Owen and Bar-Nun 1995b; Swindle 1995). The excess isotopically-normal nitrogen found in glassy inclusions from both EETA 79001 and Zagami (Wiens, Becker and Pepin 1986; Marti et al. 1995) may have been another contribution from the icy impactor that propelled these rocks into space.

We might also expect to see signs of cometary bombardment in the water that is present on these planets. Mars is the better case, because the atmosphere is so thin. The total amount of water that exchanges seasonally through the Martian atmosphere from pole to pole is just 2.9×10^{15} g, equivalent to a single comet nucleus (density $\rho = 0.5$ g/cm^3) with a radius of only 1 km. Hence the impact of a relatively small comet could have a significant effect on the surface water supply. Indications of such an effect can be found in studies of D/H in hydrous phases of the SNC meteorites. Watson et al. (1994) reported a range of values of D/H in kaersutite, biotite and apatite in Chassigny, Shergotty, and Zagami, with the highest values, 4–5.5 times the terrestrial ocean standard, occurring in Zagami apatite. Infrared spectroscopy from Earth has determined D/H in Martian atmospheric water vapor to be $5.5{\pm}2\times$ terrestrial (Owen et al. 1988; Bjoraker, Mumma and Larson 1989; Krasnopolsky, Bjoraker et al. 1997). The overlap between the SNC and Mars atmosphere values indicates that some mixing between atmospheric water vapor and crustal rocks must occur. It is therefore arresting to note that the *lowest* values of D/H measured in the SNC minerals do not cluster around 1× terrestrial, but rather around 2× terrestrial. This is also true in the most recently discovered Shergottite, QUE 94201 (Leshin, Epstein and Stolper 1996). The significance of this lies in the fact that D/H in the water in Halley's comet is also 2× terrestrial (Balsiger, Altwegg and Geiss 1995; Eberhardt et al. 1995), as is the water in Comet Hyakutake (1996 B2) (Bockelée-Morvan et al. in press), and Comet Hale–Bopp (1995 01) (Meier, Owen, Matthews et al. 1998). It thus appears likely that most of the near-surface water on Mars was contributed primarily by cometary impact, rather than by magma from the planet or by meteoritic bombardment.

This interpretation is consistent with the geochemical analysis by Carr and Wänke (1992), who concluded that Mars is much drier than the Earth, with roughly 35 ppm water in mantle rocks as opposed to 150 ppm for Earth. These authors suggested that one possible explanation for this difference is the lack of plate tectonics on Mars, which would prevent a volatile-rich veneer from mixing with mantle rocks. This is exactly what the D/H values in the SNC minerals appear to signify. In whole rock samples, D/H in the Shergottites is systematically higher than in Nakhla and Chassigny (Leshin, Epstein and Stolper 1996). This could be a result of the larger fraction of atmospheric H_2O incorporated in the Shergottites by shock, a process Nakhlites evidently avoided (Bogard, Hörz and Johnson 1986; Ott 1988; Drake et al. 1994).

Furthermore, higher values of $\Delta^{17}O$ are found in these samples of Nakhla and Chassigny compared with the Shergottites, reinforcing the idea that the Shergottites sampled a different source of water. Finally, the oxygen isotope ratios in water from whole-rock samples of the SNCs are also systematically different from the ratios found in silicates in these rocks (Karlsson et al. 1992), again suggesting a hydrosphere that is not strongly coupled to the lithosphere.

On Earth, the oxygen isotopes in sea water match those in the silicates, indicating thorough mixing for at least the last 3.5 BY (Robert, Rejon-Michel and Javoy 1992). The Earth has also lost relatively little hydrogen into space after the postulated hydrodynamic escape, so the value of D/H we measure in sea water today must be close to the original value (Yung and Dissly 1992). If Halley, Hyakutake, and Hale–Bopp are truly representative of all comets, then we can't make the oceans out of melted comets alone. This is a very different situation from Mars. It suggests that water from the inner reservoir, the rocks making up the bulk of the Earth, must have mixed with incoming cometary water to produce our planet's oceans.

Explanations for the relatively high value of O/C (12 ± 6) in the terrestrial volatile inventory also suggest such mixing (Owen and Bar-Nun 1995a). The solar value of O/C = 2.4, which was also the value found in Halley's comet (Geiss 1988; Krankowsky 1991). Impact erosion on the Earth could remove CO and CO_2 while having little effect on water in the oceans or polar caps (Chyba 1990; Chyba, Owen and Ip 1994), thereby raising the value of O/C. This process would not affect the value of D/H, however. The average value of D/H in chondritic meteorites is close to that in sea water, so mixing meteoritic and cometary water would not lead to the right result. We need a contribution from a reservoir with D/H $<1.6 \times 10^{-4}$. Lécluse and Robert (1994) have shown that water vapor in the solar nebula at 1 AU from the sun would have developed a value of D/H $\approx 0.8 \times 10^{-4}$ to 1.0×10^{-4} , depending on the lifetime of the nebula (2×10^5 to 2×10^6 years). An ocean made of roughly 35% cometary water and 65% water from the local solar nebula (trapped in planetary rocks) would satisfy the D/H constraint and would also be consistent with the observed value of O/C = 12±6.

To accept this idea, we should be able to demonstrate that water vapor from the solar nebula was adsorbed on grains that became the rocks that formed the planets. Our best hope for finding some of that original inner-nebula water appears to be on Mars, where mixing between the surface and the mantle has been so poor. The test is thus to look for water incorporated in SNC meteorites that appear to have trapped mantle gases, to see if D/H $< 1.6 \times 10^{-4}$.

The best case for such a test among the rocks we have is Chassigny, which exhibits no enrichment of ^{129}Xe and thus appears not to have trapped any atmospheric gas. However, there is no evidence of water with low D/H in this rock (Leshin, Epstein and Stolper 1996). It may be that contamination by terrestrial water has masked the Martian mantle component. This is a good project for a sample returned from Mars.

6. What about neon?

We have concentrated our analysis on the heavy noble gases: argon, krypton, and xenon. Any model for the origin of the atmosphere must also account for neon. This gas has about the same cosmic abundance as nitrogen relative to hydrogen, namely, 1.2×10^{-4} vs. 1.1×10^{-4} (Anders and Grevesse 1989). Hence we expect any atmosphere that consists of a captured remnant of the solar nebula to exhibit a ratio of Ne/$N_2 \approx 2$. On Earth, Mars, and Venus, Ne << N_2. The neon is not simply deficient in these atmospheres. Where we can measure them, the isotopes have been severely fractionated. The solar ratio of ^{20}Ne/^{22}Ne = 13.7 (Anders and Grevesse 1989), on Earth ^{20}Ne/^{22}Ne = 9.8, on Venus 11.8±0.7 (Istomin, Grechnev and Kochnev 1982), and on Mars 10.1±0.7 (Wiens, Becker and Pepin 1986). Concentrating on the Earth, we must ask how neon can be so severely fractionated, whereas nitrogen (mass 14 amu) is not.

The answer may again lie with the comets, albeit in a paradoxical manner. The laboratory work shows that neon is not trapped in ice that forms at temperatures above 20 K (Bar-Nun et al. 1985; Laufer, Kochavi and Bar-Nun et al. 1987). Because the overwhelming majority of the icy planetesimals that formed in the Solar System condensed at temperatures higher than this, we have assumed that comets carry no neon (Owen, Bar-Nun and Kleinfeld 1992;

Owen and Bar-Nun 1995a). This assumption is substantiated by the apparent absence of neon in the atmospheres of Titan, Triton, and Pluto, where the upper limits on Ne/N_2 are typically about 0.01 (Broadfoot et al. 1981, 1989; Owen et al. 1993). These three objects in the outer Solar System, especially Triton and Pluto, represent giant icy planetesimals, which can be thought of as the largest members of the Kuiper Belt. Hence the absence of neon in their atmospheres may be taken as a good indication that ice condensing in the outer solar nebula did not trap this gas, and thus we do not expect to find it in comets. This prediction is consistent with new observations of Comet Hale–Bopp carried out with the Extreme Ultraviolet Explorer satellite by Krasnopolsky, Mumma et al. (1997). These authors established an upper limit of Ne/O $<1/200\times$ solar.

If the comets don't carry neon, how did this gas reach the inner planets? Once again the meteorites don't help. Even if they brought in all the xenon, the neon they could deliver would be $<10\%$ of what we observe. Instead it seems likely that neon was brought in by the rocks. In fact, we have evidence that this was the case, because we can still find neon whose isotope abundances approach the solar ratio in rocks derived from the mantle (Craig and Lupton 1976; Honda et al. 1991). Unlike the other noble gases, neon cannot be subducted into the Earth's interior (Hiyagon 1994). Thus it is not possible to dilute the original trapped gas with highly fractionated atmospheric neon. The record of original emplacement is preserved.

If neon, which diffuses so easily through solids, was retained by the Earth's rocks from the time of the planet's accretion, we can reasonably assume that some water was also kept in the interior, to emerge after the catastrophic formation of the Moon, mixing with incoming water from comets to form the oceans we find today. This perspective supports the idea that we may yet find evidence of this original water in mantle-derived rocks on Mars.

Returning to the Earth, it appears that the atmospheric neon bears a record of an early fractionating process that sharply reduced the ratio of $^{20}Ne/^{22}Ne$ from the solar value (Zahnle, Kasting and Pollack 1991). This process must have affected all of the other species in the atmosphere at the time. It may have been the massive, hydrodynamic escape of hydrogen produced by the reduction of water by contemporary crustal iron (Dreibus and Wänke 1989; Pepin 1991). The fact that we do not see evidence of any such fractionation in atmospheric nitrogen suggests that the nitrogen in the atmosphere today reached the Earth after this process had ended. Cometary delivery of nitrogen (and other volatiles) but not neon offers an easy means of achieving this condition (Owen, Bar-Nun and Kleinfeld 1992). In this case, neon is a kind of atmospheric fossil, a remnant of conditions that existed on the Earth before the volatiles that produced the bulk of the present atmosphere were in place.

7. Conclusions

How unique is the Earth? This is a perennial question in efforts to estimate the possibilities for abundant life in the universe. We have argued here that the source of our planet's atmosphere can be found in a combination of volatiles trapped in the rocks that made the planet and a late-accreting veneer of volatile-rich material delivered by icy planetesimals. The volatiles composing the atmosphere include the carbon, nitrogen, and water essential to life. The close similarities between the elemental, isotopic, and molecular abundances found in comets and those in the interstellar medium imply that icy planetesimals that form in any planetary system originating from an interstellar cloud will carry these same biogenic materials. Hence this model for the origin of our planet's atmosphere suggests that there is nothing unique about the inventory of volatiles that was delivered to the early Earth. Current studies of the Martian atmosphere, aided by the study of the SNC meteorites, reinforce this idea by indicating a similar inventory on that planet.

Nevertheless, we are still in the stage of finding "similarities" and "indications." We do not yet have a rigorous proof of the validity of the icy impact model. In this review, we have stressed the constraints provided by the abundances and isotope ratios of the noble gases. It is clear that meteoritic delivery of volatiles cannot satisfy the constraints set by our present knowledge of noble gas abundances and isotope ratios. However, the cometary alternative that we have emphasized will remain conjectural until noble gases are actually measured in a comet. Although laboratory studies strongly suggest that comets can deliver the correct elemental abundances, trapping of gas in ice does not affect isotopic ratios. We are therefore forced to assume that comets carry xenon whose isotopes resemble the distribution found in the terrestrial and Martian atmospheres rather than in the solar wind. It is not at all obvious why this should be the case. If the cometary xenon in fact resembles solar wind xenon, it will be necessary to invoke a fractionating process that acted to produce identical results for xenon on Mars and Earth followed by subsequent selective replacement of other volatiles (Pepin 1991).

How can we move forward from this unsatisfactory situation? There are a number of possible sources of new data on the horizon:

1. MARS: The Japanese Planet B Mission to be launched in 1998 will carry instruments that may resolve the present uncertainty over the atmospheric value of $^{36}Ar/^{38}Ar$ (Section 4). It will also teach us much more about possible nonthermal escape processes on Mars, which will allow a more confident reconstruction of the early mass and composition of the atmosphere. These

parameters can then be used (again!) to test our understanding of the origin of our own atmosphere.

If present plans mature, the step forward achieved from Planet B will soon be overshadowed by information obtained from Mars Sample Return Missions, scheduled to begin in 2005. These missions will bring back samples of Martian rocks and atmosphere for analysis on Earth, enabling far more accurate measurements of isotopic ratios than we can expect from missions to the planet. These measurements will include not only the noble gases, but also isotopes of carbon, nitrogen, and oxygen, the last in both H_2O and CO_2. With the kind of precision obtainable in laboratories on Earth, great progress should be achieved in unraveling the history of the Martian atmosphere from these isotope measurements, including estimates of the size and location of contemporary reservoirs of H_2O and CO_2 (McElroy, Kong and Young 1977; Owen et al. 1988; Jakosky 1991; Owen 1992; Jakosky et al. 1994; Krasnopolsky, Bjoraker et al. 1997). Another goal of this research would be a search for low values of D/H in water from mantle-derived rocks, which should also contain neon with $^{20}Ne/^{22}Ne$ approaching the solar value of 13.7.

2. COMETS. Both NASA and ESA are planning missions that will rendezvous with comets and deploy landers to explore their nuclei. These missions will have the capability to detect and measure the abundances of the heavy noble gases and their isotopes. This will be the most definitive test of the icy impact model. It is especially important to have this information from several comets, as we already know that the composition of comets can vary, both from the laboratory work on the trapping of gas in ice (Figure 3, Bar-Nun, Kleinfeld and Kochavi 1988; Owen and Bar-Nun 1995a, b) and from observations of variations in carbon compounds in comets (Fink 1992; A'Hearn et al. 1995).

The ESA mission to Comet Wirtanen is called *Rosetta*; it should arrive at its destination in 2012. The NASA missions have not yet been approved, but are planned for approximately the same time period. There is even hope for a new atmospheric probe to Venus in this time frame, which would tell us more about the apparently anomalous noble gas abundances on this planet. We await all these new results with great interest.

References

A'Hearn, M. F., Millis, R. L., Schleicher, D. G., Osip, D. J., and Birch, P. V. 1995. The ensemble properties of comets: Results from narrowband photometry of 85 comets, 1976–1992. *Icarus* 118:223–271.

Ahrens, T. J., O'Keefe, J. D., and Lange, M. A. 1989. Formation of atmospheres during accretion of the terrestrial planets. In *Origin and Evolution of Planetary*

Satellite Atmospheres, ed. S. K. Atreya, J. B. Pollack, and M. S. Matthews, pp. 328–385. Tucson: University of Arizona Press.

Anders, E. 1968. Chemical processes in the early solar system, as inferred from meteorites. *Acc. Chem. Res.* 1:289–298.

Anders, E., and Grevesse, N. 1989. Abundances of the elements: Meteoritic and solar. *Geochim. Cosmochim. Acta* 53:197–215.

Anders, E., and Owen, T. 1977. Mars and Earth: Origin and abundances of volatiles. *Science* 198:453–465.

Atreya, S. K., Pollack, J. B., and Matthews, M. S., eds. 1989. *Origin and Evolution of Planetary and Satellite Atmospheres*. Tucson: University of Arizona Press.

Baker, V. R. 1982. *The Channels of Mars*. Austin: University of Texas Press.

Balsiger, H., Altwegg, K., and Geiss, J. 1995. D/H in Comet P/Halley. *J. Geophys. Res.* 100:5827–5834.

Bar-Nun, A., Herman, B., Laufer, D., and Rappoport, M. L. 1985. Trapping and release of gases by water ice and implications for icy bodies. *Icarus* 63:317–332.

Bar-Nun, A., Kleinfeld, I., and Kochavi, E. 1988. Trapping of gas mixtures by amorphous water ice. *Phys. Rev. B.* 38:7749–7754.

Bernatowicz, T. J., Kennedy, B. M., and Podosek, F. A. 1985. *Geochim. Cosmochim. Acta* 49:2561–2564.

Bjoraker, G. L., Mumma, M. J., and Larson, H. P. 1989. Isotopic abundance ratios for hydrogen and oxygen in the Martian atmosphere. *Bull. Amer. Astron. Soc.* 21:990 [Abstract].

Bockelée-Morvan, D., Gautier, D., Lis, D., Young, K., Keene, J., Phillips, T., Owen, T., Crovisier, J., Goldsmith, P. F., Bergin, E. A., Despois, D., and Wootten, A. In press. Deuterated water in Comet C/1996 B2 (Hyakutake) and its implications for the structure of the primitive solar nebula. *Icarus*.

Bogard, D. D., Hörz, F., and Johnson, P. H. 1986. Shock-implanted noble gases: An experimental study with implications for the origin of Martian gases in Shergottite meteorites. *Proc. Lunar and Planet. Sci. Conf. 17th*; *J. Geophys. Res.* 91: E99–E114.

Brack, A. 1996. Why exobiology on Mars. *Planet. Space Sci.* 44:1435–1440.

Broadfoot, A. L. and 15 coauthors. 1981. Extreme ultraviolet observations from Voyager 1 encounter with Saturn. *Science* 212:206–211.

Broadfoot, A. L., and 15 coauthors. 1989. Ultraviolet spectrometer observations of Neptune and Triton. *Science* 246:1459–1466.

Brown, H. 1952. Rare gases and the formation of the Earth's atmosphere. In *The Atmospheres of the Earth and Planets*, ed. G. P. Kuiper, pp 258–266. Chicago: University of Chicago Press.

Cameron, A. G. W. 1983. Origin of the atmospheres of the terrestrial planets. *Icarus* 56:195–201.

Carr, M. H. 1981. *The Surface of Mars*. New Haven: Yale University Press.

Carr. M. H. 1986. Mars: A water-rich planet? *Icarus* 68:187–216.

Carr, M. H., and Wänke, H. 1992. Earth and Mars: Water inventories as clues to accretional histories. *Icarus* 98:61–71.

Chyba, C. 1990. Impact delivery and erosion of planetary oceans in the early inner solar system. *Nature* 343:129–133.

Chyba, C. F., Owen, T., and Ip, W.-H. 1994. Impact delivery of volatiles and organic molecules to Earth. In *Hazards Due to Comets and Asteroids*, ed. T. Gehrels. Tucson: University of Arizona Press.

Clayton, R. N., and Mayeda, T. K. 1983. Oxygen and isotopes in eucrites, shergottites and nakhlites and chassignites. *Earth Planet. Sci. Lett.* 62:1–6.

Craig, H., and Lupton, J. E. 1976. Primordial neon, helium, and hydrogen in oceanic basalts. *Earth Planet Sci. Lett.* 31:369–385.
Davies, J. K., Roush, T. L., Cruikshank, D. P., Bartholomew, M. J., Geballe, T. R., and Owen, T. 1997. The detection of water ice in comet P/Hale–Bopp. *Icarus* 122:238–245.
Donahue, T. M., and Pollack, J. B. 1983. Origin and evolution of the atmosphere of Venus. In *Venus,* ed. D. M. Dunten, L. Colin, T. M. Donahue, and V. I. Moroz, pp. 1003–1036. Tucson: University of Arizona Press.
Drake, M. J., Swindle, T. D., Owen, T., and Musselwhite, D. L. 1994. Fractionated Martian atmosphere in the nakhlites. *Meteoritics* 29:854–859.
Dreibus, G., and Wänke, H. 1987. Volatiles on Earth and Mars: A comparison. *Icarus* 71:225–240.
Dreibus, G., and Wänke, H. 1989. Supply and loss of volatile constituents during the accretion of the terrestrial planets. In *Origin and Evolution of Planetary and Satellite Atmospheres,* ed. S. K. Atreya, J. B. Pollack, and M. S. Matthews, pp. 268–288. Tucson: University of Arizona Press.
Eberhardt, P., Reber, M., Krankowsky, D., and Hodges, R. R. 1995. The D/H and $^{18}O/^{16}O$ ratios in water from comet P/Halley. *Astron. Astrophys.* 302:301–316.
Fink, U. 1992. Comet Yanaka (1988r): A new class of carbon poor comet. *Science* 257:1926–1929.
Geiss, J. 1988. Composition in Halley's comet: Clues to origin and history of cometary matter. *Rev. Mod. Astron.* 1:1–27.
Heymann, D. 1971. The inert gases. In *Handbook of Elemental Abundances in Meteorites,* ed. B. Mason, pp. 29–66. New York: Gordon and Breach.
Hiyagon, H. 1994. Retention of solar helium and neon in IDPs in deep sea sediment. *Science* 263:1257–1259.
Honda, M., McDougall, I., Patterson, D., Doulgeris, A., and Clague, D. A. 1991. Possible solar noble-gas component in Hawaiian basalts. *Nature* 349:149–151.
Hunten, D. M. 1993. Atmospheric evolution of the terrestrial planets. *Science* 259:915–920.
Hunten, D. M., Pepin, R. O., and Walker, J. C. G. 1987. Mass fractionation in hydrodynamic escape. *Icarus* 69:532–549.
Ip, W. H., and Fernandez, J. A. 1988. Exchange of condensed matter among the outer and terrestrial protoplanets and the effect on surface impact and atmospheric accretion. *Icarus* 74:47–61.
Istomin, V. G., Grechnev, K. V,. and Kochnev, V. A. 1982. Preliminary results of mass spectrometric measurements on board the Venera 13 and 14 probes. *Pisma Astron. Zh.* 8:391–398.
Jakosky, B. M. 1991. Mars volatile evolution: Evidence from stable isotopes. *Icarus* 94:14–31.
Jakosky, B. M., Pepin, R. M., Johnson, R. E., and Fox, J. L. 1994. Mars atmospheric loss and isotopic fractionation by solar-wind–induced sputtering and photochemical escale. *Icarus* 111:271–288.
Jewitt, D., and Luu, J. 1993. Discovery of the Candidate Kuiper Belt object 1992 QB_1. *Nature* 362:739–732.
Jewitt, D., and Luu, J. 1997. The Kuiper Belt. In *From Stardust to Planetesimals,* ed. Y. Pendleton and A. Tieleus, pp. 335–345. San Francisco: Astron. Soc. Pac. Conference Series Vol. 122.
Karlsson, H. R., Clayton, R. N., Gibson, E. K., Jr., and Mayeda, T. K. 1992. Water in SNC meteorites: Evidence for Martian hydrosphere. *Science* 255:1409–1411.
Kasting, J. F. 1993. Earth's early atmosphere. *Science* 259:920–926.
Klein, H. P. 1996. On the search for extant life on Mars. *Icarus* 120:431–436.

Krankowsky, D. 1991. The composition of comets. In *Comets in the Post Halley Era.*, ed. R. L. Newburn, M. Neugebauer, and J. Rahe, pp. 855–878. Kluwer: Dordrecht.

Krasnopolsky, V. A., Bjoraker, G. L., Mumma, M. J., and Jennings, D. E. 1997. High-resolution spectroscopy of Mars at 3.7 and 8 μm: A search for H_2O_2, H_2CO, HC1, and CH_4, and detection HDO. *J. Geophys. Res.* 102:6525–6534.

Krasnopolsky, V. A., Mumma, M. J., Abbot, M., Flynn, B. C., Meech, K. J., Yeomans, D. K., Feldman, P. D., and Cosmovici, C. B. 1997. Detection of soft x-rays and a sensitive search for noble gases in comet Hale–Bopp (C/1995 01). *Science* 277:1488–1491.

Laufer, D., Kochavi, E., and Bar-Nun, A. 1987. Structure and dynamics of amorphous water ice. *Phys. Rev. B.* 36:9219–9227.

Lécluse, C., and Robert F. 1994. Hydrogen isotope exchange reaction rates: Origin of water in the inner solar system. *Geochim. et Cosmochim. Acta* 58:2927–2939.

Leshin, L. A., Epstein, S., and Stolper, E. M. 1996. Hydrogen isotope geochemistry of SNC meteorites. *Geochim. et Cosmochim. Acta* 60:2635–2650.

Lewis, J. S. 1974. Volatile element influx on Venus from cometary impacts. *Earth Planet. Sci. Lett.* 22:239–244.

Lewis, J. S., and Prinn, R. G. 1984. *Planets and their Atmospheres.* New York: Academic Press.

Luhmann, J. G., Johnson, R. E., and Zhang, M. H. G. 1992. Evolutionary impact of sputtering of the martian atmosphere by O^+ pickup ions. *Geophys. Res. Lett.* 19:2151–2154.

Lunine, J. I., Engel, S., Rizk, B., and Horanyi, M. 1991. Sublimation and reformation of icy grains in the primitive solar nebula. *Icarus* 94:333–344.

Marten, A., Gautier, D., Owen, T., Sanders, D. B., Matthews, H. E., Atreya, S. K., Tilanus, R. P. J., and Deane, J. 1993. First observations of CO and HCN on Neptune and Uranus at millimeter wavelengths and their implications for atmospheric chemistry. *Astrophys. J.* 406:285–297.

Marti, K., Kim, J. S., Thakur, A. N., McCoy, T. J., and Keil, K. 1995. Signatures of the Martian atmosphere in glass of the Zagami meteorite. *Science* 267:1981–1984.

McElroy, M. B., Kong, T. Y., and Yung, Y. L. 1977. Photochemistry and evolution of Mars' atmosphere: A Viking perspective. *J. Geophys. Res.* 82:4379–4388.

McKay, D. S., Gibson, E. K., Jr., Thomas-Keprta, K. S., Vali, H., Romanek, C. S., Clemett, S. J., Chillier, X. D. F., Maechling, C. R., and Zare, R. N. 1996. Search for past life on Mars: Possible relic biogenic activity in Martian meteorite ALH 84001. *Science* 273:924–930.

Meier, R., Owen, T., Jewitt, D. C., Matthews, H. E., Senay, M., Biver, N., Bockelée-Morvan, D., Crovisier, J., and Gautier, D. 1998. Deuterium in Comet C/1995 O1 (Hale Bopp): Detection of DCN. *Science* 279:1707–1710.

Meier, R., Owen, T., Matthews, H. E., Jewitt, D. C., Bockelée-Morvan, D., Biver, N., Crovisier, J., and Gautier, D. 1998. A determination of HDO/H_2O in Comet C/1995 01 (Hale Bopp). *Science* 279:842–844.

Melosh, H. J., and Vickery, A. M. 1989. Impact erosion of the primordial atmosphere of Mars. *Nature* 338:487–489.

Mumma, M. J., and six co-authors. 1996. Detection of abundant ethane and methane, along with carbon monoxide and water in comet C/1996 B2 Hyakutake: evidence for interstellar origin. *Science* 272:1310–1314.

Nealson, K. H. 1997. The limits of life on Earth and searching for life on Mars. *J. Geophys. Res.* 102:23675–23686.

Notesco, G., and Bar-Nun, A. 1996. Enrichment of CO over N_2 by their trapping in amorphous ice and implications to Comet P/Halley. *Icarus* 122:118–121.

Notesco, G., and Bar-Nun, A. 1997. The source of the high C_2H_6/CH_4 ratio in Comet Hyakutake. *Icarus*. 125:471–473
Oró, J. 1961. Comets and the formation of biochemical compounds on the primitive earth. *Nature* 190:389–390.
Ott, U. 1988. Noble gases in SNC meteorites: Shergotty, Nakhla, Chassigny. *Geochim. et Cosmochim. Acta* 52:1937–1948.
Owen, T. 1982. The composition and origin of Titan's atmosphere. *Planet. Space Sci.* 30:833–838.
Owen, T. 1992. The composition and early history of the atmosphere of Mars. In *Mars*, ed. H. H. Kieffer, B. Jakosky, C. W. Snyder, and M. Matthews, pp. 818–834. Tucson: University of Arizona Press.
Owen, T., and Bar-Nun, A. 1993. Noble gases in atmospheres. *Nature* 361:693–694.
Owen, T., and Bar-Nun, A. 1995a. Comets, impacts and atmospheres. *Icarus* 116:215–226.
Owen, T., and Bar-Nun, A. 1995b. Comets, impacts and atmospheres II: Isotopes and noble gases. *Conference Proceedings No. 341 (AIP): Volatiles in the Earth and Solar System*, ed. K. Farley, pp. 123–138.
Owen, T., Bar-Nun, A., and Kleinfeld, I. 1991. Noble gases in terrestrial planets: Evidence for cometary impacts? In *Comets in the Post-Halley Era,* ed. R. L. Newburn, Jr., M. Neugebauer, and J. Rahe, pp. 429–438. Kluwer: Dordrecht.
Owen, T., Bar-Nun, A., and Kleinfeld, I. 1992. Possible cometary origin of heavy noble gases in the atmospheres of Venus, Earth and Mars. *Nature* 358:43–46.
Owen, T., Biemann, K., Rushneck, D. R., Biller, J. E., Howarth, D. W., and Lafleur, A. L. 1977. The composition of the atmosphere at the surface of Mars. *J. Geophys. Res.* 82:4635–4639.
Owen, T., Maillard, J. P., de Bergh, C., and Lutz, B. L. 1988. Deuterium on Mars: The abundance of HDO and the values of D/H. *Science* 240:1767–1770.
Owen, T. C., Roush, T. L., Cruikshank, D. P., Elliot, J. L., Young, L. A., de Bergh, C., Schmitt, B., Geballe, T., Brown, R. H., and Bartholomew, M. J. 1993. *Science* 261:745–748.
Ozima, M., and Wada, N. 1993. Noble gases in atmospheres. *Nature* 361:693.
Pedersen, K. 1993. The deep subterranean biosphere. *Earth-Science Reviews* 34:243–260.
Pepin, R. O. 1989. Atmospheric compositions: key similarities and differences. In *Origin and Evolution of Planetary and Satellite Atmospheres,* ed. S. K. Atreya, J. B. Pollack, and M. S. Matthews, pp. 291–305 (Figure 3). Tucson: University of Arizona Press.
Pepin, R. O. 1991. On the origin and early evolution of terrestrial planet atmospheres and meteoritic volatiles. *Icarus* 92:2–79.
Robert, R., Rejon-Michel, A., and Javoy, M. 1992. Oxygen istotopic homogeneity of the Earth: new evidence. *Earth and Planet. Sci. Lett.* 108:1–9.
Shock, E. L. 1997. High temperature life without photosynthesis as a model for Mars. *J. Geophys. Res.* 102:23687–23694.
Sill, G. T., and Wilkening, L. 1978. Ice clathrate as a possible source of the atmospheres of the terrestrial planets. *Icarus* 33:13–27.
Squyres, S. W., and Kasting, J. F. 1994. Early Mars: How warm and how wet? *Science* 264:744–748.
Stevenson, D. J. 1987. Origin of the moon – The collision hypothesis. *Ann Rev. Earth Planet. Sci.* 15:271–315.
Swindle, T. D. 1995. How many Martian noble gas reservoirs have we sampled? *Conference Proceedings No. 341 (AIP): Volatiles in the Earth and Solar System*, ed. K. Farley, pp. 175–185.

Turekian, K. K., and Clark, S. P., Jr. 1969. Inhomogenous accumulation of the Earth from the primitive solar nebula. *Earth Planet. Sci. Lett.* 6:346–348.
Turekian, K. K., and Clark, S. P., Jr. 1975. The non-homogenous accumulation model for terrestrial planet formation and the consequences for the atmosphere of Venus. *J. Atmos. Sci.* 32:1257–1261.
Wacker, J. F., and Anders, E. 1984. Trapping of xenon in ice and implications for the origins of the Earth's noble gas. *Geochim. et Cosmochim. Acta* 48:2372.
Walker, J. C. G. 1986. Impact erosion of planetary atmospheres. *Icarus* 68:87–98.
Watson, L. L., Hutcheon, I. D., Epstein, S., and Stolper, E. M. 1994. Water on Mars: Clues from deuterium/hydrogen and water contents of hydrous phases in SNC meteorites. *Science* 265:86–90.
Weissman, P. 1991. Dynamical history of the Oort cloud. In *Comets in the Post-Halley Era,* ed. R. L. Newburn, M. Neugebauer, and J. Rahe, pp. 463–437. Kluwer: Dordrecht.
Wetherill, G. W. 1975. Late heavy bombardment of the moon and terrestrial planets. *Proc. Lunar Sci. Conf.* VI:1539–1561.
Wetherill, G. W. 1988. Accumulation of Mercury from planetesimals. In *Mercury*, ed. F. Vilas, C. R. Chapman, and M. S. Matthews, pp. 670–691. Tucson: University of Arizona Press.
Wetherill, G. W. 1990. Formation of the Earth. *Ann. Rev. Earth Planet. Sci.* 18:205–256.
Wetherill, G. W., and Chapman, C. R. 1988. Asteroids and meteorites. In *Meteorites and the Early Solar System*, ed. J. F. Kerridge and M. S. Matthews, pp. 35–67. Tucson: University of Arizona Press.
Wiens, R. C., Becker, R. H., and Pepin, R. O. 1986. The case for a Martian origin for the Shergottites II. Trapped and indigenous gas components in EETA 79001 glass. *Earth Planet Sci. Lett.* 77:149–158.
Yung, Y., and Dissly, R. W. 1992. Deuterium in the solar system. In *Isotope Effects in Gas-Phase Chemistry*, ed. J. A. Kaye, pp. 369–389. American Chemical Society Symposium Series No. 502.
Zahnle, K., Kasting, J. F., and Pollack, J. B. 1990. Mass fractionation of noble gases in diffusion-limited hydrodynamic hydrogen escape. *Icarus* 84:502–527.

2

The early atmosphere as a source of biogenic compounds

JAMES F. KASTING and LISA L. BROWN
Department of Geosciences
The Pennsylvania State University
University Park

1. Introduction

Any discussion of Earth's early atmosphere and ocean must be somewhat uncertain because we do not have samples of ancient air or water to analyze and because even the indirect record preserved in rocks is not very informative. We know from radiometric dating that Earth and the Solar System itself are both about 4.5 billion years old (Ga). The oldest preserved sedimentary rocks are from Isua, West Greenland, and have been dated at ~3.8 Ga. Some igneous rocks are older than this, but they contain little information about the atmosphere and oceans. Thus, conditions during the first 0.7 Ga of Earth history must be inferred largely from theoretical models. One might be tempted to forego trying to deduce what the Earth was like at that time, were it not for the fact that this is also the time interval during which life seems to have originated. If we hope to place any physical constraints on this process, we need to consider the formation and evolution of the early atmosphere and ocean.

During the first half of this century, models of the early atmosphere were shaped by the ideas of A. I. Oparin in Russia and Harold Urey in the United States. Urey, a geochemist, was aware that the atmospheres of the giant planets, Jupiter and Saturn, were dominated by H_2, CH_4, and NH_3. He reasoned that these gases had been captured from the solar nebula and that they had been retained since that time because the giant planets were too large to lose hydrogen by escape to space. He further speculated that the atmospheres of the terrestrial planets, including Earth, had a similar composition early in their histories before they had time to evolve.

Urey's ideas about early atmospheric composition received a boost in 1953 when his graduate student, Stanley Miller, showed that amino acids and other complex organic compounds could be synthesized by spark discharge (simulating lightning) in a CH_4–NH_3–H_2 atmosphere (Miller 1953, 1955). The success of these experiments helped establish Urey's model as the consensus view of the primitive atmosphere. This view remained popular among origin of life researchers for several decades afterwards.

However, even before Miller and Urey had performed their work, another American geochemist, William Rubey, published a paper in which he outlined a very different theory of early atmospheric composition (Rubey 1955). Rubey pointed out that the atmosphere was produced largely from volcanic outgassing and that modern volcanic gases are relatively oxidizing. Most of the hydrogen coming out of volcanos is released as H_2O, rather than H_2, and most of the carbon is released as CO_2, rather than CO or CH_4. Holland (1984, p. 50) lists $H_2/H_2O \approx 0.01$ and $CO/CO_2 \approx 0.03$ as typical volcanic ratios. N_2 is difficult to measure in volcanic emissions, but NH_3 is noticeably absent.

Rubey's ideas were developed further by Holland (1962), who suggested that the atmosphere passed through two distinct stages: (1) a highly reduced stage, marked by gases emitted in equilibrium with metallic iron, and (2) a more oxidized stage, dominated by gases similar to those emitted by modern volcanos. Stage 1 of this model was later abandoned when other workers (e.g., Stevenson 1983) argued that core formation was contemporaneous with accretion, so that metallic iron was removed from the upper mantle early in Earth history. Thus, the dominant view in recent years has been that the primitive atmosphere was a weakly reducing mixture of CO_2, N_2, and H_2O, combined with lesser amounts of CO and H_2 (Walker 1977; Walker et al. 1983; Holland 1984; Kasting 1993).

2. Weakly reducing atmospheres and climatically imposed constraints

A typical example of such a weakly reducing atmospheric composition is shown in Figure 2.1, from Kasting (1993). The total surface pressure is assumed to be 1 bar, although this value is somewhat arbitrary. There are few actual constraints on the surface pressure of the early Earth. More significantly, the assumed N_2 partial pressure is 0.8 bars, essentially the same as today. The partial pressure of N_2, pN_2, is equal to its volume mixing ratio, $f(N_2)$, times the total pressure. The assumption that pN_2 has remained constant is based on the idea that the atmosphere formed early as a consequence of extensive volcanic outgassing, combined with impact degassing of incoming planetesimals (Ahrens, O'Keefe and Lange 1989). Evidence for rapid atmospheric formation is provided by studies of the rates at which noble gases are released by volcanism (Holland 1984, pp. 64–76). The present release rates of these gases extrapolated back in time are too slow to account for their atmospheric inventories. Hence, it may be inferred that noble gases and other volatiles were delivered to Earth's surface early in the planet's history. Current models suggest that many of Earth's volatiles, including H_2O, CO_2,

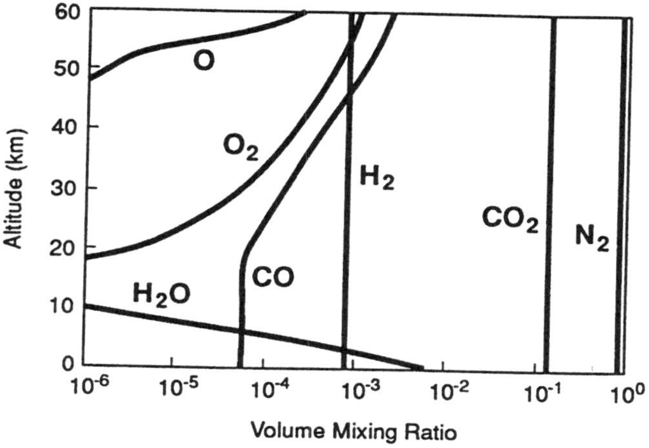

Figure 2.1. Vertical mixing ratio profiles of major atmospheric constituents in a typical weakly reducing primitive atmosphere. "Mixing ratio" is the same as "mole fraction." The assumed surface pressure is 1 bar. (From Kasting 1993.)

and N_2, were released by impact degassing of planetesimals either during the main period of accretion (Matsui and Abe 1986) or during the heavy bombardment period between 4.5 Ga and 3.8 Ga (Chyba 1987, 1989; Owen, Bar-Nun and Kleinfeld 1992). N_2, being relatively inert, has probably remained in the atmosphere throughout the entirety of Earth's history.

The CO_2 partial pressure in the model shown in Figure 2.1 is 0.2 bars. This may be compared with a present-day pCO_2 of 3.6×10^{-4} bars, or 360 ppm (parts per million by volume). A high pCO_2 is the most likely mechanism for countering the faint young Sun problem (Owen, Cess and Ramanathan 1979; Walker, Hays and Kasting 1981). Solar luminosity was ~30% lower at the time when Earth formed, as a consequence of a lower He/H ratio in the Sun's core (Gough 1981). This reduction in luminosity must have been balanced by an increased greenhouse effect in Earth's atmosphere, or the oceans would have been frozen prior to ~2.0 Ga (Sagan and Mullen 1972). We know that this did not occur, because there is evidence for liquid water back to 3.8 Ga and for life back to at least 3.5 Ga (Schopf 1983). Indeed, recent analyses of the carbon isotopic composition of apatite inclusions from Isua and from another early banded iron formation suggest that life was already present by 3.8 Ga (Mojzsis et al. 1996).

Although the faint young Sun problem has been termed a "paradox," Walker, Hays and Kasting (1981) showed that it has a natural solution. The dominant, long-term removal process for atmospheric CO_2 is silicate weathering on land, followed by deposition of carbonate sediments in the ocean. If the mean global surface temperature were to drop below freezing, silicate weathering would slow down and volcanic CO_2 would accumulate in the atmosphere. This negative feedback system involving CO_2 and weathering should have kept Earth's mean surface temperature above freezing throughout its history. This does not imply, however, that CO_2 and H_2O were the only important greenhouse gases. As will be shown later, CH_4 could also have played an important role in warming the early climate, both prior to and after the origin of life.

The 0.2 bars of CO_2 in the model atmosphere shown in Figure 2.1 represents the minimum amount of CO_2 required to offset decreased solar luminosity, assuming that CO_2 and H_2O were the only important greenhouse gases (Kasting 1987, 1992). The actual atmospheric pCO_2 could have been lower than this if methane was abundant, but it could also have been much higher, especially very early in Earth history. Walker (1985) suggested that pCO_2 could have been as high as 10 bars prior to the emergence of the continents because silicate weathering would have been inhibited and because there would have been no stable platforms on which to store carbonate rocks. If so, the early Earth could have been quite hot, perhaps 85 °C (Kasting and Ackerman 1986).

More recent analyses of Sm/Nd ratios in ancient zircons (Bowring and Housh 1995) suggest that the continents grew rapidly, so a dense, CO_2 atmosphere, if it existed at all, may have been confined to the first few hundred million years of Earth history. This might alleviate the fears of some workers, for example, Bada, Bigham and Miller (1994), that life could not have originated at high temperatures. The most probable time for the origin of life is between 3.8 Ga and ~4.0 Ga, when the chances of being hit by an ocean-vaporizing impactor become small (Sleep et al. 1989). For this time interval, an atmospheric composition similar to that in Figure 2.1 is not unreasonable.

From a photochemical standpoint, the key constituent in a weakly reducing atmosphere is H_2. H_2 controls the O_2 abundance by way of a variety of reactions that add up to

$$2H_2 + O_2 \rightarrow 2H_2O.$$

H_2 concentrations of 10^{-3}, as depicted in Figure 2.1, result in ground-level O_2 mixing ratios of $\sim 10^{-13}$, or 5×10^{-12} PAL (times the present atmospheric level). The next two sections describe how the atmospheric H_2 concentration is estimated. This discussion is central to understanding why the early atmosphere was essentially oxygen free and how our detailed simulations of such an atmosphere might change as a result of further studies.

3. Budgets of atmospheric H_2 and O_2

Almost all of the O_2 in Earth's present atmosphere is thought to have been produced by photosynthesis, followed by burial of organic carbon. Beginning with Berkner and Marshall (1965), a variety of researchers have tried to estimate how much O_2 should have been present prior to the origin of photosynthesis. The only abiotic net source for O_2 is photolysis of water vapor in the stratosphere,

$$H_2O + h\nu \rightarrow H + OH,$$

followed by escape of hydrogen to space. The OH radicals produced in this reaction can react with O atoms produced by photolysis of CO_2 to produce O_2:

$$\begin{aligned} CO_2 + h\nu &\rightarrow CO + O \\ OH + O &\rightarrow O_2 + H. \end{aligned}$$

CO_2 photolysis by itself does not provide a net source of O_2 even if the resulting O atoms recombine with each other, because any O_2 generated in this manner will eventually recombine with CO to regenerate CO_2. H_2O photolysis is unique because hydrogen atoms are light enough to escape.

The first researcher to correctly deduce how to calculate the abiotic O_2 production rate was Walker (1977). Drawing on concepts first elucidated by Hunten (1973), Walker realized that the escape rate of hydrogen from Earth's present atmosphere is limited by diffusion. In particular, the escape rate is proportional to the total hydrogen mixing ratio in the stratosphere, that is, the combined mixing ratio of hydrogen in all of its chemical forms. If we express the escape rate in units of H_2 molecules cm^{-2} s^{-1}, we can write

$$\Phi_{esc}(H_2) \approx 2.5 \times 10^{13} f_{tot}(H_2) \qquad (1)$$

where $f_{tot}(H_2)$ is the total hydrogen mixing ratio, defined by

$$f_{tot}(H_2) = 0.5 f(H) + f(H_2) + f(H_2O) + 2 f(CH_4) + \ldots \qquad (2)$$

Any stratospheric constituent that contains hydrogen will contribute to this weighted sum. Tropospheric H_2O does not count because it can be removed by condensation. The numerical constant in equation (1) represents an average of the diffusion constants for H and H_2 divided by the atmospheric scale height.

Before proceeding further, we should note that escape to space is not always limited by diffusion. Helium escape from Earth, for example, is limited not by the helium abundance in the stratosphere, but by the rate at which He atoms thermally evaporate from the exobase at 500 km. (The exobase is the altitude above which the atmosphere is effectively collisionless.) This evaporation process is termed Jeans' escape. If Jeans' escape and other, nonthermal escape processes are inefficient, which can happen if the upper atmosphere is relatively cold, then hydrogen can be "bottled up" at the exobase and its escape rate will be slower than the diffusion limit. We ignore this complication for the moment because it simply implies that equation (1) may overestimate the abiotic oxygen source. We shall return to this point at the end of the chapter, however, because it could affect some of our more detailed predictions about the nature of the primitive atmosphere, in particular, the methane abundance.

For Earth's present atmosphere, hydrogen escape is diffusion limited, so we can use equations (1) and (2) to estimate the escape rate. Earth's present lower stratosphere contains about 3 ppm H_2O and 1.6 ppm CH_4. Thus, $f_{tot}(H_2) = 3 \times 10^{-6} + 2(1.6 \times 10^{-6}) = 6.2 \times 10^{-6}$, and the diffusion-limited escape rate is $\Phi_{esc}(H_2) \simeq 1.6 \times 10^8$ H_2 molecules cm^{-2} s^{-1}. Only about half of this escaping hydrogen (8×10^7 H_2 molecules cm^{-2} s^{-1}) results in abiotic O_2 production because only half of it derives from H_2O. The abiotic O_2 production rate on the early Earth was, if anything, lower than this because the primitive stratosphere should have been even colder and drier than today (Kasting and Ackerman 1986).

Walker (1977) realized that, on an abiotic Earth, the rate at which O_2 was produced abiotically from H_2O photolysis must equal the rate at which it was destroyed. He also correctly deduced that the primary loss process for O_2 would have been its (indirect) reaction with reduced volcanic gases such as H_2 and CO. O_2 could also have been lost by rainout of photochemical oxidants (see next section) and by reaction with H_2 produced by photooxidation of ferrous iron in the primitive oceans (Braterman, Cairns-Smith and Sloper 1983), but these O_2 sinks were probably of secondary importance.

The present volcanic sink for O_2 cannot be estimated directly because it is impossible to measure the actual amounts of reduced gases coming out of volcanos. One can proceed by a more indirect route, though, that begins with an analysis of the carbonate–silicate cycle (Holland 1978, 1984). On time scales in excess of ~0.5 million years, the output of CO_2 from volcanos must be sufficient to balance the consumption rate of CO_2 by silicate weathering. The silicate weathering rate can be determined from estimates of river flow and the measured concentrations of Ca^{++}, Mg^{++}, and dissolved SiO_2 in river water (Holland 1978, pp. 270–274). Holland's estimate for the volcanic CO_2 flux is 7.5×10^{12} mol/yr. If one multiplies this number by the observed ratio of H_2O/CO_2 in volcanic gases, and by thermodynamically calculated ratios of H_2/H_2O and CO/CO_2, one obtains outgassing rates of 11×10^{11} mol/yr and

3×10^{11} mol/yr for H_2 and CO, respectively (Holland 1978, pp. 291–292). Reduced sulfur gases, SO_2 in particular, are also important constituents of volcanic emissions. The next section explains how they enter into the atmospheric redox budget.

A key point is that, from the standpoint of atmospheric redox balance, outgassing of a reduced volcanic gas such as CO is equivalent to outgassing of H_2. The reason is that CO is oxidized to CO_2 by the by-products of water vapor photolysis:

$$CO + OH \rightarrow CO_2 + H.$$

Because the OH radical was derived originally from H_2O, the net oxidation reaction for CO can be thought of as

$$CO + H_2O \rightarrow CO_2 + H_2.$$

Each molecule of CO that is oxidized in the atmosphere results in the addition of one molecule of H_2. The net source for H_2 from outgassing of H_2 and CO is thus 14×10^{11} mol/yr, or about 5×10^9 H_2 molecules cm^{-2} s^{-1}.

We can now see why the early atmosphere must have been reducing, rather than oxidizing. The abiotic source for oxygen was $\leq 8 \times 10^7$ O atoms cm^{-2} s^{-1} based on our analysis of the present hydrogen escape rate. The potential oxygen sink from volcanic gases was $\geq 5 \times 10^9$ O atoms cm^{-2} s^{-1} if early volcanos were at least as active as today. Because the potential oxygen sink overwhelms the source by a factor of 60 or more, O_2 had to have been a minor atmospheric constituent. Its concentration can therefore be calculated with a photochemical model in the same manner that ozone concentrations can be calculated today. Furthermore, the numbers that we have just calculated imply that most of the H_2 emitted by volcanos must have escaped to space, rather than reacting with abiotically produced oxygen. If we assume that the hydrogen in the stratosphere was predominantly in the form of H_2, we can invert equation (1) to find the atmospheric H_2 mixing ratio: $f(H_2) \approx 5 \times 10^9$ cm^{-2}s^{-1}/2.5 $\times 10^{13}$ cm^{-2}s^{-1} $= 2 \times 10^{-4}$, or about a factor of 5 lower than the value shown in Figure 2.1. Figure 2.1 incorporates the reasonable assumption that volcanic outgassing rates were several times higher on the early Earth as a consequence of increased internal heat flow.

4. Rainout of soluble gases and the atmospheric redox budget

Although the preceding analysis is sufficient to show that the early atmosphere was reducing, it neglects another important process that should have affected the atmospheric redox budget, namely, rainout of oxidized or reduced trace gases. An example of one such reduced gas is formaldehyde, H_2CO. H_2CO can be produced in a weakly reducing atmosphere by the reaction sequence (Pinto, Gladstone and Yung 1980)

$$H + CO + M \rightarrow HCO + M$$
$$HCO + HCO \rightarrow H_2CO + CO.$$

H and CO are derived from photolysis of H_2O and CO_2, respectively. HCO is the formyl radical, and M represents a third molecule necessary to carry off the excess energy of the collision between H and CO.

This reaction sequence is of considerable interest to prebiotic chemists because H_2CO is a key starting ingredient for the synthesis of sugars and amino acids. But it should also have affected the redox budget of the atmosphere. H_2CO is soluble in rainwater and, hence, would have been removed by precipitation. Once in the ocean, it would have been hydrolyzed to methylene glycol, $CH_2(OH)_2$, and converted to other products by processes such as the Canizzaro reaction (Pinto, Gladstone and Yung 1980). It should therefore have been permanently removed from the atmosphere. Because the H_2CO was produced by reduction of CO_2, this process constitutes a sink for atmospheric H_2. We can write the overall reaction as

$$CO_2 + 2H_2 \rightarrow H_2CO + H_2O.$$

Removal of one molecule of H_2CO results in the loss of two molecules of atmospheric H_2. The H_2O produced in this process eventually reenters the ocean, which acts as a virtually infinite source or sink for this molecule.

Just the opposite effect on the atmospheric redox budget is induced by rainout of oxidized trace gases. Hydrogen peroxide, H_2O_2, provides a good example. H_2O_2 is produced photochemically by a series of reactions that add up to

$$2H_2O \rightarrow H_2O_2 + H_2.$$

The H_2O_2 that was rained out of the atmosphere should have eventually reacted with reduced minerals in the Earth's crust or dissolved in the ocean. Thus, each H_2O_2 molecule removed from the atmosphere results in the production of one H_2 molecule. H_2O_2 was the dominant oxidized, soluble species produced in the photochemical model of Kasting, Pollack and Crisp (1984).

More recent studies that include sulfur species (Kasting et al. 1989; Kasting 1990) have shown that oxidation and reduction of sulfur gases released from volcanos should also have had a significant effect on the atmospheric redox budget. Most of the sulfur released from modern volcanos comes out as SO_2. Today, virtually all of this SO_2 is oxidized to H_2SO_4. But on the primitive Earth, volcanic SO_2 could have experienced a variety of fates, including oxidation to H_2SO_4, reduction to H_2S, or conversion to particulate S_8. Most of these sulfur gases are soluble, so they would have been removed from the atmosphere by rainout. Sulfur that was outgassed as SO_2, but exited the atmosphere in a different oxidation state, would have affected the atmospheric H_2 budget in a manner similar to rainout of H_2CO or H_2O_2.

Figure 2.2 shows the calculated fate of volcanic SO_2 as a function of atmospheric H_2 mixing ratio in a background atmosphere similar to that shown in Figure 2.1. The assumed outgassing rate is 3×10^9 SO_2 molecules $cm^{-2}\ s^{-1}$ (8.0×10^{11} mol/yr), about 3–4 times higher than the present sulfur release rate (Zehnder and Zinder 1980; Berresheim and Jaeschke 1983). In this simulation, most of the emitted SO_2 exits the atmosphere without undergoing any chemical change whatsoever. It is either removed by precipitation or it dissolves directly in the ocean. About 10% of the emitted SO_2, though, is converted to H_2SO_4. Another 20% of the emitted SO_2 is converted to H_2S and to the thermodynamically unstable product HSO, if the atmospheric H_2 concentration is high enough.

Rainout of soluble gases can be incorporated into the atmospheric redox budget in a self-consistent manner by defining "neutral" oxidation states for the various volatile elements: SO_2 for sulfur, CO_2 for carbon, H_2O for hydrogen, and N_2 for nitrogen. These definitions are arbitrary, but convenient. They represent reference points against which changes in oxidation state can be measured. If we use these definitions, then the overall atmospheric redox balance can be expressed by an equation of the form

$$\Phi_{out}(H_2) + \Phi_{out}(CO) + 3\Phi_{out}(H_2S) + \ldots = \Phi_{esc}(H_2) + \Phi_{rain}(H_2) \quad (3)$$

Here, $\Phi_{out}(i)$ represents the volcanic outgassing rate of species *i*. The left-hand side of the equation could contain additional terms if other reduced species were present in volcanic emissions, but it does not include SO_2 because SO_2 is defined as "neutral." $\Phi_{rain}(H_2)$ represents the summed rainout and surface deposition rates of reduced and oxidized gases, $\Phi_r(i)$, weighted by their contribution to the hydrogen budget:

$$\Phi_{rain}(H_2) = 2\ \Phi_r(H_2CO) + 3\ \Phi_r(H_2S) + 16\ \Phi_r(S_8) + \ldots \\ - \Phi_r(H_2SO_4) - \Phi_r(H_2O_2) - 0.5\ \Phi_r(HNO) - \ldots \quad (4)$$

The particular gases listed are those that were most abundant in the model of

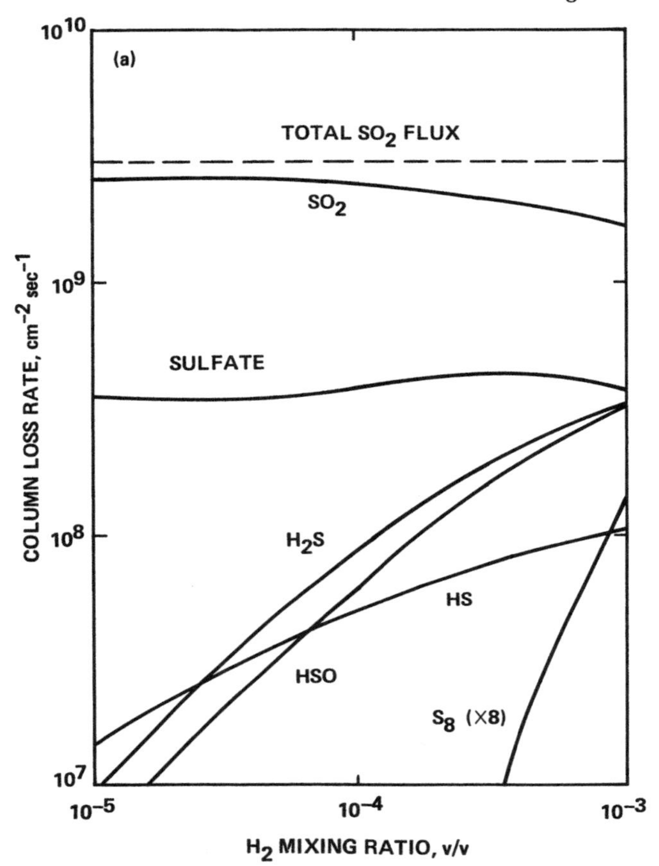

Figure 2.2. Removal rate of various sulfur species from the primitive atmosphere by rainout and by surface deposition, as a function of the assumed H_2 mixing ratio. The concentrations of reduced sulfur species become more abundant at higher H_2 levels. All of the sulfur was assumed to be outgassed initially as SO_2.

Kasting (1990). The coefficients preceding these terms, along with those multiplying the outgassing rates, are derived from redox balance relations involving reactions of "neutral" gases with H_2O and H_2. An atmospheric photochemical model written in mass conservation form should satisfy equation (3) exactly: The atmospheric H_2 concentration will adjust until this equation is balanced.

Although our inclusion of rainout processes in models of the primitive atmosphere may seem like a fine point, it has an important practical consequence: The H_2 mixing ratio in such an atmosphere can never fall below a certain value even if outgassing of reduced volcanic gases were to cease entirely.

In the model of Kasting, Pollack and Crisp (1984), that lower limit on f(H_2) was ~2×10^{-4}, or about 1/5 of the H_2 mixing ratio shown in Figure 2.1. This is more than enough H_2 to keep atmospheric O_2 at extremely low levels.

Thus, an abiotic Earth, or any abiotic planet that, like Earth, has a reduced crust and contains substantial amounts of water at its surface, cannot build up appreciable quantities of O_2 in its atmosphere. This conclusion is valid for the early Earth, and it may eventually be relevant to the detection of life on other planets. Within the next few decades, it may be possible to build a large, space-based interferometer to look for extrasolar, Earth-sized planets and to study their atmospheres spectroscopically (Angel and Woolf 1996). It appears feasible to look for ozone, O_3, which is an indirect indicator of O_2. The presence of substantial amounts of O_3 in the atmosphere of a planet located within the liquid-water habitable zone around another star would be a strong indication of the presence of life.

5. Organic synthesis in weakly reducing atmospheres

Our discussion to this point has implicitly assumed that the composition of primitive volcanic gases was similar to that of volcanic gases today. We shall argue that this was probably not the case, because the early mantle was more reduced. But, first, let us point out the problem raised by the model atmosphere previously described with respect to the origin of life.

If the earliest organisms were based on RNA or some similar molecule, then three types of compounds would be needed to form the RNA: bases (adenine, guanine, cytosine, and uracil), sugars (ribose), and phosphate. Phosphate could have been obtained from weathering of rocks, although it would have to have been converted into an activated, organic phosphate compound in order to link nucleotides together. We ignore this complication. Sugars could have been obtained from polymerization of formaldehyde. The chemical formula for ribose is $C_5H_{10}O_5$, which is just five H_2CO molecules linked together. Formaldehyde is produced in copious quantities in a weakly reducing, CO_2–N_2–H_2O atmosphere by the mechanism described in Section 4. Converting formaldehyde selectively into ribose may have been difficult but, again, we ignore this complication. From a geochemist's standpoint, the starting materials for both phosphates and sugars should have been available on the primitive Earth.

Synthesizing the four nucleic acid bases presents a more difficult problem. The chemical formula for adenine, for example, is $C_5H_5N_5$. Thus, adenine can be formed from five molecules of HCN. Synthesizing the other bases (or amino acids, for that matter) also requires HCN as a starting material. The problem is that HCN itself is difficult to form in a weakly reducing atmosphere. HCN forms readily from spark discharge in gas mixtures containing N_2 and CH_4, but it does not do so in atmospheres containing predominantly N_2 and CO_2 (Chameides and Walker 1981; Stribling and Miller 1987). In the latter case, the N atoms formed by splitting N_2 almost invariably combine with O atoms to yield NO, rather than HCN.

A way around this problem was suggested by Zahnle (1986). He pointed out that HCN could be formed photochemically in a weakly reducing atmosphere containing small amounts of CH_4. In his mechanism, N_2 molecules are broken apart in the ionosphere by ionization followed by dissociative recombination:

$$N_2 + h\nu \rightarrow N_2^+ + e$$
$$N_2^+ + e \rightarrow N + N$$

Many of the N atoms formed are in the excited $N(^2D)$ state (Fox 1993), but most of these will collisionally de-excite to ground state $N(^4S)$. These N atoms then flow down into the stratosphere where they react with the by-products of methane photolysis:

$$CH_4 + h\nu \rightarrow CH_3 + H$$
$$CH_3 + N \rightarrow HCN + H_2$$

Some of the methane will be photolyzed to CH_2, which can also react with N to form HCN.

6. Abiotic sources for methane

Zahnle's mechanism for forming HCN works quite efficiently in an atmosphere containing tens to hundreds of ppm of CH_4. Where, though, could this CH_4 have come from? CH_4 is not released in measurable quantities by surface volcanism (Holland 1984, p. 50). Instead, most of the carbon is released as CO_2, with perhaps 3% of it in the form of CO (Holland 1984). The problem is threefold: (1) The temperatures associated with surface volcanism are too high; (2) the pressures are too low; and (3) the magmas from which volcanic gases are evolved are too oxidized.

We can understand the outgassing problem better by taking advantage of the fact that the gases released from volcanos should be in an approximate state of thermodynamic equilibrium. The ratio of CH_4/CO_2 is therefore governed by the reaction

$$CO_2 + 2H_2O \leftrightarrow CH_4 + 2O_2$$

If we let K_{eq} represent the equilibrium constant for this reaction, we can express the CH_4/CO_2 ratio as

$$\frac{f_{CH_4}}{f_{CO_2}} = K_{eq} \left(\frac{f_{H_2}}{f_{O_2}} \right)^2 \tag{5}$$

Here, f_i represents the fugacity of species *i*. Fugacity is simply the partial pressure of a gas in equilibrium with some mineral assemblage. Petrologists generally represent it by the letter *f*, though, rather than *p*.

For modern surface volcanism, Holland (1984) estimates that the temperature, T, is 1,400–1,500 K, $f_{H_2O} \approx 5$ bars, and $f_{O2} \approx 10^{-8}$ bar. Temperature affects the oxidation state of the released gases because both K_{eq} and f_{O2} increase as T increases. The equilibrium constant can be determined from standard thermodynamic data (Chase et al. 1985), using $K_{eq} = \exp(-\Delta G_R/RT)$. A convenient fit to the thermodynamic data is (Deines et al. 1974)

$$\Delta G_R = -191.11478 - 5.0177493 \times 10^{-4} T_c + 7.20004174 \times 10^{-8} T_c^2 + 3.3498290 \times 10^{-10} T_c^3 - 7.8242836 \times 10^{-14} T_c^4, \tag{6}$$

where T_C is temperature in °C. The oxygen fugacity is given by Mueller and Saxena (1977) as

$$\log f_{O_2} = -\frac{A}{T} + B + \frac{C(P-1)}{T}. \tag{7}$$

Here, P is pressure in bars, and A, B, and C are constants that depend on the particular mineral assemblage with which the gases are equilibrating (see Table 2.1). Because the square of f_{O_2} increases faster with T than does K_{eq}, CH_4 is not abundant in gases released directly from molten magmas. Holland's estimate for the modern CH_4/CO_2 ratio in volcanic gases is $\sim 10^{-11}$

How, then, could there have been an abiotic source for methane when all these factors are working against it? The answer is that one does not want to rely on surface volcanism. Volcanic outgassing also occurs today at the midocean ridges. Indeed, CO_2 is currently thought to be outgassed there at a rate of $(1–2) \times 10^{12}$ mol/yr (DesMarais and Moore 1984; Marty and Jambon 1987), or about 15%–25% of the total CO_2 outgassing rate quoted earlier. Submarine volcanism provides conditions that are much more favorable to the formation of methane. Welhan (1988) has observed that submarine hydrothermal fluids typically have a CH_4/CO_2 ratio on the order of 1%–2%. These measurements were made at the East Pacific Rise, 21° north latitude, in a location where biological contamination is not thought to be a problem. Carbon isotopic ratios

Table 2.1. *Constants for calculating fo_2 of solid buffers*

Buffer	Abbreviation	A	B	C	Note
SiO_2–Fe_2SiO_4–Fe_3O_4	QFM	25,738	9.00	0.092	1
Fe–FeO	IW	27,215	6.57	0.055	2

[1] Wones and Gilbert 1969
[2] Eugster and Wones 1962

in the evolved CO_2 and CH_4 suggest that these gases last equilibrated with each other deep within the vent system at a temperature of ~900 K. By following an adiabat down from the top of the ridge (2.5 km depth), we infer that this temperature is reached at a depth of ~4 km. The hydrostatic pressure at this depth, which equals f_{H_2O}, is about 400 bars. If we assume that the oxygen fugacity is buffered by equilibrium with the quartz–fayalite–magnetite, or QFM, mineral buffer (see next section), then $f_{O_2} \approx 10^{-19.6}$ bar, and the CH_4/CO_2 ratio predicted by equation (5) is 7×10^{-3}. The reasonably close agreement between this theoretically predicted ratio and the observed ratio in vent fluids gives one some confidence that the system is being modeled correctly.

It should be noted parenthetically that methane can be generated from reactions of basalt with CO_2-bearing fluids at temperatures as low as 300 °C (Berndt, Allen and Seyfried 1996). At such low temperatures, CH_4 is thermodynamically favored over CO_2 at current mantle values of f_{O_2}. These reactions can also produce a variety of solid hydrocarbon compounds. But they are probably less important as a global methane source than the higher-temperature reaction considered here.

We conclude from this analysis that the present Earth has an abiotic methane source amounting to about 1% of the carbon flux coming out of the midocean ridges. If we average the two estimates for the mantle carbon flux, this implies that the CH_4 source is 1.5×10^{10} mol/yr, or ~5×10^7 molecules $cm^{-2}s^{-1}$. Adding a methane source of this magnitude to a weakly reducing atmospheric model similar to that shown in Figure 2.1 would generate an atmospheric CH_4 mixing ratio of ~5×10^{-7}, or 0.5 ppm (Figure 2.3). The photochemical model used to obtain this result, which has not yet been published (L. Brown thesis, in preparation), is similar to those described by Kasting, Zahnle and Walker (1983) and Zahnle (1986), except that it also includes sulfur and ammonia photochemistry.

A primitive atmosphere containing 0.5 ppm of CH_4 could provide a small source of HCN, perhaps 2×10^6 molecules $cm^{-2}s^{-1}$, using numbers from Figure 11 of Zahnle (1986). The HCN source increases roughly linearly with atmospheric CH_4 abundance, and the CH_4 abundance is roughly linear in the source

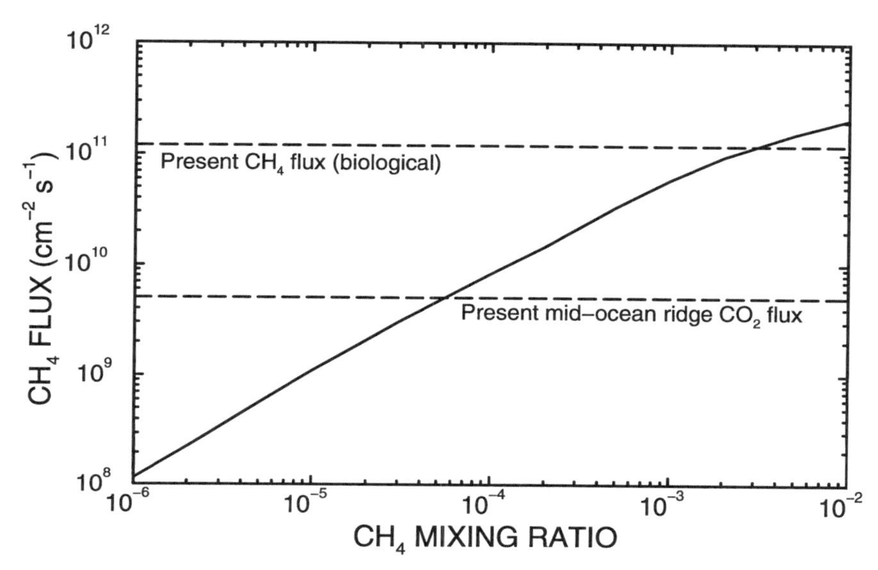

Figure 2.3. The surface CH_4 flux required to support various atmospheric CH_4 mixing ratios, as calculated by L. Brown (thesis, in preparation). In this model, CH_4 is destroyed primarily by short-wavelength solar UV radiation (especially Ly α) because most of the atmospheric OH and O reacts with H_2. If the primitive mantle were more reduced, CH_4 could have been outgassed at the midocean ridges at a rate equal to or greater than the present midocean ridge CO_2 flux.

strength (Figure 2.3), so a larger methane source would increase the rate of HCN production proportionately. In the next section, we consider whether a stronger abiotic methane source might have existed in the distant past.

7. The oxidation state of the early mantle

The third key parameter governing the CH_4/CO_2 ratio in volcanic gases (along with T and f_{H_2O}) is the oxidation state of the upper mantle. The value of f_{O_2} in volcanic emissions is assumed to be governed by equilibrium with magma (for surface volcanism) or with hot basalt (at the ridges). Today, the f_{O_2} of the upper mantle is generally agreed to lie near the QFM buffer, that is, the mineral assemblage SiO_2–Fe_2SiO_4–Fe_3O_4 (Holland 1984; Kasting, Eggler and Raeburn 1993, and references therein). This may be thought of as the boundary between ferrous iron, Fe^{+2}, and ferric iron, Fe^{+3}.

On the early Earth, the mantle oxygen fugacity could have been much lower, perhaps as low as the iron–wüstite (IW) buffer, Fe–FeO. The effective f_{O_2} in equilibrium with IW is typically about four to five orders of magnitude lower than at QFM. An f_{O_2} near IW is, after all, just what one might expect in the immediate aftermath of core formation, when metallic iron and nickel separated from the largely silicate mantle. But the argument is not straightforward, as the Earth may have been veneered by a layer of more highly oxidized material that arrived after the core formed (Turekian and Clark 1969), or it could have been oxidized by reaction of water with metallic iron, followed by loss of hydrogen to space (Dreibus and Wänke 1989). Disproportionation of FeO in the lower mantle, followed by preferential transport of ferric iron to the upper mantle, has also been suggested (Mao and Bell 1977). A discussion of the various possibilities, plus additional references, can be found in Kasting, Eggler and Raeburn (1993). From a purely theoretical standpoint, it is difficult to determine what the oxidation state of the primitive mantle must have been.

One observation that has been cited repeatedly as evidence that the early mantle was not highly reduced is the apparent overabundance of siderophile ("iron-loving") elements in the upper mantle relative to that expected at equilibrium between molten iron and silicate (e.g., Ringwood 1977, 1979; Jones and Drake 1986; Newsom 1990). But this argument may have been countered, or at least weakened, by a recent experimental study by Walter and Thibault (1995), in which they showed that the metal–silicate partition coefficients of two siderophile elements decrease with increasing degree of partial melting. High-pressure, high-temperature measurements of siderophile partition coefficients by Righter and Drake (1997) support this conclusion. According to these latter authors, if siderophile abundances were reset during the extensive melting following a Moon-forming impact, there may be little problem in reconciling them with a highly reduced upper mantle.

What evidence is there that the upper mantle was indeed more reduced in the past? Kasting, Eggler and Raeburn (1993) mentioned evidence from sulfide inclusions in 3.3 billion-year-old diamonds. Thermodynamic analyses of these inclusions indicate that they formed in an environment where f_{O_2} was 2–4 log units below QFM. It is not certain, however, that the environments where these diamonds formed were representative of the bulk upper mantle at that time.

An entirely independent line of reasoning can be developed by considering the atmospheric hydrogen budget discussed earlier in this chapter. We argued earlier that H_2 and CO are being released from modern volcanos at a combined rate of 1.4×10^{12} mol/yr. Outgassing rates on the early Earth should have been 3–4 times higher simply as a consequence of increased heat flow. If most of this hydrogen escaped to space, as seems likely, the total rate of H_2 loss would have been ~5×10^{12} mol/yr. These high hydrogen loss rates could have persisted until ~2.2 Ga, the time when free O_2 is first thought to have become abundant in the atmosphere (Walker et al. 1983; Kasting 1993; Holland 1994). Thus, the total amount of H_2 lost over the first 2.3 billion years of the Earth's history would have been of the order of 1×10^{22} moles, or about one-eighth of the hydrogen in the modern ocean. If one assumes that all of this hydrogen existed as H_2O initially, an equivalent amount of oxygen would have been left behind by the escape process. The amount of oxidizing power involved is very large. By comparison, burial of all the organic carbon in Earth's crust, 1.7×10^{21} mol (Holland 1978, p. 275), would have produced only about one-third this much oxygen.

Where did all of this oxygen go? The oxidized crustal reservoirs of carbonate, sulfate, and ferric iron (Lasaga, Berner and Garrels 1985) are large enough to have absorbed it if the carbon, sulfur, and iron were initially reduced. But, the hydrogen released from volcanos was being generated in the upper mantle. Most of the oxygen should therefore have been consumed by oxidizing ferrous iron to ferric iron in the mantle. Using numbers from Kasting, Eggler and Raeburn (1993), the upper mantle down to 700 km depth could have had its Fe^{+3}/Fe^{+2} ratio changed from 0 to 0.02 by the addition of 1.4 $\times 10^{22}$ mol O, only slightly more than the amount just estimated. This would have been sufficient to raise the mantle f_{O_2} from IW to QFM. The water needed to oxidize the mantle could have been provided by subduction of hydrated seafloor, which recycles approximately half the modern ocean downward in 2.5 Ga (Kasting, Eggler and Raeburn 1993).

This analysis demonstrates that hydrogen escape would have been an important process on the early Earth even if volcanic gases were released with their present composition, and that the mantle redox state should have evolved with time. Now, consider a more self-consistent scenario in which the mantle oxygen fugacity was initially near IW. In this case, the H_2/H_2O ratio in surface volcanic gases would have been ~1, instead of 0.01, and the CO/CO_2 ratio would have been ~3, instead of 0.03 (Holland 1984, p. 50). The effective hydrogen outgassing rate could therefore have been as much as 50 times higher than the values we have been assuming. Hydrogen escape to space should also have been faster, although at these fluxes the H_2 escape rate is likely to have been limited by the available solar EUV (extreme ultraviolet) energy, rather than by diffusion (see next section). Thus, in reality, this mechanism for oxidizing the mantle should have been even more efficient than we have just calculated.

Most importantly, for our purposes, the lower oxidation state of the early mantle would imply that the gases released from submarine hydrothermal vents should also have been more reduced. According to equation (5), these gases would become methane dominated if f_{O_2} were more than about 1 log unit lower than its current value. Thus, CH_4 could have been released from the midocean ridges at the same rate as CO_2 is released today, or perhaps even several times faster, allowing for increased geothermal heat flow. Figure 2.3 shows that such an outgassing rate could have produced atmospheric CH_4 concentrations on the order of 50–200 ppm. This is enough to allow Zahnle's HCN formation mechanism to operate with high efficiency. We conclude that a weakly reduced atmosphere supplied with methane from a highly reduced mantle could have provided a large, endogenous source of organic molecules on the early Earth.

8. Directions for further research

Although we believe that we have constructed a credible model of the primitive atmosphere, there are several different things that could be done to strengthen and extend it. The most important would be to identify another indicator of the past mantle redox state in addition to inclusions in diamonds. Several investigators have noted that the transition metals Cr and V undergo redox-dependent changes in valence state that may make them useful as mantle redox indicators (see references in Kasting, Eggler and Raeburn 1993), but these arguments have yet to be presented in a convincing manner. Ferric/ferrous ratios in ancient basalts and komatiites, which have been suggested as mantle redox indicators by some authors, are unreliable in this regard. Low-temperature reactions with water at any time during the rocks' history could have converted ferrous iron into magnetite (Berndt, Allen and Seyfried 1996, and references therein).

In addition to demonstrating that methane should have been present in the early atmosphere, further theoretical work needs to be done to understand its effect on the early climate. Preliminary calculations by one of us (L. Brown) show that methane concentrations of 100 ppm could have provided an additional 14 W/m^2 of infrared radiative forcing at the tropopause. This is approximately equivalent to four doublings of atmospheric CO_2 (each of which would itself provide ~4 W/m^2 of greenhouse forcing). Thus, the presence of 100 ppm or more of methane in the Archean/Early Proterozoic atmosphere could explain the observed factor of 20 discrepancy in atmospheric CO_2 between climate model predictions and the upper limit derived from paleosols (Rye, Kuo and Holland 1995). Perhaps more significantly, methane could provide an explanation for the anomalously warm climate of early Mars (Kasting 1991). We have yet to perform self-consistent, radiative–convective climate calculations, however, to demonstrate exactly how much greenhouse warming is possible from the combination of CH_4, CO_2, and H_2O.

Finally, additional calculations should be performed to determine the escape rate of hydrogen from the type of weakly reduced primitive atmosphere we have described. All recent photochemical model simulations have presumed that hydrogen escaped at the diffusion-limited rate, as described in Section 3. This assumption is appropriate if one is interested in estimating upper limits on atmospheric O_2 because it maximizes the abiotic oxygen production rate. But it may underestimate the actual concentrations of H_2 and CH_4 in the primitive atmosphere, particularly if early volcanic emissions were highly reduced, as suggested here. Such high rates of hydrogen input could have been balanced by escape, but only if H_2 accumulated in the atmosphere to the point where it was lost hydrodynamically (e.g., Watson, Donahue and Walker 1981). The escape rate would then have been determined by the balance between input of solar EUV energy and removal of energy by escaping hydrogen atoms and by radiation to space by CO_2, CH_4, and other infrared-active gases. It is relevant to note that the upper atmospheres of Mars and Venus are much colder than Earth's, 350–400 K compared to 1,500–2,000 K, because of their relative lack of O_2 (a good UV absorber) and their high concentrations of CO_2 (a good infrared radiator). The possible presence of CH_4 and its photolysis by-products could have depressed the exospheric temperature still further (McGovern 1969). Under these energy-limited conditions, hydrogen escape could have been severely inhibited, and both H_2 and CH_4 could have accumulated to higher atmospheric concentrations than those predicted here.

We conclude that a lot of work still needs to be done to understand the composition of the early atmosphere and its implications for the origin of life. Although the analysis presented in this chapter is far from complete, we hope that the principles outlined here will continue to guide future investigations.

References

Ahrens, T. J., O'Keefe, J. D., and Lange, M. A. 1989. Formation of atmospheres during accretion of the terrestrial planets. In *Origin and Evolution of Planetary and Satellite Atmospheres*, ed. S. K. Atreya, J. B. Pollack, and M. S. Matthews, pp. 328–385. Tucson: University of Arizona Press.

Angel, J. R. P., and Woolf, N. J. 1996. Searching for life on other planets. *Scientific American* 274 (April):60–66.

Bada, J. L., Bigham, C., and Miller, S. L. 1994. Impact melting of frozen oceans on the early Earth: implications for the origin of life. *Proc. Nat. Acad. Sci.* 91:1248–1250.

Berndt, M. E., Allen, D. E., and Seyfried, W. E. 1996. Reduction of CO_2 during serpentinization of olivine at 300° C and 500 bar. *Geology* 24:351–354.

Berkner, L. V., and Marshall, L. C. 1965. On the origin and rise of oxygen concentration in the Earth's atmosphere. *J. Atmos. Sci.* 22:225–261.

Berresheim, H., and Jaeschke, W. 1983. The contribution of volcanoes to the global atmospheric sulfur budget. *J. Geophys. Res.* 88:3732–3740.

Bowring, S. A., and Housh, T. 1995. The Earth's early evolution. *Science* 269:1535–1540.

Braterman, P. S., Cairns-Smith, A. G., and Sloper, R. W. 1983. Photooxidation of hydrated Fe^{+2} – significance for banded iron formations. *Nature* 303:163–164.

Chameides, W. L., and Walker, J. C. G. 1981. Rates of fixation by lightning of carbon and nitrogen in possible primitive terrestrial atmospheres. *Origins of Life* 11:291–302.

Chase, M. W., Jr., Davies, C. A., Downey, J. R., Jr., Frurup, D. J., McDonald, R. A., and Syverud, A. N. 1985. *JANAF Thermochemical Tables: Third Edition*. Washington: Amer. Chem. Soc.

Chyba, C. F. 1987. The cometary contribution to the oceans of primitive Earth. *Nature* 330:632–635.

Chyba, C. F. 1989. Impact delivery and erosion of planetary oceans. *Nature* 343:129–132.

Deines, P., Nafziger, R. H., Ulmer, G. C., and Woermann, E. 1974. Temperature–oxygen fugacity tables for selected gas mixtures in the system C–H–O at one atmosphere total pressure. *Bull. Earth Mineral Sci. Experimental Station* 88. University Park, PA: Penn State University.

DesMarais, D. J., and Moore, J. G. 1984. Carbon and its isotopes in mid-oceanic basaltic glasses. *Earth Planet. Sci. Lett.* 69:43–57.

Dreibus, G., and Wänke, H. 1989. Supply and loss of volatile constituents during the accretion of terrestrial planets. In *Origin and Evolution of Planetary and Satellite Atmospheres*, ed. S. K. Atreya, J. B. Pollack, and M. S. Matthews, pp. 268–288. Tucson: University of Arizona Press.

Eugster, H. P., and Wones, D. R. 1962. Stability relations of the ferruginous biotite, annite. *J. Petrol.* 3:82–125.

Fox, J. L. 1993. The production and escape of nitrogen atoms on Mars. *J. Geophys. Res.* 98:3297–3310.

Gough, D. O. 1981. Solar interior structure and luminosity variations. *Solar Phys.* 74:21–34.

Holland, H. D. 1962. Model for the evolution of the Earth's atmosphere. In *Petrologic Studies: A Volume to Honor A. F. Buddington*, ed. A. E. J. Engel, H. L. James, and B. F. Leonard, pp. 447–477. New York: Geol. Soc. Am.

Holland, H. D. 1978. *The Chemistry of the Atmosphere and Oceans*. New York: Wiley.

Holland, H. D. 1984. *The Chemical Evolution of the Atmosphere and Oceans*. Princeton: Princeton University Press.

Holland, H. D. 1994. Early Proterozoic atmospheric change. In *Early Life on Earth*, ed. S. Bengtson, pp. 237–244. New York: Columbia University Press.

Hunten, D. M. 1973. The escape of light gases from planetary atmospheres. *J. Atmos. Sci.* 30:1481–1494.

Jones, J. H., and Drake, M. J. 1986. Geochemical constraints on core formation in the Earth. *Nature* 322:221–228.

Kasting, J. F. 1987. Theoretical constraints on oxygen and carbon dioxide concentrations in the Precambrian atmosphere. *Precambrian Res.* 34:205–229.

Kasting, J. F. 1990. Bolide impacts and the oxidation state of carbon in the Earth's early atmosphere. *Origins Life Evol. Biosphere* 20:199– 231.

Kasting, J. F. 1991. CO_2 condensation and the climate of early Mars. *Icarus* 94:1–13.

Kasting, J. F. 1992. Proterozoic climates: the effect of changing atmospheric carbon dioxide concentrations. In *The Proterozoic Biosphere: A Multidisciplinary*

Study, ed. J. W. Schopf and C. Klein, pp. 165–168. Cambridge: Cambridge University Press.

Kasting, J. F. 1993. Earth's early atmosphere. *Science* 259:920–926.

Kasting, J. F., and Ackerman, T. P. 1986. Climatic consequences of very high CO_2 levels in the earth's early atmosphere. *Science* 234:1383–1385.

Kasting, J. F., Eggler, D. H., and Raeburn, S. P. 1993. Mantle redox evolution and the oxidation state of the Archean atmosphere. *J. Geol.* 101:245–257.

Kasting, J. F., Pollack, J. B., and Crisp, D. 1984. Effects of high CO_2 levels on surface temperature and atmospheric oxidation state of the early earth. *J. Atmos. Chem.* 1:403–428.

Kasting, J. F., Zahnle, K. J., Pinto, J. P., and Young, A. T. 1989. Sulfur, ultraviolet radiation, and the early evolution of life. *Origins Life Evol. Biosphere* 19:95–108.

Kasting, J. F., Zahnle, K. J., and Walker, J. C. G. 1983. Photochemistry of methane in the Earth's early atmosphere. *Precambrian Res.* 20:121–148.

Lasaga, A. C., Berner, R. A., and Garrels, R. M. 1985. An improved geochemical model of atmospheric CO_2 fluctuations over the past 100 million years. In *The Carbon Cycle and Atmospheric CO_2: Natural Variations Archean to Present*, ed. E. T. Sundquist and W. S. Broecker, pp. 397–411. Washington, DC: American Geophysical Union.

Mao, H. K., and Bell, P. M. 1977. Disproportionation equilibrium in iron-bearing systems at pressures above 100 Kbar with applications to the chemistry of the Earth's mantle. In *Energetics of Geological Processes*, ed. S. K. Saxena and S. Bhattachargi, pp. 237–249. New York: Springer-Verlag.

Marty, B., and Jambon, A. 1987. C/^{3}He in volatile fluxes from the solid Earth: implications for carbon geodynamics. *Earth Planet. Sci. Lett.* 83:16–26.

Matsui, T., and Abe, Y. 1986. Evolution of an impact-induced atmosphere and magma ocean on the accreting Earth. *Nature* 319:303–305.

McGovern, W. E. 1969. The primitive Earth: thermal models of the upper atmosphere for a methane-dominated environment. *J. Atmos. Sci.* 26:623–635.

Miller, S. L. 1953. A production of amino acids under possible primitive Earth conditions. *Science* 117:528–529.

Miller, S. L. 1955. Production of some organic compounds under possible primitive Earth conditions. *J. Am. Chem. Soc.* 77:2351–2361.

Mojzsis, S. J., Arrhenius, G., McKeegan, K. D., Harrison, T. M., Nutman, A. P., and Friend, C. R. L. 1996. Evidence for life on Earth before 3,800 million years ago. *Nature* 384:55–59.

Mueller, R. F., and Saxena, S. K. 1977. *Chemical Petrology.* New York: Springer-Verlag.

Newsom, H. E. 1990. Accretion and core formation in the Earth: evidence from siderophile elements. In *Origin of the Earth*, ed. H. E. Newsom and J. H. Jones, pp. 273–288. New York: Oxford University Press.

Owen, T., Bar-Nun, A., and Kleinfeld, I. 1992. Possible cometary origin of heavy noble gases in the atmospheres of Venus, Earth, and Mars. *Nature* 358:43–46.

Owen, T., Cess, R. D., and Ramanathan, V. 1979. Early Earth: An enhanced carbon dioxide greenhouse to compensate for reduced solar luminosity. *Nature* 277:640–642.

Pinto, J. P., Gladstone, C. R., and Yung, Y. L. 1980. Photochemical production of formaldehyde in the earth's primitive atmosphere. *Science* 210:183–185.

Righter, K., and Drake, M. J. 1997. Metal/silicate equilibrium in a homogeneously accreting Earth: new results for Re. *Earth Planet. Sci. Lett.* 146:541–553.

Ringwood, A. E. 1977. Composition of the core and implications for origin of the earth. *Geochem. J.* 11:111–135.

Ringwood, A. E. 1979. *The Origin of the Earth and Moon*. New York: Springer-Verlag.
Rubey, W. W. 1955. Development of the hydrosphere and atmosphere, with special reference to probable composition of the early atmosphere. In *Crust of the Earth*, ed. A. Poldervaart, pp. 631–650. New York: Geol. Soc. Am.
Rye, R., Kuo, P. H., and Holland, H. D. 1995. Atmospheric carbon dioxide concentrations before 2.2 billion years ago. *Nature* 378:603–605.
Sagan, C., and Mullen, G. 1972. Earth and Mars: Evolution of atmospheres and surface temperatures. *Science* 177:52–56.
Schopf, J. W. 1983. *Earth's Earliest Biosphere: Its Origin and Evolution*. Princeton: Princeton University Press.
Sleep, N. H., Zahnle, K. J., Kasting, J. F., and Morowitz, H. J. 1989. Annihilation of ecosystems by large asteroid impacts on the early Earth. *Nature* 342:139–142.
Stevenson, D. J. 1983. The nature of the Earth prior to the oldest known rock record: the Hadean Earth. In *Earth's Earliest Biosphere: Its Origin and Evolution*, ed. J. W. Schopf, pp. 32–40. Princeton: Princeton University Press.
Stribling, R., and Miller, S. L. 1987. Energy yields for hydrogen cyanide and formaldehyde syntheses: the HCN and amino acid concentrations in the primitive ocean. *Origins of Life* 17:261–273.
Turekian, K. K., and Clark, S. P. 1969. Inhomogeneous accretion of the Earth from the primitive solar nebula. *Earth Planet. Sci. Lett.* 6:346–348.
Walker, J. C. G. 1977. *Evolution of the Atmosphere*. New York: Macmillan.
Walker, J. C. G. 1985. Carbon dioxide on the early Earth. *Origins of Life* 16:117–127.
Walker, J. C. G., Hays, P. B., and Kasting, J. F. 1981. A negative feedback mechanism for the long-term stabilization of Earth's surface temperature. *J. Geophys. Res.* 86:9776–9782.
Walker, J. C. G., Klein, C., Schidlowski, M., Schopf, J. W., Stevenson, D. J., and Walter, M. R. 1983. Environmental evolution of the Archean-Early Proterozoic Earth. In *Earth's Earliest Biosphere: Its Origin and Evolution*, ed. J. W. Schopf, pp. 260–290. Princeton: Princeton University Press.
Walter, M. J., and Thibault, Y. 1995. Partitioning of tungsten and molybdenum between metallic liquid and silicate melt. *Science* 270:1186–1189.
Watson, A. J., Donahue, T. M., Walker, J. C. G. 1981. The dynamics of a rapidly escaping atmosphere: applications to the evolution of Earth and Venus. *Icarus* 48:150–166.
Welhan, J. A. 1988. Origins of methane in hydrothermal systems. *Chem. Geol.* 71:183–198.
Wones, D. R., and Gilbert, M. C. 1969. The fayalite–magnetite–quartz assemblage between 600° and 800° C. *Amer. J. Sci.* 267-A:480–488.
Zahnle, K. J. 1986. Photochemistry of methane and the formation of hydrocyanic acid (HCN) in the Earth's early atmosphere. *J. Geophys. Res.* 91:2819–2834.
Zehnder, A. J. B., and Zinder, S. H. 1980. The sulfur cycle. In *The Natural Environment and the Biogeochemical Cycles*, ed. O. Hutzinger, pp. 105–145. Berlin: Springer-Verlag.

Part II

Organic molecules on the early Earth

3

The endogenous synthesis of organic compounds

STANLEY L. MILLER
Department of Chemistry and Biochemistry
University of California, San Diego
La Jolla

Introduction

It is now generally accepted that life arose on the Earth early in its history. The sequence of events started with the synthesis of simple organic compounds by various processes. These simple organic compounds reacted to form polymers, which in turn reacted to form structures of greater and greater complexity until one was formed that could be called living. This is a relatively new idea, first expressed clearly by Oparin (1938) with contributions by Haldane (1929), Urey (1952), and Bernal (1951). It is sometimes referred to as either the Oparin–Haldane or the Heterotrophic Hypothesis (Horowitz 1945). Older ideas that are no longer regarded seriously include the seeding of the Earth from another planet (panspermia), the origin of life at the present time from decaying organic material (spontaneous generation), and the origin of an organism early in the Earth's history by an extremely improbable event. This last process assumed that the primitive Earth was essentially the same as at present, except possibly for the absence of molecular oxygen, and that the first organism would have to have been autotrophic. That is, it would have to have been photosynthetic, using light energy to synthesize all its organic compounds from CO_2 and H_2O.

One of Oparin's major contributions was to propose that the first organisms were heterotrophic, that is, organisms that utilized prebiotically produced organic compounds available in the environment. They still had to build proteins, nucleic acids, and so forth, within their cells, but they did not have to synthesize *de novo* the amino acids, purines, pyrimidines, and sugars that make up these biopolymers. Oparin as well as Urey also proposed that the early Earth had a reducing atmosphere composed of methane (CH_4), ammonia (NH_3), water vapor (H_2O), and molecular hydrogen (H_2), and that organic compounds might be synthesized in such an atmosphere. On the basis of Urey's and Oparin's ideas, it was shown that amino acids could be synthesized in surprisingly high yield from the action of spark, simulating lightning discharges, on just such a strongly reducing atmosphere (Miller 1953). There is now a large literature dealing with the prebiotic synthesis of organic compounds and polymers, which is too extensive to review here, so only the highlights will be covered (for reviews, see Kenyon and Steinman 1969; Lemmon 1970; Miller and Orgel 1974; Miller 1987; Oró, Miller and Lazcano 1990; Mason 1991; Chyba and McDonald 1995).

In addition to these prebiotic syntheses under various assumed prebiotic conditions, there is a large literature on organic syntheses on Jupiter, Saturn, and Saturn's satellite, Titan. There is also organic synthesis in interstellar space, as well as on asteroids and on interstellar dust particles.

This chapter will examine these syntheses and try to evaluate their relative importance in the origin-of-life process.

The composition of the primitive atmosphere

There is only limited agreement regarding the probable constituents of the primitive atmosphere. It is to be noted that because no rocks older than 3.8 billion years in age are known, there is no geological evidence concerning the conditions on the Earth from the time of its formation, 4.5 billion years ago, to the time of deposition of these oldest known rocks. Even the 3.8-billion-year-old Isua rocks in Greenland are not sufficiently well preserved to provide much evidence about the atmosphere at that time. Proposed atmospheres and the reasons given to favor them will not be discussed here. However, as will be shown, the more-reducing atmospheres are more favorable for the synthesis of organic compounds, both in terms of the yield and the variety of compounds obtained. However, such considerations cannot prove that the Earth had a particular type of primitive atmosphere. There is considerable opinion that strongly reducing conditions were never present on the primitive Earth, but this would mean that the organic compounds would have to be brought in on comets and meteorites, and this assumption has its own set of problems.

Temperature of the primitive Earth

The necessity to accumulate organic compounds for the synthesis of the first organism requires a low temperature Earth; otherwise, the organic compounds would decompose. The half-lives for decomposition at 25 °C vary from several billion years for the amino acid alanine, to a few million years for the amino acid serine, to thousands of years for hydrolytic decomposition of peptides and polynucleotides, to a few hundred years for sugars. Because the temperature coefficients for these decompositions are large, the half-lives would be much less at temperatures of 50 °C or 100 °C. Conversely, the half-lives would be longer at 0 °C. The rates of hydrolysis of peptide and polynucleotide polymers, and of decomposition of sugars, are so large that it seems impossible that such compounds could have accumulated in aqueous solution and been used in the first organism, unless the temperature was low.

These considerations argue strongly in favor of low-temperature synthesis and accumulation of organic compounds, but there is some opinion that favors a high-temperature origin of life. A high-temperature origin would have to be very rapid, that is, not much greater than the half-life for decomposition of the essential organic building blocks.

Energy sources

A wide variety of energy sources has been utilized with various gas mixtures since the first experiments using electric discharges. The importance of a given energy source is determined by the product of the energy available and its efficiency for organic compound synthesis. Since neither factor can be evaluated with precision, only a qualitative assessment of the energy sources can be made. It should be emphasized that neither any single source of energy, nor any single process, is likely to account for all the organic compounds on the primitive Earth (Miller, Urey and Oró 1976). An estimate (Miller and Urey 1959) of the sources of energy on the Earth at the present time is given in Table 3.1. A somewhat different estimate is given by Chyba and Sagan (1991).

Ultraviolet light was probably the largest source of energy on the primitive Earth. The wavelengths absorbed by the primitive atmospheric constituents are all below 2,000 Å except for ammonia (<2,300 Å) and hydrogen sulfide, H_2S (<2,600 Å). Whether ultraviolet light was the most effective energy source for production of organic compounds is not clear. Most of the photochemical reactions would occur in the upper atmosphere, and the products formed would, for the most part, absorb the longer wavelengths and therefore be decomposed before they reached the oceans, in which they would be dissolved and protected from photochemical destruction. The yield of amino acids from the photolysis of CH_4, NH_3, and H_2O at wavelengths of 1,470 Å and 1,294 Å is quite low (Groth and von Weyssenhoff 1960), probably due to the low yields of hydrogen cyanide. The synthesis of amino acids by the photolysis of CH_4, C_2H_6, NH_3, H_2O, and H_2S mixtures by ultraviolet light of wavelengths greater than 2,000 Å (Khare and Sagan 1971; Sagan and Khare 1971) is also a low-yield synthesis, but the amount of energy is much greater

Table 3.1. *Present sources of energy averaged over the Earth.*

Source	ENERGY (cal/cm²/yr)	ENERGY (J/cm²/yr)
Total radiation from sun	260,000	1,090,000
Ultraviolet light		
<3,000Å	3,400	14,000
<2,500Å	563	2,360
<2,000Å	41	170
<1,500Å	1.7	7
Electric discharges	4[a]	17
Cosmic rays	0.0015	0.006
Radiactivity (to 1.0 km depth)	0.8	3.0
Volcanoes	0.13	0.5
Shock waves	1.1[b]	4.6

[a] 3 cal/cm²/yr of corona discharge + 1 cal/cm²/yr of lightning.
[b] 1 cal/cm²/yr of this is in the shockwave of lightning bolts and is also included under electric discharges.

in this region of the Sun's spectrum. Only H_2S absorbs the ultraviolet light, but the photodissociation of H_2S results in a hydrogen atom having a high kinetic energy, which activates or dissociates the methane, ammonia, and water.

A photochemical source of HCN using very short wavelengths (796–912 Å) has been proposed by Zahnle (1986). The nitrogen atoms produced diffuse lower into the atmosphere and react with CH_2 and CH_3, producing HCN. The process depends on the N atoms reacting with nothing else before they encounter the CH_2 and CH_3. Whether this proposal is valid remains to be determined.

The most widely used sources of energy for laboratory syntheses of prebiotic compounds are electric discharges. These include sparks, semicorona, arc, and silent discharges, with the spark being the most frequently used type. The ease of handling and high efficiency of electric discharges are factors favoring their use, but the most important reason is that electric discharges are very efficient in synthesizing hydrogen cyanide, whereas ultraviolet light is not, except for the very short wavelengths. Hydrogen cyanide is a central intermediate in prebiotic synthesis, being needed for amino acid synthesis via the Strecker reaction or via self-polymerization, and most importantly for the prebiotic synthesis of the purines adenine and guanine.

Use of any of these energy sources for prebiotic syntheses requires activation of molecules in a local area, followed by quenching of the activated mixture, and then protection of the synthesized organic compounds from further influence of the energy source. The quenching and protective steps are critical because the organic compounds will be destroyed if they are subjected continuously to the energy source.

Prebiotic synthesis of organic compounds

Amino acids: synthesis in strongly reducing atmospheres

Mixtures of CH_4, NH_3, and H_2O, with or without added hydrogen (H_2), are considered strongly reducing atmospheres. The atmosphere of Jupiter contains these species, with the H_2 in large excess over the CH_4. The first successful prebiotic amino acid synthesis was carried out using an electric discharge as the energy source and a strongly reducing atmosphere (Figure 3.1; Miller 1953, 1955). The result was a large yield of amino acids, together with hydroxy acids, short aliphatic acids, and urea (Table 3.2). One of the surprising results of this experiment was that the products were not a random mixture of organic compounds; rather, a relatively small number of compounds were produced in substantial yield. In addition, the compounds were, with a few exceptions, of biological importance.

The mechanism of synthesis of the amino and hydroxy acids was investigated (Miller 1957). It was shown that the amino acids were not formed directly in the electric discharge, but were the result of solution reactions of smaller molecules produced in the discharge, in particular, reactions of hydrogen cyanide and aldehydes. The reactions are summarized as follows:

$$\underset{\text{aldehyde}}{RCHO} + HCN + NH_3 \rightleftharpoons \underset{\text{amino nitrile}}{RCH(NH_2)CN} \xrightarrow{H_2O} RCH(NH_2)\overset{O}{\overset{\|}{C}}-NH_2 \xrightarrow{H_2O} \underset{\text{amino acid}}{RCH(NH_2)COOH}$$

$$\underset{\text{aldehyde}}{RCHO} + HCN \rightleftharpoons \underset{\text{hydroxy nitrile}}{RCH(OH)CN} \xrightarrow{H_2O} RCH(OH)\overset{O}{\overset{\|}{C}}-NH_2 \xrightarrow{H_2O} \underset{\text{hydroxy acid}}{RCH(OH)COOH}$$

These reactions were later studied in detail, and their equilibrium and rate constants were measured (Miller and Van Trump 1981). The results show that amino and hydroxy acids can be synthesized at high dilutions of HCN and aldehydes in a simulated primitive ocean. It is also to be noted that the rates of these reactions are rather rapid on the geological time scale; the half-lives for the hydrolysis of the intermediate products in the reactions, amino and hydroxy nitriles, are less than a thousand years at 0 °C.

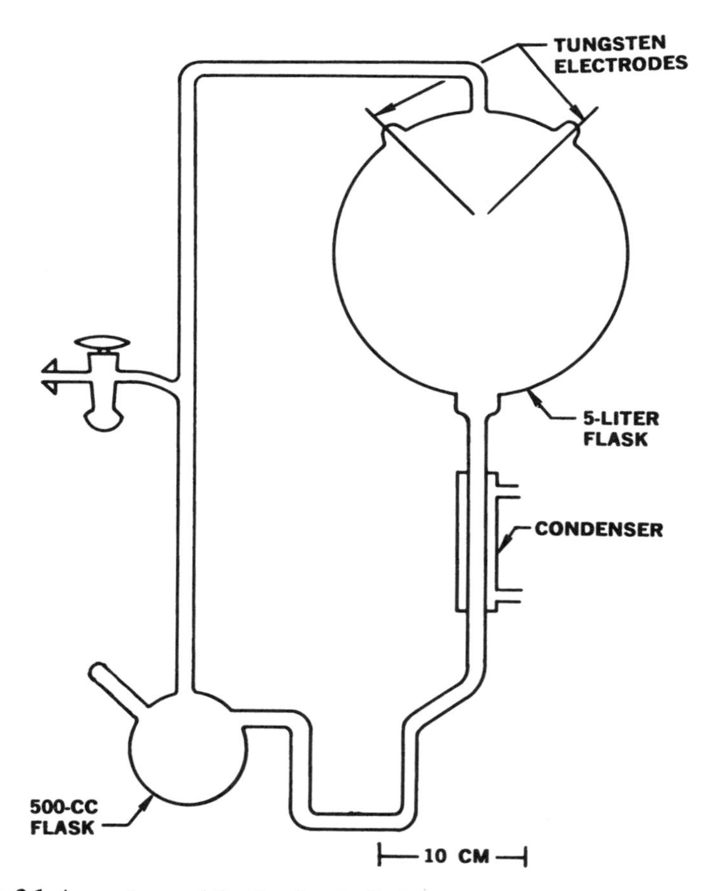

Figure 3.1. Apparatus used for the electric discharge synthesis of amino acids and other organic compounds in a reducing atmosphere of methane (200 torr), ammonia (200 torr), water, and hydrogen (100 torr). The small flask contained about 250 ml of H_2O, and since the temperature near the electrodes was about 65 °C, the p_{H2O} was 190 torr. Results from experiments carried out in this apparatus are summarized in Table 3.2.

This synthesis of amino acids, called the Strecker synthesis, requires the presence of the ammonium ion, NH_4^+ (and therefore the presence of NH_3), in the primitive environment. On the basis of the experimentally determined equilibrium and rate constants it can be shown (Miller and Van Trump 1981) that equal amounts of amino and hydroxy acids are obtained when the NH_4^+ concentration in the simulated primitive ocean is about 0.01 M at pH 8 and 25 °C, with this NH_4^+ concentration being insensitive to temperature and pH. This translates into a low partial pressure of NH_3 in the atmosphere (2×10^{-7} atm at

Table 3.2. *Yields from sparking a mixture of CH_4, NH_3, H_2O, and H_2. The percentage yields are based on carbon (59 mmoles [710mg] of carbon was added as CH_4).*

Compound	YIELD (μmoles)	%
Glycine	630	2.1
Glycolic acid	560	1.9
Sarcosine	50	0.25
Alanine	340	1.7
Lactic acid	310	1.6
N-Methylalanine	10	0.07
α-Amino-*n*-butyric acid	50	0.34
α-Aminoisobutyric acid	1	0.007
α-Hydroxybutyric acid	50	0.34
β-Alanine	150	0.76
Succinic acid	40	0.27
Aspartic acid	4	0.024
Glutamic acid	6	0.051
Iminodiacetic acid	55	0.37
Iminoacetic-propionic acid	15	0.13
Formic acid	2,330	4.0
Acetic acid	150	0.51
Propionic acid	130	0.66
Urea	20	0.034
N-Methyl urea	15	0.051
Total		15.2

0 °C, and 4×10^{-6} atm at 25 °C). Thus, at least a small amount of atmospheric NH_3 would seem necessary for amino acid synthesis. A similar estimate of the NH_4^+ concentration in the primitive ocean can be obtained from the equilibrium decomposition of aspartic acid, a prebiotically produced amino acid (Bada and Miller 1968). Ammonia would have been decomposed in the early environment by ultraviolet light, but mechanisms for its resynthesis are also known. The details of the ammonia balance on the primitive Earth remain to be worked out.

In a typical electric discharge experiment, the partial pressure of CH_4 is 0.1–0.2 atm. This pressure is used for convenience, and it is likely (but has never been demonstrated) that organic compound synthesis would work at much lower partial pressures of methane. There are no estimates available for pCH_4 on the primitive Earth, but low levels (10^{-5} to 10^{-3} atm) seem plausible.

Higher pressures are not reasonable because the sources of energy would convert the CH_4 to organic compounds too rapidly for higher pressures of CH_4 to build up.

Ultraviolet light acting on a strongly reducing mixture of gases is not efficient in producing amino acids, as just discussed. However, ultraviolet light can be efficient in producing aldehydes, as shown experimentally (Ferris and Chen 1975; Bar-Nun and Hartman 1978; Bar-Nun and Chang 1983) and in atmospheric model calculations (Pinto, Gladstone and Young 1980).

Heating reactions (pyrolysis) of CH_4 and NH_3 give very low yields of amino acids. The pyrolysis conditions are from 800 °C to 1,200 °C with contact times of a second or less (Lawless and Boynton 1973). However, the pyrolysis of CH_4 and other hydrocarbons gives good yields of benzene (C_6H_6), phenylacetylene (C_8H_6), and many other hydrocarbons. It can be shown that phenylacetylene would be converted to the amino acids phenylalanine and tyrosine in the primitive ocean (Friedmann and Miller 1969). Pyrolysis of hydro-carbons in the presence of NH_3 gives substantial yields of indole, which can be converted to the amino acid tryptophan in the primitive ocean (Friedmann, Haverland and Miller 1971).

Because NH_3 would have dissolved in the ocean, rather than accumulating in large amounts in the early atmosphere, a mixture of CH_4, N_2, and traces of NH_3, and H_2O is a more realistic atmosphere for the primitive Earth. Such an atmosphere, however, would nevertheless be strongly reducing. Experimental studies (Figure 3.2) have demonstrated that this mixture of gases is quite effective with an electric discharge in producing amino acids (Ring et al. 1972; Wolman, Haverland and Miller 1972). The yields are somewhat lower than with higher partial pressures of NH_3, but the products are more diverse (Table 3.3). Hydroxy acids, short aliphatic acids, and dicarboxylic acids are produced along with the amino acids (Peltzer and Bada 1978; Peltzer et al. 1984). Ten of the 20 amino acids that occur in the proteins of present-day organisms are produced directly in this experiment. With addition of the amino acids asparagine and glutamine, which are formed in the experiment but are hydrolyzed before analysis, and methionine, which is formed when H_2S is added to the reaction mixture (Van Trump and Miller 1972), one can say that 13 of the 20 amino acids occurring in modern proteins can be formed in this single experiment. The amino acid cysteine has been produced prebiotically via photolysis of CH_4, NH_3, H_2O, and H_2S (Khare and Sagan 1971). And the pyrolysis of hydrocarbons, as previously discussed, leads to the amino acids phenylalanine, tyrosine, and tryptophan. This leaves only the three basic amino acids, lysine, arginine, and histidine, unaccounted for. Although there are, so far, no established prebiotic syntheses of these amino acids, there is no fundamental reason that these

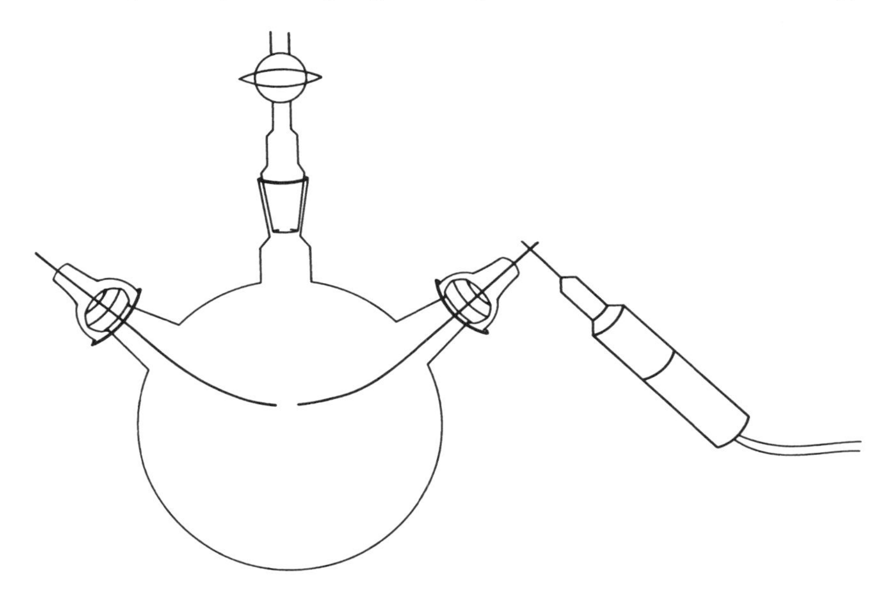

Figure 3.2. Apparatus used for the electric discharge synthesis of amino acids in a strongly reducing atmosphere containing traces of NH_3. The 3-liter flask is shown with the two tungsten electrodes and a spark generator. The second electrode is usually not grounded. In the experiments summarized in Table 3.3, the flask contained 100 ml of 0.05 M NH_4Cl brought to pH 8.7, giving p_{NH_3} of 0.2 torr. The p_{CH_4} was 200 torr and p_{N_2} was 80 torr. Since the temperature was about 30 °C during the 48 hours of sparking, p_{H_2O} was 32 torr.

compounds cannot be synthesized, and their prebiotic syntheses may be accomplished before too long.

Amino acids: synthesis in mildly reducing and nonreducing atmospheres

There has been less experimental work with gas mixtures containing CO and CO_2 as carbon sources instead of CH_4. Spark discharges have been the source of energy most extensively investigated (Abelson 1965; Schlesinger and Miller 1983a, b; Stribling and Miller 1987). Figure 3.3 compares amino acid yields using CH_4, CO, or CO_2 as a carbon source with various amounts of H_2. Separate experiments were performed with and without added NH_3. It is clear from the figure that CH_4 is the best source of amino acids using a spark discharge, but that CO and CO_2 are almost as good if a high H_2-to-C ratio is used. Without added H_2, however, the amino acid yields are very low, especially with CO_2 as the carbon source. The amino acids produced in

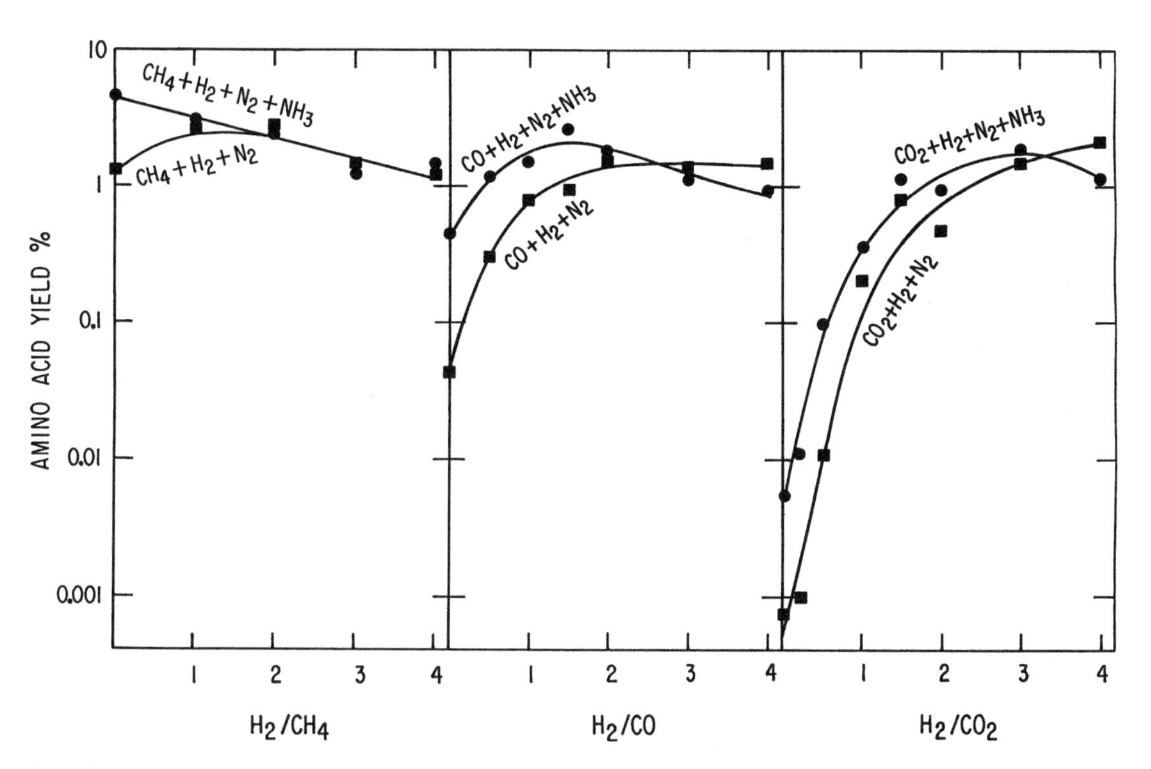

Figure 3.3. Amino acid yields based on initial carbon. In all experiments, p_{N_2} was 100 torr, and p_{CH_4}, pCO, or p_{CO_2} was 100 torr. The flask contained 100 ml H_2O for the curves with N_2 but no NH_3, and it contained 100 ml of 0.05 M NH_4Cl for the curves with $N_2 + NH_3$ (0.2 torr). The flask was kept at room temperature, and the spark generator was operated continuously for 48 hours.

Table 3.3. *Yields from sparking CH_4 (336 mmoles), N_2, and H_2O, with traces of NH_3* *

	μ moles
Glycine	440
Alanine	790
α-Amino-*n*-butyric acid	270
α-Aminoisobutyric acid	~30
Valine	19.5
Norvaline	61
Isovaline	~5
Leucine	11.3
Isoleucine	4.8
Alloisoleucine	5.1
Norleucine	6.0
tert-leucine	<0.02
Proline	1.5
Aspartic acid	34
Glutamic acid	7.7
Serine	5.0
Threonine	~0.8
Allothreonine	~0.8
α,γ-Diaminobutyric acid	33
α-Hydroxy-γ-aminobutyric acid	74
α,β-Diaminopriopionic	6.4
Isoserine	5.5
Sarcosine	55
N-Ethylglycine	30
N-Propylglycine	~2
N-Isopropylglycine	~2
N-Methylalanine	~15
N-Ethylalanine	<0.2
β-Alanine	18.8
β-Amino-*n*-butyric acid	~0.3
β-Aminoisobutyric acid	~0.3
γ-Aminobutyric acid	2.4
N-Methyl-β-alanine	~5
N-Ethyl-β-alanine	~2
Pipecolic acid	0.05

* Yield based on the carbon added as CH_4. Glycine = 0.26%; alanine = 0.71%; total yield of amino acids in the table = 1.90%.

the CH_4 experiments are similar to those shown in Table 3.3. With CO and CO_2, glycine was the predominant amino acid, with little else besides some alanine being produced.

The implication of these results with regard to the composition of the atmosphere of the primitive Earth is that CH_4 is the best carbon source for prebiotic synthesis, especially for amino acid synthesis. Although glycine was essentially the only amino acid synthesized in the spark discharge experiments with CO and CO_2, other amino acids (e.g., serine, aspartic acid, alanine) would probably have been formed from this glycine, H_2CO, and HCN as the primitive ocean matured. Since we do not know which amino acids made up the first organism, we can say only that CO and CO_2 are less favorable than CH_4 for amino acid synthesis, but that the amino acids produced from CO and CO_2 may have been adequate. The synthesis of purines and sugars, to be described later, would not be greatly different with CH_4, CO, or CO_2 as long as sufficient H_2 were available. Although the spark discharge yields of amino acids, HCN, and H_2CO are about the same with CH_4, with $H_2/CO>1$, and with $H_2/CO_2>2$, for the last two reaction mixtures it is not clear how such high molecular hydrogen-to-carbon ratios could have been maintained in the primitive atmosphere, because H_2 escapes gravitationally from the Earth's atmosphere into outer space. These problems are poorly understood and are beyond the scope of this brief review.

Amino acids from HCN polymerization and other sources

The large number of prebiotic amino acid syntheses reported are far beyond this review to cover. Most of these involve electric discharges or ionizing radiation or the reactions of simple prebiotic organic compounds, for example, formaldehyde, formamide, and so forth. However, the amino acids from HCN polymerizations need to be mentioned.

The first prebiotic amino acid synthesis from HCN was by Oró (1961a), confirmed by Lowe, Rees and Markham (1963), and intensively investigated by Ferris and co-workers (Ferris, Donner and Lobo 1973; Ferris et al. 1974a; Ferris et al. 1978). It is also claimed that peptides are produced in the HCN polymerizations (Matthews 1975), but the evidence is weak. The amino acids produced are mostly glycine, along with alanine, aspartic acid, and α-aminoisobutyric acid. Urea, guanidine, and oxalic acid are also products. The yields are not particularly high except for glycine (~1%). The intense interest in HCN polymerization is due to its production of adenine.

Purines and pyrimidines

Hydrogen cyanide plays an important role in the synthesis of purines as well as amino acids. This is illustrated in a remarkable synthesis of the purine adenine. If concentrated solutions of ammonium cyanide are refluxed for a few days, adenine is obtained in up to 0.5% yield along with 4-aminoimidazole-5-carboxamide and the usual cyanide polymer (Oró 1960; Oró and Kimball 1961; 1962).

The mechanism of adenine synthesis in these experiments is probably the following:

H—C≡N (Hydrogen cyanide) + −C≡N (Cyanide) → H—C(=NH)—C≡N —HCN→ N≡C—CH(NH₂)—C≡N (Aminomalonitrile (HCN trimer)) —HCN→

Diaminomaleonitrile (HCN tetramer) —HN=CH—NH₂ (Formamidine)→ Aminoimidazole carbonitrile —HN=CH—NH₂→ Adenine

The difficult step in the synthesis of adenine as summarized is the reaction of the HCN tetramer with formamidine. This step may be bypassed by the photochemical rearrangement of the HCN tetramer to aminoimidazole carbonitrile as shown next, a reaction that proceeds readily in present-day sunlight (Sanchez, Ferris and Orgel 1967, 1968).

Diaminomaleonitrile (HCN tetramer) —hv→ Diaminofumaronitrile —hv→ Aminoimidazole carbonitrile → Adenine

A further possibility is that tetramer formation may have occurred in an eutectic solution of HCN–H_2O. High yields of tetramer (>10%) can be obtained by cooling dilute cyanide solutions to between -10 °C and -30 °C for a few months (Sanchez, Ferris and Orgel 1966a). The production of adenine by a cyanide polymerization is accelerated by the addition of formaldehyde and other aldehydes as catalysts (Schwartz and Goverde 1982; Voet and Schwartz 1983).

Guanine and the additional purines, hypoxanthine, xanthine, and diaminopurine could have been synthesized in the primitive environment by variations of the adenine synthesis, as shown next, using aminoimidazole carbonitrile and aminoimidazole carboxamide (Sanchez, Ferris and Orgel 1968):

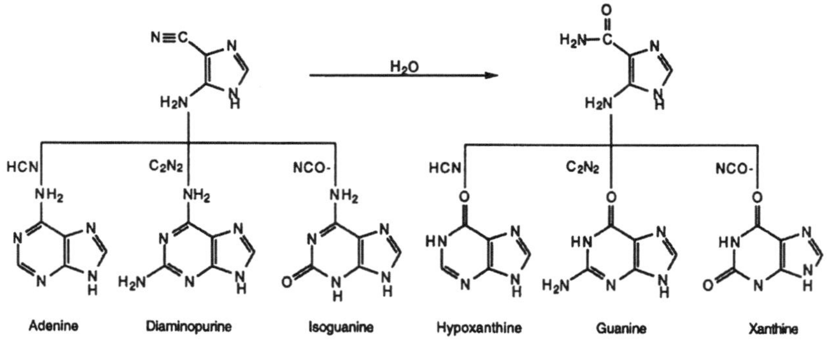

Adenine Diaminopurine Isoguanine Hypoxanthine Guanine Xanthine

The prebiotic synthesis of the pyrimidines of nucleic acids involves cyanoacetylene, which is synthesized in good yield by sparking mixtures of CH_4 and N_2. As shown, cyanoacetylene reacts with cyanate (NCO^-) to give the pyrimidine cytosine (Sanchez, Ferris and Orgel 1966b; Ferris, Sanchez and Orgel 1968):

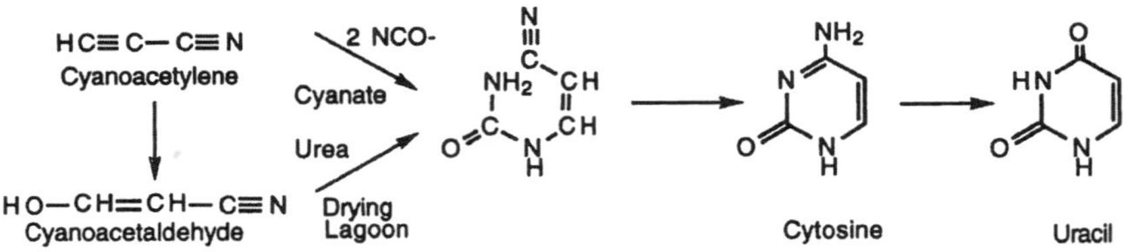

Cyanoacetylene Cyanoacetaldehyde Cyanate Urea Drying Lagoon Cytosine Uracil

As shown, the cytosine, in turn, can be converted to uracil. Cyanate can come from cyanogen (NCCN) or by the decomposition of urea (H_2NCONH_2), compounds that are both produced readily in prebiotic syntheses. A more efficient synthesis of cytosine and uracil starts with cyanoacetaldehyde (derived from the hydration of cyanoacetylene) and urea. These react under the drying lagoon or beach conditions, under which the nonvolatile compounds are concentrated (Robertson and Miller 1995a). A related synthesis uses cyanoacetaldehyde and guanidine to give diaminopyrimidine (Ferris et al. 1974b) with very high yields under drying lagoon conditions (Robertson et al. 1996b). Diaminopyrimidine hydrolyzes to cytosine and uracil.

Very small amounts of pyrimidines are produced in hydrogen cyanide polymerizations (Ferris et al. 1978). It is likely that 5-substituted uracils are prebiotic compounds (Robertson and Miller 1995b) that may have played a major role in the pre-RNA world.

Sugars

The synthesis of sugars from formaldehyde under alkaline conditions was discovered long ago (Butlerow 1861). However, the Butlerow synthesis, or "formose reaction," is very complex and incompletely understood. It depends on the presence of a suitable inorganic catalyst, with calcium hydroxide or calcium carbonate being the most commonly used. In the absence of such mineral catalysts, little or no sugar is obtained. At 100 °C, clays such as kaolin serve to catalyze formation of simple sugars, including ribose, in small yields from dilute (0.01 M) solutions of formaldehyde (Gabel and Ponnamperuma 1967; Reid and Orgel 1967).

The reaction is autocatalytic and proceeds in a series of stages through glycolaldehyde, glyceraldehyde, and dihydroxyacetone, four-carbon sugars, and five-carbon sugars, to give finally hexoses, or six-carbon sugars, including the biologically important sugars glucose and fructose. One proposed reaction sequence is the following:

REVERSE ALDOL

CH_2O (Formaldehyde) $\xrightarrow{CH_2O}$ $CHO{-}CH_2OH$ (Glycoaldehyde) $\xrightarrow{CH_2O}$ $CHO{-}CHOH{-}CH_2OH$ (Glyceraldehyde) $\rightleftharpoons$ $CH_2OH{-}C{=}O{-}CH_2OH$ (Dihydroxyacetone) $\xrightarrow{CH_2O}$ $CH_2OH{-}C{=}O{-}CHOH{-}CH_2OH$ $\rightleftharpoons$ $CHO{-}CHOH{-}CHOH{-}CH_2OH$

Glycoaldehyde + Glyceraldehyde → PENTOSE

Glyceraldehyde + Dihydroxyacetone → HEXOSE

There are two problems with the formose reaction as a source of sugars on the primitive Earth. The first is that the formose reaction gives a wide variety of sugars, both straight chain and branched. Indeed, more than 40 different sugars have been separated from one reaction mixture (Decker, Schweer and Pohlman 1982). Ribose occurs in this mixture, but is not particularly abundant. It is difficult to envision how the relative yield of ribose, needed for formation of RNA, could be greatly increased in this reaction, or how any prebiotic reaction producing sugars could give mostly ribose, although considerable selectivity in the synthesis of ribose-2,4-diphosphate has been accomplished (Müller et al. 1990).

The second problem with ribose is its instability. Its half-life for decomposition is 73 minutes at 100 °C pH7 and 44 years at 0 °C pH7 (Larralde, Robertson and Millér 1995). Therefore, ribose cannot be considered a prebiotic reagent unless it is used immediately after its prebiotic synthesis. The three other pentoses and eight hexoses are similarly unstable. It therefore has become apparent that ribonucleotides could not have been the first components of prebiotic nucleic acids (Shapiro 1988). A number of substitutes for ribose (on the left) have been proposed (Joyce et al. 1987) that include the two compounds on the right:

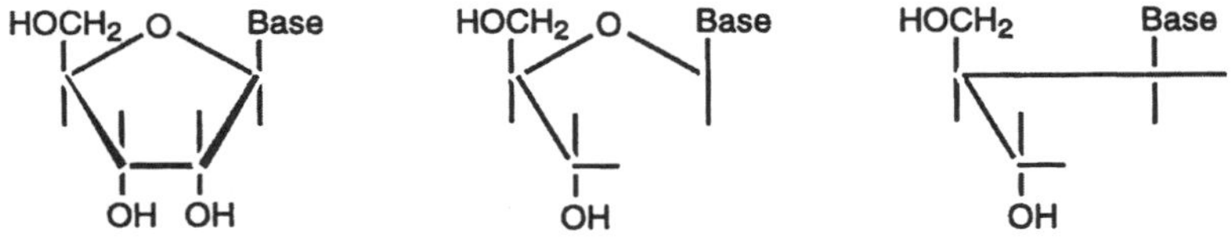

There are many other possible substitutes for ribose, but the two shown here are attractive because they are open-chain, flexible molecules that do not have an asymmetric carbon. The prebiotic synthesis of these compounds has not yet been demonstrated.

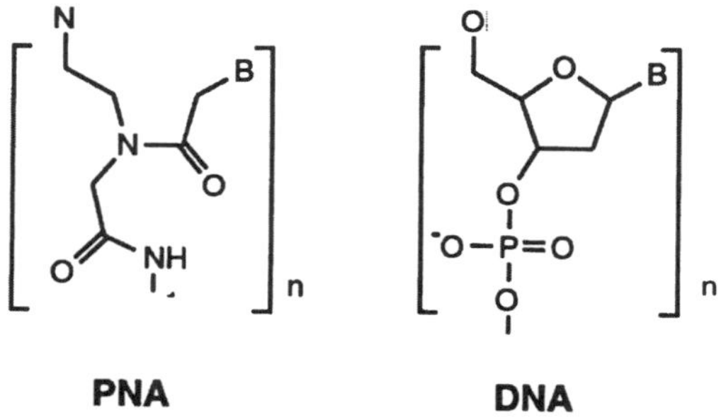

PNA **DNA**

An even more attractive precursor to the ribose phosphate backbone is that in peptide nucleic acids (PNA) (Nielsen 1993). The backbone is a polymer of ethylenediamine monoacetic acid (2-aminoethylglycine). The bases (B) are attached by an acetic acid. This polymer binds DNA very strongly, and double helices of DNA–PNA are stable. The prebiotic synthesis of PNA has not been accomplished, but the monomer ethylenediamine monoacetic acid appears to be a prebiotic compound.

Other prebiotic compounds

In addition to the foregoing, numerous other compounds have been synthesized under primitive Earth conditions, including the following:

dicarboxylic acids
tricarboxylic acids
fatty acids C_2–C_{10} (branched and straight)
fatty alcohols (straight chain via Fischer–Tropsch reaction)
porphin
nicotinonitrile and nicotinamide
triazines
imidazoles

Other prebiotic compounds that may have been involved in polymerization reactions include:

cyanate [NCO^-]
cyanamide [H_2NCN]
cyanamide dimer [$H_2NC(NH)NH$–CN]
dicyanamide [NC–NH–CN]
cyanogen [NC–CN]
HCN tetramer
diiminosuccinonitrile
acylthioesters
phosphate polymers

Compounds that have not been synthesized prebiotically

It is a matter of opinion as to what constitutes a plausible prebiotic synthesis. In some syntheses, the conditions are so forced (e.g., by use of anhydrous solvents) or the concentrations are so high (e.g., 10 M formaldehyde) that the syntheses could not be expected to have occurred extensively (if at all) on the primitive Earth. Reactions under these and other extreme conditions cannot be considered plausibly prebiotic.

There have been many reported "prebiotic syntheses" in which the compound claimed to have been produced has not been properly identified. At present, the best method for unequivocal identification is gas chromatography–mass spectrometry of a suitable derivative, although melting points and mixed melting points can sometimes be used. Identification of a compound cannot be established unequivocally based solely on the use of an amino acid analyzer, or via paper chromatography in multiple solvent systems.

Some important biological compounds that do not yet have adequate prebiotic syntheses are the following:

arginine	porphyrins	riboflavin
lysine	pyridoxal	folic acid
histidine	thiamine	lipoic acid
straight-chain fatty acids		biotin

It is probable that plausible prebiotic syntheses will become available before too long for some of these compounds. In other cases, the compounds may not have been synthesized prebiotically, their occurrence in living systems being a result of intracellular biochemical evolution that occurred after the origin of life.

Production rates of organic compounds on the primitive Earth

A definitive statement cannot be made about the production rates and the concentrations of compounds in the primitive ocean because the atmospheric composition, ambient temperature, and ocean size are unknown. In addition, the mechanisms of destruction of organic compounds in the early environment are poorly understood. Nevertheless, quantitative estimates have been made (Stribling and Miller 1987). The calculations are complicated and uncertain, but the results give an amino acid concentration of 3×10^{-4} M and an adenine concentration of about 1.5×10^{-5} M. This can be considered as a relatively concentrated prebiotic soup.

Other proposed sources of organic compounds

Pyrite

An elaborate autotrophic theory of the origin of life that has been proposed by Wächtershäuser (1992) involves the absorption of anions on pyrite (FeS_2) and the use of $FeS + H_2S$ as a reducing agent. The reaction $FeS + H_2S \rightarrow$ '$FeS_2 + H_2$ constitutes a strong reducing agent, and it will produce H_2 from H_2O, and will reduce alkynes, alkenes, and thiols to hydrocarbons, and ketones to thiols.

Central to any autotrophic theory of the origin of life is the reduction of CO_2 to amino acids, purines, pyrimidines, and other organic compounds. This was shown not to take place (yields of less than 10^{-4}%) under the conditions tried by Keefe et al. (1995). There may be some conditions that will give adequate yields, but this does not seem promising. Thus FeS/H_2S may have been an important prebiotic reducing agent, but any other role in the origin of life process remains to be demonstrated.

Submarine vents

The discovery of the hydrothermal vents at the oceanic ridge crests and the appreciation of their importance in the element balance of the oceans is one of the most important advances in oceanography (Corliss et al. 1979). It is likely that the vents were present in the oceans of the primitive Earth because the process of hydrothermal circulation probably began early in the Earth's history. Large amounts of water now pass through the vents, with the whole ocean passing through the vents every 10 million years (Edmond et al. 1982. The flow of ocean water may have been greater on the early Earth because the heat flow from the Earth's interior was greater at that time.

A theory of the origin of life in the vents followed shortly after discovery of the vents (Corliss, Baross and Hoffman 1981); it was proposed that the amino acids and other organic compounds are produced during passage through the temperature gradient of the 350 °C vent waters to the 0 °C ocean waters. Polymerization was also proposed to take place in this gradient.

Although there are attractive features to this theory, there are no relevant experiments to support it. The problems are that organic compounds are not produced from heating the low concentrations of CH_4 found in the vent fluids. Even worse, vent temperatures of 350 °C decompose most organic compounds in a time span ranging from seconds to a few hours. In short, submarine vents do not synthesize organic compounds, they decompose them (Miller and Bada 1988). These statements are not contradicted by reports of the synthesis of amino acids under "hydrothermal conditions" (Hennet, Holm and Engel 1992; Marshall 1994). These experiments heated various combinations of H_2CO, NH_4^+, and CN^- at high concentrations (0.19 M) and at temperatures between 150 °C and 275 °C for 2–54 hours. The products were mostly glycine, and the yields were as high as 10%. No explanation was given as to where the 0.19 M reagents could come from. These were clearly not prebiotic experiments, because they started with the components of the Strecker synthesis. In other words, starting with H_2CO and NH_4CN is equivalent to starting with glycine. The Strecker synthesis works better at lower temperatures, although the rate is then slower. The amino acids synthesized survived because of the short heating times and because the temperatures were considerably lower than the 350 °C in the submarine vents.

It should be kept in mind that the submarine vents are of considerable prebiotic importance, but it is in their destructive role rather than in synthesis. In spite of these considerations, the vent hypothesis has continued popularity, with a whole issue of the journal *Origins of Life and Evolution of the Biosphere* devoted to it (Holm 1992).

Prebiotic synthesis in clouds

There appears to have been little thought given to the prebiotic chemistry that could have taken place in clouds and their raindrops. At first sight, clouds might be considered to be a minor reaction zone because of the short time available before rain-out. However, there may be a few prebiotic reactions that could have taken place exclusively in raindrops because of conditions different from those in the ocean, such as pH and concentrations of reactants. Most of this type of chemistry in present-day clouds is related to oxidized sulfur species that would not be applicable to the primitive Earth (Warneck 1988; Wayne 1991). However, consideration needs to be given to the possible prebiotic cloud chemistry of HCN, formaldehyde, cyanoacetylene, and others.

Equilibrium calculations

Over the years, there has been a series of attempts to do prebiotic chemistry by thermodynamic calculations, which can give misleading results (Miller and Bada 1991; Bada, Miller and Zhao 1995). The first problem with such calculations is that organic reactions that involve making or breaking carbon–carbon bonds rarely come to equilibrium below about 700 °C. Dehydrogenations and ester/ether formation come to quasi equilibrium at lower temperatures, but no carbon–carbon bonds are broken in these reactions. Another way of stating this is that thermodynamics tells us which reactions are possible or impossible, but it does not tell us which reactions actually take place.

The second problem is that prebiotic chemistry is largely based on the synthesis of "high energy" compounds by electric discharges or ultraviolet light. Examples are hydrogen cyanide, cyanoacetylene, and formaldehyde. These compounds are unstable to hydrolysis or other decompositions at room temperature, but at elevated temperatures they are stable or they are photochemical products. The production of amino acids, purines, pyrimidines, and sugars is a downhill process from these "high energy" compounds, and the reactions actually take place. However, these products are uphill reactions from atmospheric constituents such as CH_4, CO_2, N_2, NH_3, H_2O, and H_2.

Extraterrestrial sources of organic material

The discovery of large amounts of organic material in meteorites, comets, planetary atmospheres, and interstellar space has led some to propose these as the sources of prebiotic organic compounds (Anders 1989; Chyba 1990; Chyba et al. 1990; Chyba and Sagan 1992). There is little doubt that some organic compounds came to the Earth from these sources, but the question is, how much and will the organic compounds survive impact? There have been a number of discussions of these extraterrestrial sources, as well as a symposium in Eau Claire, Wisconsin, in 1991 and a symposium volume (Thomas 1992; Thomas, Chyba and McKay 1996).

The major sources of exogenous organics would appear to be comets and dust, with asteroids and meteorites being minor contributors. Asteroids would have impacted the Earth frequently in prebiotic times, but the amount of organic material brought in would seem to be small, even if the asteroid is a Murchison meteorite type object. Murchison contains 1.8% organic carbon, but most of this is a polymer, and there are only about 100 parts per million of amino acids. Thus the 10-km asteroid that hit the Earth 65 million years ago at the Cretaceous/Tertiary boundary left amino acids at these boundary sediments that are detectable only by the most sensitive modern methods (Zhao and Bada 1989). The same considerations would apply to carbonaceous chondrites, although the survival of the organics on impact would be much better than with asteroids.

Comets are the most promising source of exogenous organics. The first proposal for the prebiotic importance of comets was by Oró (1961b). Comets contain about 80% water, 1% HCN, and 1% formaldehyde, as well as CO_2 and CO. Thus a 1-km-diameter comet (density = 1 gm cm^{-3}) would contain 2×10^{11} moles of HCN or 40 nmoles cm^{-2} of the Earth's surface (5×10^{18} cm^2). This is comparable to the yearly HCN production in a reducing atmosphere from electric discharges and N atoms (10–500 nmoles $cm^{-2} yr^{-1}$). This does not seem like an enormous contribution. On the other hand, it would be quite important if the Earth had a nonreducing atmosphere.

This calculation assumes complete survival of the HCN on impact. There is little understanding of what happens during an impact of a kilometer-sized object, but much of it would be heated to temperatures above 300 °C and this would decompose HCN, formaldehyde, and many other organic compounds. Even heating HCN to temperatures a little above 100 °C would hydrolyze HCN to formate. Thus the survival of the bulk of the organics in a comet on impact is problematical. However, it is likely that a few percent of a comet's organic compounds would survive the impact without excessive heating.

The input from dust may also have been important. The present dust infall is 40×10^6 kg per year or 8×10^{-9} gm cm^{-2} yr^{-1} (Love and Brownlee 1993). However, the input on the primitive Earth may have been greater by a factor of 100 or 1,000.

Unfortunately, the organic composition of this cosmic dust is very poorly known. The only individual molecules that have been detected are a few polycyclic aromatic hydrocarbons (Gibson 1992; Clemett et al. 1993), but much of the dust could be organic polymers called tholins, which are produced by electric discharges, ionizing radiation, and ultraviolet light. These tholins are intractable polymers and are difficult to hydrolyze, although a few percent of amino acids are released on acid hydrolysis.

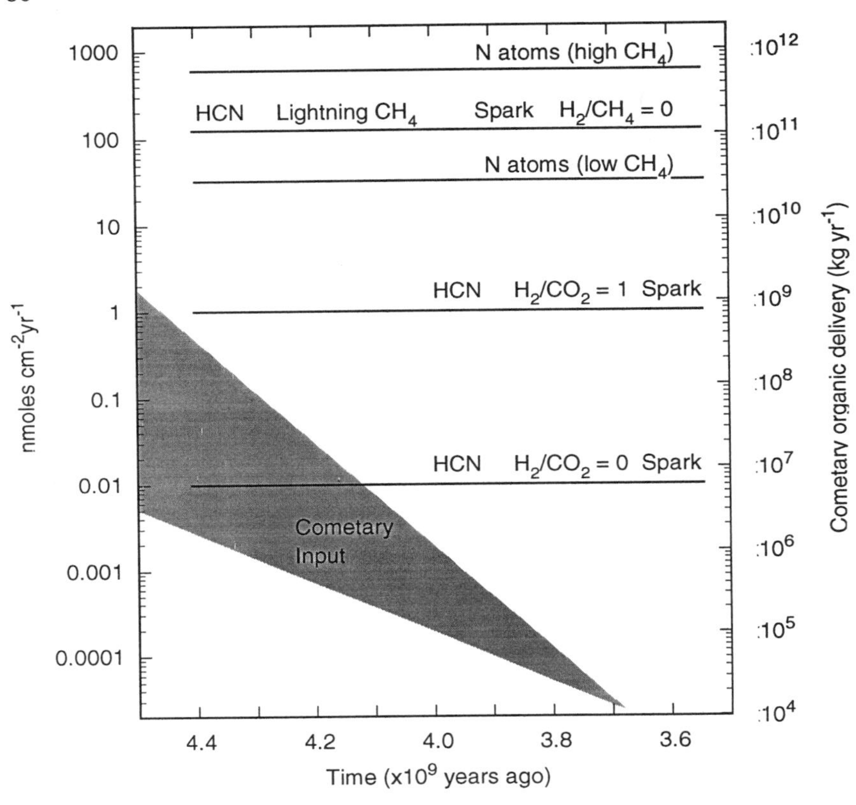

Figure 3.4. Estimated inputs of HCN from various energy sources and organic compounds from comets. The cometary inputs are from Chyba and Sagan (1992). The area is shaded because of the uncertainty of the fall-off of dust input on the early Earth. The cometary input has been converted to nmoles cm^{-2} yr^{-1} assuming a molecular weight of 100. The HCN production rates are discussed in Stribling and Miller (1987).

A more promising role for the tholins in the dust would be as a source of prebiotic molecules such as HCN, cyanoactetylene, and aldehydes. On entry to the Earth's primitive atmosphere, the dust particle would be heated and the tholin pyrolyzed, creating the HCN and other products (Mukhin, Gerasimov and Safonova 1989; Chyba et al. 1990). This does not happen in our present atmosphere because the tholins are burned up by the O_2 rather than pyrolyzed. It would be useful to determine experimentally what compounds are produced by the pyrolytic process during entry of an organic dust particle.

There are considerable differences in opinion on the sources of organic compounds that led to the origin of life. It is clear that there were multiple sources of organic compounds, and the question to decide is the relative importance of each.

Figure 3.4 shows my estimate of production and input rates from various sources. It should be emphasized that there is considerable dispute on these numbers as well as on the size of the energy sources. However, the overall conclusions seem clear:

1. Reducing conditions give very high production rates of HCN and other organic compounds.
2. Neutral atmospheres give very much lower ($\sim 10^{-4}$) rates.
3. The exogenous sources become important only if the Earth has a neutral atmosphere.
4. If the only sources of organics are exogenous or production in a neutral atmosphere, the amounts produced maybe too low for the origin of life.

This last point is debatable because it is not known what concentrations are necessary for the origin of life. But there is a lower limit on the concentrations because at some point the dilute organics in the ocean cannot be concentrated by evaporation and other processes due to absorption. Thus the reducing atmosphere could produce a steady-state concentration of amino acids of 3×10^{-4} M but a neutral atmosphere or exogenous sources would give only 3×10^{-8} M. Some thought needs to be given to whether such low concentrations are adequate.

References

Abelson, P. H. 1965. A biogenic synthesis in the Martian environment. *Proc. Natl. Acad. Sci. USA* 54:1490–1494.

Anders, E. 1989. Pre-biotic organic matter from comets and asteroids. *Nature* 342:255–257.

Bada, J. L., and Miller, S. L. 1968. Ammonium ion concentration in the primitive ocean. *Science* 159:423–425.

Bada, J. L., Miller, S. L., and Zhao, M. 1995. The stability of amino acids at submarine hydrothermal vent temperatures. *Origins Life Evol. Biosphere* 25:111–118.

Bar-Nun, A., and Chang, S. 1983. Photochemical reactions of water and carbon monoxide in Earth's primitive atmosphere. *J. Geophys. Res.* 88:6662–6672.

Bar-Nun, A., and Hartman, H. 1978. Synthesis of organic compounds from carbon monoxide and water by UV photolysis. *Origins Life Evol. Biosphere* 9:93–101.

Bernal, J. D. 1951. *The Physical Basis of Life*. London: Routledge and Kegan Paul.

Butlerow, A. 1861. Formation synthétique d'une substance sucrée. *C. R. Acad. Sci.* 53:145–147.

Chyba, C. F. 1990. Impact delivery and erosion of planetary oceans in the early inner Solar System. *Nature* 343:129–133.

Chyba, C. F., and McDonald, G. D. 1995. The origin of life in the Solar System: Current issues. *Ann. Rev. Earth Planet. Sci.* 23:215–249.

Chyba, C. F., and Sagan, C. 1991. Electrical energy sources for organic synthesis on the early Earth. *Orig. Life Evol. Biosphere* 21:3–17.

Chyba, C. F., and Sagan, C. 1992. Endogenous production, exogenous delivery and impact-shock synthesis of organic molecules: an inventory for the origins of life. *Nature* 355:125–132.

Chyba, C. F., Thomas, P. J., Brookshaw, L., and Sagan, C. 1990. Cometary delivery of organic molecules to the early Earth. *Science* 249:366–373.

Clemett, S. J., Maechling, C. R., Zare, R. N., Swan, P. D., and Walker, R. M. 1993. Identification of complex aromatic molecules in individual interplanetary dust particles. *Science* 262:721–725.

Corliss, J. B., Baross, J. A., and Hoffman, S. E. 1981. An hypothesis concerning the relationship between submarine hot springs and the origin of life on Earth. *Oceanologica Acta* 4 Suppl., 59–69.

Corliss, J. B., Dymond, J., Gordon, L. I., Edmond, J. M., von Herzen, R. P., Ballard, R. D., Green, K., Williams, D., Bainbridge, A., Crane, K., and van Andel, T. H. 1979. Submarine thermal springs on the Galapagos Rift. *Science* 203:1073–1083.

Decker, P., Schweer, H., and Pohlmann, R. 1982. Identification of formose sugars, presumable prebiotic metabolites, using capillary gas chromatography/gas chromatography–mass spectrometry of *n*-butoxime trifluoroacetates on OV-225. *J. Chromatography* 225:281–291.

Edmond, J. M., Von Damm, K. L., McDuff, R. E., and Measures, C. I. 1982. Chemistry of hot springs on the east Pacific Rise and their effluent dispersal. *Nature* 297:187–191.

Ferris, J. P., and Chen, C. T. 1975. Chemical evolution. XXVI. Photochemistry of methane, nitrogen, and water mixtures as a model for the atmosphere of the primitive Earth. *J. Am. Chem. Soc.* 97:2962–2967.

Ferris, J. P., Donner, D. B., and Lobo, A. P. 1973. Possible role of hydrogen cyanide in chemical evolution: Investigation of the proposed direct synthesis of peptides from hydrogen cyanide. *J. Mol. Biol.* 74:499–510.

Ferris, J. P., Joshi, P. D., Edelson, E. H., and Lawless, J. G. 1978. HCN: A plausible source of purines, pyrimidines and amino acids on the primitive Earth. *J. Mol. Evol.* 11:293–311.

Ferris, J. P., Sanchez, R. A., and Orgel, L. E. 1968. Studies in prebiotic synthesis. III. Synthesis of pyrimidines from cyanoacetylene and cyanate. *J. Mol. Biol.* 33:693-704.

Ferris, J. P., Wos, J. D., Nooner, D. W., and Oró, J. 1974a. Chemical evolution. XXI. The amino acids released on hydrolysis of HCN oligomers. *J. Mol. Evol.* 3:225–231.

Ferris, J. P., Zamek, O. S., Altbuch, A. M., and Frieman, H. 1974b. Chemical evolution. XVIII. Synthesis of pyrimidines from guanidine and cyanoacetaldehyde. *J. Mol. Evol.* 3:301–309.

Friedmann, N., Haverland, W. J., and Miller, S. L. 1971. In *Chemical Evolution and the Origin of Life*, ed. R. Buret and C. Ponnamperuma, pp. 123–135. Amsterdam: North Holland.

Friedmann, N., and Miller, S. L. 1969. Phenylalanine and tyrosine synthesis under primitive Earth conditions. *Science* 166:766–767.

Gabel, N. W., and Ponnamperuma, C. 1967. Model for origin of monosaccharides. *Nature* 216:453–455.

Gibson, E. K., Jr. 1992. Volatiles in interplanetary dust particles: A review. *J. Geophys. Res.* 97:3865–3875.

Groth, W., and von Weyssenhoff, H. 1960. Photochemical formation of organic compounds from mixtures of simple gases. *Planet. Space Sci.* 2:79–85.

Haldane, J. B. S. 1929. The origin of life. *Rationalist Annual* 148:3–10.

Hennet, R. J.-C, Holm, N. G., and Engel, M. H. 1992. Abiotic synthesis of amino acids under hydrothermal conditions and the origin of life: a perpetual phenomenon? *Naturwissenschaften* 79:361–365.

Holm, N. G. (ed.). 1992. *Marine Hydrothermal Systems and the Origin of Life.*

Dordrecht: Kluwer Academic. (Also a special issue of *Origins Life Evol. Biosphere* 22:1–241.)

Horowitz, N. H. 1945. On the evolution of biochemical synthesis. *Proc. Natl. Acad. Sci. USA* 31:153–157.

Joyce, G. F. 1987. Nonenzymatic template-directed synthesis of informational macromolecules. *Cold Spring Harbor Symp. Quant. Biol.* 52:41–51.

Joyce, G. F., Schwartz, A. W., Miller, S. L., and Orgel, L. E. 1987. The case for an ancestral genetic system involving simple analogues of the nucleotides. *Proc. Natl. Acad. Sci. USA* 84:4398–4402.

Keefe, A. D., Miller, S. L., McDonald, G., and Bada, J. 1995. Investigation of the prebiotic synthesis of amino acids and RNA bases from CO_2 using FeS/H_2S as a reducing agent. *Proc. Natl. Acad. Sci. USA* 92:11904–11906.

Kenyon, D. H., and Steinman, G. 1969. *Biochemical Predestination.* New York: McGraw-Hill.

Khare, B. N., and Sagan, C. 1971. Synthesis of cystine in simulated primitive conditions. *Nature* 232:577–578.

Larralde, R., Robertson, M. P., and Miller, S. L. 1995. Rates of decomposition of ribose and other sugars: Implications for chemical evolution. *Proc. Natl. Acad. Sci. USA* 92:8158–8160.

Lawless, J. G., and Boynton, C. G. 1973. Thermal synthesis of amino acids from a simulated primitive atmosphere. *Nature* 243:405–407.

Lemmon, R. M. 1970. Chemical evolution. *Chem. Rev.* 70:95–109.

Love, S. G., and Brownlee, D. E. 1993. A direct measurement of the terrestrial mass accretion rate of cosmic dust. *Science* 262:550–553

Lowe, C. J., Rees, M. W., and Markham, R. M. 1963. Synthesis of complex organic compounds from simple precursors: Formation of amino acids, amino-acid polymers, fatty acids and purines from ammonium cyanide. *Nature* 199:219–222.

Marshall, W. L. 1994. Hydrothermal synthesis of amino acids. *Geochim. Cosmochim. Acta* 58:2099–2106.

Mason, S. F. 1991. *Chemical Evolution.* Clarendon: Oxford.

Matthews, C. N. 1975. The origin of proteins: Heteropolypeptides from hydrogen cyanide and water. *Origins of Life* 6:155–163.

Miller, S. L. 1953. Production of amino acids under possible primitive Earth conditions. *Science* 117:528–529.

Miller, S. L. 1955. Production of some organic compounds under possible primitive Earth conditions. *J. Am. Chem. Soc.* 77:2351–2361.

Miller, S. L. 1957. The formation of organic compounds on the primitive Earth. *Ann. N.Y. Sci.* 69:260–274 . (Also in *The Origin of Life on the Earth,* 1959, ed. A. Oparin, pp. 123–135. Oxford: Pergamon Press.)

Miller, S. L. 1987. Which organic compounds could have occurred on the prebiotic Earth? *Cold Spring Harbor Symp. Quant. Biol.* 52:17–27.

Miller, S. L., and Bada, J. L. 1988. Submarine hot springs and the origin of life. *Nature* 334:609–611.

Miller, S. L., and Bada, J. L. 1991. Extraterrestrial synthesis. *Nature* 350:388–389.

Miller, S. L., and Orgel, L. E. 1974. *The Origins of Life on the Earth.* Englewood Cliffs, N.J.: Prentice Hall.

Miller, S. L., and Urey, H. C. 1959. Organic compound synthesis on the primitive Earth. *Science* 130:245–251.

Miller, S. L., Urey, H. C., and Oró, J. 1976. Origin of organic compounds on the primitive Earth and in meteorites. *J. Mol. Evol.* 9:59–72.

Miller, S. L., and Van Trump, J. E. 1981. The Strecker synthesis in the primitive ocean. In *Origin of Life,* ed. Y. Wolman, pp. 135–141. Dordrecht: Reidel.

Mukhin, L. M., Gerasimov, M. V., and Safonova, E. N. 1989. Origin of precursors of organic molecules during evaporation of meteorites and mafic terrestrial rocks. *Nature* 340:46–48.
Müller, D. von, Pitsch, S., Kittaka, A., Wagner, E., Wintner, C. E., and Eschenmoser, A. 1990. Chemie von α-Aminonitrilen. *Helv. Chim. Acta* 73:1410–1468.
Nielsen, P. E. 1993. Peptide nucleic acid (PNA): a model structure for the primordial genetic material? *Origins Life Evol. Biosphere* 23:323–327.
Oparin, A. I. 1938. *The Origin of Life*. New York: McMillan
Orgel, L. E. 1987. Evolution of the genetic apparatus: A review. *Cold Spring Harbor Symp. Quant. Biol.* 52:9–16.
Oró, J. 1960. Synthesis of adenine from ammonium cyanide. *Biochem. Biophys. Res. Commun.* 2:407–412.
Oró, J. 1961a. Amino-acid synthesis from hydrogen cyanide under possible primitive earth conditions. *Nature* 190:442–443.
Oró, J. 1961b. Comets and the formation of biochemical compounds on the primitive earth. *Nature* 190:389–390.
Oró, J., and Kimball, A. P. 1961. Synthesis of purines under primitive Earth conditions. I. Adenine from hydrogen cyanide. *Arch. Biochem. Biophys.* 94:221–227.
Oró, J., and Kimball, A. P. 1962. Synthesis of purines under possible primitive Earth conditions. II. Purine intermediates from hydrogen cyanide. *Arch. Biochem. Biophys.* 96:293–313.
Oró, J., Miller, S. L., and Lazcano, A. 1990. The origin and early evolution of life on Earth. *Annu. Rev. Earth Planet. Sci.* 18:317–356.
Peltzer, E. T., and Bada, J. L. 1978. α-Hydroxy carboxylic acids in the Murchison meteorite. *Nature* 272:443–444.
Peltzer, E. T., Bada, J. L., Schlesinger, G., and Miller, S. L. 1984. The chemical conditions on the parent body of the Murchison meteorite: Some conclusions based on amino, hydroxy and dicarboxylic acids. *Adv. Space Res.* 4(No.12): 69–74.
Pinto, J. P., Gladstone, C. R., and Yung, Y. L. 1980. Photochemical production of formaldehyde in the Earth's primitive atmosphere. *Science* 210:183–185.
Reid, C., and Orgel, L. E. 1967. Synthesis of sugar in potentially prebiotic conditions. *Nature* 216:455.
Ring, D., Wolman, Y., Friedmann, N., and Miller, S. L. 1972. Prebiotic synthesis of hydrophobic and protein amino acids. *Proc. Natl. Acad. Sci. USA* 69:765–768.
Robertson, M. P., Levy, M., and Miller, S. L. 1996. Prebiotic synthesis of diaminopyrimidine and thiocytosine. *J. Mol. Evol.* 43:543–550.
Robertson, M. P., and Miller, S. L. 1995a. An efficient prebiotic synthesis of cytosine and uracil. *Nature* 375:772–774.
Robertson, M. P., and Miller, S. L. 1995b. Prebiotic synthesis of 5-substituted uracils: A bridge between the RNA world and the DNA-protein world. *Science* 268:702–705.
Sagan, C., and Khare, B. N. 1971. Long-wavelength ultraviolet photoproduction of amino acids on the primitive Earth. *Science* 173:417–420.
Sanchez, R. A., Ferris, J. P., and Orgel, L. E. 1966a. Conditions for purine synthesis: Did prebiotic synthesis occur at low temperatures? *Science* 153: 72–73.
Sanchez, R. A., Ferris, J. P., and Orgel, L. E. 1966b. Cyanoacetylene in prebiotic synthesis. *Science* 154:784–786.
Sanchez, R. A., Ferris, J. P., and Orgel, L. E. 1967. Studies in prebiotic synthesis. II. Synthesis of purine precursors and amino acids from aqueous hydrogen cyanide. *J. Mol. Biol.* 30:223–253.

Sanchez, R. A., Ferris, J. P., and Orgel, L. E. 1968. Studies in prebiotic synthesis. IV. The conversion of 4-aminoimidazole-5-carbonitrile derivatives to purines. *J. Mol. Biol.* 38:121–128.

Schlesinger, G., and Miller. S. L. 1983a. Prebiotic synthesis in atmospheres containing CH_4, CO, and CO_2. I. Amino acids. *J. Mol. Evol.* 19:376–382.

Schlesinger, G., and Miller, S. L. 1983b. Prebiotic synthesis in atmospheres containing CH_4, CO, and CO_2. II. Hydrogen cyanide, formaldehyde and ammonia. *J. Mol. Evol.* 19:383–390.

Schwartz, A. W., and Goverde, M. 1982. Acceleration of HCN oligomerization by formaldehyde and related compounds: Implications for prebiotic syntheses. *J. Mol. Evol.* 18:351–353.

Shapiro, R. 1988. Prebiotic ribose synthesis: A critical analysis. *Origins Life Evol. Biosphere* 18:71–85.

Stribling, R., and Miller, S. L. 1987. Energy yields for hydrogen cyanide and formaldehyde syntheses: The HCN and amino acid concentrations in the primitive ocean. *Origins Life Evol. Biosphere* 17:261–273.

Thomas, P. J. 1992. Comets and the origin and evolution of life. *Orig. Life Evol. Biosphere* (Special issue) 22:265–434.

Thomas, P. J., Chyba, C. F., and McKay, C. P. 1996. *Comets and the Origin and Evolution of Life*. New York: Springer.

Urey, H. C. 1952 On the early chemical history of the earth and the origin of life. *Proc. Natl. Acad. Sci.* 38:351–363.

Van Trump, J. E., and Miller, S. L. 1972. Prebiotic synthesis of methionine. *Science* 178:859–860.

Voet, A. B., and Schwartz, A. W. 1983. Prebiotic adenine synthesis from HCN: Evidence for a newly discovered major pathway. *Bioinorganic Chem.* 12:8–17.

Wächtershäuser, G. 1992. Groundworks for an evolutionary biochemistry: the iron–sulphur world. *Prog. Biophys. Mol. Biol.* 58:85–201.

Warneck, P. 1988. *Chemistry of the Natural Atmosphere*. San Diego, Calif.: Academic Press.

Wayne, R. P. 1991. *Chemistry of Atmospheres*, 2nd Ed. Oxford: Clarenden Press.

Wolman, Y., Haverland, W. J., and Miller, S. L. 1972. Nonprotein amino acids from spark discharges and their comparison with the Murchison meteorite amino acids. *Proc. Natl. Acad. Sci. USA* 69:809–811.

Zahnle, K. J. 1986. The photochemistry of hydrocyanic acid (HCN) in the Earth's early atmosphere. *J. Geophys. Res.* 91:2819–2834.

Zhao, M., and Bada, J. L. 1989. Extraterrestrial amino acids in Cretaceous/Tertiary boundary sediments at Stevns Klint, Denmark. *Nature* 339:463–465.

4

Hydrothermal systems

NILS G. HOLM and EVA M. ANDERSSON
Department of Geology and Geochemistry
Stockholm University, Sweden

The conditions for potential abiotic formation of organic compounds from inorganic precursors have important implications for understanding past and present global carbon budgets. With respect to the early Earth, much of the current discussion has focused on whether the Earth's inventory of organic compounds was introduced from space or was a natural consequence of reactions taking place in the atmosphere, hydrosphere, and geosphere. It is, of course, of great interest in geochemistry to determine plausible pathways for abiotic synthesis of organic compounds. It is, however, equally important to identify natural settings that would be conductive for such reactions, as well as for the accumulation of simple organic compounds and for their subsequent reaction to form more complex macromolecules. The geosphere with its submarine hydrothermal systems offers a type of environment that in principle has not changed during the four billion years that have passed since the Earth's crust was formed.

Formation of oceanic crust and hydrothermal systems

New oceanic crust is created mainly by basalt production at ocean ridge spreading centers (Figure 4.1). Some basalt is also produced at oceanic hotspots, like the one that has created the Hawaiian Island Chain and is now overlain by the Loihi Seamount. During the process of basaltic crust formation, convecting water acts as a cooling fluid. The convecting water carries thermal energy away toward the relatively cold rock surface, thus creating hydrothermal systems. Geologists differentiate hydrothermal systems on the basis of their tectonic settings. In Figure 4.1, the most common types are shown: sediment-free on-axis systems on plate tectonic spreading centers, and off-axis systems on the flanks of the spreading centers. On-axis systems are driven by "forced convection" due to the steep temperature gradients caused by a magma chamber underneath the ridge axis. Off-axis systems, on the other hand, are driven by "free convection" due to the cooling of the new oceanic crust. The mean temperature of such hydrothermal systems is around 150 °C,

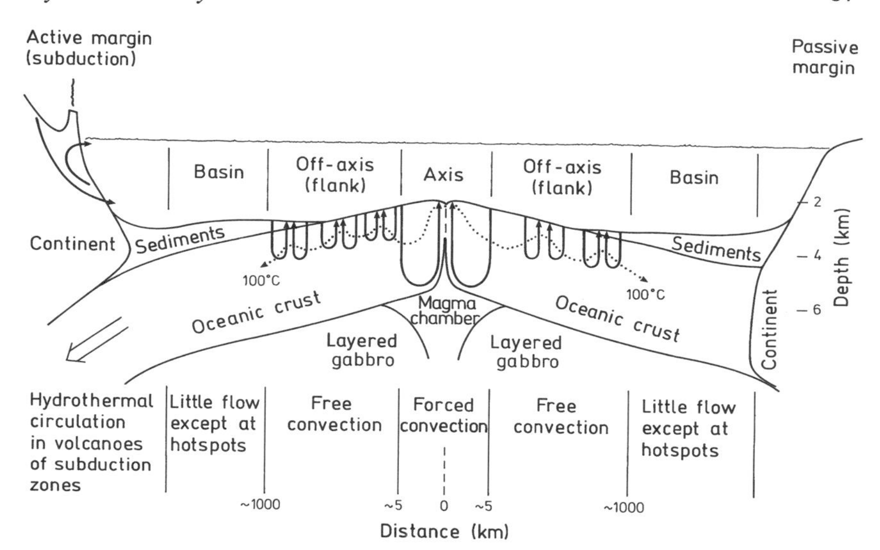

Figure 4.1. Outline map showing the major plate tectonic boundaries of the Earth's lithosphere. Most of the hydrothermal activity occurs at divergent plate margins (spreading centers). (From Holm 1992. Reprinted by permission of Kluwer Academic Publishers.)

whereas on-axis systems may have water temperatures of 350–370 °C. The COSOD II Report (1987) estimated the flow of hydrothermal water through the ridge flanks at an average temperature of 150 °C to be on the order of 560 km^3 a^{-1}. This volume should be compared to the estimated volume of water circulating at an average temperature of 350 °C through the hot on-axis systems. For that type of system the COSOD II Report arrived at the value 24 km^3 a^{-1}, that is, a volume 20–25 times smaller than the off-axis hydrothermal fluids. In a more recent paper, Cathles (1990) has published modified values for the two types of circulation. He estimated free convection of water in oceanic crust to be >240 km^3 a^{-1}, exceeding forced convection by a factor of >6 (40 km^3 a^{-1}).

The prevailing minerals that control the redox conditions at the surface of the young basaltic ocean floor down to about 300–1,300 m below the seafloor are pyrite (FeS_2), pyrrhotite (FeS), and magnetite (Fe_3O_4) – the PPM mineral assemblage (Shock 1990). Deeper in the circulation systems, the redox conditions are buffered by the FMQ mineral assemblage, that is, fayalite (Fe_2SiO_4), magnetite, and quartz (SiO_2) (Shock 1990). The major mass of midocean ridge basalts are believed to have crystallized near the redox conditions of the FMQ mineral assemblage (Christie, Carmichael, and Langmuir 1986). The buffering ability of the mineral assemblages may be represented by the following reactions:

$$2FeS + 4/3H_2O \rightarrow FeS_2 + 1/3Fe_3O_4 + 4/3H_2$$
pyrrhotite pyrite magnetite

$$3Fe_2SiO_4 + 2H_2O \rightarrow 2Fe_3O_4 + 3SiO_2 + 2H_2.$$
fayalite magnetite quartz

The buffer capacity of the assemblage can thus be expressed in terms of the fugacity of hydrogen set by the minerals at different temperatures and pressures. Values can be calculated from empirical fits of experimental data and from known thermodynamic data. The capacity can, of course, equally well be expressed as the fugacity of oxygen set by the mineral assemblage:

$$3Fe_2SiO_4 + O_2 \rightarrow 2Fe_3O_4 + 3SiO_2.$$
fayalite magnetite quartz

When plotting the log fH_2 or log fO_2 set by the mineral assemblages as a function of the temperature, a set of buffer curves is obtained (Figure.4 2). In Figure 4.2, the expression

$$CO_2 + 4H_2 \rightarrow CH_4 + 2H_2O$$

has also been included, calculated from data and equations of Helgeson and coworkers (1978). It is evident from the predominance fields of the carbon compounds in Figure 4.2 that the two buffering mineral assemblages in oceanic crust discussed previously are compatible with ratios of CO_2 to CH_4>1 above temperatures of 350–400 °C, and ratios <1 at temperatures less than about 250 °C. This means that at high temperatures in oceanic crust, CO_2 degassing from Earth's mantle is the dominating carbon compound. As a hydrothermal solution in the surface of ocean crust cools below 250 °C, it passes into the predominance field of CH_4. The carbon system then strives to adjust itself to the new equilibrium conditions but is obviously hindered by kinetic barriers. The nitrogen system behaves similarly, with N_2 being the predominant species at high temperatures and NH_3 at low temperatures. Isotopic data indicate that CO_2/CH_4 equilibration occurs only at temperatures >500 °C (Shock 1992). At temperatures <500 °C, equilibration between CO_2 and CH_4 is prevented by kinetic barriers. It appears, however, that the existence of kinetic barriers allows the formation of metastable equilibrium states like alkanes, alkenes, carboxylic acids, amino acids, and so forth For instance, Berndt, Allen and Seyfried (1996) have carried out abiotic synthesis experiments at 300 °C and 500 bar under FMQ buffered conditions. After 69 days of the experiment, 1% of the CO_2 initially present had been converted to hydrocarbon gases, primar-

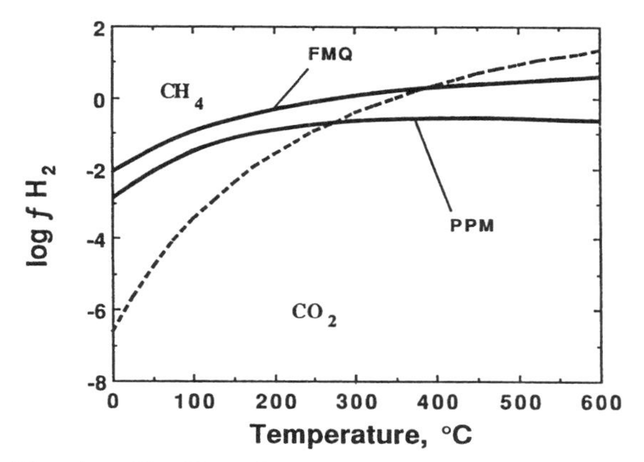

Figure 4.2. The plot of logfH_2 against temperature. Solid curves correspond to the pyrite–pyrrhotite–magnetite (PPM) buffer assemblage (lower curve), and the fayalite–magnetite–quartz (FMQ) buffer assemblage (upper curve). The dotted line indicates equal fugacities of CO_2 and CH_4 in equilibrium with H_2O, and the fields of relative predominance are labeled. (From Shock 1990. Reprinted by permission of Kluwer Academic Publishers.)

ily CH_4 but also C_2H_6 and C_3H_8. A much larger fraction (>75%) had, however, been reduced and converted to "carbonaceous" material consisting of predominantly aliphatic C constituents with a distinct aromatic component. Theoretical calculations (Shock 1990, 1992) of metastable equilibria of aqueous organic compounds like carboxylic acids in the lithosphere reveal activity maxima in the temperature range 150–250 °C (Figure 4.3). These predictions have been combined with the knowledge of the dominant water circulation at 150 °C in off-axis hydrothermal systems (see Holm 1992). The general interest with regard to potential geochemical environments for abiotic production of organic compounds has therefore recently been focused on the ridge flanks instead of the ridge crests. The studies by Kelley (Kelley 1996; also see Evans 1996) of fluid inclusions in the plutonic rocks (gabbro) of the slow-spreading Southwest Indian Ridge (SWIR) support such hypotheses. The fluid inclusions record concentrations of about 15–40 mol% CH_4, that is, at least 15–40 times those of fast-spreading hydrothermal vent fluids (Lilley et al. 1993). The results indicate that fracturing of cooling, brittle ocean floor is important for weathering of the rocks and, as a result, reduction of CO_2 to CH_4 and, potentially, metastable organic compounds. We will discuss this more in detail later.

Fischer–Tropsch type reactions

Organic compounds percolating through hydrothermal systems will be adsorbed and desorbed many times on and off mineral surfaces. Some organ-

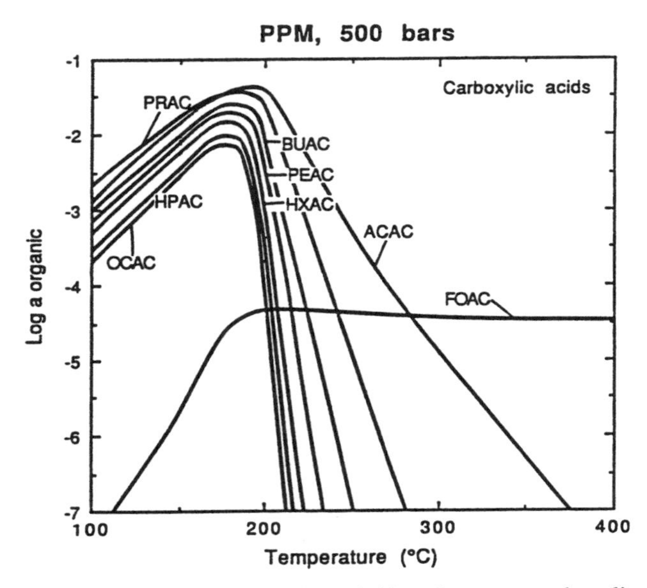

Figure 4.3. Plot of the logarithms of the activities of aqueous carboxylic acids as functions of temperature at 500 bars, in equilibrium with an initial log $fCO_2 = 2.6$, with hydrogen fugacities set by the PPM assemblage. (From Shock 1990. Reprinted by permission of Kluwer Academic Publishers.)

ics may adhere to the mineral until they undergo a chemical transformation to another compound that does not bind. The mineral surfaces may catalyze the transformation of some of these organics to higher molecular weight compounds or it may affect their decomposition to simpler derivatives. In the previous discussion we noted that thermodynamic calculations indicate the potential for the formation of metastable organic compounds in oceanic crust. We know little, however, of possible reaction pathways. One class of reaction mechanisms that would have high priority for investigation would be the FTT (Fischer–Tropsch type) syntheses on different classes of minerals. Industrial FTT reactions are normally optimized for the synthesis of hydrocarbons from CO and H_2 on iron catalysts. Reduction of CO_2 to organic compounds in the presence of H_2 or Fe(II) in olivine in laboratory experiments is also considered to follow a FTT pathway (Berndt, Allen and Seyfried 1996), probably with CO as an intermediate stage. FTT synthesis may provide a route to linear hydrocarbons and fatty acids. Formation of some of the linear hydrocarbons and fatty acids has been reported to take place in laboratory experiments using montmorillonite or metal sulfides as catalysts (Ferris 1992). It is also possible to prepare amino acids and heterocyclic compounds with limited chain branching by the FTT reaction of H_2, CO, and NH_3 (Hayatsu and Anders 1981). Even though organic chemistry nomenclature would normally restrict the use of the term *FTT synthesis reactions* to carbon-chain compounds, geochemists have described the reduction from CO_2 to CH_4 in the lithosphere as a process belonging to the FTT class of reaction pathways (see, for instance, Tedesco and Sabroux 1987; Charlou et al. 1991; Bougalt et al. 1993). Standard FTT reactions are usually performed using different metal catalysts under anhydrous reaction conditions (see Ferris 1992). Miller and Bada (1988) have, therefore, concluded that the FTT reactions could not be involved in the synthesis of organic compounds in hydrothermal systems. The chemical process in such systems, they claimed, would be inhibited by the presence of both H_2S and H_2O. The inhibition of iron FTT catalysts by H_2S is known to be permanent, which is not to say that other catalysts necessarily are. In addition, in hydrothermal systems fresh mineral surfaces are constantly being generated deep in the oceanic crust (Siskin and Katritzky 1991). Water, on the other hand, is a reversible inhibitor in some FTT reactions. It is not known, however, whether water in its supercritical form binds to the potential catalysts strongly enough to inhibit CO_2/CO conversion to organic compounds. An interesting problem to solve would be whether Fischer–Tropsch type syntheses can occur in water under certain circumstances with, for instance, sulfide minerals as catalysts. Metal sulfides like chalcopyrite, pyrite, pyrrhotite, and sphalerite are formed in great quantities in hydrothermal systems. Molybdenum sulfide (Storch, Golumbic and Anderson 1951), nickel sulfide, and tungsten sulfide (Asinger 1968) were long ago shown to have catalytic properties in FTT reactions. Magnetite is, as far as we are aware, the only oxide mineral that has a demonstrated ability to catalyze methanogenesis from CO_2. Experimental data by Berndt and co-workers (1996) suggest that magnetite's catalyzing effect on Fischer–Tropsch synthesis is not lost during reaction under high water pressures. In addition, Madon and Taylor (1981) have shown that magnetite is much less susceptible to poisoning by H_2S than is metallic iron.

Serpentinization of oceanic crust

Serpentinization of oceanic crust is probably the most likely geochemical process that would result in FTT synthesis of organic compounds. Serpentinites from oceanic areas are often associated with fracture zones and trenches (Christensen 1972). They have been formed by oxidation processes from peridotite, an ultramafic rock (<45% silica) consisting of olivine and pyroxene. Oxidation of Fe(II) of the olivine $(Mg,Fe)_2SiO_4$, in particular, in peridotite to magnetite would lead to the reduction of water and the formation of H_2. The reaction may be registered by the pH increase due to the occurrence of free OH^-. Serpentinization of mantle and deep, hot crustal rocks appears to be a significant process, especially in crustal development in zones of relatively low magmatic supply. Fracturing of the rocks in such zones enables fluids to migrate along deeply penetrating fault systems, thus intensifying hydration of deep-seated mafic mantle material and leading to significant H_2 generation. The precise reaction mechanisms that are responsible for the H_2 generation have until recently not been demonstrated in controlled experiments. The laboratory work by Berndt and co-workers (1996) has been briefly referred to previously. They carried out experiments for the reduction of CO_2 during serpentinization of olivine. The olivine was reacted with a CO_2-bearing NaCl fluid at 300 °C and 500 bar. H_2 gradually increased from essentially nothing to 158 mmol/kg by the end of the experiment (69 days). At the same time, magnetite was an abundant product of the experimental runs. The CO_2 decreased from 8.0 mmol/kg initially to 2.3 mmol/kg simultaneously with systematic increases in the concentrations of CH_4, C_2H_6, C_3H_8 (84, 26, and 12 μmol/kg, respectively), and solid "carbonaceous" material. Stevens and McKinley (1995) have detected free H_2 and communities of autotrophic methanogenic bacteria (that use the dissolved hydrogen gas as an electron donor during carbon fixation) in deep aquifers of the Columbia River basalts. The ground water aquifers of the Columbia River basalts may contain 60 μM H_2 and 160 μM CH_4 at a pH varying from 8.0 to 10.5 (Stevens and McKinley 1995). Fe(II) minerals are not as abundant in basalts as in peridotites. Nevertheless, Stevens and McKinley were able to show in the laboratory that H_2 was produced by reactions between crushed basalt and anaerobic water at room temperature. No H_2, however, was produced in assays with rocks that contained little Fe(II). The FTT synthesis is likely to be most significant in fractured rock in which degassing of CO_2 from the mantle occurs and percolation of water leading to serpentinization of olivine is efficient. This is also the type of environment where fresh surfaces of potential mineral catalysts like magnetite, native metals, and the low-sulfur sulfides are common (Moody 1976; Frost 1985). Studies of supraaquatic ultramafic rocks of the Oman Ophiolite by Neal and Stanger (1983) have suggested that H_2 generation in mantle source rocks at depth may be more widespread than has been hitherto realized and not restricted to the marine environment. They found that H_2, associated with OH-rich groundwaters (pH 10–12), is emanating from partly serpentinized and deeply fractured peridotites of the Oman Ophiolite. The gases vary in chemical composition from almost pure H_2 (99%) to almost pure N_2 (96%). Small amounts of CH_4 (<5%) occur together with trace amounts of CO and saturated hydrocarbons like ethane, propane, isobutane, and *n*-butane. Sherwood Lollar and co-workers (1993) analyzed the composition of mixed H_2, CH_4 and C_2–C_4 alkane gas in continental crystalline rocks of the Canadian and Fennoscandian Shields associated with ultramafic intrusions. Some of the gas mixtures were characterized by unexpectedly high concentrations of H_2, ranging from several volume percent up to 30 volume percent. On the basis of the widespread occurrence of serpentinized and altered ultramafic rocks at the measured sites on both the Canadian and Fennoscandian Shields, Sherwood Lollar and co-workers (1993) suggested that the serpentinization process is the likely mechanism responsible for both the H_2 and hydrocarbon production. The measured concentrations of CH_4, C_2H_6, C_3H_8, and C_4H_{10} were <82 , <11, <3.9, and <0.29 vol%, respectively.

A couple of decades ago, Pineau, Javoy and Bottinga (1976) analyzed for the stable carbon isotope composition of the gas recovered from "popping" tholeiitic basalt rocks of the Mid-Atlantic Ridge. Gas contents of about 0.9 mg g^{-1} have been found in such rocks (Hekinian, Chaigneau and Cheminée 1973; Moore, Batchelder and Cunningham 1977). The gas is composed mainly of CO_2 (95%–99%), with traces of CO, and has in general a $\delta^{13}C$ value of -7.6. This value is in the range of isotopic compositions usually found for carbonatites and diamonds and has been interpreted to indicate a deep-seated (mantle) origin of the carbon (Pineau, Javoy and Bottinga 1976). An interesting source of information is also the article by Sakai and co-workers (1990) on the hydrothermal fields of the Okinawa backarc spreading center. These researchers observed the release of bubbles of liquid CO_2 from the seafloor of active hydrothermal sites. The approximate composition of the bubbles was 86% CO_2, 3% H_2S, and 11% residual gas, mostly CH_4 and H_2. The isotopic ratios of C and of S, as well as of He, indicated that the CO_2-rich fluid had a magmatic origin. Charlou and co-workers (1991) reported intense degassing of CH_4 associated with H_2 and CO_2 at 15°05′ north latitude on the Mid-Atlantic Ridge that was related to an axial mount consisting of serpentinized peridotites. Bougalt and co-workers (1993) described the occurrence of the same type of environment at several sites in the axial region of the Mid-Atlantic Ridge, with CH_4 outputs at exposures of serpentinized ultramafic rocks. The most apparent potential evidence for FTT synthesis in subaquatic oceanic crust (as opposed to, for instance, the terrestrial Oman Ophiolite) has been provided by Mottl (1992) from holes drilled by ODP (the Ocean Drilling Program) Leg 125 at Conical Seamount in the Mariana forearc. He reported the occurrence of methane concentrations up to 37 mmol/kg in serpentine-associated sediment pore fluids, along with ethane and propane and a pH of 12.6. Geological evidence indicates that Conical Seamount is an active mud volcano, with a central conduit of low-density serpentine that rises buoyantly through the crust and produces periodic cold serpentine mud and debris flows (Mottl 1992). The high pH indicates that serpentinization was going on when the fluids were sampled. In the same serpentine-associated pore fluids sampled during ODP Leg 125, Haggerty and Fisher (1992) were able to show the presence of some of the simple carboxylic acids. The maximum measured concentration of formate was 2.3 mmol/kg, whereas acetate amounted to 210 μmol/kg. In addition, propionate and malonate were detected. CH_4 was also found in fluid inclusions of cold seep carbonate chimneys at the summit of Conical Seamount, along with aromatic compounds, long-chain paraffins, naphtenes, and acetate ions (Haggerty 1989, 1991). Mottl (1992) suggested that the presence of acetate limits the temperature range in the source region of the organics to 150 °C or less; that is, the source region would maintain a temperature at which the activities of metastable organic phases like the organic acids are close to their theoretical maximum values (Figure 4.3). Mottl (1992) hypothesized that the organic compounds of the pore fluids were thermogenic. The organic carbon content of the serpentine sediments was, however, only 0.1%–0.3%. Since little organic material was present in the pore fluids when peridotite was absent, some other mechanism for the formation of organic substances is likely to have been active. Unfortunately, Mottl did not provide stable isotope data that would shed light on the origin of the organic compounds. Mottl also discussed whether or not the low salinity values of Conical Seamount pore fluids depends on the melting of gas hydrates that could be present at depth in the drilled hole. Such gas hydrates might continuously melt, supplying fresh water and hydrocarbon gases that may upwell to the summit. No direct evidence of hydrocarbon gas hydrates was, however, found in any of the drilled cores of Conical Seamount, nor was there any seismic evidence of a bottom-simulating reflector that might indicate the presence of such gas hydrates (Mottl 1992).

Some analysis data of fossil material that may support the hypotheses of abiotic formation of organic compounds during serpentinization processes on the early Earth have been presented by de Ronde and Ebbesen (1996). They noted that $\delta^{13}C$ data of 3.2 Ma Fig Tree Group shale of South Africa compared favorably with isotopic data from serpentinized ultramafic samples associated with regionally extensive seafloor alteration zones in the Barberton greenstone belt. The Fig Tree shale overlies the ultramafic rocks in the central part of the greenstone belt and contains between 8 and 14 wt% organic C. It exhibits a great variety of organic compounds, with atomic mass numbers ranging from 15 to about 450 (as determined by electron ionization mass spectroscopy). De Ronde and Ebbesen (1996) suggested on the basis of the $\delta^{13}C$ signatures that (1) either were chemoautotrophic bacteria related to seafloor hydrothermal activity the dominant source of the Fig Tree Group carbonaceous material, or (2) were organisms that existed in the ambient Barberton ocean water of the same general carbon isotopic composition as the biota with concomitant hydrothermal activity, and/or (3) postdepositional alteration may have modified the Barberton carbon isotopic ratios.

Both low- and high-temperature generation of CH_4

We discussed previously that Stevens and McKinley (1995) were able to show in the laboratory that H_2 may be formed at room temperature from reduction of water by Fe(II) in basalt. This indicates that FTT formation of CH_4 from H_2 and CO/CO_2 may be possible at low temperature. The temperature–fractionation relationship for H_2–H_2O vapor of the Oman Ophiolite indicates a formation temperature range of 20–50 °C (Neal and Stanger 1983). Low-temperature serpentinization has been suggested as the mechanism for reduction of water and formation of H_2 and CH_4 in other ophiolites as well. Abrajano and co-workers (1988) reported, for instance, the escape of CH_4-rich gas from seeps in serpentinized ultramafic rock in the Zambales Ophiolite, Philippines. The major components of the gas were CH_4 (55 mol%) and H_2 (42 mol%). About 0.1% ethane was also present. The $\delta^{13}C$ value of the CH_4 was -7.0% (PDB), similar to values commonly attributed to mantle carbon. Abrajano and co-workers (1988), using a variety of methods, estimated the possible serpentinization temperature range of the Zambales Ophiolite as 30–350 °C. Charlou and co-workers (1991) have suggested that hydrothermal methane and unsaturated hydrocarbons may be derived either from high-temperature (150–400 °C) Fischer–Tropsch type synthesis or from reduction of CO_2 through a modified FTT reaction at relatively low temperatures (20–100 °C). It was mentioned earlier that Kelley (1996) found that fluid inclusions in gabbros recovered from the slow-spreading Southwest Indian Ridge record CH_4 concentrations 15–40 times those of hydrothermal vent fluids of, for instance, the fast-spreading East Pacific Rise and Juan de Fuca Ridge. Mass spectrometric analyses of volatiles released from olivine and plagioclase mineral separates indicated at least two phases of CH_4 generation. The volatiles of the first phase were released at temperatures of 900 °C. They contained up to 30–50 mol% CO_2 and up to 33 mol% CH_4. Isotopic compositions of these CO_2–CH_4–H_2O fluids were generally consistent with a magmatic origin for the CO_2 and H_2O. Data on the CH_4 showed no evidence of a biogenic origin, suggesting that the CH_4 was either magmatically or rock derived. The carbon isotope data suggested that the CO_2 and CH_4 had reequilibrated at temperatures of about 400–600 °C. If, on the other hand, the inclusions had been formed under nonequilibrium conditions, the CH_4-rich compositions may reflect respeciation of CO_2-rich fluid inclusions by inward diffusion of H_2. The second phase of CH_4 generation is marked by CH_4-H_2O±H_2±graphite-bearing fluids. Volatiles released from the inclusions separate at temperatures of about 300–500 °C. The fluids of the inclusions contain up to 40 mol% of CH_4, with average concentrations of about 10–15 mol%; they may also contain measurable amounts of H_2. Kelley (1996) suggested two possible origins for H_2 within the system: either magmatic degassing or interaction of aqueous fluids with olivine to form magnetite. The phase equilibria indicate that the fluids were trapped under equilibrium conditions in the presence of graphite at close to fayalite-magnetite-quartz (FMQ) buffered conditions. Fluids released during heating reveal distinct gas release profiles indicative of CH_4 and H_2O, as well as C_2–C_5 hydrocarbons (Kelley 1996).

Supercritical fluids

Supercritical fluids are the focus of a new research field that has opened up in lithospheric geochemistry. Observed and postulated supercritical fluids in hydrothermal systems are CO_2, H_2O, CH_4, and perhaps NH_3. The divergence between the characteristics of supercritical fluids and of liquids at room temperature and atmospheric pressure has been reviewed by Shaw and co-workers (1991). Supercritical fluids are excellent solvents and extractors of organic compounds (Hawthorne 1990); they have much better mass transfer characteristics than liquid solvents (Hawthorne 1990) – the heat capacity of water, for instance, approaches infinity at the critical point (Shaw et al. 1991); both oxidative and reductive reactions can easily be carried out in modified supercritical fluids. Much fundamental research needs to be done, however, in order to achieve an improved understanding of the thermodynamic and molecular properties in the supercritical region.

Future research efforts

A general characterization of the type of fluids circulating in the Earth's crust would be of great interest. Both aqueous and nonaqueous phases do occur. The magnitude of circulation of supercritical CO_2 or other supercritical nonaqueous fluids is virtually unknown. It is, therefore, necessary to develop sampling technology designed for collecting the nonaqueous fluids.

The difficulties of carrying out high-temperature experiments should not be underestimated. It may be necessary to explore an appreciable temperature and pressure range (probably from 40 °C to 300 °C and from 1 to 300 atmospheres), as well as to try to control the redox and pH of the system throughout this range. The pH and redox buffering will be crucial. To date it is not possible to measure pH directly much above 120 °C. We know, however, that pH is likely to have a large influence on stability to hydrolytic degradation. Additionally, the mineral assemblages likely to play an important role have not been fully characterized. Among these are the potential catalysts made of metal sulfides discussed previously in conjunction with Fischer–Tropsch type reactions.

Much more information is needed on the temperature stability of the small organic molecules (amino acids, sugars, nucleotides) as well as the larger biomolecules (polypeptide chains). It is currently, for example, an open question whether, among all the protein structures available, there exist proteins of a nature that will resist the temperatures to be found in some hydrothermal systems (e.g., 150 °C and upwards) and that will also retain a structure capable of functioning catalytically as an enzyme. The evidence we have on protein stability is mostly based on proteins that have evolved to function well below 105 °C. These are presumably highly adapted to these conditions and to their particular functions, so the results may not be applicable to much more stable proteins. In general, even high-temperature stability experiments have not been carried out with much pH or redox buffering.

Few organic substances that are considered unequivocal evidence of abiotic synthesis have yet been found in hydrothermal systems. Potential abiotic synthesis products are normally masked by an overwhelming amount of pyrolysates from alteration of organic matter. One group of abiotic key substances would probably be the heterocyclic sulfur compounds (tetrathiolane, pentathiane, pentathiepane, hexathiepane), which have few known biological formation mechanisms. Heterocyclic sulfur compounds have been reported to occur in pore waters of the Guyamas Basin hydrothermal field in the Gulf of California (Kawka and Simoneit 1987; Simoneit 1992). Their existence can be interpreted as a result of abiotic synthesis from formaldehyde and sulfur or HS^- present in the hydrothermal fluids.

Hydrothermal activity has continued uninterrupted for more than four billion years since the Earth's crust was formed (see de Ronde and Ebbesen 1996). A tantalizing possibility is, therefore, that "ancient" chemical processes of great potential to the primitive organic geochemistry on Earth are still active in modern geological environments.

References

Abrajano, T. A., Sturchio, N. C., Bohlke, J. K., Lyon, G. L., Poreda, R. J., and Stevens, C. M. 1988. Methane–hydrogen gas seeps, Zambales Ophiolite, Philippines: Deep or shallow origin? *Chemical Geology* 71:211–222.

Asinger, F. 1968. *Paraffins Chemistry and Technology.* Oxford: Pergamon Press.

Berndt, M. E., Allen, D. E., and Seyfried, Jr., W. E. 1996. Reduction of CO_2 during serpentinization of olivine at 300°C and 500 bar. *Geology* 24:351–354.

Bougalt, H., Charlou, J. -L., Fouquet, Y., Needham, H. D., Vaslet, N., Appriou, P., Baptiste, Ph. J., Rona, P. A., Dmitriev, L., and Silantiev, S. 1993. Fast and slow spreading ridges: Structure and hydrothermal activity, ultramafic topographic highs, and CH_4 output. *Journal of Geophysical Research* 98:9643–9651.

Cathles, L. M. 1990. Scales and effects of fluid flow in the upper crust. *Science* 248:323–329.

Charlou, J. L., Bougalt, H., Appriou, P., Nelsen, T., and Rona, P. 1991. Different TDM/CH_4 hydrothermal plume signatures: TAG site at 26°N and serpentinized ultrabasic diapir at 15°05'N on the Mid-Atlantic Ridge. *Geochimica et Cosmochimica Acta* 55:3209–3222.

Christensen, N. I. 1972. The abundance of serpentinites in the oceanic crust. *Journal of Geology* 80:709–719.

Christie, D. M., Carmichael, I. S. E., and Langmuir, C. H. 1986. Oxidation states of mid-ocean ridge basalt glasses. *Earth and Planetary Science Letters* 79:397–411.

COSOD II. 1987. *Report of the Second Conference on Scientific Ocean Drilling.* Washington, DC/Strasbourg: Joint Oceanographic Institutions for Deep Earth Sampling/European Science Foundation.

de Ronde, C. E. J., and Ebbesen, T. W. 1996. 3.2 billion years of organic compound formation near seafloor hot springs. *Geology* 24:791–794.

Evans, W. C. 1996. A gold mine of methane. *Nature* 381:114–115.

Ferris, J. P. 1992. Chemical markers of prebiotic chemistry in hydrothermal systems. *Origins Life Evol. Biosphere* 22:109–134.

Frost, B. R. 1985. On the stability of sulfides, oxides, and native metals in serpentinite. *Journal of Petrology* 26:31–63.

Haggerty, J. A. 1989. Fluid inclusion studies of chimneys associated with serpentinite seamounts in the Mariana Forearc. *PACROFI II* 2:29.

Haggerty, J. A. 1991. Evidence from fluid seeps atop serpentine seamounts in the Mariana Forearc: Clues for emplacement of the seamounts and their relationship to forearc tectonics. *Marine Geology* 102: 293–309.

Haggerty, J. A., and Fisher, J. B. 1992. Short-chain organic acids in interstitial waters from Mariana and Bonin forearc serpentinites: Leg 125. *Proceedings of the Ocean Drilling Program, Scientific Results* 125:387–395.

Hawthorne, S. B. 1990. Analytical-scale supercritical fluid extraction. *Analytical Chemistry* 62:633A–642A.

Hayatsu, R., and Anders, E. 1981. Organic compounds in meteorites and their origins. *Topics in Current Chemistry* 99:1–37.

Hekinian, R., Chaigneau, M., and Cheminée, J. L. 1973. Popping rocks and lava tubes from the Mid-Atlantic Rift Valley at 36°N. *Nature* 245:371–373.

Helgeson, H. C., Delaney, J. M., Nesbitt, H. W., and Bird, D. K. 1978. Summary and critique of the thermodynamic properties of rock-forming minerals. *American Journal of Science* 271:1–229.

Holm, N. G. 1992. *Marine Hydrothermal Systems and the Origin of Life*. Dordrecht: Kluwer Academic Publishers.

Kawka, O. E., and Simoneit, B. R. T. 1987. Survey of hydrothermally-generated petroleums from the Guaymas Basin spreading center. *Organic Geochemistry* 11:311–328.

Kelley, D. S. 1996 Methane-rich fluids in the oceanic crust. *Journal of Geophysical Research* 101:2943–2962.

Lilley, M. D., Butterfield, D. A., Olson, E. J., Lupton, J. E., Macko, S. A., and McDuff, R. E. 1993. Anomalous CH_4 and NH_4^+ concentrations at an unsedimented mid-ocean–ridge hydrothermal system. *Nature* 364:45–47.

Madon, R. J., and Taylor, W. F. 1981. Fischer–Tropsch synthesis on a precipitated iron catalyst. *Journal of Catalysis* 69:32–43.

Miller, S. L., and Bada, J. L. 1988. Submarine hot springs and the origin of life. *Nature* 334:609–611.

Moody, J. B. 1976. Serpentinization: a review. *Lithos* 9:125–138.

Moore, J. G., Batchelder, J. N., and Cunningham, C. G. 1977. CO_2-filled vesicles in mid-ocean basalt. *Journal of Volcanology and Geothermal Research* 2:309–327.

Mottl, M. J. 1992. Pore waters from serpentinite seamounts in the Mariana and Izu-Bonin forearcs, Leg 125: Evidence for volatiles from the subducting slab. *Proceedings of the Ocean Drilling Program, Scientific Results* 125:373–385.

Neal, C., and Stanger, G. 1983. Hydrogen generation from mantle source rocks in Oman. *Earth and Planetary Science Letters* 66:315–320.

Pineau, F., Javoy, M., and Bottinga, Y. 1976. $^{13}C/^{12}C$ ratios of rocks and inclusions in popping rocks of the Mid-Atlantic Ridge and their bearing on the problem of isotopic composition of deep-seated carbon. *Earth and Planetary Science Letters* 29:413–421.

Sakai, H., Gamo, T., Kim, E-S., Tsutsumi, M., Tanaka, T., Ishibashi, J., Wakita, H., Yamano, M., and Oomori, T. 1990. Venting of carbon dioxide–rich fluid and hydrate formation in mid–Okinawa Trough backarc basin. *Science* 248:1093–1096.

Shaw, R. W., Brill, T. B., Clifford, A. A., Eckert, C. A., and Franck, E. U. 1991. Supercritical water: a medium for chemistry. *Chemical and Engineering News* 69:26–39.

Sherwood Lollar, B., Frape, S. K., Weise, S. M., Fritz, P., Macko, S. A., and Welhan, J. A. 1993. Abiotic methanogenesis in crystalline rocks. *Geochimica et Cosmochimica Acta* 57:5087–5097.

Shock, E. L. 1990. Geochemical constraints on the origin of organic compounds in hydrothermal systems. *Origins Life Evol. Biosphere* 20:331–367.

Shock, E. L. 1992. Chemical environments of submarine hydrothermal systems. *Origins Life Evol. Biosphere* 22:67–107.

Simoneit, B. R. T. 1992. Aqueous organic geochemistry at high temperature/high pressure. *Origins Life Evol. Biosphere* 22:43–65.

Siskin, M., and Katritzky, R. R. 1991. Reactivity of organic compounds in hot water: Geochemical and technological implications. *Science* 254:231–237.

Stevens, T. O., and McKinley, J. P. 1995. Lithoautotrophic microbial ecosystems in deep basalt aquifers. *Science* 270:450–454.

Storch, H. H., Golumbic, N., and Anderson, R. B. 1951. *The Fischer–Tropsch and Related Syntheses*. New York: John Wiley & Sons.

Tedesco, D., and Sabroux, J. C. 1987. The determination of deep temperatures by means of the CO–CO_2–H_2–H_2O geothermometer: an example using fumaroles in the Campi Flegrei, Italy. *Bulletin of Volcanology* 49:381–387.

5

Cosmic origin of the biosphere

ARMAND H. DELSEMME
The University of Toledo
Toledo, Ohio

Cosmic abundances of the elements

All chemical elements are present in about the same proportions in the Sun, in the nearby stars, and in the interstellar medium. Standing in contrast, planets have been deeply transformed by gravity and igneous differentiation. For this reason, their surfaces do not represent a fair sample of the cosmic abundances of the elements. However, there are still minor bodies in the Solar System (in particular in the asteroid belt) that have not been differentiated. Some of their fragments often reach the Earth; these primitive meteorites are called chondrites. Chondrites display the same elemental abundance ratios present in the Sun, at least for 60 elements of low volatility (loosely called "metals" by astronomers). The most volatile gases, such as hydrogen, helium, oxygen, and nitrogen, depleted in chondrites, were assumedly present in the same proportions as in the Sun, in the primitive mixture that formed the Sun and the planetary system 4.6 billion years ago.

The "cosmic abundances" are not essentially different elsewhere. About 98% of the total mass of the Universe seems to be 3/4 hydrogen and 1/4 helium. The next three most abundant elements are carbon, nitrogen, and oxygen; taken together, they make about one single percent of the mass. The other 87 elements, often called "metals" by astronomers, to the great dismay of chemists, represent a total of less than one percent of the mass.

The previous statement is an average, and there are small fluctuations. For instance, Snow and Witt (1995) review the evidence that carbon is more abundant by a factor of two in the Sun than in the nearby stars. Another important qualifier is that the cosmic abundances refer to the present epoch. When we go backwards in time, there was less and less C, N, and O, and less and less "metals." This is apparent from the spectra of very old stars, such as those in globular clusters. These stars contain 100 to 1,000 times less "metals" than the Sun (Arnett 1995).

This seems to imply that C, N, O, and "metals" are still being made by a continuous process. Such a process has been identified with the hot nuclear syntheses that take place in the stellar cores and that explain the stars' luminosities (Clayton 1968). The presence of these new elements in the interstellar medium results from different processes of stellar evolution (stellar winds, novas, and supernovas) (Chiosi and Maeder 1986; Chiosi, Bertelli and Bressan 1992) that scatter the results of stellar nucleosynthesis into space.

Stars differ mostly by their masses; most of them range from 0.1 to 100 solar masses. The mass of the star limits the possible compression of its core and hence its highest possible temperature. The great variety of stellar masses (Shu, Adams and Lizano 1987) leads to a large number of bifurcations in the nuclear syntheses of the heavier elements. The heaviest stars (for instance, more than 15 solar masses) will not stop after the "burning" of their helium into carbon, nitrogen, and oxygen near 100 million degrees. Their further compression eventually results in the nuclear ignition of these elements into metals, such as silicon and magnesium. Finally, near 4 billion degrees, silicon itself will ignite and "burn" into iron. When the temperature rises again, near 6 billion degrees, gamma rays photodissociate iron. This is an endothermic reaction that induces the core *implosion*. The rebound of this implosion makes the whole star explode into what is called a *supernova*.

A supernova is a big ball of fire that shines like one hundred billion stars (a whole galaxy). Expanding into space, it scatters the previous products of nuclear reactions to interstellar space. The supernova explosion also ignites new nuclear reactions. Those neutrons that have been freed by the iron photodissociation help the synthesis of the heaviest metals and radioactive elements.

The supernova phenomenon is rather well understood; this is made clear by the prediction of the observed abundance ratios of most heavy metals and radioactive elements (Figure 5.1). The only large discrepancy left by the theory of nucleosyntheses in stellar cores and supernovas comes from helium; helium produced in nucleosyntheses from the stellar cores cannot explain the large abundance of observed helium. This discrepancy can be explained only if the bulk of helium is primordial.

A primordial ignition of hydrogen, between 10 and 100 billion degrees, cooling off very fast to quench the thermonuclear equilibrium between hydrogen and helium, would indeed explain the proportions of 3/4 hydrogen and 1/4 helium. It also predicts the very small abundances of the very light elements, namely lithium, beryllium, and boron, plus the primeval abundance of deuterium, the heavy isotope of hydrogen. The observed abundances of the very light elements are generally accepted as one of the proofs of the existence of the hot "Big Bang" that started the Universe (Walker et al. 1991). The Hubble Space Telescope has recently confirmed that primordial helium indeed existed at the cosmological distances of the quasars, that is, too early to have been produced by any process other than the Big Bang.

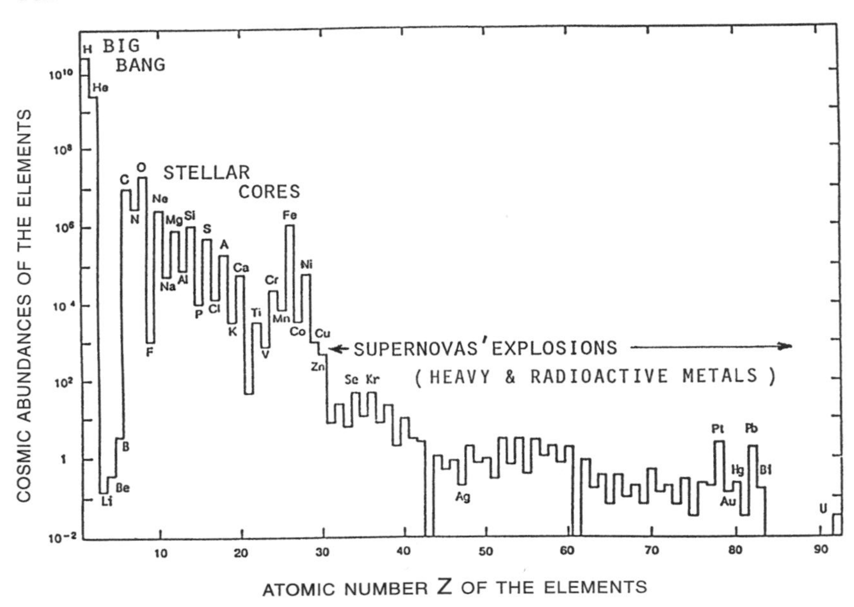

Figure 5.1. The cosmic abundances of the elements are normalized to 10^6 for the silicium Si. These abundances are found in the Sun, in interstellar space, and in most of the neighboring stars. They are believed to come from three separate processes: (1) the ratio of H to He, as well as a few traces of light elements like Li, Be, and B, come from the Big Bang; (2) the atomic numbers from 6 to 28 have been and are still steadily produced in the cores of stars; and (3) the heavy metals and radioactive elements are produced during the explosions of supernovas.

Interstellar chemistry

In our Galaxy, explosions of hundreds of millions of supernovas have already produced numberless gas bubbles expanding and cooling in space. The heaviest stars eject somewhat more oxygen than carbon; after the formation of CO in the cooling bubble, the extra amount of oxygen maintains oxidizing conditions; metal-oxide and silicate grains condense, and water vapor covers the submicroscopic grains with frost.

Less massive stars make more carbon than oxygen (Figure 5.2); after formation of CO, carbon prevails, condensing into submicroscopic grains of graphite, diamond, polyaromatic hydrocarbons, and so forth.

The millions of gas bubbles that have appeared along the arms of the Galaxy mix imperfectly and never reach chemical equilibrium. In the deep cold of space (from 3 to 30 K), chemical kinetics is controlled by fast charge–exchange reactions between ions and molecules. The source of energy is provided by the ultraviolet light of faraway stars, which ionizes the most abundant gases, namely hydrogen and helium.

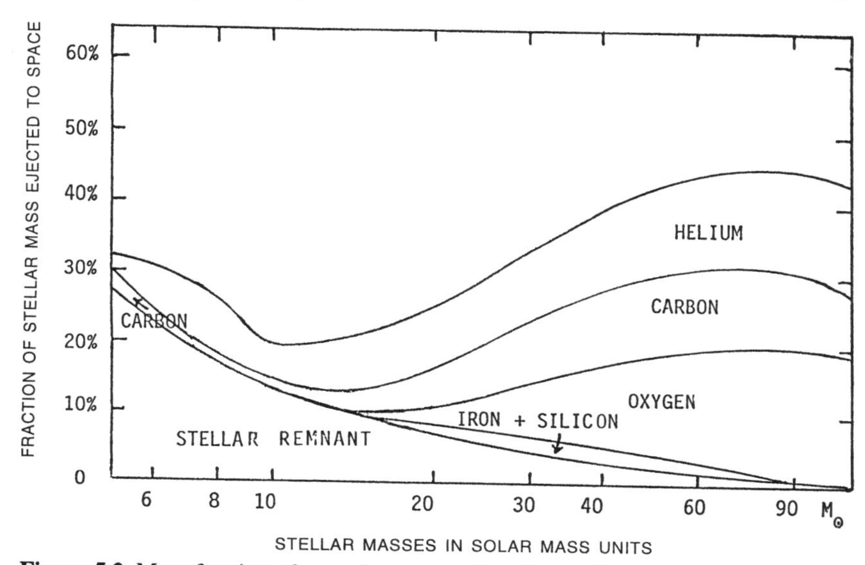

Figure 5.2. Mass fraction of new elements produced in stellar cores and eventually ejected into interstellar space (from Chiosi and Maeder 1986). Supernova explosions are the major contributing factor at left and in the center; at right, stellar winds prevail. At right, there is more oxygen than carbon, hence oxides and silicates will be formed in circumstellar bubbles. At left, carbon is more abundant than oxygen, hence reduced iron and carbon particles form in circumstellar bubbles.

Many different types of molecules appear in this highly variable environment; therefore there is no standard composition for the interstellar medium. Only in dense clouds does a complex chemistry develop, because the clouds' opacity protects large molecules from further ionization or dissociation. Submicroscopic grains of carbon or silicates, present in the interstellar medium, also play an important catalytic action, because in the high vacuum of space, triple collisions of atoms or molecules are nonexistent except on the surface of solid grains.

Radioastronomy has played a fundamental role in identifying almost one hundred interstellar molecules. Their pure rotation transitions can be measured with many significant figures, leaving no doubt for the identifications. The existence of numerous molecular clouds became obvious in the 1970s, in particular after the first three polyatomic molecules were identified as H_2O (water), NH_3 (ammonia), and H_2CO (formaldehyde) in 1968. Later, it became clear that dense molecular clouds are associated with regions of star formation (Genzel and Stutzki 1989). Three-fourths of the identified interstellar molecules are organic, and the list is still growing steadily year after year. Since large variations exist from cloud to cloud, we will include only the most abundant molecules of an average cloud. Abundances are given here in parts per million (ppm) of the overwhelmingly abundant hydrogen atom:

10 ppm:	CO	N_2	H_2O	H_2CO	
1 ppm:	HCN	HNC	NH_3	CO_2	CH_3OH
0.1 ppm:	CH_4	$CH_3{-}0{-}CH_3$			
0.01 ppm:	CS	SO_2	SO		

It is remarkable that the most abundant triatomic molecule present in the Universe is water, and that formaldehyde (H_2CO) and cyanhydric acid (HCN) are the next most abundant polyatomic organic molecules. Hundreds of the lines detected by radiotelescopes are still unidentified, mostly because of missing data on the pure rotation spectra of complex molecules. The kinetics of ion–molecular reactions is well understood, but predictions meet two obstacles: the lack of thermodynamic data for large polyatomic ions, and the lack of rate constants for their kinetics. The largest molecule identified so far contains 13 atoms ($HC_{11}N$). Among the molecular species identified, the whole gamut of organic chemistry is present, including hydrocarbons, alcohols, acids, aldehydes, ketones, amides, esters and ethers, organo-sulfur, acetylene derivatives, paraffin derivatives with CN, and so forth.

A pure rotation line of glycine, the lightest of the amino acids, had been previously reported at least twice, but a single line was not deemed sufficient to confirm this important result. Then Snyder, Kuang and Miao (1994) searched in Sagittarius B2, a massive molecular cloud of star-forming gas and dust, where 93 of the almost 100 molecular species known in space had already been found. In a dense nodule of the cloud, where methyl cyanide (CH_3CN) had already been identified, two spectral lines of glycine were found. This discovery suggests that glycine and other amino acids may form from simpler molecules by catalytic reactions at the surface of frosty dust grains.

Star formation

In the constellation of Orion, there is a giant molecular cloud; its center is visible with binoculars as an irregular nebula. It contains about one million solar masses and is sufficiently dense and opaque to have cooled by radiating in infrared, down to 10 or 20 K. This cooling diminishes the gas pressure in such a way that the numerous clumps are no longer able to resist their self-gravitation. They are right now collapsing into thousands of new stars. Already, several hundreds of them are hot enough to be seen: They are in the T Tauri stage of their evolution (from the name of the first star of this type) (Genzel and Stutzki 1989). In 1993, circumstellar disks of dust were identified surrounding about 50% of these T Tauri stars, suggesting their evolution toward planetary systems.

The existence of circumstellar disks had been proposed, before these discoveries, as an unavoidable consequence of the collapse of a gas nodule when it makes a single star. During stellar contraction, the disk sheds its excess angular momentum to its expanding margin, and mass accretes onto the protostar (Pringle 1981). The process ends when the infalling matter slows down and eventually stops. At this stage, the turbulence subsides in the disk because it is no longer fed by the kinetic energy of infalling matter. No longer supported by gas turbulence, the interstellar dust grains start sedimenting toward the midplane of the disk, where they aggregate into larger and larger clumps. As soon as its accretion is finished, the central T Tauri star starts blowing a stellar wind that quickly sweeps away the gaseous remnants of the disk, leaving only the dust that sediments into a circular ring, larger than, but not unlike, Saturn's rings (Levy and Lunine 1993).

Formation of the Solar System

In our Solar System, planets have been too much differentiated by thermal processes to show telltale signs of their formation. In contrast, the very low gravity of the smallest asteroids has preserved signs of dust sedimentation. Primitive meteorites (chondrites) that have come to Earth from the asteroid belt are clearly sedimentary rocks, in which grains of widely different origins, barely compressed together, are still recognizable. Reduced and oxidized grains touch each other, as do anhydrous and hydrated grains; refractory and volatile grains are in close contact.

Liquid water did not exist either in space or on small asteroids, hence the relevant sedimentation was not a separation of solids from a liquid phase. The only epoch in the meteorite history when a sedimentation took place was the separation of dust from the nebular gas.

In the primitive meteorites, different isotopic anomalies are also recognizable from grain to grain; this implies that the grains are individual interstellar dust grains, coming from different stars. Their identity has been sufficiently preserved during their compaction into meteorites (Levy and Lunine 1993).

Two classes of chondrites (carbonaceous chondrites and ordinary chondrites) show different rates of depletion in their volatile metals (namely, Pb, Bi, Ti, and In), whereas their 60-odd less-volatile elements are strictly in solar (that is, cosmic) abundances. This is a telltale sign that a thermal fractionation took place just before dust accretion into larger bodies, that is, at the time of dust sedimentation in the accretion disk. All observed depletions in the volatile metals agree on the fractionation temperature (Kerridge and Matthews 1988).

Besides, all carbonaceous chondrites come from the dark asteroids (called

C), whereas light, ordinary chondrites come from the light asteroids (called S). In the asteroid belt, the C asteroids outnumber the S asteroids only beyond 2.6 astronomical units (Morrison 1977). It has been concluded that, at such a distance, the dust temperature at sedimentation was close to 450 K, in agreement with other deductions, but the depletion of volatile metals seems to have been the best "cosmothermometer" at the time when dust separated from gas.

This fact also establishes the accretion temperature of the Earth's dust at the same epoch, because the radial distribution of temperature in the midplane of the accretion disk is rather well understood. The reason is that the temperature distribution is linked to the shape of the gravitational potential by the virial theorem. To put it simply, dust and gas, after spiraling from outside, settle into circular orbits. Their orbital velocity transforms half of the gravitational potential energy into kinetic energy, and the other half is transformed into heat. As soon as there is a large central mass (the protosun), the gravitational energy at distance r is proportional to r^{-1}, and so is the temperature (assuming no great amount of latent heat intervenes).

More sophisticated models of the viscous accretion disk, reported by Boss, Morfill and Tscharnuter (1989), show that the radial dependence of temperature on distance in the midplane may vary slightly, but remains close to r^{-1} at the epoch of sedimentation. Temperature plateaus are introduced by the latent heat of vaporization of silicates at 0.4 AU, that is, near Mercury, and by ice condensation from water vapor at 5.2 AU, that is, near Jupiter. No significant latent heat intervenes in between, in particular, between the asteroid belt and the Earth.

It is therefore possible to show the temperature distribution in the midplane of the accretion disk at the time of dust sedimentation (Figure 5.3). This epoch represents the state of affairs just before the accretion into larger and larger objects got started. The dust that was going to accrete in the Earth's zone (from 0.8 to 1.25 AU) was already very hot, near 1,000 K; therefore, its grains were totally degassed, since in that temperature range, water was in steam, nitrogen was entirely in gaseous N_2, and carbon was in CO.

Only when the Earth grows large and massive enough will it be able to deflect large planetesimals coming from further-away orbits and possibly having retained some volatile materials. However, the further the planetesimal comes from, the larger its impact velocity, hence the larger its impact heat. This is still a delayed effect of the virial theorem, which indicates that most volatile materials will be vaporized by the impact. Even if some volatiles are trapped by gravitation, the amount kept cannot be significant enough to explain the bulk of the oceans, the large amount of carbonates and organic compounds, or the atmosphere of the Earth.

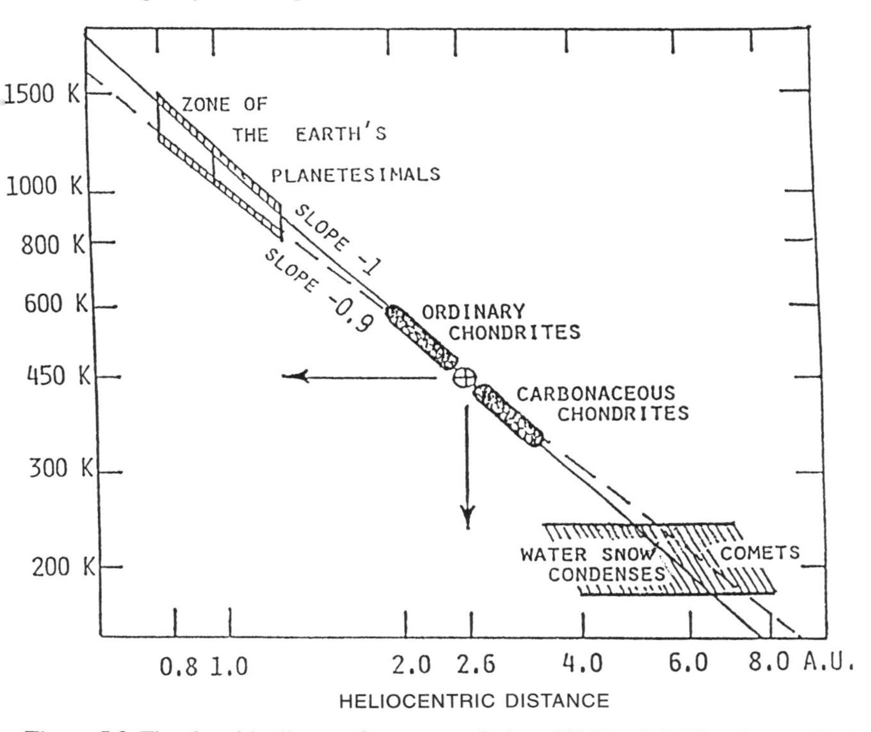

Figure 5.3. The chondrite "cosmothermometer" gives 450 K at 2.6 AU at the epoch of dust sedimentation from the nebular gas. The virial theorem predicts a temperature gradient of -l, given by the -1 slope of this log–log diagram. The midplane gradient is unlikely to be less steep than -1; however, even with -0.9, the zone of the Earth's planetesimals lies in the general range of 1,000 K, implying that the grains were degassed and that water was in steam, nitrogen in N_2, and carbon in CO. Note that water snow condenses at Jupiter's distance; this is not likely to be a coincidence.

Origin of the Oort cloud of comets

In the accretion disk, the solid fraction that sediments to the midplane has a composition that depends on the temperature. Solid grains hotter than 450 K are mostly degassed silicates and reduced iron grains. From 450 K to 225 K, the grains are less degassed and contain a growing fraction of organic and inorganic carbon. Grains cooler than 225 K do not lose their water snow, and their cold surfaces act as a condensing site for water vapor coming from warmer places.

If this last effect did not exist, cosmic abundances predict that carbonaceous icy grains should be four times heavier than degassed grains. With the cumulative effect of water vapor condensation, Figure 5.3 suggests that a large planetary embryo containing snows should accrete at a 5.2-AU distance. Theory predicts that any planetary embryo greater than 10 terrestrial masses will rapidly capture a thick atmosphere of several hundred terrestrial masses. It has been concluded that it is not a coincidence that the giant planet Jupiter lies precisely at that distance from the Sun, and still has an "atmosphere" of hydrogen and helium of almost 300 terrestrial masses. Of course, this implies that Jupiter's embryo accreted *before* the gas of the accretion disk was lost by the action of the T Tauri wind.

The early presence of this fast-growing mass soon deflected the so-far circular orbits of the planetesimals that had accreted in the same zone, that is, roughly from 4 to 7 AU. Those planetesimals are icy; more than half of their mass is water ice and ices of other volatile materials – in short, they are comets. Their orbital diffusion by the growing proto-Jupiter has two major consequences: The giant planet captures some comets by direct hits, and ejects the others in all directions at high velocities. As Jupiter's mass grows, comet velocities soon reach the escape velocity of the Solar System. A large fraction of the comets are lost forever.

However, there is also a fraction of these comets that do not quite reach the escape velocity from the Solar System. Because they go very far (up to half-way to the nearest stars), the faint action of the nearby stars widens their orbits and stores them, more or less permanently, in a huge sphere 1,000 times as large as our planetary system. They constitute the Oort cloud, where perhaps 10 billion comets are kept in their pristine state by the deep cold of space. The other giant planets also feed the Oort cloud by the same mechanism.

The Oort cloud is constantly slightly perturbed by galactic tides and by the close passage of single stars, which explains the trickle of one or two comets per year that come back from it into visibility. These "new" comets (as they are called) have been used by the Dutch astronomer Jan Oort to demonstrate the existence of the cloud that now bears his name.

Origin of volatiles on terrestrial planets

Another fundamental consequence of the orbital scattering of comets by the giant planets' growing embryos is the origin of the water and carbon compounds on the terrestrial planets. This origin comes from a cometary bombardment of the inner Solar System. Comets have been ejected at random by the giant planets, and a large fraction of them went first at high velocity through the inner planetary system, often colliding with any obstacle present on their path. The carbon dioxide atmospheres of Venus and Mars, the polar ice caps of Mars, and the atmosphere and oceans of the Earth have no other possible explanation known (Delsemme 1995a, b).

In 1972, the Russian astronomer V. S. Safronov showed that there is a direct relationship between the mass accreted on the giant planets' embryos and the total mass of the comets ejected at random. This relationship allowed him to compute the total mass of comets present in the Oort cloud, which is consistent with that deduced from the rate of "new" comets coming back from it. But the mass of the giant planets' embryos is well known; it corresponds to the solid cores detected inside these planets. In particular, the American astronomer W. B. Hubbard (1984) says the solid core of Jupiter contains 29 terrestrial masses. It must be emphasized here that even if the capture of a large atmosphere of hydrogen and helium started with an embryo of 10 terrestrial masses, the capture of comets went on during the growth of Jupiter to its present size. In a way, the recent capture of Comet Shoemaker–Levy 9 can be considered as the faint tail of this process still going on.

In 1991, I used Safronov's results to compute a model of the amount of volatiles brought about by the cometary bombardment on the terrestrial planets. In particular, this model (Delsemme 1991) shows that comets brought to the Earth 10 times more water than needed to fill the oceans, and 1,000 times more gases than are needed for our atmosphere. These excesses are welcome because of the inefficiency of the process. Different investigators have confirmed these values within a factor of 2 or 3. The origin of the Moon has been explained (Cameron 1988) by the grazing collision with the Earth of a Mars-sized protoplanet. This also explains very well the further depletion of the Earth's atmosphere. Such a depletion did not occur on Venus; this clarifies why our sister planet has an atmosphere two orders of magnitude denser than ours.

Mercury is too small and too hot to maintain a sizable atmosphere, and Mars (whose mass is only 10% of ours) has steadily lost more than 99.9% of its primitive atmosphere during the last 4 billion years. Its large, dry river beds testify that a much thicker atmosphere kept its temperature above the freezing point of water as recently as 2 billion years ago.

The history of the Solar System as we have just described it, is at variance with what was believed in geology until recently. Therefore, it is imperative to review all critical evidence about the early Earth.

The first billion years of the Earth

There is no geological record for the first billion years of the Earth's evolution. "During that first billion years, all the volatiles of the Earth could have been derived and recycled many times, while the evidence for the exact mechanism of supply was obliterated completely" (Turekian 1972).

Nowadays, a large fraction of the volcanic gases are constantly recycled by known geological processes. Recent processes include plate tectonics, which induces subduction zones; these zones heat carbonates and water enough to maintain volcanic activity. In want of a better explanation, geologists often assumed that this recycling was initiated by an early outgassing of primitive volatiles coming from the Earth's mantle. Astronomical evidence suggests that this is not so; the Earth formed from completely outgassed dust, and the volatiles were brought about by an intense bombardment of comets. This bombardment "gardened" the crust and the upper mantle, and it buried volatiles deep enough to get the whole volcanic process started. Because of the traditional explanation, the *total* outgassing of the primitive Earth remains controversial (Atreya et al. 1989) but the cometary bombardment is not, because of the lunar evidence.

Because the Moon and the Earth are close enough to have shared the same bombardment, lunar craters are a testimony of its existence. The much smaller gravity of the Moon was not able to retain any trace of the water or of the volatiles. The radioactive ages of the lunar rocks brought back by the astronauts set the age of solidification of different places on the Moon (Figure 5.4).

The ages of the rocks are interpreted as setting the epoch at which each region was flooded by lava. Such a flooding has erased the trace of the previous impact craters. Therefore, the cumulative flux of the craters still visible in the region can be traced back to the age of each rock (Basaltic Volcanism Study Project 1981). The formation age of the Moon has been assessed recently (Der-Chuan Lee et al. 1997). It occurred 50–60 million years later than the formation of the terrestrial core.

Figure 5.4 shows that the early bombardment of the Moon lasted for a very long time. It subsided significantly only after the first billion years. This long duration implies a very slow rate of orbital scattering for the comets. This can be explained only by using the very long orbital periods of the giant planets. Hence the objects involved in the bombardment were formed in the cold and faraway zones of the Solar System, implying that they were comets, containing large quantities of water and other volatile snows. The previous statement was quantified in a model published by the author (Delsemme 1991, 1995a). It is remarkable that the previous model predicts a deuterium enrichment larger in the Oort cloud than in seawater (Delsemme 1998). The observation of recent comets confirms this fact (Meier et al. 1998).

The total number of comets available in a given zone drops exponentially with time; the four characteristic times of decay induced by the four giant planets are in proportion to their periods of revolution. The contribution of the zone of each giant planet is a straight line on the logarithmic plot of Figure 5.4, but the cumulative curve is not. The comparison of the theoretical curve with the observational data from the Moon shows that the early cratering rate is essen-

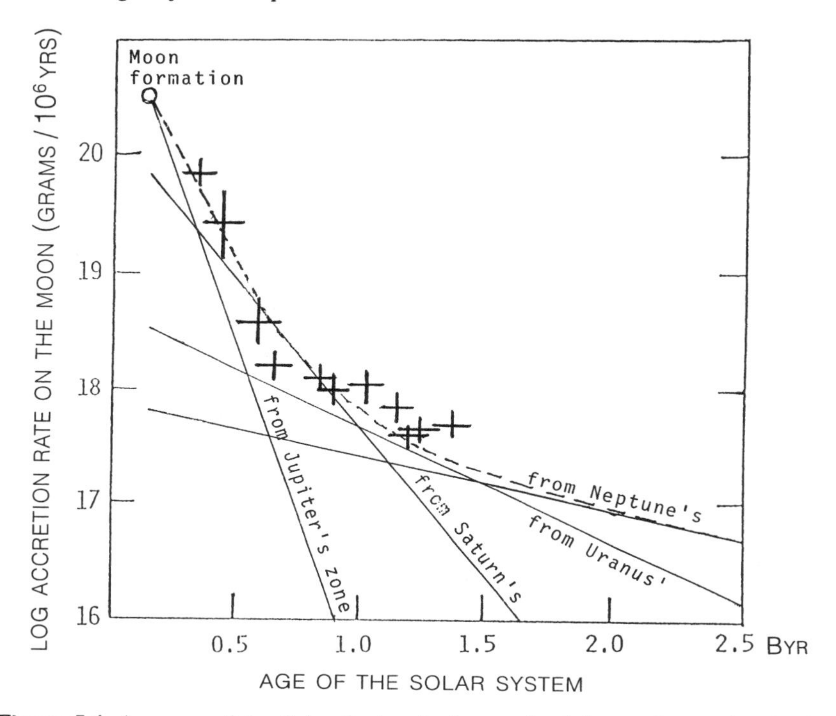

Figure 5.4. *Age zero* of the Solar System is the epoch of dust sedimentation in the accretion disk. The cumulative flux of lunar craters has been computed for different regions of the Moon from the observed crater density. The age of each region is deduced from the solidification age of those lunar rocks that have been brought back to Earth (previous craters were erased by molten lava). The data come from the Basaltic Volcanism Study Project (1981). The uncertainties are shown by the size of the crosses. The dotted line is the sum of the four exponential decay times for the number of comets coming from the four giant planets.

tially due to comets. A more recent, moderate contribution from asteroids detached from the asteroid belt by resonances with the period of Jupiter is probable, but it has not played a significant role during the first billion years.

Cometary populations

Two cometary populations can be distinguished: comets coming from the Oort cloud, and comets coming from the Kuiper Belt. The first population originated in the zones of the giant planets and was responsible for the early bombardment of the inner Solar System. The comets' cores were never warmer than 225 K for Jupiter's comets, 150 K for Saturn's, 75 K for Uranus's, and 50 K for Neptune's.

Nowadays, these comets are all intermingled in the Oort cloud and, when a "new" comet comes back into visibility, we do not know any more which accretion zone it came from. This was not the case during the cometary bombardment of the terrestrial planets. Figure 5.4 shows that Jupiter's comets came first, and Neptune's comets came last, implying that the most volatile gases and the most pristine prebiotic molecules were brought to the Earth in the late half-billion years of the cometary bombardment, that is, beginning 4.1 billion years ago.

During the last three centuries, 174 short-period comets have also been observed. As a group, their orbits are very much flattened on the ecliptic plane, which proves that they do not come from the Oort cloud, whose symmetry is more spherical. It has been demonstrated that they come from the Kuiper Belt (Weissman 1995). This belt of comets extends beyond Neptune, in the plane of the ecliptic, and has to be a remnant of the accretion disk that has never formed a planet. The belt's existence had been predicted by the Dutch astronomer G. Kuiper. Several dozen very large cometary nuclei have now been discovered in the belt, which must contain billions of them, too faint to be identified in our largest telescopes.

The mechanism that detaches the short-period comets from the Kuiper Belt is nothing else but the tail of the orbital diffusion of Neptune's comets. Neptune's zone has of course already been depleted of its comets, but there are resonances with Neptune's period that slowly transform the orbits in the Kuiper Belt. Hence the collision with Jupiter of short-period comet Shoemaker–Levy 9 in July 1994 is just a contemporary example of the end of the cometary bombardment, which never completely stopped.

Even if interstellar prebiotic molecules were kept in their pristine state (say, at less than 100 K) during their accretion into cometary nuclei, it is often assumed that the heat of their impact on the Earth destroyed these fragile molecules. This was completely true when the Earth's atmosphere did not exist. During the first million years, all molecules were destroyed by the heat of the impacts, which was often intense enough to produce large ponds or seas of molten lava (this is well demonstrated by the Moon).

When a substantial atmosphere exists, the situation changes dramatically because of another phenomenon, namely the large range of the sizes of cometary particles that are going to be slowed down to a complete stop in the atmosphere (Chyba and Sagan 1992). The reason is that when they come into the inner Solar System, cometary nuclei lose their volatile materials by sublimation and they decay into dust. This is the reason why comets display such spectacular tails when they come near the Earth.

The fact that cometary dust is braked gently without much heating in the upper atmosphere has been demonstrated by capturing it with the U2 plane of NASA, and discovering that it still contains many organics, including prebiotic molecules.

Cometary chemistry

Some 15 years ago, the author (1981) calculated a table (Table 5.1) that summarizes the relative abundances of elements in comets, interstellar frost, and living organisms. The elemental composition of living matter is much less similar to that of the Earth (silicates of the rocks) than to that of interstellar frost or of the volatile fraction of comets. The first two columns show that calcium was a recent addition by evolved animals, assumedly because they needed a skeleton. Only phosphorus was somewhat concentrated in primitive bacteria (only by a factor of 1.5 from cosmic abundances), presumably because phosphates brought them the first energy source before the use of photosynthesis.

In 1983, the outstanding British astrophysicist Fred Hoyle wrote that "Professor Delsemme has concluded from his results that cometary material must be the feedstuff of life. It is a conclusion that might seem remarkable enough, but one which I think is too cautious. Cometary material is *life,* I would say, not simply its precursor." Sir Fred's arguments to put the origin of life in comets were never confirmed.

The perihelion passage of Comet Halley in 1986 brought great progress (Greenberg and Shalabiea 1993). The quantitative analysis of major compounds has been summarized in Delsemme (1991). First, the abundances of the 17 elements measured in Comet Halley (including C, N, O, and S) were strictly in solar proportions (as in the previous five bright comets observed from 1957 to 1976) except hydrogen, which was depleted by a factor of 500 (confirming the same sedimentation process for comets and planets in the primitive accretion disk). The dust-to-gas mass ratio of Halley was 0.8.

The abundances in Table 5.2 remain somewhat controversial because they try to incorporate discordant data (see discussion in Delsemme 1991). Another interesting feature emerges from the elemental abundance dispersion of the submicroscopic grains. They have been classified in four families: Group A is carbon rich, Group B is oxygen rich, Group C is rich in reduced Mg + Si, and Group D is rich in reduced iron.

This suggests that individual grains, coming from different circumstellar environments, were mixed together during their sedimentation from the nebular gas. Numerous isotopic anomalies suggest the same scenario. In the volatile fraction, mass spectra taken during the spacecraft flyby have revealed $(H_2CO)_n$. Finally, pyrimidine, purine, and other heterocycles with N have been identified in the organic grains dubbed CHON by astronomers.

Our qualitative and quantitative knowledge of molecular compounds in comets remains very incomplete. In particular, no amino acids were seen in Halley; it is unlikely that any could have been detected with the equipment used

Table 5.1. *Comparison of elemental abundances*

	Bacteria	Mammals	Interstellar frost	Volatile fraction of comets
Hydrogen	63.1%	61.6%	55%	56%
Oxygen	29.0%	26.0%	30%	31%
Carbon	6.4%	10.5%	13%	10%
Nitrogen	1.4%	2.4%	1%	2.7%
Sulfur	0.06%	0.13%	0.8%	0.3%
Phosphorus*	0.12%	0.13%	–	(0.08%)
Calcium	–	0.23%	–	–

* Phosphorus has not been observed in comets; the figure in parentheses represents its cosmic abundance, which is reasonably expected to be present in comets.

during the few minutes' flyby with the Giotto spacecraft. It may be useful to recall here that the extensive analysis of primitive meteorites brought much more detailed results. In particular, carbonaceous chondrites contain not only saturated and unsaturated hydrocarbons (including cyclo-compounds), but also fatty acids, purines, pyrimidines, and amino acids. This large variety of organic compounds almost certainly formed by inorganic pathways. Because they were present in the asteroid belt, they may have formed near 400 K (Figure 5.3) by reactions of the Fischer–Tropsch type, catalyzed by silicate grains. Of course, this is not applicable in comets where pristine interstellar grains may have been kept in their primitive state because of low temperatures.

In spite of the significant prebiotic chemistry found in carbonaceous chondrites, we must conclude here that it is not relevant to the supply of organic matter to the primitive Earth; interstellar grains, however, are relevant because they sedimented into cometary nuclei in zones of the Solar System where they were never heated as much as chondrites. Finally, cometary dust, blown away in the cometary tails before its entry into the atmosphere, is the vehicle best able to carry organic prebiotic matter to the oceans and to the ground, as initially suggested by Oró (1961).

The early Earth

During the end of its accretion, the growing gravitation of the Earth increased the impact velocities of the accreting planetesimals. Their energy was transformed into more and more heat, melting the surface rocks and starting the large-scale differentiation that was soon going to separate the mantle from the iron core. The largest impacts occurred somewhat later, in particular the grazing impact of a protoplanetary body, 10 times less massive than the Earth, that

Table 5.2. *Chemical composition of Comet Halley*

DUST: 44% BY MASS	VOLATILE FRACTION: 56% BY MASS	
Inorganics: 29.5%	With oxygen: 51.5%	Hydrocarbons: 1.2%
22.7% silicates	44.0% H_2O	0.8% C_2H_2
2.6% FeS troilite	2.5% HCO·OH	0.3% CH_4
1.3% C graphite	2.2% H_2CO	0.1% C_3H_2
0.4% S sulfur	2.0% CO_2	–
2.4% H_2O (hydration)	0.8% CO	–
Organics: 14.5%	With nitrogen: 3.1%	Sulfur compounds: 0.1%
7.0% unsat. hydrocarbons	1.5% N_2	0.04% H_2S
2.3% with oxygen	0.6% HCN	0.03% CS_2
2.0% with nitrogen	0 4% NH_3	0.03% S_2
0.1% with sulfur	0.4% N_2H_2	–
2.4% water	0.2% $C_4H_4N_2$	–

is likely to have formed the Moon. The growing mass of the Earth (as well as resonances with the orbital period of Jupiter) also perturbed large asteroidal bodies, but the contribution of their impacts during the first billion years was insignificant when compared with the flux of comets.

After 500 million years, 90% of the cometary bombardment was finished; the Earth surface had cooled down, and the large amount of steam produced by the cometary impacts had formed a thick atmosphere. This steam soon started cooling off and condensing into a hot rain that formed very hot oceans. The Moon had started spiraling away from the Earth, but during this epoch, it was still at least three times closer than now, hence it turned in 5 days about the Earth (which rotated twice as fast as now). Tides were gigantic, with an amplitude about 30 times as large as now; therefore, there were extensive tidal ponds of hot water everywhere.

At that time, the atmospheric pressure was still presumably several tens of atmospheres, and it produced a greenhouse effect capable of maintaining extremely hot oceans in spite of the low radiation of the Sun. This radiation was at a minimum at that time, about 25%–30% less than now. In spite of the considerable losses in volatiles due to the great impacts, the atmospheric pressure could have gone up to 60 or 80 atmospheres, if a new phenomenon had not transformed a large amount of CO_2 into solid carbonates. CO_2 is increasingly soluble in rain water with higher pressure in the atmosphere. Rain becomes rich in acid, H_2CO_3, which attacks silicates in rocks and transforms them into SiO_2 through the formation of solid carbonates (mostly limestone and dolomite). This phenomenon was possible because eventually the Earth cooled off more than Venus, since is is farther away from the Sun. On Venus, carbonates were never able to form because of the heat, and its atmospheric CO_2 induced a runaway greenhouse effect that stopped any possible cooling of the ground.

Standing in contrast, the acid rain process worked for millions of years on the Earth until CO_2 reached a much lower pressure than nitrogen, which became the major atmospheric constituent only for this reason. Sedimentary rocks 3.8 billion years old exist in the southwest of Greenland. In order for these rocks to have been made, water and CO_2 were needed to alter previous rocks. In these ancient rocks, small amounts of dolomite are rich in reduced iron compounds, which establishes that there was very little, if any, oxygen in the air.

The chemical composition of the early atmosphere is suggested by Table 5.2. In particular, as soon as the atmosphere was cool enough to condense water as rain into the oceans, the rest of the gases must have remained for very long periods in an intermediate state of oxidation, since the incoming comet flux was compensating for the losses and changes happening in the upper atmosphere, like ultraviolet photochemistry and exospheric dissipation.

However, the major prebiotic molecules are likely to have been brought to Earth by the slow fallout of cometary dust smoothly braked by the atmosphere. This dust has been freed in immense amounts in the inner Solar System for hundreds of millions of years. The faint remnants of this cometary dust is still visible in the inner Solar System and is called the zodiacal light. Stopped by the progressively cooling surface of very hot oceans, this dust either was trapped in microscopic foam bubbles at the surface of the oceans, or sank immediately down to the bottom of sea shallows. Millions of sites with variable physicochemical conditions were available, in particular in the hot tidal ponds produced everywhere by the immense lunar tides.

Conclusion

It is still early to describe how life appeared on the Earth. However, the existence of amino acids in interstellar space, and of purines and pyrimidines in Comet Halley, suggest that these prebiotic molecules were brought about by cometary dust and do not need ad hoc scenarios to be made again in a terrestrial environment.

References

Arnett, D. 1995. Explosive nucleosynthesis in stars. *Ann. Rev. Astron. Astrophys.* 33:113–132.

Atreya, S. K. et al. (eds). 1989. *Origin and Evolution of Planetary and Satellite Atmospheres.* Tucson: University of Arizona Press.

Basaltic Volcanism Study Project. 1981. *Basaltic Volcanism of the Terrestrial Planets.* New York: Pergamon Press.

Boss, A. P., Morfill, G. E., and Tscharnuter, W. M. 1989. Models of the formation and evolution of the solar nebula. In *Origin and Evolution of Planetary and Satellite Atmospheres,* ed. S. K. Atreya, J. B. Pollack, and M. S. Matthews, pp. 35–79. Tucson: University of Arizona Press.
Cameron, A. C. W. 1988. The origin of the solar system. *Ann. Rev. Astron. Astrophys.* 26:441–472.
Chiosi, C., Bertelli, G., and Bressan, A. 1992. New developments in understanding stellar evolution. *Ann. Rev. Astron. Astrophys.* 30:235–283.
Chiosi, C., and Maeder, A. 1986. The evolution of massive stars with mass loss. *Ann. Rev. Astron. Astrophys.* 24:329–373.
Chyba, C., and Sagan, C. 1992. Endogenous production, exogenous delivery and impact shock synthesis of organic matter. *Nature* 355:125–132.
Clayton, D. D. 1968. *Principles of Stellar Evolution and Nucleosyntheses.* New York: McGraw Hill.
Delsemme, A. H. 1981. Are comets connected to the origin of life? In *Comets and the Origins of Life,* ed. C. Ponnamperuma, pp. 141–159. Dordrecht: Reidel.
Delsemme, A. H. 1991. Nature and history of the organic compounds in comets. In *Comets in the Post-Halley Era,* vol. I, ed. R. L. Newburn, Jr., M. Neugebauer, and J. Rahe, pp. 387–428. Dordrecht: Kluwer Academic Publ.
Delsemme, A. H. 1995a. Cometary origin of the biosphere. *Adv. Space Res.* 15(3):349–357.
Delsemme, A. H. 1995b. Giant planets needed, to make biospheres on terrestrial planets. In *Progress, Search for Extraterrestrial Life,* ed. G. S. Shostak, pp. 31–50. San Francisco: Astron. Soc. Pacific, Conf. Series vol. 74.
Delsemme, A. H. 1998. The deuterium enrichment observed in recent comets is consistent with the cometary origin of seawater. *IAU Colloquium 168 Abstract.* Nanjing, 18–22 May 1998.
Genzel, R., and Stutzki, J. 1989. Orion molecular cloud and star-forming region. *Ann. Rev. Astron. Astrophys.* 27:41–85.
Greenberg, J. M., and Shalabiea, O. M. 1993. Comets as a reflection of the interstellar medium chemistry. In *Asteroids, Comets, Meteors,* ed. A. Milani, M. Di Martino, and A. Cellino, pp. 327–342. Dordrecht: IAU-Kluwer Acad.
Hoyle, F. 1983. *The Intelligent Universe.* New York: Holt, Rinehart and Winston.
Hubbard, W. B. 1984. The Jovian planets. In *Planetary Interiors,* pp. 234–295. New York: Van Nostrand-Reinhold.
Kerridge, J. F., and Matthews, M. S. (eds.). 1988. *Metéorites and the Early Solar System.* Tucson: University of Arizona Press.
Lee, D.-C. Halliday, A. H., Snyder, G. A., and Taylor, L. A. 1997. Age and origin of the Moon. *Science* 278:1098–1103.
Levy, E. H., and Lunine, J. I. (eds.). 1993. *Protostars and Planets III.* Tucson: University of Arizona Press.
Meier, R., Owen, T. C., and eight co-authors. A determination of the HDO/H_2O ratio in Comet C/1995 01 (Hale–Bopp).1998. *Science* 279:842–844.
Morrison, D. 1977. In *Comets, Asteroids, Meteorites,* ed. A. H. Delsemme, pp. 177–184. Toledo, Ohio: The University of Toledo.
Oró, J. 1961. Comets and the formation of biochemical compounds on the primitive Earth. *Nature* 190:389–390.
Pringle, J. E. 1981. Accretion disks in astrophysics. *Ann. Rev. Astron. Astrophys.* 19:137–162.
Safronov, V. S. 1972. Ejection of bodies from the Solar System in the course of the accretion of the giant planets and the formation of the cometary cloud. In *Motions, Evolution of Orbits and Origin of Comets,* ed. Chebotarev et al., pp. 329–334. Dordrecht: IAU-Reidel Publ.

Shu, T. P., Adams, F. C., and Lizano, S. 1987. Star formation in molecular clouds. *Ann. Rev. Astron. Astrophys.* 25:23–81.
Snow, T. P., and Witt, A. N. 1995. The interstellar carbon budget and the role of carbon in dust and large molecules. *Science* 276:1455–1460.
Snyder, L., Kuang, Y. J., Miao, Y. 1994. Hints of first amino acid outside solar system. *Science* 264:1668.
Turekian, K. K. 1972. *Chemistry of the Earth.* New York: Holt-Rinehart-Winston.
Walker, T. P., Steigman, G., Schramm, D. N., Olive, K. A., and Kang, H. S. 1991. Primordial nucleosynthesis redux. *Astrophys. J.* 376:51–69.
Weissman, P. R. 1995. The Kuiper Belt. *Ann. Rev. Astron. Astrophys.* 33:327–357.

6

Clues from the origin of the Solar System: meteorites

JOHN R. CRONIN
Department of Chemistry and Biochemistry
Arizona State University, Tempe

A puzzle, particularly a jigsaw puzzle, can serve as a useful metaphor for the origin of life problem, although the figurative puzzle must be unusually challenging. Ordinarily, a jigsaw puzzle provides well-defined pieces that are just sufficient to complete a picture or design; however, in this case, it is complicated by missing pieces (some irretrievably lost and others perhaps yet to be found), pieces that no longer have their original shapes, and other pieces that are perhaps irrelevant, that is, that come from other puzzles and have accidentally been added to the mix. Consequently, the puzzlers (origin of life scientists) work, hoping that these difficulties will not, in the end, be insurmountable – that the irrelevant pieces can be culled, and that those remaining can be properly arranged and will be sufficient to show the outlines of the overall picture or design.

One of the puzzle pieces is a record of early Solar System organic chemical evolution found in a class of meteorites called carbonaceous chondrites. It is a well-defined piece, with fairly sharp edges and much of its original color still showing. It seems to interlock with adjacent pieces – but, is it relevant? The nature of the organic matter and its accretion by the early Earth suggest that it is.

Theories of the origin of terrestrial life, with the exceptions of panspermia (Arrhenius 1908) and those that posit an inorganic, that is, "clays first," origin (Cairns-Smith 1982), require preexisting organic (reduced carbon) compounds as raw material for self-assembly of the progenote, the earliest life form (Haldane 1929; Oparin 1938). They differ with respect to the source of this organic matter, that is, whether it was formed on the primitive Earth (endogenous formation) (Miller 1953), was delivered by late-accreting bodies (exogenous delivery) (Anders 1989), or involved both of these processes. The paucity of the early rock record makes it difficult to know conditions on the early Earth and, consequently, to confidently evaluate the possibilities for endogenous synthesis of organic compounds before the origin of life. Recent attempts to model the early atmosphere (Kasting 1993) suggest that it may not have provided the reduced gases necessary for the abundant production of organic matter, as had once been thought (Urey 1952). On the other hand, studies of interstellar clouds (Irvine in press), interplanetary dust (IDPs) (Clemett et al. 1993), micrometeorites (Maurette et al. 1994), meteorites, comets (Mumma et al. 1996), and terrestrial impacts (Chyba and Sagan 1992) have led to a growing awareness of the possibilities for exogenous delivery, and of the nature of the organic matter brought to the primitive Earth.

Understanding in detail the organic chemistry of IDPs, comets, and meteorites is essential if we are to understand the role of exogenous delivery in the origin of life. Chapters by Delsemme and Maurette deal with comets and micrometeorites, respectively; however, the most detailed record of organic chemistry in the early Solar System has come from analyses of meteorites, specifically the carbonaceous chondrites. In this chapter, the organic chemistry of these meteorites is described. Readers are referred to the following cited reviews for more detailed and comprehensive treatments of the subject: Cronin and Chang 1993; Cronin, Pizzarello and Cruikshank 1988; Mullie and Reisse 1987.

What are carbonaceous chondrites?

Meteorites are classified as belonging to one of three categories: stones, irons, and stony-irons. Stones are the most numerous and account for more than 90% of meteorite falls. About 90% of the stones can be further classified as chondrites based on the presence of chondrules, millimeter-sized or smaller spherical bodies distributed throughout the stone. The chondrules evidence a pervasive thermal event that occurred prior to agglomeration of the bodies from which the chondritic meteorites derived, their so-called parent bodies. The various parent bodies required to account for the properties of chondritic meteorites are believed to have originated in the asteroid belt, about 2–4 astronomical units (AU) from the Sun. The chondrites are considered to be primitive Solar System materials on the basis of their chemical composition. With the exception of the volatile elements (hydrogen, helium, carbon, nitrogen, and oxygen), they have an elemental composition equivalent to that of the Sun and, since more than 99% of the Solar System mass resides in the Sun, equivalent to that of the solar nebula from which the Sun and planets formed. Thus the chondrites did not experience processes that led to extensive chemical fractionation, such as partial evaporation or gravitational separation from a melt.

Chondrites can be subclassified on the basis of small compositional differences and the extent to which they have been altered by reactions with water (aqueous alteration), heating, or shock. The least altered are the carbonaceous

(C) chondrites, so named because of their relatively high content of carbon (0.3% to >3%) and the other volatile elements, H, N, O, and S. The least altered carbonaceous chondrites are the CI and CM types. These are unique in having experienced aqueous alteration of their original anhydrous silicate matrix to mineral assemblages dominated by hydrous, layer lattice silicates (clay minerals) and in being rich in organic compounds. It should be emphasized that abundant organic matter is observed *only* in those chondrites in which the original anhydrous minerals were converted to hydrous assemblages by the action of liquid water.

The unusual properties of the CI and CM chondrites have led to the suggestion that they are not asteroidal in origin, but rather fragments of devolatilized cometary nuclei, a proposition that Anders (1975, 1978) has convincingly argued against. On the other hand, an origin from cometlike material cannot be excluded. Bunch and Chang (1980) have pointed out that accretion of icy cometesimals into the parent body/bodies could have provided both organic compounds and, upon melting, the liquid water for aqueous alteration.

Books by Heide and Wlotzka (1995) and by Dodd (1986) are recommended to those who might wish a broader, although concise, overview of the field of meteoritics.

The state of carbon in carbonaceous chondrites

The organic-rich CI and CM chondrites have mean total carbon contents of, respectively, 3.54% and 2.46% (Mason 1963). The carbon of these meteorites is dominated by the organic fraction but also includes carbonate minerals and "exotic" carbon phases, such as diamond, graphite, and silicon carbide. These last three materials bear witness to direct stellar inputs to the presolar nebula. An approximate quantitative distribution of carbon among the various organic and inorganic forms in the Murchison meteorite, a CM chondrite, is given in Table 6.1.

The organic matter of CM chondrites can be fractionated on the basis of solubility. The soluble fraction is obtained by extracting powdered meteorite samples sequentially with a series of solvents of varying polarity. These extracts contain a complex mixture of discrete compounds, such as amino acids and hydrocarbons, accounting for perhaps 10%–20% of the total carbon. Hayes (1967) has summarized the results of several early experiments in which the mass extracted by various solvents was determined. The bulk of the organic matter is retained in the residual insoluble fraction. It is composed primarily of a poorly characterized, structurally heterogeneous, macromolecular material, which is commonly referred to as either meteorite "polymer" or "kerogen-like"

Table 6.1. *Carbon distribution in the Murchison meteorite*

Form		Amount	
Total carbon		2.12%, 1.96% (1, 2)	
Interstellar grains			
Diamond		400 ppm (3)	
Silicon carbide		7 ppm (4)	
Graphite		<2 ppm (5)	
Carbonate minerals		2%–10% total C (6)	
Macromolecular carbon		70%–80% total C	
Organic compounds		10%–20% total C	
Aliphatic hydrocarbons	++	Dicarboxylic acids	++
Aromatic hydrocarbons	++	Sulfonic acids	+++
Polar hydrocarbons	+++	Phosphonic acids	+
Volatile hydrocarbons	+	N-heterocycles	+
Aldehydes & ketones	++	Purines & Pyrimidines	+
Alcohols	++	Carboxamides	++
Amines	+	Hydroxy acids	++
Monocarboxylic acids	+++	Amino acids	++

+++ >100 ppm ++ >10 ppm + >1 ppm

1. Jarosevich 1971. 2. Fuchs, Olsen and Jensen 1973. 3. Lewis et al. 1987; Blake et al. 1988. 4. Tang et al. 1989. 5. Amari et al. 1990. 6. Grady et al. 1988.

material. This macromolecular carbon, along with the exotic carbon phases, can be obtained by digesting the insoluble fraction with HF–HCl mixtures, a process that dissolves most inorganic minerals, including the carbonates.

Macromolecular carbon

The macromolecular carbon of the Murchison meteorite has been shown by both chemical degradation (Hayatsu et al. 1977, 1980) and ^{13}C NMR (Cronin, Pizzarello and Frye 1987) to have both aliphatic and aromatic character and to be similar to the more aromatic (type III) terrestrial kerogens (Miknis et al. 1984). High resolution electron microscopy (HREM) has shown this material to occur predominantly in irregular clumps having both amorphous and spiral layered structures (Lumpkin 1986). The clumps are composed of particles in the range 0.01 to <0.1 μm (Reynolds et al. 1978). Some portion of the macromolecular carbon may occur in fluorescent, μm-sized inclusions of various morphologies. Rossignol-Strick and Barghoorn (1971) thoroughly surveyed the particles liberated by HF-digestion of the Orgueil meteorite (CI) and identified numerous hollow spheres, as well as less abundant irregular objects having membranous and spiraled structures. Similar fluorescent particles have been observed in Murchison sections (Deamer 1985). The fluorescent material extracted with a chloroform–methanol mixture forms droplets when dispersed in aqueous solution, an observation that has led to the proposal of a possible role for this material in prebiotic membrane formation (Deamer 1986).

Organic compounds

Historically, attempts to establish the composition of carbonaceous chondrites with respect to specific organic compounds have been plagued by the problem of terrestrial contamination; but, by 1969, the year of the fall of the Murchison meteorite, the problem was widely appreciated and appropriate care generally was taken to avoid contamination artifacts in the handling, sampling, and analysis of the Murchison stones. In what follows, the emphasis will be on the results of analyses made since 1969, mainly on the Murchison meteorite.

Higher (nonvolatile) hydrocarbons

The hydrocarbons of CM meteorites have been the subject of considerable interest and analytical effort. Two distinct suites of hydrocarbons differing in molecular weight and volatility have been found. The higher-molecular-weight, nonvolatile hydrocarbons can be extracted from meteorite powders using various organic solvents or solvent mixtures, for example, benzene–methanol, and the extract then separated into aliphatic, aromatic, and polar hydrocarbon fractions by silica gel chromatography (Meinschein 1963).

The nature of the Murchison higher aliphatic hydrocarbons has been controversial. Analyses made soon after the meteorite fell gave somewhat discordant results. One group reported the predominance of aliphatic polycyclic compounds (Kvenvolden et al. 1970) and another claimed predominance of straight-chain alkanes (Studier, Hayatsu and Anders 1972). The latter result has often been cited in support of a nebular Fischer–Tropsch type (FTT) process as the formation mechanism for the meteoritic organic compounds (see, for example, Hayatsu and Anders 1981).[1]

[1] The Fischer–Tropsch reaction is the catalyzed synthesis of hydrocarbons (and other organic compounds) by reduction of carbon monoxide by hydrogen. The possibility that such reactions occurred in the cooling solar nebula is an attractive idea, but one that is not supported by analyses of meteoritic organic matter.

Recently, analyses of the Murchison aliphatic hydrocarbons showed that when interior samples were carefully obtained and the analyses were made under conditions that minimized environmental contaminants, the principal aliphatic components of the meteorite were not straight-chain alkanes, but rather C_{15} to C_{30} branched alkyl-substituted mono-, di-, and tricyclic alkanes of great structural diversity (Cronin and Pizzarello 1990). In addition, carbon isotope analyses have shown the straight-chain alkanes of a Murchison extract to have $\delta^{13}C$ values in the range of terrestrial hydrocarbons (Gilmour and Pillinger 1993).[2] The *n*-alkanes, along with the methyl alkanes and isoprenoid alkanes found in some extracts, thus appear to be terrestrial contaminants and the argument for an FTT synthesis based on the dominance of straight-chain compounds is not valid.

Aromatic hydrocarbons

Several analyses of the Murchison aromatic hydrocarbons have given results in generally good agreement. In a recent analysis, the most abundant compounds found were pyrene, fluoranthene, phenanthrene, and acenaphthene in ratios of about 10:10:5:1, respectively (Krishnamurthy et al. 1992). Numerous other unsubstituted aromatic compounds were found in lower amounts, along with alkyl-substituted and partially hydrogenated forms of these polycyclic aromatic hydrocarbons (PAHs).

It should be noted that, due to evaporative losses of the lower PAHs and the involatility of the higher PAHs, the solvent extraction–GC analytical method gives valid results for only a fairly narrow range of aromatic compounds, that is, over the molecular weight range of about 178–252 daltons (tri- through pentacyclic compounds and their alkyl derivatives).

Recently, the laser desorption/ionization–mass spectrometry (L^2MS) technique has been applied to analyses of meteorite aromatic hydrocar-

[2] Stable isotope ratios are commonly expressed in relation to a standard as δ-values, where

$$\delta\ (‰) = \frac{F(H/L)_{sample} - (H/L)_{standard}}{(H/L)_{standard}} \times 1000$$

and H/L represents the abundance ratio of the higher mass isotope to the lower mass isotope. δ-values range from -1,000 (heavy isotope absent) to ∞ (light isotope absent). When the isotopic ratios of sample and standard are equal, δ = 0. A positive δ-value is indicative of enrichment, and a negative δ-value of depletion of the sample in the heavy isotope with respect to the standard. The isotopic enrichment (depletion) can be calculated from δ as follows:

$$\text{Enrichment} = \frac{(H/L)_{sample}}{(H/L)_{standard}} = \frac{\delta\ (‰)}{1000} + 1$$

The commonly used standards and their H/L ratios are hydrogen, standard mean ocean water (SMOW), D/H = 1.557×10^{-4}; carbon, Peedee belemnite limestone (PDB), $^{13}C/^{12}C = 1.12372 \times 10^{-2}$; nitrogen, atmospheric N_2, $^{15}N/^{14}N = 3.6765 \times 10^{-3}$.

bons. Through this method, naphthalene, phenanthrene/anthracene, fluoranthene/pyrene, and various C_1, C_2, and C_3 alkyl derivatives were observed in powdered Murchison samples (Hahn et al. 1988). The aromatic hydrocarbons in intact, freshly exposed interior fragments from Murchison have been analyzed by the SALI technique (surface analysis by laser ionization) (Tingle, Becker and Malhotra 1991). Unsubstituted PAHs were the dominant aromatic compounds observed in Murchison. The molecular weight range of aromatic hydrocarbons was extended to encompass coronene and its alkyl-substituted derivatives by applying the L^2MS technique to sublimates obtained under high vacuum at 300, 450, and 600 °C from Murchison residues that had been demineralized by HCl and HF treatments (de Vries et al. 1993). The sublimation process favors retention of the higher molecular weight compounds and loss of the more volatile, lower molecular weight species. Unidentified PAHs up to 750 daltons were revealed when Fourier transform ion cyclotron resonance mass spectrometry was used to analyze a dried toluene extract of an interior Murchison sample. Although it was not observed, an upper limit of 2 ppb was set for Buckminsterfullerene, C_{60}.

Polar hydrocarbons

The bulk of the hydrocarbons extractable from Murchison (~70% of the total benzene–methanol extracted carbon) are polar hydrocarbons (Krishnamurthy et al. 1992). Four classes of compounds were identified on the basis of mass spectra: (1) several series of alkyl aryl ketones; for example, alkyl naphthyl-, alkyl biphenyl-, and alkyl anthracyl/phenanthryl ketones; (2) aromatic ketones and diketones; for example, fluoren-9-one, anthracenone, and anthracenedione; (3) nitrogen heterocycles that include $C_{13}H_9N$ (benzoquinoline isomers) and $C_{15}H_9N$ (azapyrene/azafluoranthene); and (4) sulfur heterocycles, of which dibenzothiophene was identified. Using different extraction methods, others have found several water-soluble N-heterocyclic compounds, such as substituted pyridines, quinolines, isoquinolines, purines, and pyrimidines, in Murchison extracts (Hayatsu et al. 1975; Stoks and Schwartz 1982). Although the identified compounds were mainly aromatic, mass spectra suggested the presence, in addition, of aliphatic compounds with polar substituents (for example, heteroalicyclic compounds) and aliphatic-substituted heteroaromatic compounds.

Lower (volatile) hydrocarbons

Light hydrocarbons presumably remain in CM chondrites only because they are immobilized (for example, within crystals or between crystal boundaries), firmly adsorbed to grain surfaces, or dissolved in the macromolecular carbon phase (Belsky and Kaplan 1970). Nevertheless, these compounds have been obtained from CM chondrites (Belsky and Kaplan 1970; Studier, Hayatsu and Anders 1972; Yuen et al. 1984). The work of Yuen and co-workers is particularly significant in that individual members of the lower hydrocarbon series were isolated for $^{13}C/^{12}C$ measurements. These researchers found methane and its homologues through the butanes to be isotopically heavier than their terrestrial counterparts, clearly indicating an extraterrestrial origin. More interestingly, they discovered a trend in which $^{13}C/^{12}C$ decreased with increasing carbon chain length (also observed for the monocarboxylic acids). These results have important implications with respect to the production mechanism (see "Monocarboxylic acids").

The lower aliphatic hydrocarbons trapped in carbonaceous chondrites may represent only a fraction of the parent body's original content of such compounds. It was the impression of those who collected Murchison stones immediately after the fall that the fresh meteorite had a high volatile hydrocarbon content. Given the amounts of nonvolatile hydrocarbons found in Murchison and the exponential decline in amount with increasing carbon number that characterizes several other classes of compounds, it would follow that the primitive meteorite parent body contained very large quantities of the volatile lower hydrocarbons. From this it might be inferred that comets also contain large amounts of these compounds.

Alcohols, aldehydes, and ketones

Functionalized light hydrocarbons are also trapped in the carbonaceous chondrite matrix. Series of lower alcohols, aldehydes, and ketones have been found in aqueous extracts of the Murchison and Murray (CM) meteorites (Jungclaus et al. 1976). Alcohols and aldehydes through C_4 and ketones through C_5 were observed. Declining concentrations with increasing carbon number were seen in the straight-chain homologous series of each class, and all possible isomers were present through C_4.

Amines

Initially, 10 aliphatic amines were tentatively identified in aqueous extracts of Murchison using GC as well as ion exchange chromatography (Jungclaus et al. 1979). These analyses were recently confirmed by GC–MS and 10 additional aliphatic amines were positively identified (Pizzarello et al. 1994). The amines comprise a mixture that includes both primary and secondary isomers through C_5. As commonly observed for meteoritic aliphatic compounds, almost all isomers are present; the concentrations within homologous series decrease with increasing chain length; and branched-chain isomers are more abundant than the straight-chain compounds. As with the amino acids, their concentration is about doubled by acid hydrolysis of the extract.

Monocarboxylic acids

The monocarboxylic acids of the Murchison meteorite have been the subject of a series of detailed analyses by Yuen and co-workers (Yuen and Kvenvolden 1973; Lawless and Yuen 1979; Yuen et al. 1984; Epstein et al. 1987). This work has provided several significant insights into the origin of these compounds and, by inference, into the origin of meteorite organic compounds in general.

Yuen positively identified 17 saturated aliphatic carboxylic acids with chain lengths from 1 (Kimball 1988; Yuen, unpublished results) to 8 (Yuen and Kvenvolden 1973). All isomers through C_5 were identified, as were at least 6 of the 8 possible C_6 carboxylic acids. Similar results were obtained (21 carboxylic acids identified through C_{12}) with a CM chondrite from the Japanese Antarctic collection (Shimoyama et al. 1989). The amounts decline in the straight-chain series by about 60% with each additional carbon atom, decreases similar to those seen in the amino acids. The branched-chain isomers are abundant, at least through C_6. The acids presumably exist in the meteorite largely as carboxylate salts. The slightly alkaline pH of the water extract of a Murchison powder suggests similar pH conditions for the final aqueous phase experienced by the parent body. Furthermore, if the acids had been present in the protonated carboxyl form, the lower members of the series would have been lost by evaporation unless sequestered in a special environment like methane, for example. A portion of the carboxylic acids occurs as carboxamides and acetyl amino acids, which yield carboxylic acids and acetic acid, respectively, upon acid hydrolysis (Cooper and Cronin 1995). The presence of carboxamides suggests the possibility of formation of the carboxylic acids by hydrolysis of nitriles.

Stable isotope ratios have been obtained for the C_2–C_5 branched- and straight-chain monocarboxylic acids along with CO and CO_2 (Yuen et al. 1984). These results parallel those described for the lower aliphatic hydrocarbons, and show (1) a decline in $\delta^{13}C$ from acetic acid through valeric acid, and (2) C_4 and C_5 branched-chain isomers that are isotopically slightly heavier than their straight-chain counterparts. The data suggest a kinetically controlled chain elongation mechanism in which the higher homologues are built up by the addition of one-carbon species.

Recently, δD (deuterium) values have been obtained for the Murchison monocarboxylic acids (Epstein et al. 1987; Krishnamurthy et al. 1992). They are D rich, although not to the degree of the amino acids. The enrichment is believed to be indicative of an interstellar origin (see "Isotopic characteristics of meteorite organic matter"). If so, the intermolecular and intramolecular differences in carbon isotopic composition among the carboxylic acids discovered by Yuen and his co-workers represent previously unrecognized characteristics of carbon chain formation under interstellar cloud conditions.

Dicarboxylic acids

Dicarboxylic acids are found in hot-water extracts of the Murchison meteorite (Lawless et al. 1974). Initially, 17 compounds in this series were identified with carbon numbers through C_9. Three have chiral centers and in the case of methylsuccinic acid, both enantiomers were identified and shown to be present in about equal amounts. Fifteen had saturated aliphatic carbon chains.

Recently, the Murchison dicarboxylic acids prepared for isotopic analysis were analyzed and 42 compounds through C_8 were recognized by GC–MS (Cronin et al. 1993). All of the 8-chain isomers through C_5 were recognized and many (25 of 33) at C_6 and C_7. These compounds, like the amines and monocarboxylic acids, constitute a suite of great, if not complete, structural diversity.

We assume that the dicarboxylic acids exist in the meteorite as the carboxylate dianions for the reasons discussed previously for the monocarboxylic acids. Oxalic acid has been shown to occur as whewellite, its calcium salt (Fuchs, Olsen and Jensen 1973). It is interesting to note that series of pyrrolidiones and piperidiones have recently been recognized in meteorite extracts (Cooper and Cronin 1995). These compounds are converted, respectively, to various β- and γ-dicarboxylic acids upon hydrolysis. It is not known whether such compounds are precursors of all meteoritic dicarboxylic acids or represent an alternative pathway to their formation.

Sulfonic and phosphonic acids

Recent analyses of water extracts of Murchison have disclosed analogous suites of alkyl sulfonates and alkyl phosphonates (Cooper, Onwo and Cronin 1992). The alkyl phosphonates represent the first organophosphorous compounds recognized in meteorites. In the case of the alkyl sulfonates, seven of the eight possible compounds through C_4 were recognized; five of the corresponding alkyl phosphonates were observed. Both series appear to show the decrease in homologue concentration with increasing carbon number that has been seen in other aliphatic acid series from Murchison.

Although isotopic analyses have not yet been done with sulfonic and phosphonic acids, these acids are thought to be derived from interstellar precursors. An abundant sulfur-containing carbon chain series, C_nS, has been identified in molecular clouds (Wilson et al. 1971; Saito et al. 1987; Yamamoto et al. 1987), and the radical of the first member of a possible phosphorous-containing series, CP, has also been observed (Guélin et al. 1990). Deuterium analyses of the meteoritic compounds would be particularly interesting since they could be indicative of the site of hydrogenation, that is, whether it was interstellar or parent body.

Nitrogen heterocycles

The central role of purines and pyrimidines in biological information storage and transfer has evoked a strong interest in whether these compounds occur in Murchison and other carbonaceous chondrites. The early literature in this area is contradictory; however, later investigations by Schwartz and his colleagues (Van der Velden and Schwartz 1977; Stoks and Schwartz 1979, 1981a, b, 1982) clarified the situation considerably. A summary of this work was given in an earlier review (Cronin, Pizzarello and Cruikshank 1988), and complete references are given there and in the papers by Stoks and Schwartz. The current view is that Murchison contains the purines xanthine, hypoxanthine, guanine, and adenine, as well as the pyrimidine uracil, at a total concentration of about 1.3 ppm.

A suite of basic N-heterocyclic compounds was also identified in the Murchison meteorite (Stoks and Schwartz 1982). Positive identifications were made of 2,4,6-trimethylpyridine, quinoline, isoquinoline, and 2-methyl- and 4-methylquinoline. In addition, N-methyl-aniline, a suite of 12 additional alkyl pyridines, and a suite of 14 methylquinolines and/or isoquinolines were identified. Higher members of the quinoline/isoquinoline series were found among the polar hydrocarbons described previously (Krishnamurthy et al. 1992).

In summary, Murchison appears to contain several classes of basic and neutral N-heterocycles, including purines, pyrimidines, quinolines/isoquinolines, and pyridines. The last two groups are structurally diverse and contain a large number of isomeric alkyl derivatives. Taken together they total about 7 ppm. In contrast, only one pyrimidine and four purines are found, and only in smaller amounts. All of the purines and pyrimidines found are biologically common and no biologically unknown or unusual analogues seem to accompany them. Isotopic measurements have not been reported, and stereoisomeric criteria are not applicable since the compounds are achiral. Thus, although blank runs testify to contamination-free procedures, the possibility that these compounds originated in terrestrial microorganisms should be kept in mind.

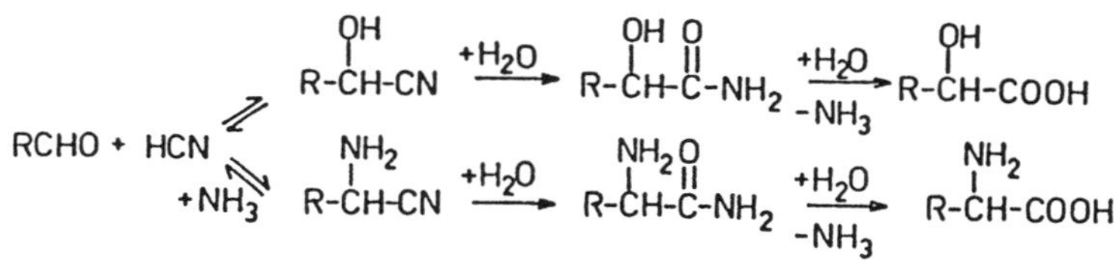

$$\mathrm{RCHO + HCN \rightleftharpoons R\text{-}CH(OH)\text{-}CN \xrightarrow{+H_2O} R\text{-}CH(OH)\text{-}C(=O)\text{-}NH_2 \xrightarrow[-NH_3]{+H_2O} R\text{-}CH(OH)\text{-}COOH}$$

$$\mathrm{RCHO + HCN + NH_3 \rightleftharpoons R\text{-}CH(NH_2)\text{-}CN \xrightarrow{+H_2O} R\text{-}CH(NH_2)\text{-}C(=O)\text{-}NH_2 \xrightarrow[-NH_3]{+H_2O} R\text{-}CH(NH_2)\text{-}COOH}$$

Figure 6.1. Strecker–cyanohydrin reaction.

Hydroxycarboxylic acids

Initially, seven α-hydroxycarboxylic acids, the complete set of α-isomers having five or fewer carbon atoms, were identified in the Murchison meteorite (Peltzer and Bada 1978). Volatile diastereomeric derivatives were prepared for four of the five chiral hydroxy acids, and enantiomer ratios (D/L) ranging from 0.82±0.1 to 0.93±0.3 were determined by GC. The ratios were interpreted as showing the α-hydroxy acids to be essentially racemic, and therefore of abiotic origin. The presence of α-hydroxy- and α-amino acids suggests the formation of both classes of compounds by a Strecker–cyanohydrin synthesis. The molar ratios of three structurally analogous pairs of α-hydroxy- and α-amino acids are consistent with their equilibration at a common ammonia concentration, as would the case if they had been formed by a common Strecker synthesis (Peltzer et al. 1984).

A recent reanalysis of the hydroxy acids has shown them to comprise a much more extensive suite than was originally observed (Cronin et al. 1993). A total of 51 hydroxy acids were identified, including β- and γ-substituted isomers, with carbon chains through C_8. The distribution of α-isomers is similar to that observed for the α-amino acids, that is, they appear to comprise structurally analogous sets of compounds. This more extensive structural correlation reinforces the earlier suggestion of a Strecker synthesis, although the presence of β- and γ-amino-position isomers indicates that at least one additional formation process was operative as well.

Isotopic analyses of the total hydroxy acids gave a large, positive δD value (Cronin et al. 1993). The fact that the δD and $\delta^{13}C$ values are both lower than those of the amino acids is, at face value, inconsistent with a common Strecker synthesis; however, it should be noted that the preparation of hydroxy acids analyzed lacked the hydroxy dicarboxylic acids, whereas the analogous amino acids were present in the sample analyzed. In other words, the two sets of compounds isotopically analyzed were not exactly comparable in composition. Compound-specific carbon isotopic analyses of analogous α-amino and α-hydroxy acid pairs would be of considerable interest.

Hydroxydicarboxylic acids

Analysis of the Murchison dicarboxylic acids prepared for isotopic analysis disclosed an extensive suite of hydroxydicarboxylic acids, compounds which had previously escaped detection (Cronin et al. 1993). More than 50 hydroxydicarboxylic acids were recognized, of which 5 α-substituted compounds were positively identified, including malic acid, α-hydroxy glutaric acid, and α-hydroxy-2-methyl succinic acid. The last 3 correspond to 3 meteoritic α-amino acids: aspartic acid, glutamic acid, and 2-methyl aspartic acid, respectively. Many β-, γ-, and other hydroxydicarboxylic acids were found, for which corresponding amino dicarboxylic acids have not yet been identified. This finding suggests that a search for additional members of the last class of compounds might be productive.

Amino acids

The presence of amino acids in meteorites has been a matter of considerable interest. They provide unequivocal evidence for the natural abiotic synthesis of biologically important compounds, and thus seem to support the chemical evolution hypothesis, that is, that prebiotic processes gave rise to the organic compounds necessary for the origin of life. Because amino acids are essential for all terrestrial life, it has been widely assumed that they were also essential for the origin of life; however, the discovery that RNA can function catalytically (Cech, Zaug and Grabowski 1981), as well as in information transfer, has led to the concept of an RNA world in which proteins might have had only a secondary role, or perhaps no role at all (Gilbert 1986). The difficulty in demonstrating plausible prebiotic routes to RNA suggests the prior existence of a pre-RNA world based on as yet unimagined chemistry. An early role for amino acids and proteins remains a strong possibility, particularly in view of the abundance of amino acids in carbonaceous chondrites and the ease with which they are produced abiotically from simple precursors (Miller 1953; Oró 1963).

Early attempts to find amino acids in meteorites were compromised by contamination of the available samples with amino acids of biological origin (Hamilton 1965; Oró and Skewes 1965) and it remained for analyses of the then recently fallen Murchison meteorite to convincingly show that amino acids are indigenous to a carbonaceous chondrite (Kvenvolden et al. 1970). The occurrence of chiral amino acids as nearly 1:1 mixtures of D- and L-enantiomers and the discovery of several amino acids not found in proteins, and thus unlikely to be terrestrial contaminants, provided compelling evidence for their extraterrestrial origin. By 1975, 22 amino acids had been positively identified in Murchison (Kvenvolden et al. 1970; Kvenvolden, Lawless and Ponnamperuma 1971; Oró et al. 1971; Lawless 1973; Buhl 1975). These included 8 of the protein amino acids, 10 of more restricted biological occurrence, and 4 that were considered to be very rare, if not nonexistent, in the biosphere. Since then, more than 50 additional amino acids have been identified, bringing the total to more than 70 (Cronin, Gandy and Pizzarello 1981; Cronin, Pizzarello and Yuen 1985; Cronin and Pizzarello 1986). Most of the more recently identified amino acids are biologically unknown and thus may be unique to extraterrestrial matter.

Structurally, the meteoritic amino acids are of two types: monoamino monocarboxylic acids and monoamino dicarboxylic acids. Two variations on these basic structures have been observed: N-alkyl derivatives, and cyclic amino acids in which an alkyl side chain forms a ring connecting the α-carbon and the α-amino group. Within these constraints, complete structural diversity prevails. For example, there are 14 chain isomers for the acyclic 7-carbon α-amino acids, 4 of which have two chiral centers and therefore exist as two diastereomeric forms. Thus, counting diastereomers but excluding enantiomers, 18 isomeric forms exist that are, in principle, separable by ordinary chromatographic methods. All 18 isomers have been identified in a Murchison extract (Cronin and Pizzarello 1986). Other structural characteristics of the meteoritic amino acids are as follows: (1) at each carbon number the abundance order is α- >γ- >β-isomers; (2) within isomeric sets, branched-carbon-chain isomers are more abundant than the straight-chain isomers; (3) within homologous series, for example, the straight-chain α-amino acids, there is an exponential decline in amount with increasing carbon number (slope ~ -0.7).

The amino acids found in greatest abundance are usually glycine, the simplest α-amino acid, and α-aminoisobutyric acid, the lowest-carbon-number branched-chain α-amino acid. The abundance of this 4-carbon amino acid clearly illustrates the preference for branched-chain structures previously noted. The most abundant amino acids are found in concentrations of about 100 nmol g^{-1}, and the total amino acids may reach a concentration approaching 700 nmol g^{-1}. In discussing amino acid composition, it is important to note that there is significant quantitative heterogeneity among Murchison specimens. For example, in some samples the composition is dominated by α-methyl-α-amino acids like α-aminoisobutyric acid and isovaline (Cronin and Pizzarello 1983).

The hydroxy amino acids serine and threonine are commonly found in meteorite extracts; however, since these amino acids, serine in particular, are easily acquired contaminants, their status as meteorite constituents is questionable. Samples have been obtained by drilling inward from the surface of a Murchison stone in increments of a few millimeters and these then analyzed for serine and threonine enantiomers by HPLC. Although small amounts of both amino acids were found in the interior samples, enantiomeric analyses indicated a biological origin (Pizzarello and Cronin, unpublished).

Amino acid precursors

Amino acids occur in a hot-water extract of the Murchison meteorite both as free amino acids and as derivatives or precursors that can be converted to amino acids by acid hydrolysis (Cronin and Moore 1971). The total free amino acid content of an extract can be approximately doubled by hydrolysis, with increases in individual amino acids ranging from about 40% to 800% (Cronin 1976). The acid-labile compounds that account for these increases have been partially characterized by ion exchange chromatography. About 70% were found to be acidic compounds, suggesting that they may be derivatives in which the basicity of the amino group has been lost, for example, as in N-acyl amino acids (Cronin 1976).

A recent investigation of the composition of the neutral and acidic fraction of a Murchison extract revealed an extensive series of 2-carboxy-γ-lactams (pyroglutamic acid and alkyl-substituted pyroglutamic acids) and 2-carboxy-δ-lactams, as well as alkyl-substituted and unsubstituted γ- and δ-lactams, compounds that on hydrolysis would yield, respectively, glutamic acid and higher γ-carboxy amino acid homologues; α-amino adipic acid and higher δ-carboxy-α-amino adipic acid homologues; γ-amino butyric acid and higher γ-amino acid homologues; and δ-amino valeric acid and higher δ-amino acid homologues (Cooper and Cronin 1995). In addition, N-acetyl derivatives of glycine, alanine, α-aminoisobutyric acid, and aspartic acid were found in small amounts. The carboxy-γ- and δ-lactams and N-acetyl amino acids have the properties of the acidic amino acid precursors previously identified. It remains to be determined whether the amounts of these compounds are sufficient to account for the amino acid increases seen on acid hydrolysis.

Amino acid chirality

A fundamental question with respect to the origin of life is whether enantiomeric selection took place before or after the origin of life, that is, whether it was a feature of chemical or biological evolution. Consequently, since meteoritic organic matter represents prebiotic organic chemistry, the enantiomeric ratios of the chiral amino acids from Murchison have been of considerable interest. Early GC analyses showed them to be racemic, within experimental error, when extracted from uncontaminated interior samples (Kvenvolden et al. 1970; Oró et al. 1971; Pollock et al. 1975). These results provided a powerful argument for the extraterrestrial origin of the amino acids, especially those that also are common to terrestrial organisms. Subsequently there were conflicting reports of excesses of the L-enantiomers in five Murchison amino acids (all protein amino acids) (Engel and Nagy 1982). This work was criticized by others in the field, who attributed the result to terrestrial contamination of the sample (Bada et al. 1983). More recently, isotopic data were used to support the claim of an excess of the L-enantiomer in alanine from the Murchison meteorite (Engel, Macko and Silfer 1990). The $\delta^{13}C$ values of both L- and D-enantiomers showed enrichments in ^{13}C relative to terrestrial organic matter and, more significantly, the difference between them was too small ($\Delta\delta^{13}C = 3‰$) to be consistent with the terrestrial contamination necessary to account for the L-enantiomer excess. Although the isotopic data support the earlier claim, the observation that the enantiomeric excess occurred in the biologically preferred enantiomer (L) of a common protein amino acid, the rather large L-excess and its contradiction of earlier results obtained with a fresher and presumably more pristine meteorite sample (Kvenvolden et al. 1970), and the possibility that alanine coeluted with other meteoritic amino acids, all gave reason for doubt, or at least to reserve judgment, regarding indigenous enantiomeric excesses.

Recently, a new approach to this problem has been taken that is not subject to some of the criticisms of the earlier work (Cronin and Pizzarello 1997). In these studies, the targets for analysis were Murchison amino acids that (1) either have no known terrestrial source or are of very restricted occurrence in the biosphere, and (2) have chiral centers that are resistant to racemization (epimerization). Consequently, it is likely that the chiral centers of these amino acids retained their original configuration through the aqueous and mild thermal processing experienced in the meteorite parent body and that the original enantiomer ratios were not compromised by terrestrial contamination. The amino acids analyzed were 2-amino-2,3-dimethylpentanoic acid, which has two chiral centers and occurs as four stereoisomers (DL-α-methylisoleucine/DL-α-methylalloisoleucine), 2-amino-2-methylbutanoic acid (isovaline), and 2-amino-2-methylpentanoic acid (α-methylnorvaline). An L-enantiomeric excess was observed in each case: α-methylisoleucine (7.0%), α-methylalloisoleucine (9.1%), isovaline (8.4%), and α-methylnorvaline (2.8%).[3] Interestingly, the α-H analogues of these last two amino acids, that is, α-amino-*n*-butyric acid and α-aminopentanoic acid, were found to be racemates. The failure to observe enantiomeric excesses in the α-amino acids

[3] Enantiomer excess $= \frac{|L - D|}{L + D} \times 100 = |\%L - \%D|$

with α-H substituents contrasts with the reported L-enantiomer excess in alanine, an α-H amino acid (Engel, Macko and Silfer 1990).

It was suggested (Cronin and Pizzarello 1997) that these enantiomeric excesses may have originated in the presolar cloud as a result of its irradiation by circularly polarized ultraviolet light (UVCPL) from a neutron star, an idea originally put forward by others (Rubenstein et al. 1983; Bonner and Rubenstein 1987), and more recently discussed in the context of interstellar grains and cometary organic matter (Greenberg 1996). The role of interstellar chemistry in the formation of meteorite organic matter (discussed in the following section) and the magnitude of experimentally observed enantiomeric excesses obtained when leucine was irradiated with UVCPL (Flores, Bonner and Massey 1977) seem consistent with this possibility.

Isotopic characteristics of meteoritic organic matter

The preceding sections have summarized mainly molecular analyses of the organic compounds found in the Murchison meteorite. Substantial efforts have also been made to obtain isotopic data for these compounds. Several generalizations can be drawn from the data, which are collected elsewhere (Cronin and Chang 1993).

1. The indigenous organic matter, including the macromolecular (insoluble) material, is generally enriched in D and found to have δD values in the range +100‰, the heavy end of the terrestrial range, to +2,500‰.
2. The indigenous soluble organic compounds are enriched in ^{13}C relative to terrestrial matter or, at minimum, lie at the heavy end of the terrestrial range.
3. The indigenous soluble organic compounds are also significantly enriched in ^{15}N relative to terrestrial matter, showing $\delta^{15}N$ values in the range +90‰ to +100‰
4. Five isotopically distinct components, exclusive of the "exotic" phases, can be recognized in Murchison:
   a. amino acids: δD = +1,000‰ to +2,500‰; $\delta^{13}C$ = +5‰ to +40‰; $\delta^{15}N$ = +90‰
   b. other organic compounds: δD = +100‰ to +1,400‰; $\delta^{13}C$ = -10‰ to +10‰; $\delta^{15}N \leq$ +98‰
   c. insoluble carbon: δD +500‰ to +1,000‰; $\delta^{13}C \cong$ -15‰; $\delta^{15}N$ = +18‰
   d. carbonate: $\delta^{13}C$ = +35‰ to +45‰
   e. phyllosilicates (clay minerals): δD ≅ -100‰
5. The monocarboxylic acids and light alkanes show a decline in $\delta^{13}C$ as carbon number increases.

The isotopic enrichments, especially in D, in the Murchison organic matter suggest its formation from interstellar organic compounds. These D-enrichments have been attributed (Kolodny, Kerridge and Kaplan 1980; Geiss and Reeves 1981; Robert and Epstein 1982) to isotopic fractionation effects in the formation of organic compounds by low-temperature (<50 K), zero-activation-energy gas-phase ion–molecule reactions (Watson 1976) and, possibly, grain-mediated reactions (Tielens 1983) in the interstellar cloud from which the Solar System formed. Even greater D-enrichments are observed in the organic molecules of dark interstellar clouds (Millar, Bennett and Herbst 1989). Ordinary neutral–neutral chemical reactions occurring in the cooling solar nebula seem to be ruled out. Reactions at the low temperatures (<200 K) required to achieve comparable D-enrichments through thermodynamic equilibration with nebular H_2 would be too sluggish to be effective over the theoretical lifetime of the solar nebula (Geiss and Reeves 1981). The lower ΔE values for ion–molecule ^{12}C–^{13}C and ^{14}N–^{15}N exchange reactions dictate the lower enrichments observed for ^{13}C and ^{15}N (Wannier 1980).

The survival of interstellar organic compounds once seemed to be precluded by the extreme temperatures thought to have been attained in the solar nebula and the consequent conversion of preexisting molecules and minerals to a well-mixed gas of simple atomic or molecular composition. Under such circumstances the nebula would have retained no structural or isotopic "memory" of its initial organic raw materials. This view has gradually given way to models of the solar nebula in which temperatures where meteorites formed, about 3 AU from the nascent Sun, never exceeded more than a few hundred K (Wood and Morfill 1988). At these temperatures, interstellar grains would retain their integrity and their aggregates could be incorporated into CI chondrites and the matrices of CM chondrites (Cameron 1973).

The first evidence suggesting that the organic matter of carbonaceous chondrites had a presolar origin came from the discovery that the D-enrichment observed in bulk samples is concentrated in the organic fraction (Kolodny, Kerridge and Kaplan 1980; Robert and Epstein 1982). This suggested the incorporation of interstellar molecules (Kolodny, Kerridge and Kaplan 1980; Geiss and Reeves 1981; Robert and Epstein 1982), an interpretation that has subsequently been widely accepted because large, positive δD values have been obtained for various classes of soluble organic compounds, as well as for the macromolecular carbon. The range of the isotopic variations is too large to be attributable to ordinary Solar System processes but is consistent with interstellar processes. In the latter case, variations in "local" conditions such as temperature or fractional ionization can influence the extent of isotopic fractionation associated with gas-phase ion–molecule and grain surface reactions. Thus, the organic compounds of Murchison may largely represent an interstellar inheritance.

Formation of meteoritic organic compounds

From the preceding sections a number of puzzle pieces can be gleaned that allow us to understand the formation of the organic compounds of the Murchison meteorite, a small domain within the larger origin of life puzzle.

1. The compounds are *isotopically unusual*, typically enriched in D. Minimally, this indicates formation by low-temperature ion–molecule chemistry, and probably that the compounds, per se, or their precursors, were formed in an interstellar cloud.
2. Many classes of meteoritic compounds are *common to the interstellar medium*, for example, carboxylic acids, amides, amines, alcohols, aldehydes, ketones, aliphatic hydrocarbons, and PAHs.
3. Many classes of compounds show essentially *complete structural diversity*, apparently reflecting a random synthesis of carbon chains from single carbon units independent of directing influences, such as surface catalysis.
4. The *exponential decline in abundance with increasing carbon chain length and decreasing ^{13}C content with increasing chain length* suggest kinetically controlled chain build-up from single carbon units.
5. The *predominance of branched carbon chains* suggests their formation by a process in which radical or ion stability may have been influential.

These characteristics suggest, or are at least consistent with, an interstellar origin for many of the meteoritic organic compounds; however, in the case of the amino acids, it has seemed reasonable to propose a role for Solar System (parent body) chemistry in their formation as well. As previously noted, the coexistence in Murchison of analogous suites of α-hydroxy- and α-amino acids suggests their formation by a common Strecker synthesis. A volatile-rich parent body should have had the necessary precursors in abundance (HCN, NH_3, and carbonyl compounds are well-known interstellar molecules), and the liquid water present during aqueous processing of the parent body would have provided appropriate reaction conditions. Consequently, the amino acids have been viewed as the products of a two-stage formation process in which the precursors were made in the presolar cloud and the amino acids, per se, were formed later in the early Solar System.

The presence of enantiomeric excesses in the Murchison amino acids are difficult to reconcile with this hypothesis. If UVCPL was the asymmetric agent that gave rise to the amino acid enantiomeric excesses, its effects must have been manifest after (or during) formation of the amino acids; however, if this occurred in the early Solar System it seems likely that they would have been shielded by the parent body. On the other hand, if the exposure to UVCPL occurred in the presolar cloud, only the precursor ketones of the α-methyl amino acids would have been irradiated and there would have been no effect on the chiral α-carbons of the amino acids that were formed later. The C_7 amino acid, 2-amino-2,3-dimethyl pentanoic acid, is interesting in this regard. The precursor of the carbon chain of this amino acid in a Strecker synthesis is the chiral ketone, 3-methyl-2-pentanone (asymmetric C3). In this case, if the presolar cloud had been exposed to UVCPL, the ketone could have carried an enantiomeric excess when incorporated into the primitive parent body. Formation of the amino acid by Strecker reactions would then have given all four stereoisomers, but because of the preexisting enantiomeric excess in the ketone, an L-excess would have been observed in one pair of enantiomers, and a D-excess in the other.[4] In contrast, an L-excess was observed in both pairs. Consequently, if the asymmetric influence was, in fact, stellar UVCPL, formation of the nonracemic α-methyl-α-amino acids seems likely to have been entirely a presolar process.

Two explanations seem plausible for the observation of enantiomeric excesses *only* in the α-substituted-α-amino acids. One is that all the amino acids originally were nonracemic, but during aqueous processing of the parent body, those with an α-hydrogen atom racemized, whereas those with an α-substituent, which are resistant to racemization, did not. In this case, analyses of the state of chiral compounds in cometary samples should be of interest as enantiomeric excesses would be expected in a much wider range of compounds than in the aqueously altered carbonaceous chondrites. The second possibility is that the two classes of amino acids represent the products of different formation processes: for example, presolar formation of the α-substituted amino acids, but parent body formation of those with α-hydrogen atoms. This is consistent with observations that the two classes of amino acids vary in relative amount among different Murchison specimens (Cronin and Pizzarello 1983) and differ in the ease of extraction from

[4] This result is analogous to the α-epimerization of L-isoleucine (2*S*, 3*S*) which gives D-alloisoleucine (2*R*, 3*S*); however, in the case of a Strecker synthesis, the 2*S*–2*R* equilibrium is achieved by formation of the chiral center rather than by inversion of configuration at an existing chiral center. As envisioned here, the C3 of 2-a-2,3-dmpa would not necessarily have been homochiral, as in the case of biological isoleucine, but might have had only a modest enantiomeric excess. Consequently, excesses of the D- and L-enantiomers of the respective diastereomers would also have been small.

meteorite powders (Pizzarello, unpublished). Both observations suggest that the two types of amino acids may be associated with different phases within the meteorite matrix.

Meteoritic organic compounds and the origin of life

Our puzzle piece seems to fit comfortably into a space bounded on one side by pieces denoting the nucleosynthesis of the biogenic elements (C, H, N, O, P, and S) and the formation of organic compounds in interstellar clouds, and, on the other side, by a piece denoting the delivery of organic matter to the early Earth by the infall of asteroidal and cometary matter. Lying beyond, are less distinct pieces showing the further chemical evolution of this organic matter on the early Earth, and its self-organization into structural and functional macromolecules; however, in this part of the puzzle many pieces are missing. Beyond this void lie pieces of increasing clarity indicating an RNA world, the common ancestor with its modern biochemistry, divergence of the phylogenetic tree, and extant life.

The challenge facing origin of life puzzlers is filling the empty space between the products of chemical evolution as exemplified by the organic chemistry of carbonaceous chondrites and the biochemistry of first life, whatever that may have been. An RNA world is likely at some point, but imagining it as the basis for first life is difficult. How the molecular and stereoisomeric demands of RNA biochemistry might have been met by prebiotic chemistry is unclear. The prebiotic synthesis of ribose is problematic, to say nothing of the specific assembly of β-nucleosides, their regiospecific phosphorylation and activation for polymerization. Nonenzymatic polymerization of activated nucleotides is strongly dependent on consistently correct nucleotide stereoisomerism (Joyce et al. 1984). An earlier pre-RNA world based on "low-tech" biochemistry, to borrow a useful term (Cairns-Smith 1982), seems probable.

Could extrapolation from the chemical evolution side of the gap provide useful clues to the nature of this earliest biochemistry? As we have seen, the organic chemical inventory of carbonaceous chondrites includes many interesting compounds, such as amino acids, purines, and pyrimidines, and common metabolic intermediates like the hydroxy acids and dicarboxylic acids. The presence of vesicle-forming components and possible light-harvesting membrane pigments (Deamer and Pashley 1989) is also intriguing. Nevertheless, the lack of structural specificity and the preponderance of the low molecular weight homologues among the meteorite organics make it difficult to imagine this mixture as the ingredients list in a recipe for life.

The recent discovery of enantiomeric excesses in the α-substituted amino acids of the Murchison meteorite adds a new dimension to arguments for exogenous delivery as an important contributor to prebiotic chemistry. If it was the exogenous delivery of amino acids and perhaps other compounds with enantiomeric excesses that "solved" the stereoisomerism problem, perhaps more serious consideration should be given to the organic matter of comets and meteorites as important ingredients of the prebiotic milieu. The α-methyl amino acids are a case in point. These amino acids, which have not been viewed as important for the origin of life because they are generally unimportant in life, are abundant in Murchison and may have been well suited for a role in further amplification of the small initial enantiomeric excesses. Polymerization accompanied by formation of regular secondary structure, for example, α-helices and β-sheets, has been shown to be an effective way to amplify modest initial enantiomeric excesses (Bonner, Blair and Dirbas 1981; Brack and Spach 1981), and α-methyl amino acids are known to have strong helix-inducing and stabilizing effects (Altman et al. 1988). Although a transition to α-H amino acids would obviously have been required at some later stage, the α-substituted amino acids could have played a significant early role in the development of homochirality. The polymerization of these amino acids and the conformation, conformational stability, and catalytic activity of their oligomers might be fruitful lines of research in prebiotic chemistry.

Clay minerals have long been proposed to have had a central role in prebiotic chemistry (Bernal 1967) or even in early biochemistry (Cairns-Smith 1982). In the matrix of carbonaceous chondrites, a rich organic chemistry is joined with abundant clay minerals. The further chemical evolution of the meteorite organics on the surfaces or within the layers of the associated clay minerals is another line of research that might yield interesting insights. The clay minerals of carbonaceous chondrites could, in principle, have provided both catalysis and specificity for subsequent steps in the development of the organic chemical complexity necessary for evolution of the first biochemistry.

References

Altman, E., Altman, K. H., Nebel, K., and Mutter, M. 1988. Conformational studies on host–guest peptides containing chiral α-methyl-α-amino acids. *Int. J. Pept. Protein Res.* 32:344–351.

Amari, S., Anders, E., Virag, A., and Zinner, E. 1990. Interstellar graphite in meteorites. *Nature* 345:238–240.

Anders, E. 1975. Do stony meteorites come from comets? *Icarus* 24:363–371.

Anders, E. 1978. Most stony meteorites come from the asteroid belt. In *Asteroids: An*

Exploration Assessment, ed. D. Morrison and W. C. Wells, Washington, D.C.: NASA CP-2053, U.S. Govt. Printing Office.

Anders, E. 1989. Pre-biotic organic matter from comets and asteroids. *Nature* 342:255–256.

Arrhenius 1908. *Worlds in the Making*. New York: Harper and Row.

Bada, J. L., Cronin, J. R., Ho, M.-S., Kvenvolden, K. A., Lawless, J. G., Miller, S. L., Oró, J., and Steinberg, S. 1983. On the reported optical activity of amino acids in the Murchison meteorite. *Nature* 301:494–497.

Belsky, T., and Kaplan, I. R. 1970. Light hydrocarbon gases, ^{13}C, and origin of organic matter in carbonaceous chondrites. *Geochim. Cosmochim. Acta* 34:257–278.

Bernal, J. D. 1967. *The Origin of Life*. Cleveland: The World Publishing Company.

Blake, D. F., Freund, F., Krishnan, K. F. M., Echer, C. J., Shipp, R., Bunch, T. E., Tielens, A. G., Lipari, R. J., Hetherington, C. J. D., and Chang, S. 1988. The nature and origin of interstellar diamond. *Nature* 332:611–613.

Bonner, W. A., Blair, N. E., and Dirbas, F. M. 1981. Experiments on the abiotic amplification of optical activity. *Origins of Life* 11:119–134.

Bonner, W. A., and Rubenstein, E. 1987. Supernovae, neutron stars and biomolecular chirality. *BioSystems* 20:99–111.

Brack, A., and Spach, G. 1981. Enantiomer enrichment in early peptides. *Origins of Life* 11:135–142.

Buhl, P. 1975. An investigation of organic compounds in the Mighei meteorite. Ph.D. Thesis, University of Maryland, College Park.

Bunch, T. E., and Chang, S. 1980. Carbonaceous chondrites: II. Carbonaceous chondrite phyllosilicates and light element geochemistry as indicators of parent body processes and surface conditions. *Geochim. Cosmoshim. Acta* 44:1543–1577.

Cairns-Smith, A. G. 1982. *Genetic Takeover and the Mineral Origins of Life*. Cambridge: Cambridge University Press.

Cameron, A. G. W. 1973. Interstellar grains in museums? In *Interstellar Dust and Related Topics*, ed. J. M. Greenberg and H. C. Van de Hulst, pp. 545–547. Dordrecht: D. Reidel Publishing Co.

Cech, T. R., Zaug, A. J., and Grabowski, P. J. 1981. *In vitro* splicing of the ribosomal RNA precursor of Tetrahymena: Involvement of a guanosine nucleotide in the excision of the intervening sequence. *Cell* 27:487–496.

Chyba, C. F., and Sagan, C. 1992. Endogenous production, exogenous delivery and impact–shock synthesis of organic molecules: an inventory for the origins of life. *Nature* 355:125–132.

Clemett, S. J., Maechling, C. R., Zare, R. N., Swan, P. D., and Walker, R. M. 1993. Identification of complex aromatic molecules in individual interplanetary dust particles. *Science* 262:721–725.

Cooper, G. W., and Cronin, J. R. 1995. Linear and cyclic aliphatic carboxamides of the Murchison meteorite: Hydrolyzable derivatives of amino acids and other carboxylic acids. *Geochim. Cosmochim. Acta* 59:1003–1015.

Cooper, G. W., Onwo, W. M., and Cronin, J. R. 1992. Alkyl phosphonic acids and sulfonic acids in the Murchison meteorite. *Geochim. Cosmochim. Acta* 56:4109–4115.

Cronin, J. R. 1976. Acid-labile amino acid precursors in the Murchison meteorite. I. Chromatographic fractionation. *Origins of Life* 7:337–342.

Cronin, J. R., and Chang, S. 1993. Organic matter in meteorites: Molecular and isotopic analyses. In *The Chemistry of Life's Origins,* ed. J. M. Greenberg, C. X.

Mendoza-Gómez, and V. Pirronello, pp. 209–258. Dordrecht: Kluwer Academic Publishers.

Cronin, J. R., Gandy, W. E., and Pizzarello, S. 1981. Amino acids of the Murchison meteorite: I. Six carbon acyclic primary α-amino alkanoic acids. *J. Mol. Evol.* 17:265–272.

Cronin, J. R., and Moore, C. B. 1971. Amino acid analyses of the Murchison, Murray, and Allende carbonaceous chondrites. *Science* 172:1327–1329.

Cronin, J. R., and Pizzarello, S. 1983. Amino acids in meteorites. *Adv. Space Res.* 3:3–18.

Cronin, J. R., and Pizzarello, S. 1986. Amino acids of the Murchison meteorite. III. Seven carbon acyclic primary α-amino alkanoic acids. *Geochim. Cosmochim. Acta* 50:2419–2427.

Cronin, J. R., and Pizzarello, S. 1990. Aliphatic hydrocarbons of the Murchison meteorite. *Geochim. Cosmochim. Acta* 54:2859–2868.

Cronin, J. R., and Pizzarello, S. 1997. Enantiomeric excesses in meteoritic amino acids. *Science* 275:951–955.

Cronin, J. R., Pizzarello, S., and Cruikshank, D. P. 1988. Organic matter in carbonaceous chondrites, planetary satellites, asteroids, and comets. In *Meteorites and the Early Solar System,* ed. J. F. Kerridge and M. S. Matthews, pp. 819–857. Tucson: University of Arizona Press.

Cronin, J. R., Pizzarello, S., Epstein, S., and Krishnamurthy, R. V. 1993. Molecular and isotopic analyses of the hydroxy acids, dicarboxylic acids, and hydroxydicarboxylic acids of the Murchison meteorite. *Geochim. Cosmochim. Acta* 57:4745–4752.

Cronin, J. R., Pizzarello, S., and Frye, J. S. 1987. ^{13}C NMR spectroscopy of the insoluble carbon of carbonaceous chondrites. *Geochim. Cosmochim. Acta* 51:299–303.

Cronin, J. R., Pizzarello, S., and Yuen, G. U. 1985. Amino acids of the Murchison meteorite. II. Five carbon acyclic primary β-, γ-, and δ-amino alkanoic acids. *Geochim. Cosmochim. Acta* 49:2259–2265

Deamer, D. W. 1985. Boundary structures are formed by organic components of the Murchison carbonaceous chondrite. *Nature* 317:792–794.

Deamer, D. W. 1986. Role of amphiphilic compounds in the evolution of membrane structure on the early Earth. *Origins of Life* 17:3–25.

Deamer, D. W., and Pashley, R. M. 1989 Amphiphilic components of the Murchison carbonaceous chondrite: Surface properties and membrane formation. *Origins Life Evol. Biosphere* 19:21–38.

Dodd, R. T. 1986. *Thunderstones and Shooting Stars.* Cambridge: Harvard University Press.

Engel, M. H., and Nagy, B. 1982. Distribution and enantiomeric composition of amino acids in the Murchison meteorite. *Nature* 296:837–840.

Engel, M. H., Macko, S. A., and Silfer, J. A. 1990. Carbon isotope composition of individual amino acids in the Murchison meteorite. *Nature* 348:47–49.

Epstein, S., Krishnamurthy, R. V., Cronin, J. R., Pizzarello, S., and Yuen, G. U. 1987. Unusual stable isotope ratios in amino acid and carboxylic acid extracts from the Murchison meteorite. *Nature* 326:477–479.

Flores, J. J., Bonner, W. A., and Massey, G. A. 1977. Asymmetric photolysis of (*R,S*)-leucine with circularly polarized light. *J. Am Chem. Soc.* 99:3622–3625.

Fuchs, L. H., Olsen, E., and Jensen, K. J. 1973. Mineralogy, crystal chemistry and composition of the Murchison (C2) meteorite. *Smithsonian Contrib. Earth Sci.* 10:1–39.

Geiss, J., and Reeves, H. 1981. Deuterium in the early solar system. *Astron. Astrophys.* 93:189–199.

Gilbert, W. 1986. The RNA world. *Nature* 319:618.

Gilmour, I., and Pillinger, C. 1993. Extraction and isotopic analysis of medium molecular weight hydrocarbons from Murchison using supercritical carbon dioxide. *Proc. Lunar Planet. Sci. Conf.* 24:535–536.

Grady, M. M., Wright, I. P., Swart, P. K., and Pillinger, C. T. 1988. The carbon and oxygen isotopic composition of meteoritic carbonates. *Geochim. Cosmochim. Acta* 52:2855–2866.

Greenberg, J. M. 1996. Chirality in interstellar dust and in comets: Life from dead stars. In *Physical Origin of Homochirality in Life*, ed. D. B. Cline, pp. 185–210. Woodbury, N.Y.: American Institute of Physics.

Guélin, M., Cernicharo, J., Paubert, G., and Turner, B. E. 1990. Free CP in IRC$^+$ 10216. *Astron. Astrophys.* 230:L9–L11.

Hahn, J. H., Zenobi, R., Bada, J. L., and Zare, R. N. 1988. Application of two-step laser mass spectrometry to cosmogeochemistry: direct analysis of meteorites. *Science* 239:1523–1525.

Haldane, J. B. S. 1929. The origin of life. Reprinted in J. D. Bernal, *The Origin of Life,* 1967, pp. 242–249. London: Weidenfeld and Nicolson.

Hamilton, P. B. 1965. Amino acids on hands. *Nature* 205:284–285.

Hayatsu, R., and Anders, E. 1981. Organic compounds in meteorites and their origins. *Topics Curr. Chem.* 99:1–37.

Hayatsu, R., Matsuoka, S., Scott, R. G., Studier, M., and Anders, E. 1977. Origin of organic matter in the early solar system. VII. The organic polymer in carbonaceous chondrites. *Geochim. Cosmochim. Acta* 41:1325–1339.

Hayatsu, R., Studier, M. H., Moore, L. P., and Anders, E. 1975. Purines and triazines in the Murchison meteorite. *Geochim. Cosmochim. Acta* 39:471–488.

Hayatsu, R., Winans, R. E., Scott, R. G., McBeth, R. L., Moore, L. P., and Studier, M. H. 1980. Phenolic ethers in the organic polymer of the Murchison meteorite. *Science* 207:1202–1204.

Hayes, J. M. 1967. Organic constituents of meteorites: A review. *Geochim. Cosmochim. Acta* 31:1395–1440.

Heide, F., and Wlotzka, F. 1995. *Meteorites: Messengers from Space.* Berlin:Springer-Verlag.

Irvine, W. In press. Extraterrestrial organic matter: a review. *Origins Life Evol. Biosphere.*

Jarosevich, E. 1971 Chemical analysis of the Murchison meteorite. *Meteoritics* 6:49–52.

Joyce, G. F., Visser, G. M., van Boeckel, C. A. A., van Boom, J. H., Orgel, L. E., and van Westrenen, J. 1984. Chiral selection in poly(C)-directed synthesis of oligo(G). *Nature* 310:602–604.

Jungclaus, G., Cronin, J. R., Moore, C. B., and Yuen, G. U. 1979. Aliphatic amines in the Murchison meteorite. *Nature* 261:126–128.

Jungclaus, G. A., Yuen, G. U., Moore, C. B., and Lawless, J. G. 1976. Evidence for the presence of low molecular weight alcohols and carbonyl compounds in the Murchison meteorite. *Meteoritics* 11:231–237.

Kasting, J. F. 1993. Earth's early atmosphere. *Science* 259:920–926.

Kimball, B. A. 1988. Determination of formic acid in chondritic meteorites. M.S. Thesis, Arizona State University, Tempe.

Kolodny, Y., Kerridge, J. F., and Kaplan, I. R. 1980. Deuterium in carbonaceous chondrites. *Earth Planet. Sci. Lett.* 46:149–158.

Krishnamurthy, R. V., Epstein, S., Cronin, J. R., Pizzarello, S., and Yuen, G. U. 1992. Isotopic and molecular analyses of hydrocarbons and monocarboxylic acids of

the Murchison meteorite. *Geochim. Cosmochim. Acta* 56:4045–4058.
Kvenvolden, K., Lawless, J., Pering, K., Peterson, E., Flores, J., Ponnamperuma, C., Kaplan, I. R., and Moore, C. 1970. Evidence for extraterrestrial amino acids and hydrocarbons in the Murchison meteorite. *Nature* 228:923–926.
Kvenvolden, K., Lawless, J. G., and Ponnamperuma, C. 1971. Nonprotein amino acids in the Murchison meteorite. *Proc. Natl. Acad. Sci. USA* 68:486–490.
Lawless, J. G. 1973. Amino acids in the Murchison meteorite. *Geochim. Cosmochim. Acta* 37:2207–2212.
Lawless, J. G., and Yuen, G. U. 1979. Quantification of monocarboxylic acids in Murchison carbonaceous meteorite. *Nature* 282:396–398.
Lawless, J. G., Zeitman, B., Pereira, W. E., Summons, R. E., and Duffield, A. M. 1974. Dicarboxylic acids in the Murchison meteorite. *Nature* 251:40–41.
Lewis, R. S., Tang, M., Wacker, J. F., Anders, E., and Steel, E. 1987. Interstellar diamonds in meteorites. *Nature* 326:160–162.
Lumpkin, G. R. 1986. Electron microscopy of carbonaceous matter in Allende acid residues. *Proc. Lunar Planet. Sci. Conf.* 12B:1153–1166.
Mason, B. 1963. The carbonaceous chondrites. *Space Sci. Rev.* 1:621–646.
Maurette, M., Brack, A., Kurat, G., Perreau, M., and Engrand, C. 1994. Were micrometeorites a source of prebiotic molecules on the early earth? *Adv. Space Res.* 15:113–126.
Meinschein, W. G. 1963. Hydrocarbons in terrestrial samples and the Orgueil meteorite. *Space Sci. Rev.* 2:653–679.
Miknis, F. P., Lindner, A. W., Gannon, J., Davis, M. F., and Maciel, G. E. 1984. Solid state ^{13}C NMR studies of selected oil shales from Queensland, Australia. *Org. Geochem.* 7:239–248.
Millar, T. J., Bennett, A., and Herbst, E. 1989. Deuterium fractionation in dense interstellar clouds. *Ap. J.* 340:906–920.
Miller, S. L. 1953. Production of amino acids under possible primitive Earth conditions. *Science* 117:528–529.
Mullie, F., and Reisse, J. 1987. Organic matter in carbonaceous chondrites. *Topics in Current Chemistry* 139:85–117.
Mumma, M. J., DiSanti, M. A., Dello Russo, N., Fomenkova, M., Magee-Sauer, K., Kaminski, C. D., and Xie, D. X. 1996. Detection of abundant ethane and methane along with carbon monoxide and water, in Comet C/1996 B2 Hyakutake: Evidence for interstellar origin. *Science* 272:1310–1314.
Oparin, A. I. 1938. *The Origin of Life*. New York: MacMillan. Reprinted, New York: Dover, 1953.
Oró, J. 1963. Studies in experimental organic cosmochemistry. *Ann. N.Y. Acad. Sci.* 108:464–481.
Oró, J., Gibert, J., Lichtenstein, H., Wikstrom, S., and Flory, D. A. 1971. Amino acids, aliphatic, and aromatic hydrocarbons in the Murchison meteorite. *Nature* 230:105–106.
Oró, J., and Skewes, H. B. 1965. Free amino acids on human fingers: the question of contamination in microanalysis. *Nature* 207:1042–1045.
Peltzer, E. T., and Bada, J. L. 1978. α-Hydroxycarboxylic acids in the Murchison meteorite. *Nature* 272:443–444.
Peltzer, E. T., Bada, J. L., Schlesinger, G., and Miller, S. L. 1984. The chemical conditions on the parent body of the Murchison meteorite: Some conclusions based on amino, hydroxy, and dicarboxylic acids. *Adv. Space Res.* 4:69–74.
Pizzarello, S., Feng, X., Epstein, S., and Cronin, J. R. 1994. Isotopic analyses of nitrogenous compounds from the Murchison meteorite: ammonia, amines, amino acids, and polar hydrocarbons. *Geochim. Cosmochim. Acta* 58:5579–5587.

Pizzarello, S., Krishnamurthy, R. V., Epstein, S., and Cronin, J. R. 1991. Isotopic analyses of amino acids from the Murchison meteorite. *Geochim. Cosmochim. Acta* 55:905–910.
Pollock, G. E., Chang, C.-N., Cronin, S. E., and Kvenvolden, K. E. 1975. Stereoisomers of isovaline in the Murchison meteorite. *Geochim. Cosmochim. Acta* 39:1571–1573.
Reynolds, J. H., Frick, U., Neil, J. M., and Phinney, D. L. 1978. Rare-gas–rich separates from carbonaceous chondrites. *Geochim. Cosmochim. Acta* 42:1775–1797.
Robert, F., and Epstein, S. 1982. The concentration and isotopic composition of hydrogen, carbon, and nitrogen in carbonaceous meteorites. *Geochim. Cosmochim. Acta* 46:81–95.
Rossignol-Strick, M., and Barghoorn, E. 1971. Extraterrestrial abiogenic organization of organic matter: The hollow spheres of the Orgueil meteorite. *Space Life Sci.* 3:89–107.
Rubenstein, E., Bonner, W. A., Noyes, H. P., and Brown, G. S. 1983. Supernovae and life. *Nature* 306:118.
Saito, S., Kawaguchi, K., Yamamoto, S., Ohishi, M., Suzuki, H., and Kaifu, N. 1987. Laboratory detection and astronomical identification of a new free radical, CCS ($^3S^-$). *Astrophys. J.* 317:L115–L119.
Shimoyama, A., Naroka, H., Komiya, M., and Harada, K. 1989. Analyses of carboxylic acids and hydrocarbons in Antarctic carbonaceous chondrites, Yamato-74662 and Yamato-793321. *Geochem. J.* 23:181–193.
Stoks, P. G., and Schwartz, A. W. 1979. Uracil in carbonaceous meteorites. *Nature* 282:709–710.
Stoks, P. G., and Schwartz, A. W. 1981a. Nitrogen-heterocyclic compounds in meteorites: Significance and mechansim of formation. *Geochim. Cosmochim. Acta* 45:563–569.
Stoks, P. G., and Schwartz, A. W. 1981b. Nitrogen compounds in meteorites: A reassessment. In *Proceedings of the 6th International Conference on the Origin of Life,* ed. Y. Wolman, pp. 59–64. Dordrecht: D. Reidel.
Stoks, P. G., and Schwartz, A. W. 1982. Basic nitrogen-heterocyclic compounds in the Murchison meteorite. *Geochim. Cosmochim. Acta* 46:309–315.
Studier, M. H., Hayatsu, R., and Anders, E. 1972. Origin of organic matter in the early solar system. V. Further studies of meteoritic hydrocarbons and a discussion of their origin. *Geochim. Cosmochim. Acta* 36:189–215.
Tang, M., Anders, E., Hoppe, P., and Zinner, E. 1989. Meteoritic silicon carbide and its stellar sources: implications for galactic chemical evolution. *Nature* 339:351–354.
Tielens, A. G. G. M. 1983. Surface chemistry of deuterated molecules. *Astron. Astrophys.* 119:177–184.
Tingle, T. N., Becker, C. H., and Malhotra, R. 1991. Organic compounds in the Murchison and Allende carbonaceous chondrites studied by photoionization mass spectrometry. *Meteoritics* 26:117–127.
Urey, H. C. 1952. On the early chemical history of the earth and the origin of life. *Proc. Natl. Acad. Sci. USA* 38:351–363.
Van der Velden, W., and Schwartz, A. W. 1977. Search for purines and pyrimidines in the Murchison meteorite. *Geochim. Cosmochim. Acta* 41:961–968.
de Vries, M. S., Reihs, K., Wendt, H. R., Golden, W. G., Hunziker, H. E., Fleming, R., Peterson, E., and Chang, S. 1993. Search for C_{60} in the Murchison meteorite. *Geochim. Cosmochim. Acta* 57:933–938.
Wannier, P. G. 1980. Nuclear abundances and evolution of the interstellar medium. *Ann. Rev. Astr. Astrophys.* 18:399–437.

Watson, W. D. 1976. Interstellar molecule reactions. *Rev. Mod. Phys.* 48:513–552.

Wilson, R. W., Solomon, P. M., Penzias, A. A., and Jefferts, K. B. 1971. Millimeter observations of CO, CN, and CS emission from IRC^+ 10216. *Astrophys. J.* 169:L35–L37.

Wood, J. A., and Morfill, G. E. 1988. Solar nebula models. In *Meteorites and the Early Solar System,* ed. J. F. Kerridge and M. S. Matthews, pp. 329–347. Tucson: University of Arizona Press.

Yamamoto, S., Saito, S., Kawaguchi, K., Kaifu, K., Suzuki, H., and Ohishi, M. 1987. Laboratory detection of a new carbon-chain molecule C_3S and its astronomical identification. *Astrophys. J.* 317:L119–L121.

Yuen, G., Blair, N., Des Marais, D. J., and Chang, S. 1984. Carbon isotope composition of low molecular weight hydrocarbons and monocarboxylic acids from Murchison meteorite. *Nature* 307:252–254.

Yuen, G., and Kvenvolden, K. A. 1973. Monocarboxylic acids in Murray and Murchison carbonaceous meteorites. *Nature* 246:301–302.

7

Micrometeorites on the early Earth

MICHEL MAURETTE
Astrophysique du Solide
C.S.N.S.M., Orsay-Campus, France

I. Introduction

Since July 1984, "large" micrometeorites with sizes ranging from ≈50 to 500 μm were successfully collected in large number on the Greenland and Antarctica ice sheets (Maurette et al. 1987, 1991, 1994). It was shown that they represent by far the dominant source of extraterrestrial materials that "survive" upon their hypervelocity capture by the Earth.

The cleanest collections of micrometeorites have been recovered from Antarctica ices. These Antarctica micrometeorites (AMMs) are mostly carbonaceous objects, related mainly to a relatively rare class of carbonaceous meteorites (CM-type chondrites). They contain high concentrations of carbonaceous material (average C-content ≈7 wt%), including complex organics such as amino acids (Brinton et al. in press) and polycyclic aromatic hydrocarbons (Clemett et al. in press).

The characteristics of the AMMs' constituent grains were acquired in the early Solar System, before their accretion in the micrometeorite parent bodies (comets and/or asteroids). But some of them have been since altered as a result of interactions effective either in these parent bodies or in the interplanetary medium and the terrestrial environment.

Such interactions have to be understood, first, to identify the primitive characteristics of micrometeorites (which can be used either to relate them to other types of Solar System bodies, or to find new constraints on the early history of the Solar System), and secondly, because they might have produced new features improving the functioning of micrometeorites as microscopic "chondritic chemical reactors" to synthesize prebiotic molecules on the early Earth.

The major objective of this chapter is to present this "early micrometeorites" scenario, as well as the major difficulties and speculations still facing it. But first we have to know about both the characteristics of Antarctica micrometeorites (including their carbon chemistry) and their fast interactions with the terrestrial environment that follow their capture by the Earth.

The smaller micrometeorites collected since 1975 in the stratosphere by NASA (IDPs) will be only briefly mentioned in this chapter. Much information about their collection and characteristics can be found in a book edited by Zolensky et al. (1994). The IDPs are even more "exotic" than Antarctica micrometeorites because they show marked H and N isotopic anomalies, and unique components such as GEMS (Glass Embedded with Metals and Sulfides). However, their contribution to the bulk accretion rate of the Earth is negligible, being about 100 times smaller than that of the larger micrometeorites. Moreover, their smaller size precludes both the search for important components of the micrometeorite flux (e.g., chondrules) and the formation of shooting stars upon atmospheric entry. (Information from shooting stars has helped formulate useful inferences about the parent bodies of large micrometeorites; see Section II.3.3.) However, it should be kept in mind that they might deliver to the Earth unique chondritic reactors, not to be found in the flux of the larger micrometeorites.

II. Micrometeorites today

II.1. Controlling all steps in micrometeorite studies

There is plenty of dust in the Solar System; it is concentrated in the "zodiacal" cloud, fed by dust grains ejected from comets and asteroids. The Earth moves through this cloud and captures (as micrometeorites) its constituent dust at a rate of about 100 tons per day (see Section II.2.2). In spite of its ubiquity we have only recently identified its nature and we still poorly know its sources .

Several major difficulties explain this situation. The small size of the grains, when coupled to their scarcity, considerably complicates their collection and their analysis, which is frequently performed with instruments exploited at the limits of their sensitivity. Furthermore, the grains can easily be contaminated as a result of their high surface-to-volume ratio, their marked porosity , and their chemical reactivity when exposed to gases and waters (see Sections II.2.3 and II.2.4). Consequently, all steps in the study of micrometeorites have to be strictly controlled, from their collection from the cleanest terrestrial "sediments" (e.g., Antarctic ices) to their preparation in ultraclean conditions and their microanalysis. Very few research groups have developed the required expertise to tackle such difficulties.

We shall just describe the Antarctic collections that have rejuvenated the study of large (>50 μm) micrometeorites and have allowed, for the first time, an investigation into their carbon chemistry. These collections were made since December 1988 in the blue ice fields of Cap-Prudhomme, at 6 km from the French station of Dumont d'Urville. With a micrometeorite "factory" (see Figure 7.1), three pockets of melt ice water are made on each day of good

Figure 7.1. Micrometeorite "factory" used at Cap-Prudhomme during the Antarctic summer of 1993–94. The caravan hosts three steam generators that deliver jets of reheated melt ice water at ≈70 °C that are injected into three drill holes with an initial diameter and depth of about 30 cm and 1.5 m, respectively (the water quickly reaches an equilibrium temperature of a few °C). After 8 hours of functioning, each hole has generated a pocket of melt ice water with a volume of about 3–5 m^3. The chimney of the factory was equipped with a smoke filter that stopped all fly ashes with a size ≥36 μm.

weather (about 50% of the time). The glacial sand deposited on the bottom of the holes is vacuumed up with a water pump and filtered on a stack of stainless steel sieves, yielding four distinct size fractions (25–50 μm; 50–100 μm; 100–400 μm; ≥400 μm).

This sand is amazingly rich in micrometeorites, with its 50–100-μm size fraction being the best. In this fraction, where up to ≈50% of the grains are unmelted micrometeorites, a good daily collection typically yields about 500 melted micrometeorites (cosmic spherules) and 2,000 unmelted to partially "dehydrated" micrometeorites (see Section II.2.1 for this "textural" classification of micrometeorites). After a rapid counting with a field microscope of the total number of cosmic spherules in the 100–400-μm size fraction (this number is used as an index of "quality" of the daily collection), the grains are refrozen in a small aliquot of their original melt ice water and transported in deep-freeze conditions to our laboratory freezers.

The best collections were made during the 1993–94 field operations. We drastically improved the micrometeorite factory already used in 1988 and 1991. In particular, all parts exposed to hot water were made of either a brand of stainless steel used in the tubings of the French nuclear reactors or Teflon, in order to minimize the formation of corrosion and leach products in water, which could end up contaminating the grains. Furthermore, during the same operations, the heaviest snowfalls ever recorded since the creation of the French station of Dumont d'Urville in 1950 very efficiently shielded the blue ice fields from contaminations generated by human activities, thus completing the effect of the very clean wind blowing almost constantly from the center to the margin of the ice cap, which transforms the working area into a gigantic dust-free room.

These Antarctica collections allowed, for the first time, the recovery of "giant" micrometeorites (size ≈200 μm) with a mass sufficient both to investigate their carbon chemistry (Section II.3.2) and to assess the concentration of important objects, such as chondrules and refractory inclusions, in the micrometeorite flux (Section II.3.3).

A new and important collection was made in January 1996 by Taylor, Lever and Harvey (1996) on the bottom of the water well of the South Pole station. It gives the unique opportunity to sample micrometeorites over time windows of about 300 years, corresponding to the ice thickness melted each year (about 10 m).

II.2. Fast interactions with the terrestrial environment

Micrometeorites recovered from Antarctic ices have suffered only "fast" interactions with the terrestrial environment during their capture by the atmosphere and their short exposure to melt ice water. These interactions are already relevant to exobiology. Indeed, in the new "early micrometeorites" scenario presented in Section III.5, some of the key characteristics of micrometeorites, which can be thought of as chemical reactors, were acquired during such fast interactions.

II.2.1. Frictional heating: inadequacy of the modeling and textural classification

Upon atmospheric entry, micrometeorites suffer a frictional heating during their aerodynamical braking with air molecules in the upper layers of the atmosphere, which can trigger their complete melting, generating cosmic spherules (CSs).

After the pioneering work of Whipple (1951), many theoretical models of this heating have been reported over the last 30 years. One of the first unexpected results of the study of the 50–100-μm-sized AMMs was to show that the theories are inadequate (Maurette et al. 1990, 1995). For example, they predict that around a size of ≈100 μm, the ratio of micrometeorites to cosmic spherules is less than 1:100 (Brownlee 1985), whereas we constantly found a ratio of 4:1 during several random samplings of these micrometeorites! This corresponds to a micrometeorite survival rate about 400 times larger than the value predicted by the models. In the larger (100–400 μm) size fraction, the ratio of AMMs to CSs drops by a factor of about 20, as frictional heating increases with the size of the impacting bodies. But it is still much larger than the predicted values .

We still poorly understand the inadequacy of these models. Philippe Bonny (Bonny et al. 1988; Bonny 1990), in the most detailed and up-to-date model reported yet, which was developed at the Department of Thermophysics of ONERA, obtained new clues. He applied to micrometeorites, after many improvements, a computer code (COKE) initially developed to describe the effects of frictional heating on meter-sized thermal blankets used to shield satellites upon atmospheric entry.

These meter-sized blankets were made of glass fibers (70%) impregnated with a phenolic resin (30%). Considering this mixture as a model of a "synthetic" micrometeorite, Bonny showed that organics partially degrade upon pyrolysis in the upper layers of the atmosphere through a complex network of about 10 distinct endothermic reactions that cool down micrometeorites from the "inside" (Figure 7.2). The synthetic mixture happens to just correspond to the average ratio of organics to silicates inferred from the in situ analysis of grains from Comet Halley (Krueger and Kissel 1987).

So micrometeorites could survive better than predicted because they carry their own thermal shield that would get partially "consumed" upon atmospheric entry! From this viewpoint it is important to check whether the partial dehydration of the constituent hydrous minerals of micrometeorites further contributes to this shielding. But even with his improved modeling, Bonny predicted only a slightly higher value (a few percent) for the micrometeorites-to-cosmic-spherules ratio for AMMs 100 μm in size, still quite short of the

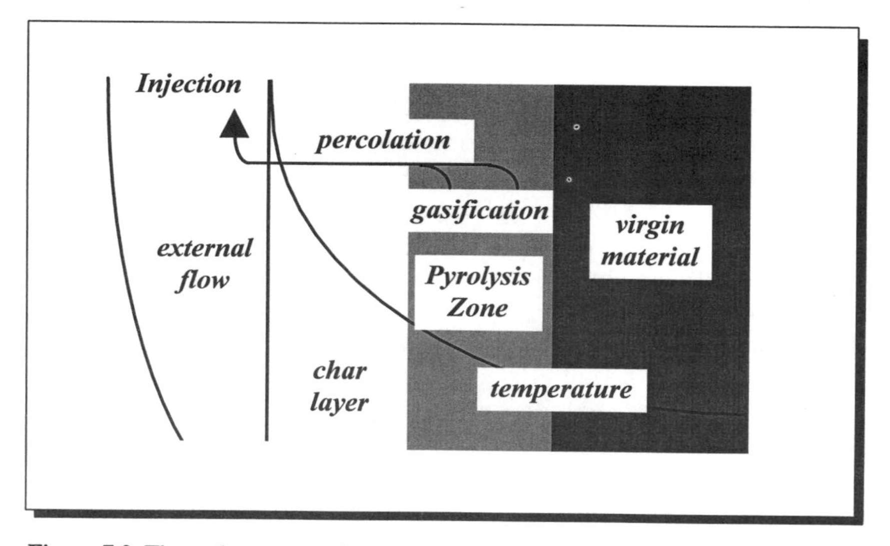

Figure 7.2. Thermal response of a synthetic charring material upon frictional heating during atmospheric entry (from Bonny et al. 1988).

real value. All previous models are inadequate because their mathematics cannot take into consideration the very brief duration (about 1 s) of the flash heating suffered by micrometeorites, their complex composition, and their encapsulation into a thin shell (shield?) of magnetite. Moreover, according to Bonny (Bonny et al. 1988; Bonny 1990), the gas "gasket" formed on the leading edge of the incoming micrometeorite is still poorly described.

Fortunately, we have a much larger number of micrometeorites to investigate than predicted by the modelings!

Observations of polished sections of AMMs and cosmic spherules from Antarctica yield a simple textural classification (Figure 7.3) related to the degree of thermal metamorphism upon atmospheric entry, which is estimated by looking at the abundance of tiny vesicles resulting from the loss of the structural "water" (either H_2O or OH) of their constituent hydrous silicates.

In the richest 50–100-μm size fraction, about 20% of the chondritic grains are melted cosmic spherules (Figure 7.3, micro-77, top left). An amazing feature recently discovered during ion microprobe analyses (Engrand et al. 1996a, b) is that spherules previously considered as the most severely heated up (e.g., the chondritic barred spherules) still contain 1%–2% extraterrestrial water (as characterized by D/H measurements) originating from their initial hydrous minerals, as well as nuggets of a variety of "dirty" iron hydroxide called ferrihydrite, containing a high concentration of Ni (indicative of the nuggets' extraterrestrial origin), (Figure 7.3, micro-77). These characteristics cannot be predicted by the modelings.

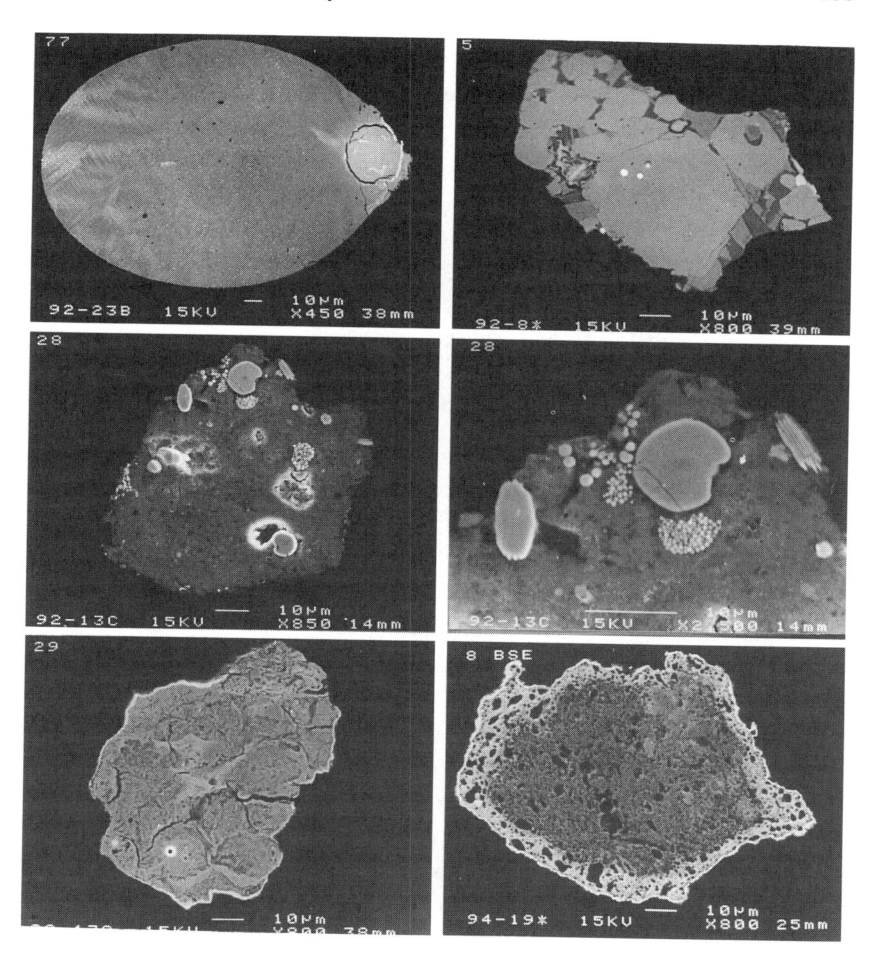

Figure 7.3. Polished sections of micrometeorites with sizes of about 100 µm, observed with a scanning electron microscope (each micrograph is identified with a number on the upper left corner) and illustrating the major textural groups of micrometeorites. From top to bottom one thus finds: **micro-77**, cosmic spherule showing a bright nugget of COPS phase (on the right), very rich in C, O, P, S, and Fe; **micro-5**, unmelted crystalline micrometeorite, essentially made of olivine crystals containing bright nodules of primitive Fe/Ni metal (e.g., with a low Ni content); **micro-28**, fine-grained unmelted micrometeorite of a relatively rare class (about 5% of the micrometeorites) related to CI-type carbonaceous chondrites, which is rich in phyllosilicates and in small magnetite framboids; it can be observed at a higher magnification on the right-hand micrograph; **micro-29**, fine-grained unmelted micrometeorite rich in hydrous silicates; **micro-8**, partially dehydrated micrometeorite (scoriaceous type), which was initially rich in hydrous silicates that started loosing their constituent water upon frictional heating, thus developing a 10-µm-thick "rind" of vesicles. In spite of their dehydration, these grains still contain high concentrations of carbonaceous material and extraterrestrial water (as identified from D/H ratios). (Courtesy of C. Engrand and Mineralogische Abteilung, Naturhistorisches Museum, Vienna.)

In the same size fraction, the remaining AMMs are divided into unmelted (48%) and partially dehydrated (32%) chondritic grains. The unmelted AMMs show two major subgroups: (1) The crystalline micrometeorites (16%) are made of a few single crystals of olivine and pyroxene containing metallic globules (Figure 7.3, micro-5); (2) the fine-grained AMMs (32%) are dominated by a highly unequilibrated assemblage of hydrous and anhydrous minerals (Figure 7.3, micro-28 and micro-29).

The partially dehydrated AMMs appear as typical scoriaceous fine-grained particles loaded with tiny vesicles (Figure 7.3, micro-8), which start invading the grains from their external surfaces. They still contain several percent extraterrestrial water (characterized by D/H measurements), a high concentration of carbonaceous material, and, frequently, high levels of implanted solar neon.

II.2.2. Transmitted flux

An extraterrestrial body (meteoroid) is fully decelerated in the atmosphere when it has impacted its own mass of air molecules. Because the total thickness of the atmosphere is about 10 m of liquid water equivalent, only rocks smaller than ≈5 m can be fully decelerated. Larger bodies cannot be decelerated, and they explode upon impact with the Earth's surface. For example, the most recent modeling of the research group of Dieter Stoffler shows that a 1-km-sized meteoroid penetrates the oceans and/or the continents up to a depth comparable to its original diameter. This body is then transformed into a giant ball of buried gas, at a pressure of 5 million bars and at a temperature of ≈10,000 °C, that finally triggers a cataclysmic explosion. The explosion products fall back on the Earth, because their speed is lower than the escape velocity (≈11 km/s), and they dominate the bulk accretion rate of extraterrestrial material on the Earth. However, this domination, which can be assessed from the meteoroid mass spectrum (as determined from lunar impact craters), is already limited to about 10 times the total mass delivered by the transmitted micrometeorite flux (see later). Moreover, the material thus cycled throughout the explosion can hardly contain the delicate complex organics required for the origin of life. Consequently, only stratospheric IDPs, micrometeorites, and meteorites that survive upon atmospheric entry should be considered in the delivery of complex organics to the early Earth.

The meteorite flux reaching the whole Earth surface after atmospheric entry has been estimated by several groups of investigators, who relied on two very different methods. (1) Wilson and collaborators (see Zolensky et al. 1992), after years of slow and careful walking in an arid zone of New Mexico (Roosevelt County) collected all meteorites standing on an area of about 2,000 acres. From a measurement of the exposure age of this zone to the meteorite infall (about 50,000 years), the value of the meteorite flux for the whole Earth was estimated to be about 100 tons per year. (2) Halliday, Blackwell and Griffin (1989), from a completely different study of "meteors" observed in the Canadian prairies (e.g., the light phenomenon that decorates the trajectory of meteorites in the atmosphere), inferred a smaller value of the meteorite flux. (3) Recent counts of meteorites recovered from the Sahara Desert and the Nullarbor Plain in Australia, when associated with both the data of Wilson and collaborators and a clever estimate of the lifetimes of meteorites in such deserts, yielded a reliable estimate of the meteorite flux reaching the Earth's surface of ≤10 tons per year (Bland et al. 1996), in good agreement with the value determined by Halliday, Blackwell and Griffin (1989).

Eric Robin (see Maurette et al. 1987) measured the total mass of micrometeorites with a size >50 μm that accumulate in dark sediments (cryoconite) deposited on the bottom of the largest cryoconite holes found on the melt zone of the West Greenland ice sheet (Figure 7.4). In such "mature" holes, the concentration of micrometeorites per kilogram of sediment reaches a saturation value reflecting a maximum lifetime, Δ (max), of the holes on the ice surface (and consequently a maximum exposure time to the micrometeorite infall), against destruction when they flow through a crevasse or a fast river. From the ice flow model developed specifically by Niels Reeh, a value of Δ (max) ≈ 250 years was estimated, yielding in turn a value of ≈20,000 tons a year for the "transmitted" micrometeorite flux, F_t, reaching the whole Earth's surface after atmospheric entry (Maurette et al. 1991; Hammer and Maurette 1996).

This value was recently reconfirmed by two direct methods that have the great advantage of being free of uncertainties related to the use of an ice flow model: (1) At the South Pole station, where the snow accumulates at a steady rate of about 8 cm per year, Taylor, Lever and Harvey (1996) deduced a rather similar value of F_t from a direct count of cosmic spherules recovered from the bottom of the station water well; (2) Rasmussen, Clausen, and Kallemeyn (1995) measured the total iridium concentration of a well-defined annual ice layer (precipitation rate of about 40 cm/year) sampled in central Greenland, from which a value of F_t ≥20,000 tons/year can be inferred (Hammer and Maurette 1996).

The comparison of these transmitted fluxes shows first that micrometeorites found in the 50–500-μm size range represent by far the dominant source of extraterrestrial material reaching the Earth's surface. Indeed, they deliver about 2,000 times more extraterrestrial material to the Earth's surface than meteorites.

The micrometeorite flux *before atmospheric entry* was determined from a

Figure 7.4. Typical cryoconite hole used as a micrometeorite collector, investigated in July 1987 at the latitude of Sondrestromfjord, at 20 km from the ice margin, on the melt zone of the West Greenland ice sheet. The density of the dark sediments deposited on the bottom of the holes reaches a saturation concentration of about 2 kg m^{-2}, which reflects their limited lifetime against various destruction processes. The disaggregation of the sediments on a stainless steel sieve with an opening of 50 μm yielded a glacial sand, in which the concentration of iridium was determined by INAA (instrumental neutron activation analysis). From the average iridium content measured separately in individual cosmic spherules and micrometeorites (about 20,000 times larger than that in terrestrial material), the total mass of micrometeorites deposited per m^2 of cryoconite hole could be determined.

direct count of impact craters on metallic plates of the LDEF satellite, from which Love and Brownlee (1993) inferred a flux value of about 40,000 tons (± 20,000 tons) per year for the whole Earth, which was recently revised down to 30,000 tons per year (Taylor, Lever and Harvey 1996). This value, which is quite compatible with that inferred from the ice, shows independently that micrometeorites will survive during their hypervelocity capture by the Earth.

Love and Brownlee (1993) also noted that the amount of extraterrestrial material delivered outside the size range 50–500 μm (including meteorites and IDPs) is about 100 times smaller, and that most of the mass is found around a size of 200 μm. Consequently, we selected ≈ 200-μm-sized AMMs for most of the studies of carbon chemistry reported in Section II.3.2.

II.2.3. Interactions with the upper atmosphere

All types of micrometeorites, down to the smallest size fraction collected in Antarctica (25–50 μm), and including both unmelted and partially dehydrated grains, are encapsulated into a very thin shell of magnetite that appears as a bright, 1-μm-thick rim around polished sections of the grains (Figure 7.5). This shell increases the mechanical strength of micrometeorites (like the shell of an egg) and, consequently, their durability in the terrestrial environment (in particular, magnetite is one of the most durable minerals in deep sea sediments).

This shell is not a primary characteristic of micrometeorites, and its formation on the Earth is still poorly understood. First, Bonny (1990) noted that frictional heating cannot build up a thermal gradient smaller than 10 μm on the edge of micrometeorites. Consequently, the small thickness of the magnetite shell (≤1 μm) would argue against an ultrathin fusion "crust" generated by some type of in situ melting and/or decomposition of micrometeorite matter in a thermal gradient. On the other hand, the 10-μm-thick scoriaceous rinds of vesicles observed around some micrometeorites (see Figure 7.3, micro-8) bear the print of this "minimum" thermal gradient.

The magnetite rim is observed in particular on top of the external surface of dehydrated micrometeorites. Indeed, its thickness is independent of the nature of the host AMMs and/or the host particles in a given AMM. This characteristic signature suggests that the magnetite shell possibly resulted from some accretionary process effective near the end of the deceleration range of micrometeorites in the upper atmosphere.

The statistics of shooting star trails demonstrate that most micrometeorites slow down between 120 and 80 km of elevation, right into the so called *E-layer,* which contains high concentrations of ions, including iron, which is the dominant metallic species. The source of these ions is currently attributed to those of the micrometeorites and meteorites that get the most extensively ablated during atmospheric entry. The surviving micrometeorites would scavenge the metallic ashes of "dead" extraterrestrial material to build up their own magnetite shell. One major difficulty with this model is the low column density of iron presently measured in the E-layer, from which a 1-μm-thick shell could hardly originate.

Evidence supporting this formation process of the magnetite shell was found

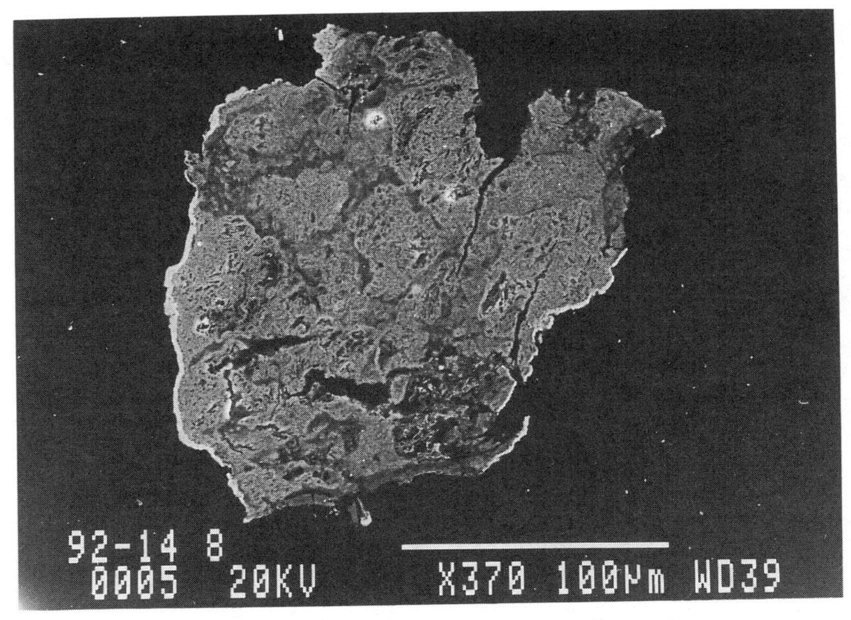

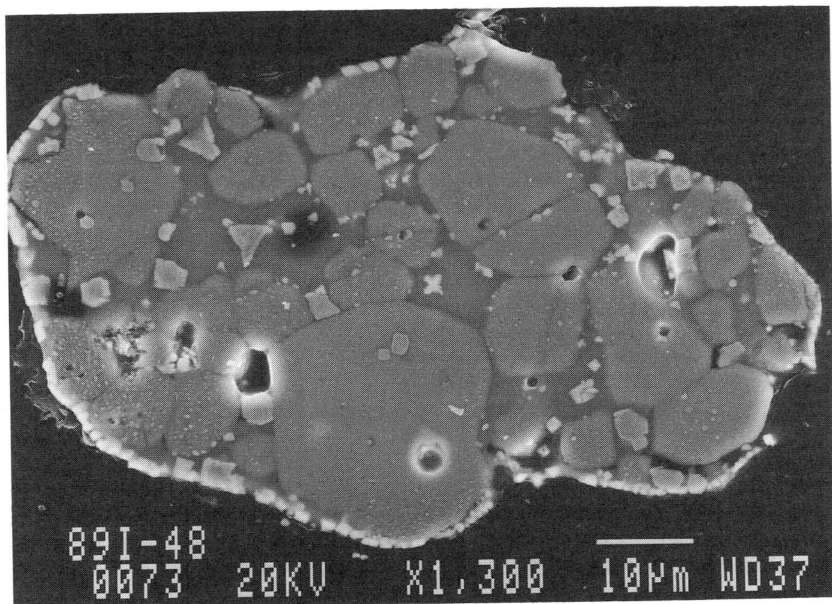

Figure 7.5. Polished sections of ≈100-µm-sized crushed fragments of two distinct types of micrometeorites (top: "fine-grained hydrous"; bottom: "crystalline"), observed with a scanning electron microscope. The ≈1-µm-thick shell of magnetite is observed as a discontinuous bright rim, because it was originally formed on the external surfaces of these fragments now showing a few "freshly crushed" internal surfaces without rim. (Courtesy of C. Engrand and Mineralogische Abteilung, Naturhistorisches Museum, Vienna.)

during a study of the fusion crust of carbonaceous chondrites. The meteorite fusion crust, which has a thickness of a few 100 μm, looks rather similar to a fully dehydrated scoriaceous-type micrometeorite with large vesicles. But there is no magnetite rim on top of the first layer of vesicles. Due to their much larger masses, meteorites slow down at smaller elevations than micrometeorites, typically between 80 and 30 km. Near the end of their deceleration range, they are well below the E-layer and can hardly accrete a magnetite shell.

After this fast deceleration, micrometeorites start their free fall to the Earth's surface. During their sedimentation gravitational time, Δ_g, which can be perturbed by strong winds, they interact with the stratosphere. The first evidence of such interactions was the observation of a high sulfur content (two to three times that of chondritic) observed on some of the small micro- meteorites collected in the stratosphere (IDP), which exhibit the highest value of Δ_g due to their small mass and/or their fluffiness. These high S-contents could reflect the scavenging of sulfuric acid aerosols in the stratosphere. Then Br and Zn, two volatile elements, were found to be enriched in IDPs up to 1,000× and up to a few times the chondritic values, respectively (Jessberger et al. 1992). Both excesses were interpreted as resulting from the exposure of IDPs to the stratospheric environment.

To assess the effects of an ≈8-hour exposure of AMMs to melt ice water (see next section) we drastically reduced this exposure to about 15 minutes, just melting (in a microwave oven) and filtering quickly a ≈50-kg chunk of Antarctica ice. We thus found on the external surface of cosmic spherules tiny deposits of K and Na sulfates (Figure 7.6) that likely resulted from the scavenging of stratospheric aerosols.

The UV photolysis of polycyclic aromatic hydrocarbons (PAHs) in the lower atmosphere generates nitro-PAHs. We searched for them in individual crushed fragments of about 20 large AMMs with sizes of 200 μm and have found none as yet. But nitro-PAHs were reported in two IDPs with much smaller sizes, and interpreted as being primordial (Clemett et al. 1993).

These few examples indicate that micrometeorites interact with the stratosphere, acquiring characteristics that might assist their functioning as chemical reactors.

II.2.4. Weathering in melt ice water

Gero Kurat (see Maurette et al. 1992; Kurat, Koeberl et al. 1994) calculated that the depletion in AMMs of S, Ca, and Ni (up to a factor of 5 with regard to chondritic composition) could be due to the fast dissolution of their constituent soluble salts (sulfates and carbonates) during their ≈8 hours' exposure in their host pocket of melt ice water. In the "mass balance" computations supporting this conclusion, Kurat assumed that the initial contents of these salts in

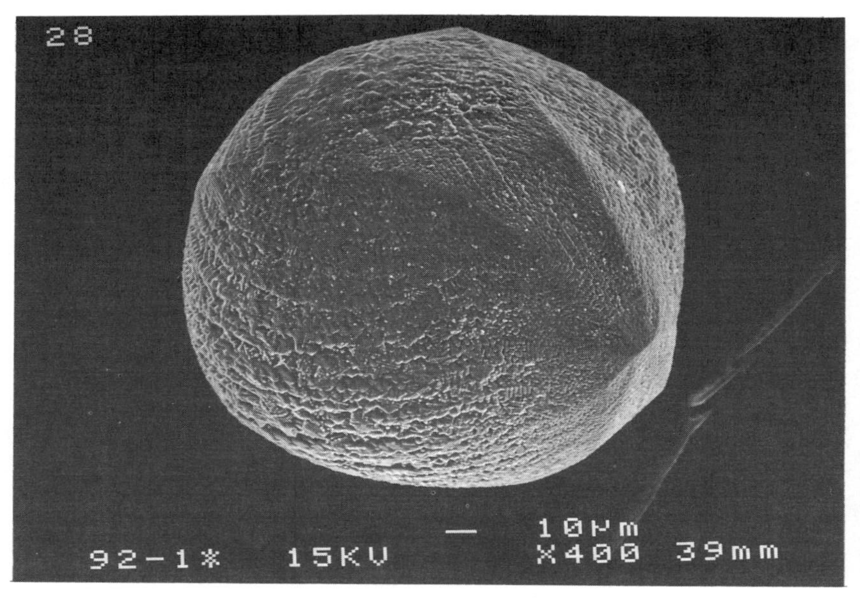

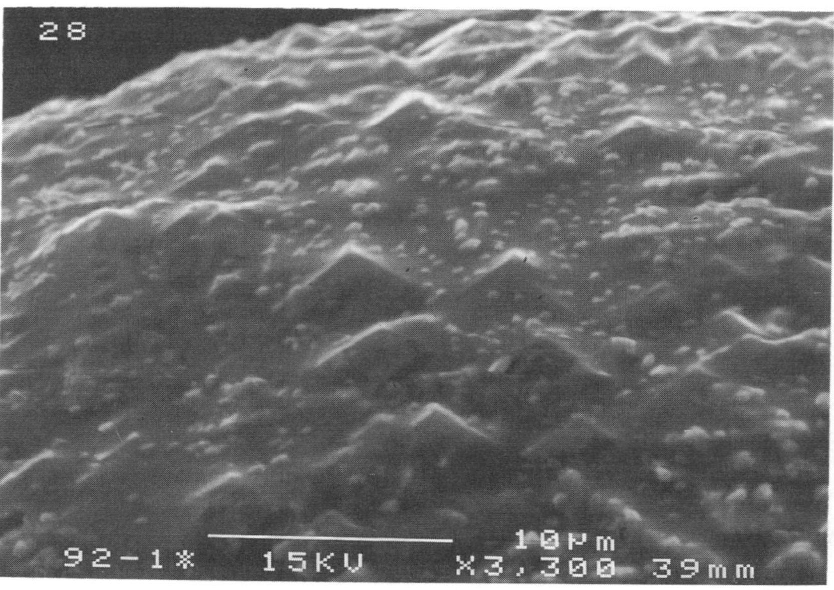

Figure 7.6. Scanning electron microscope observations of sulfate deposits on a cosmic spherule (size ≈ 100 μm), exposed less than 15 minutes in melt ice water. The tiny sulfates can be best observed on the bottom micrograph, representing an 8× enlargement of a small zone of the upper surface of the spherule, located at "11 o'clock." The parent ice of the grains was collected in January 1991 in Antarctica by R. Harvey and R. M. Walker. The grains were extracted from the ice in the ultraclean facilities of the Cold Region Research and Engineering Laboratory (Hanover, N.H.), in collaboration with J. Cragin and S. Taylor. (Courtesy of G. Boivin, Office des Matériaux, ONERA.)

micrometeorites were similar to those measured in CM-type meteorites.

AMMs were quickly washed out from their soluble salts and this has several implications: (1) Because AMMs recovered from melt ice water still show excess contents of Zn and Br, similar to those observed in IDPs, the carrier phases of these two elements in the stratosphere are not soluble. (2) In carbonaceous chondrites, soluble carbonates and organics carry about one-third and two-thirds of the total carbon, respectively. But because carbonates have been washed out from AMMs, their bulk C-content directly yields that of their carbonaceous matter. (3) Carbonates can trap complex organics in meteorites (Carter 1978; Thomas, Clouse and Longo 1993) but, because they have been lost from AMMs, they cannot be considered as a possible host phase of AMMs' complex organics. (4) The magnetite shell of a micrometeorite is not a waterproof "vesicle" for soluble salts.

II.3. A new population of Solar System objects

II.3.1. Relationships with CM carbonaceous chondrites

We summarize here results obtained during analyses of the chemical and mineralogical composition of about 500 Antarctic micrometeorites (AMMs) selected at random in the glacial sand collected at Cap-Prudhomme. All studies were made in collaboration with Gero Kurat in Vienna.

Most micrometeorites are complex aggregates composed of millions of individual grains embedded into an abundant carbonaceous component. They all have a chondritic composition of their nonvolatile major elements (e.g., the composition of the solar atmosphere), which already relates them to the most abundant stony meteorites, which are called chondrites and are divided into *ordinary* and *carbonaceous* chondrites, representing about 80% and 4% of the meteorite falls, respectively. (See Heide and Wlotzka 1995 for a simple description of all types of meteorites, and Cronin and Chang 1993 for a technical description of carbonaceous chondrites.)

More detailed mineralogical studies, as well as measurements of the isotopic composition of hydrogen (Engrand et al. 1996a), show further that AMMs are mostly related to a specific group of carbonaceous chondrites, the so-called CM-type carbonaceous chondrites, which represent only ≈ 2% of the meteorite falls. Indeed, about 80% of them, classified as fine grained, are made of a highly unequilibrated and primitive assemblage of hydrous silicates, anhydrous phases, and carbonaceous matter like the CM chondrites. The remaining 20%, classified as crystalline, are also unequilibrated aggregates of a few anhydrous silicates, related to those observed in CM chondrites (pyroxenes and olivines).

We have not found as yet any AMMs related to the groups of "differentiated" meteorites, which include iron and stony-iron meteorites and the stony meteorites called achondrites.

These findings are quite amazing: Micrometeorites constitute by far the dominant source of extraterrestrial material accreted by the Earth today (see Section II.2.2), and this most abundant matter is not represented either in the great diversity of *the most abundant meteorites,* the ordinary chondrites (≈80% of the meteorite falls), or in the group of differentiated meteorites (14% of the meteorite falls). This in turn reflects a drastic difference between the parent bodies of meteorites and micrometeorites (see Section II.3.3).

But the list of surprises is far from being closed! In spite of similarities, there are also marked differences between micrometeorites and CM chondrites, with micrometeorites showing, in particular, much higher C/O ratios, reflecting higher carbon contents; a very strong depletion of chondrules, which constitute about 15% of the mass of CM chondrites (their size distribution peaks around 200–300 μm); and a ratio of pyroxene to olivine that is about 10 times larger than that found in CM chondrites.

Micrometeorites thus represent a new population of primitive Solar System objects not represented as yet in meteorite collections (Section II.3.3).

II.3.2. Carbon chemistry

A knowledge of the carbon chemistry of individual micrometeorites is required for any application of the study of micrometeorites to exobiology.

To avoid carbon contamination, the initial micrometeorite and/or a tiny chunk of meteorite (used as a standard) is first crushed into several fragments between two quartz plates in ultraclean conditions. One of these fragments is polished in an epoxy resin mount for mineralogical and textural classification (see previous section).

A second fragment is crushed into much smaller grains (sizes up to a few micrometers) directly onto a clean, gold electron microscope grid held between two quartz plates. Then, at least 30 of these grains, partially embedded in the grid and selected at random, can be analyzed with an analytical transmission electron microscope (TEM) (see next section).

A third fragment is crushed onto a gold foil for analysis either by the μL^2MS (microscopic double laser mass spectrometry) technique at Stanford University (Section II.3.2.B) or with an ion microprobe (SIMS) at the Centre de Recherches Géochimiques et Pétrographiques, Nancy.

Then several fragments are kept for additional analyses and/or future use by the Star Dust and Rosetta cometary missions, which will require comparisons between the cometary grains, well-characterized IDPs and AMMs, and primitive meteorites.

Table 7.1. *C/O atomic ratios measured both in 2 hydrous carbonaceous chondrites (Orgueil and Murchison) and in a random selection of 18 Antarctic micrometeorites with a size of ≈200 μm. Each upper bold number refers to the average value obtained for about 30 distinct micron-sized fragments (selected at random) of a given meteorite and/or micrometeorite; the lower numbers, within the parentheses, indicate the highest C/O ratio measured in a given object. (From Engrand 1995.)*

Orgueil	**0.27** (01.24)					
Murchison	**0.20** (1.20)					
AMMs 100-400 μm	**1.63** (2.27)	**1.14** (3.13)	**1.08** (2.44)	**1.08** (1.54)	**1.00** (2.27)	**0.88** (3.33)
	0.87 (2.56)	**0.61** (1.22)	**0.55** (2.17)	**0.46** (1.49)	**0.36** (1.22)	**0.36** (0.89)
	0.32 (1.63)	**0.24** (0.91)	**0.18** (1.22)	**0.15** (0.71)	**0.12** (0.65)	**0.05** (0.25)
Average value for 18 AMMS		**0.62**				

A. FROM COMETARY TYPE C/O RATIOS TO HIGH CARBON CONTENTS. At the Laboratoire d'Etudes des Microstructures (CNRS/ONERA), Cecile Engrand and Michel Perreau have used an analytical TEM equipped with both an electron energy loss spectrometer (EELS) and an energy dispersive X-ray spectrometer (EDS) to analyze small volumes of material with a size of ≈0.1 μm.

The EELS has the best sensitivity for light elements (C, N, O) and, in particular, yields C/O atomic ratios with an accuracy of ≈1% when the sample is sufficiently thin (e.g., about 0.1 μm thick) to produce a double peak structure in the electron energy loss spectra of O.

The EDS analysis is not good for C and N determinations, but it gives the chemical composition of the major nonvolatile elements in the sample volume analyzed with the EELS (with an accuracy of ≈5%), which is related in turn to the mineralogical composition of the host phase of carbon.

Table 7.1 gives both the average and the maximum values of the C/O ratios for large Antarctica micrometeorites (100–400-μm size fraction) and for two carbonaceous chondrites used as standards, Orgueil (CI type) and Murchison (CM type).

On the average, AMMs show much higher C/O ratios than meteorites, about 2.5 and 3.0 times larger than in Orgueil and Murchison, respectively.

This adds to the list of major differences between micrometeorites and meteorites. But these ratios are now rather comparable to those measured during the in situ analyses of dust grains from Comet Halley in March 1986 , about 3–10 times larger than in Murchison (Kissel, personal communication 1996). Indeed, the ratio of 10 is *a maximum value*, to be compared with the corresponding value of ≈8 inferred from the *average* of *all* the numbers within parentheses listed in Table 7.1 for 18 micrometeorites. These analyses, pertaining to masses of cometary dust grains (≈10^{-14} g) quite comparable to those analyzed with the TEM, thus reveal another relationship between large micrometeorites and cometary dust grains.

The exact meaning of C/O ratios measured at a scale of 0.1 μm is still not fully understood. With the exception of the tiny vesicles reported in Figure 7.8d, all C-rich grains show an oxygen peak. This suggests that organics are closely associated with the mineral phases mostly responsible for the O peak. In both micrometeorites and cometary dust grains, this close association between organics and minerals would be more predominant than in meteorites, as a result of higher C-contents, possibly coupled to a peculiar process of "impregnation" of organics in the fine-grained matrix, which could function as a "cosmochromatograph."

The ratio of the average C/O values measured for Orgueil and Murchison gives exactly the ratio of their average bulk carbon contents, determined on much larger volumes of material with the best techniques of geochemistry. We extrapolated this trend to micrometeorites, scaling their carbon contents directly from the C/O values, relative to those measured in Orgueil and/or Murchison. Thus, in Antarctica micrometeorites, C-contents would be about three times higher on the average than in CM chondrites.

Graham et al. (1996) recently reported on studies of the isotopic composition of carbon during the individual pyrolysis of six AMMs with a size of ≈200 μm. The isotopic composition of carbon released above 400 °C looks quite similar to that observed in CM meteorites. But in this case, the total contents of C inferred from these difficult measurements performed near the limits of sensitivity of the technique would be even smaller than the value measured in Murchison.

At the present time, more work is needed to understand these discrepancies between these very different techniques. In this chapter we shall only cite the high carbon contents of micrometeorites measured with the EELS. Another independent piece of evidence for these high C-contents was obtained during the direct observation of ultramicrotomed sections of AMMs with the TEM: In the field of view reported in Figure 7.8b, the relative total cross-sectional area of the C-rich grains relative to the other grains is at least 25%.

We have difficulty detecting nitrogen in micrometeorites. In carbonaceous chondrites, this element is already ≈10 times less abundant than C. Furthermore, an additional depletion by a factor of 3 was noted for the Halley dust grains (Kissel, personal communication 1996). If the C/N ratio of micrometeorites is similar to the cometary value, then N is still below the limit of detection of EELS.

B. SEARCH FOR COMPLEX ORGANICS. The most sensitive techniques of geochemistry now available to detect complex organics in micrometeorites include μL^2MS and HPLC (high performance liquid chromatography), as presently used in the groups of Richard Zare (Department of Chemistry, Stanford University) and Geoffrey Bada (NESCORT for Exobiology, Scripps Institution of Oceanography, La Jolla). Again, in all analyses of AMMs, the Murchison meteorite was used as a standard. The major results of these studies will be published shortly (Brinton et al. in press; Clemett et al. in press).

For the μL^2MS analyses small (≈50-μm-sized) chunks of micrometeorites and/or meteorites are crushed on gold foils. An IR laser gently desorbs organics from the grains. This forms a "plume," which is irradiated with a UV laser to specifically ionize the constituent polycyclic aromatic hydrocarbons (PAHs) of the samples, and a time-of-flight mass spectrometer yields their mass spectra.

A comparison of the PAHs spectra of meteorites representing three major groups of carbonaceous chondrites (Orgueil, CI type; Murchison, CM type; Allende, CV type) reveals the following features (Clemett et al. 1992): (1) The meteoritic PAHs show similarities, because they yield simple mass spectra dominated by four peaks (Figure 7.7, top spectra) corresponding to core PAHs observed at m/z values of 128, 178, 202, and 228. (2) There are very few fragmentation products observed at masses ≤100 amu, as well as few PAHs with masses ≥ 300 amu. (3) During laboratory pyrolysis experiments, the degree of alkylation (related to the number of substitutions on the core PAHs), which is reflected in the tails of the high mass peaks, increases with temperature. (4) The content of PAHs in meteorites decreases from Allende to Murchison and then to Orgueil. It has been suggested that this trend is inversely correlated with a scale of hydrous alteration (on parent bodies) that increases from Allende (no hydrous minerals) to Murchison (≈50% hydrous minerals) to Orgueil (≈100% hydrous minerals).

The PAHs mass distributions of Antarctic micrometeorites (see Figure 7.7) show the following differences with regard to carbonaceous chondrites (see Clemett et al. in press): (1) a much larger variety of PAHs spectra, which range from simple spectra rather similar to those of carbonaceous chondrites

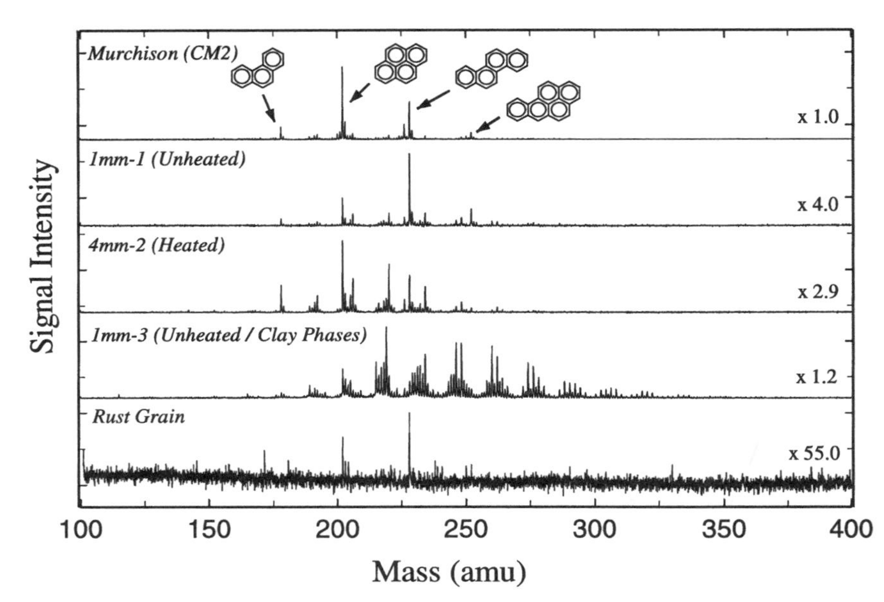

Figure 7.7. Mass distribution of polycyclic aromatic hydrocarbons (PAHs) observed in the Murchison meteorite; three Antarctic micrometeorites showing very different PAHs spectra, ranging from the most "simple" (1mm-1) to the most "complex" (1mm-3) ones; and a highly porous rust grain. The three micrometeorites and the rust grain were recovered from the same water pocket in 1991 and were subsequently handled in the same way. The multiplication factors reported on the right-hand side of each spectrum were used to get a similar height for the major peak observed in each spectrum, relative to Murchison. They do not give the relative yields of desorption of PAHs, which would require a summation of all peak intensities (in particular, the desorption yield of 1mm-3 is found to be about 20 times higher than that of Murchison). However, they already show that the yield of PAHs observed in the rust grain corresponds to the background level that is frequently observed on very clean surfaces and that is related to the "memory" of the instrument for previous extraterrestrial PAHs analyses. (Courtesy of S. Clemett and R. Zare.)

(Figure 7.7, grain 1mm-1), to complex spectra with many peaks (Figure 7.7, grain 1mm-3), extending to higher mass numbers and showing predominant tails around the major peaks; (2) barely detectable peaks at m/z = 128 and m/z = 252); (3) yields that vary between the values observed for Allende to a few times those for Orgueil, with most values clustering around the "intermediate" yield observed for Murchison; (4) a much higher degree of alkylation of the PAHs, which clearly increases with the extent of thermal metamorphism, as inferred from the degree of "invasion" of AMMs texture by tiny vesicles (see Section II.2.1 and Figure 7.3d); and (5) vinyl PAHs (e.g., containing a CH_2=CH– group), not observed yet in meteorites, and which can be best observed in the mass range 250–290 amu (see 1mm-3, Figure 7.7).

We can exclude the possibility that the micrometeorite PAHs result from terrestrial contamination. The mass distribution of PAHs in a sample of Antarctic ice has been measured with a similar technique of L²MS (Becker, Glavin and Bada 1996). This distribution is much different from that of Antarctic micrometeorites, which do not show major peaks at m/z = 166 and 252 and which have a much higher degree of alkylation, and the content of PAHs is ≥100,000 times higher in micrometeorites than in the ice. The great diversity of PAHs is observed in particular for micrometeorites recovered from the same pocket of melt ice water, the same day (compare 1mm-1 and 1mm-3 in Figure 7.7), and this strongly argues against a common source of terrestrial contamination. Furthermore, rust grains extracted from the same pocket of melt ice water do not show PAHs besides the normal background level (Figure 7.7, rust) that is attributed to the "memory" of the instrument for previous analyses and that is also observed on very clean surfaces. This demonstrates that the PAHs observed in AMMs do not result from the terrestrial contamination observed in the ice.

Although the HPLC technique has been utilized in the search for amino acids, the sensitivity of the method is still not sufficient to analyze individual micrometeorites. Thus, an aliquot of ≈30–35 micrometeorites and/or a ≈5-mg chunk of the fine-grained matrix of the Murchison meteorite has to be run. So far, five distinct aliquots of AMMs (corresponding to two distinct daily collections made in 1991 and in 1994) have been analyzed.

The major results of these investigations (see Brinton et al. in press) are: (1) The most abundant amino acid is AIB; (2) the ratio of AIB to isovaline is much (≥10 times) higher than in CM chondrites; (3) the content of AIB is highly variable in these five aliquots, ranging from 280 ppm to the blank level (<<1 ppm); (4) the average content of AIB in the five aliquots (≈ 180 AMMs), about 80 ppm, is thus ≈10 times higher than the value measured in a ≈5-mg sample of the Murchison meteorite (5–10 ppm) used as a standard during these runs; (5) alanine would be the next most abundant extraterrestrial amino acid, with a similar abundance of L- and D-enantiomers; (6) there is a clear contamination of terrestrial amino acids showing a strong dominance of the L-enantiomers, which looks quite unique (see next section).

C. PRELIMINARY IMPLICATIONS TO EXOBIOLOGY. Micrometeorites show a huge diversity of PAHs spectra, which has to be contrasted with the amazing uniformity of the simple PAHs spectra observed in carbonaceous chondrites.

From their thermodynamic computations, Shock and Schulte (1990) suggested that the catalyzed hydrolysis of PAHs in the "damp" regolith of the parent bodies of carbonaceous chondrites could generate the rich mixture of

Table 7.2. *Concentration of α-aminoisobutyric acid (in ppm) in five samples of 100–200-μm-sized Antarctic micrometeorites collected in 1991 (samples 1–3) and 1994 (samples 4 and 5). For comparison, ≈5-mg chunks of Murchison analyzed as "standards" during the same runs yielded values ranging from 5 to 10 ppm. (Courtesy of G. Bada.)*

NSCORT/EXOBIOLOGY

AIB in Antarctic Micrometeorites

Sample	Size (μg)	AIB (ppm)
A91	50	280
I91	175	78
III91	310	22
IV94	166	<<1
V94	259	20

Scripps Institution Of Oceanography
University Of California at San Diego

amino acids observed in these meteorites. Because the variety and chemical reactivity of the micrometeorite PAHs are much larger than in meteorites, this mechanism should be even more efficient with micrometeorites, especially with the very reactive vinyl-PAHs.

Exobiologists have listed several processes to synthesize the amino acids found in CM meteorites (see Cronin and Chang 1993). In meteorites, the ratio of AIB to isovaline (about 1:1) would be typical of a Strecker type synthesis, which is known to produce a large number of amino acids (Wolman, Haverland and Miller 1972; Lerner et al. 1993). In micrometeorites, the much higher value of this ratio (≥20:1) would imply a very different process, one related to the polymerization of HCN precursors (Wolman, Haverland and Miller 1972; Yuasa et al. 1984), which are thought to be abundant at least in comets showing CN^- ions in their tails.

The high variability of the AIB content in aliquots of about 30–35 micrometeorites suggests that AIB is concentrated in peculiar families of micrometeorites that are quite rare (a few percent at most). Consequently, these AIB-rich micrometeorites can be missed during a random selection of ≈30 AMMs. In this case, these rare grains would be extremely rich in amino acids, and their identification is one of the most exciting prospects of micrometeorite studies in exobiology.

The contamination of AMMs by terrestrial amino acids looks odd because it involves only very few amino acids (L-alanine; glycine; D,L-glutamate; L-serine; L-aspartate). But it is similar to the unique contamination already observed in about 50 distinct samples of ice recovered from Greenland and Antarctica (see Figure 3 in McDonald and Bada 1995). This contamination was thus likely introduced into AMMs during their short exposure of 8 hours in their host pocket of melt ice water.

However, the concentration of these amino acids in AMMs is ≥100,000 times higher than in the ices. Clearly, some peculiar constituent phases of micrometeorites function as very efficient "cosmochromatographs" to scavenge amino acids (and probably other complex organics) initially present in the ice and released in the pockets of melt ice water from which micrometeorites were recovered.

This property of cosmochromatography of AMMs greatly complicates the study of their extraterrestrial amino acids found at lower concentrations than AIB, and it is unlikely that we can further reduce this terrestrial contamination, because AMMs have been recovered from the cleanest zones of the Earth. Our only hope is to extract ≈30 large micrometeorites from ≈300 kg of clean Antarctic ice by a dry liophilization technique, which represents a very difficult technical challenge. On the other hand, this property would be one of the key characteristics of micrometeorites, considered as tiny chemical reactors, which can interact with both gases and waters.

D. SEARCH FOR THE HOST MINERAL PHASES OF ORGANICS. The analytical TEM yields the C/O ratios of the C-rich grains, as well as some clues about their mineralogy, as inferred from both EDS analyses, electron diffraction patterns, and textural features.

The four TEM micrographs reported in Figure 7.8 shows several zones of an ultrathin section of a partially dehydrated scoriaceous type micrometeorite, in which hydrous silicates started to lose their constituent water. But this micrometeorite still contains abundant C-rich grains, which can be easily distinguished from the epoxy resin used to prepare the thin section, as a result of their darker contrast, peculiar texture, and distinct C/O ratios.

The C-rich grains can be classified into four major textural groups, identified with the dark arrows in Figure 7.8, and which appear as (a) a coating of COPS phase (see next paragraph) on the external surface of the grain,

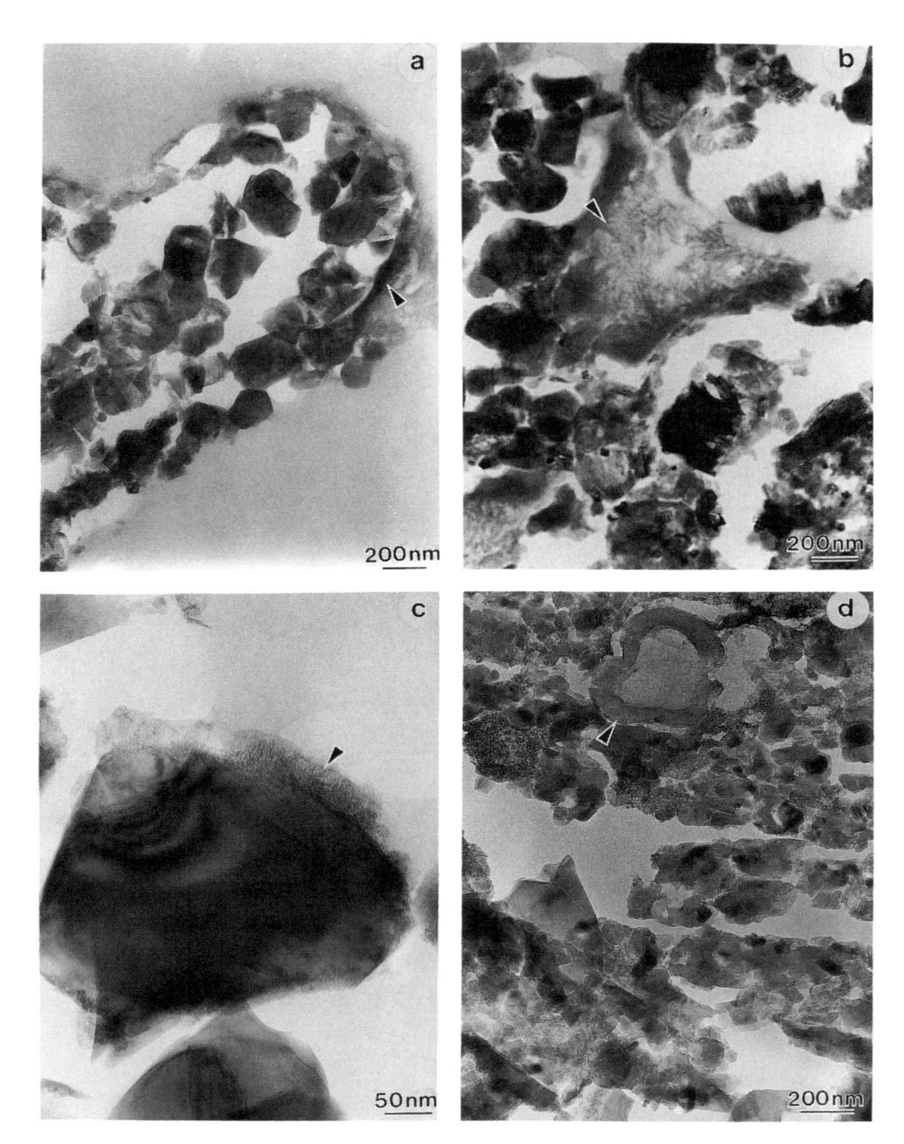

Figure 7.8. Electron micrographs obtained with a 400-kV analytical transmission electron microscope at a high magnification (see scale bars ranging from 50 to 200 nanometers). They show some details of an ultramicrotomed section of a partially dehydrated scoriaceous micrometeorite. Four distinct zones can be observed, directly illustrating the presence of abundant carbonaceous components, which can be easily distinguished from the epoxy resin used to prepare the ultrathin section, as a result of their darker contrast, specific texture, and different C/O ratios. These components (see dark arrows) can be classified into four major textural groups: (a) ≈1-μm-thick coating of a variety of poorly crystallized ferrihydrite, the so-called COPS phase,

which is thus somewhat encapsulated into a "vesicle"; (b) an interstitial COPS particle; (c) ultrathin coatings on individual constituent particles of the micrometeorite; and (d) a small vesicle of pure organics.

All the C-rich grains are very poorly crystallized phases, which are very difficult to identify from electron diffraction patterns. The only one that was clearly identified is a variety of iron hydroxide, called ferrihydrite ($5Fe_2O_3$, $9H_2O$), which can be observed in Figure 7.8a and which represents about 10% of the C-rich grains (Engrand et al. 1993). This phase is also very rich in Ni and S, and in other elements (Si, Mg, Al, etc.). It thus bears a clear relationship with the variety observed in CI chondrites (Tomeoka and Buseck 1988), and in the unique type 3 Kakangari chondrite (Brearley 1989), and this relationship supports its extraterrestrial origin.

This phase looks quite promising for exobiology. It associates, in a very small volume, important elements, such as C, P, and S, together with Fe and Ni, and sometimes detectable amount of impurities such as Zn. It has already been suggested that ferrihydrite and other varieties of Fe-rich "poorly characterized phases" (PCPs) could be the host phase of complex organics such as amino acids in the hydrous carbonaceous chondrites. The same property could explain, in turn, the high concentrations of terrestrial amino acids naturally "extracted" from melt ice water by micrometeorites functioning now as "cosmochromatographs." Moreover, the AMMs ferrihydrite is very different from the variety reported in carbonaceous chondrites with regard to its *much higher P-content,* and for this reason was coined as the COPS phase. Phosphorus is a very important element for the chemistry of life (Schidlowski 1988). It might also constitute a unique tracer to identify the parent bodies of micrometeorites.

The CM chondrites, to which most of the AMMs are related, do not contain ferrihydrite, but the other varieties of PCPs (Bunch and Chang 1980; McKinnon and Zolensky 1984; Tomeoka and Buseck 1985). On the other hand, ferrihydrite would be a major component of CI meteorites such as Orgueil, containing up to ≈60% of the iron of the fine-grained matrix (Tomeoka and Buseck 1988). All these findings indicate that the host phases of complex organics might be quite different in AMMs and CM chondrites, thus adding an important characteristic to their list of "major" differences.

---

which delineates here a tiny vesicle; (b) another textural variety (e.g., a "large" grain) of COPS phase; this micrograph helps visualize directly the high abundance of the carbonaceous material, which represents in this field of view about 1/4 of the total volume of all grains; (c) ≈0.05-μm-thick ultrathin coating observed on the external surface of individual mineral grains (not a COPS phase); and (d) small vesicles of pure organics with a size of about 0.5 μm, which represent about 1% of the C-rich grains. (Courtesy of C. Engrand and M. Perreau, and Laboratoire d'Etudes des Microstructures CNRS/ONERA.)

In terrestrial clay mineralogy, ferrihydrite is frequently associated with clays. Because it very efficiently adsorbs minor and trace elements, ferrihydrite "masks" the primary chemical composition of the clays, which has to be known to get information about their fabric. Clay mineralogists have thus learned to wash out this phase in a mixture of acetic acid and other solvents (Jeanroy 1990). This procedure could be applied both to carbonaceous (and hydrous) meteorites and to micrometeorites, and could be used to look at the residual distribution of both PAHs and amino acids after the washing of ferrihydrite. That might help demonstrate that ferrihydrite is the "magic" phase that transforms micrometeorites into very efficient microscopic cosmochromatographs.

II.3.3. Parent bodies in the early solar nebula

Meteorites and micrometeorites have short lifetimes in the interplanetary medium (up to about 1 my and 100 my, respectively), which are limited by destructive collisions with other meteoroids and/or by perturbations of their orbits. Consequently, those objects that we collect today had to have been shielded for periods comparable to the age of the Solar System within much larger parents bodies such as comets and asteroids, formed in the early Solar System, and from which they were released in recent times.

When the speed of the incoming parent meteoroids can be accurately determined, the light and/or ion trails delineating the atmospheric entry of micrometeorites (shooting stars) and meteorites (meteors) in the atmosphere allow the determination of their orbits. From these studies, applicable only to large micrometeorites, it has been convincingly argued, ever since the early work of Whipple (1951) and Whipple and Hughes (1955), that the parent bodies of meteorites are asteroids from the main asteroidal belt, whereas micrometeorites originate predominantly from comets, formed at much larger heliocentric distances.

Our recent comparison of meteorites and large (≥100 μm) AMMs independently supports such earlier inferences. In studies of carbonaceous chondrites, it is frequently stated that either an increase in carbon content or a decrease in the concentration of chondrules with regard to the most abundant ordinary chondrites indicates that the meteorite parent bodies did form at larger heliocentric distances. We can tentatively apply the same scaling to micrometeorites. Because they show both much higher carbon content (3×) and a very strong depletion of chondrules (by at least a factor of 20) with regard to CM carbonaceous chondrites, their parent bodies were likely formed at much larger heliocentric distances, in a zone of the outer Solar System where differences between asteroids and comets probably vanish.

But our comparison also yields new clues about the early history of the

Solar System. Besides chondrules, carbonaceous chondrites contain refractory inclusions (CAIs), composed mostly of mixtures of Mg, Al, Ca, and Ti oxides. It is generally assumed that the formation of chondrules and the formation of CAIs are related, with CAIs being the first high-temperature phases to condense during the cooling of a hot solar nebula at temperatures $\gtrsim$2,000 K, and chondrules being formed shortly thereafter at lower temperatures. In the classical hot nebula models, both processes would occur within the formation zone of the terrestrial planets (including the main asteroidal belt), in order to get the high temperatures required to produce CAIs. Then further away, beyond the "frost line" (separating the terrestrial from the giant planets), where icy bodies could survive, both processes would be inoperative.

AMMs force us to modify this classical view because they appear to be "chondrites without chondrules." In spite of being strongly depleted in chondrules, they contain CAIs showing striking similarities with those observed in CM carbonaceous chondrites with regard to their abundance (a few percent of the grains), their anomalous oxygen isotopic composition (e.g., excess of ^{16}O), and their mineralogical characteristics (Kurat, Hoppe et al. 1994; Hoppe et al. 1995).

In the formation zones of AMMs, the production of chondrules and that of CAIs were clearly uncoupled, because the formation of chondrules was severely suppressed, whereas that of CAIs extended with the same efficiency to the outer nebular disk. This drastically narrows the choice of models for the formation of the Solar System to those derived from the work of Frank Shu and collaborators (Shu, Adams and Lizano 1987) based on studies of young T Tauri stars. In this work, both CAIs and chondrules are thought to be formed near the Sun, at about the orbit of Mercury. Then, they can be somewhat fired on ballistic trajectories by the so-called "extraordinary" wind emitted by the early Sun, up to the outer fringes of the nebular disk. The fact that micrometeorites and their parent comets show only CAIs would just imply that comets were formed slightly earlier than asteroids (the parent bodies of meteorites), at a time when the temperature in the formation zone of CAIs was still too high to allow the survival of chondrules, which form at lower temperatures than CAIs.

Thus, AMMs yield not only unexpected constraints about the formation of the Solar System, but also a new model of comets, in which most of the dust grains trapped into cometary ices would be related to a CM-type chondritic matter strongly depleted in chondrules. This challenges the widespread belief that cometary dust grains should be much more primitive than carbonaceous chondrites. One has to make sure that the future Star Dust and Rosetta cometary missions can distinguish between these two distinct types of cometary models.

III. Synthesis of prebiotic molecules during the "accretionary tail"

III.1. Key ingredients, at the right time, at the right place

Any scenario dealing with the origin of life on the early Earth requires the following ingredients: ponds and/or oceans of liquid water, in which the key organic molecules for life can accumulate to form a prebiotic "soup"; a sufficient amount of these molecules, as expressed in terms of their total amount of carbon, which should exceed the mass of carbon present in the present-day biosphere (about 10^{18} g); and a set of efficient processes, probably involving some key catalyzers, to synthesize at least the organic molecules considered as the building blocks of life, such as amino acids, sugars, nucleotides, and lipids.

When this prebiotic soup was ready, a fantastic event occurred: a complex organic molecule, which originated from reactions between these building blocks, started to propagate "information" to its localized environment, thus initiating a very primitive form of life, which very quickly spread out.

Such a prebiotic synthesis of complex organics occurred on the early Earth, probably in a narrow time interval of a few hundred million years, somewhere between 4.2 and 3.9 billion years ago, during a period of much enhanced bombardment by meteoroids known as the accretionary tail (see Section III.4). As described by Sleep et al. (1990), before 4.2 billion years ago, the intensity of this bombardment was sufficiently high as to "sterilize" the whole Earth surface, thus preventing the development of life. Near the end of the accretionary tail, liquid water was already present because the oldest terrestrial rocks (3.8 billion years old) show typical sediments that can be formed only in water.

III.2. Difficulties with earlier prebiotic "soups"

The Miller–Urey electrical discharge experiment did convince exobiologists for about 30 years that the Earth's atmosphere did generate these complex organics in a very simple way. At that time, the terrestrial atmosphere was thought to be reducing, being made mostly of a mixture of CH_4, NH_3, and H_2, in contact with water vapor. With this type of atmosphere it is very easy to synthesize a primitive soup of complex organics, including amino acids, by depositing any form of energy, for example through electric discharges simulating lightning, in the primitive atmosphere.

But over the last 15 years this model has had to be replaced by an atmosphere dominated by CO_2, N_2, and H_2O, somewhat similar to the atmosphere of Venus and Mars. Indeed, exobiologists and planetologists were progressively accumulating a lot of evidence against the strongly reducing atmosphere. For example, laboratory measurements of the lifetime of CH_4 and NH_3 against photodissociation by UV light demonstrated that these two key molecules in the scenario of Miller and Urey would have been quickly destroyed over periods ranging from weeks to a few years.

But in this new atmosphere, the synthesis of organics is very inefficient, whatever the source of energy might be. Exobiologists then started to look at other scenarios, and the two most plausible ones are still in competition today. The first one originates from a suggestion made in 1961 by Juan Oró (1961), attributing to comets the delivery of complex organics to early seas, and thus the formation of a primitive soup of extraterrestrial origin. The second one, proposed after the discovery of hot hydrothermal sources on the oceans floor in 1977, postulates that the primitive soup was formed in these sources (see Section III.5).

III.3. Extraterrestrial scenarios

The Oró scenario (1961) was supported until 1990 by theoretical models, such as those proposed by Chyba (1987), Chyba et al. (1990), and Oberbeck and Aggarwal (1992), suggesting that some fraction of an incoming comet could survive the cataclysmic impact with the Earth. But in 1992, Chyba and Sagan (1992) gave up supporting the role of comets and accepted earlier arguments (Anders 1989; Maurette et al. 1990) indicating that micrometeorites should mainly be responsible for the delivery of organics to the early Earth.

Indeed, improved modelings failed to prove that a sufficient amount of "intact" organics could survive cataclysmic impacts (Section II.2.2), in contrast to cometary water, which was likely added as a minor component to early oceans. However, the finding of fullrenes in breccias generated during the impact responsible for the formation of the giant Sudbury basin (diameter ≈200 km) suggests that very refractory organics can still be generated during such cataclysmic explosions (Becker, Poreda and Bada 1996).

Before Chyba and Sagan (1992), several authors had already suggested, and/or even supported by computations, the role of micrometeorites (Brownlee 1981; Krueger and Kissel 1987; Anders 1989; Krueger and Kissel 1989; Oró and Mills 1989; Greenberg and Hage 1990; Maurette et al. 1990, 1991). In particular, soon after their impressive in situ analyses of ≤1-μm-sized dust grains from Comet Halley during the Giotto and Vega missions (March 1986), revealing up to 30 wt% of carbonaceous material (coined as CHON) in cometary dust, Krueger and Kissel (1987, 1989) quoted thermodynamic computations to stress the *"central importance of heterocatalytic thermodynamic properties needed for self organisation of matter to life."* In brief, they proposed a "combined scenario" in which cometary dust grains of the Halley type, when added to a preexisting terrestrial soup of organics, trigger the self-organization of complex organics such as nucleic acids. Subsequently, Anders (1989), relying on the characteristics of tiny interplanetary dust particles (size 10–20 μm) collected in the stratosphere (SIDPs), argued that the micrometeorite flux could have played a major role in the delivery of complex organics to the Earth.

In all these models, the characteristics of very small cometary (≤1 μm) and/or tiny interplanetary (10–20 μm) dust particles (IDPs) had to be extrapolated to the much larger size range (100–200 μm) that corresponds to the dominant extraterrestrial material accreted by the Earth and that delivers at least 100 times more material than objects found outside this size range, such as meteorites and the much smaller IDPs. These were bold extrapolations. Indeed, all models of the frictional heating of such giant micrometeorites upon atmospheric entry predict that around a size of ≈100 μm, ≥99% of them should be melted (Brownlee 1985); even ≈10-μm-sized IDPs would be heated sufficiently to destroy amino acids. Furthermore, it has been discovered that all micrometeorites get altered to some extent during their gravitational settling in the atmosphere (see Section II.2.3), which is particularly slow for the tiny IDPs. The resulting changes in properties have to be considered in any scenario.

Thus the development of any realistic "micrometeorite scenario" involving the "giant" micrometeorites required their successful collection in the cleanest conditions (Section II.1), a minimum understanding of their interactions with the terrestrial environment (Sections II.2.3 and II.2.4), and the difficult task of investigating their carbon chemistry (including the search for complex organics) with the most sensitive techniques, frequently exploited near their limit of sensitivity (Section II.3.2). Then, as first quoted by Ponchelet (1989) and Bonny (1989), we could propose that micrometeorites might have been functioning as microscopic chemical reactors on the early Earth during their interactions with gases and waters (Maurette et al. 1990).

III.4. Accretionary tail: delivery of total carbon

If the infall of extraterrestrial materials was effective in the synthesis of prebiotic molecules, it is not necessary to speculate either on the survival of a minute fraction of a cometary nucleus impacting the Earth, or on a delivery of organics by meteorites.

Direct flux measurements (see Section II.2.3) show that micrometeorites from the 50–500-μm size fraction represent today the dominant source of extraterrestrial materials that reach the Earth's surface without suffering too much damage. Micrometeorites in this specific size range deliver about 2,000 times more material than meteorites.

But this result is not sufficient for exobiology, because most micrometeorites might have been completely melted upon frictional heating in the upper atmosphere, thus losing their carbonaceous matter during a natural pyrolysis. One should assess how much of this material survives unmelted, and how much carbon it delivers to the Earth.

Cecile Engrand (1995) made a detailed "experimental" inventory of the characteristics of Antarctica micrometeorites that led both to their classification in three major groups as indicated in Section II.2.1 (i.e., unmelted; partially dehydrated; and melted) and to the measurement of the average carbon content of each group (Section II.3.2.A). From this inventory, the present-day delivery of carbon to the Earth was inferred to be ≈500 tons per year, a value about 50,000 times higher than that expected from carbonaceous chondrites.

Then, from the average content of AIB (≈80 ppm) and the average yield of PAHs (similar to that of Murchison) in micrometeorites, we can estimate that the micrometeorite delivery rate of complex organics to the Earth is ≈500,000 and ≈50,000 times higher for AIB (and the associated amino acids) and PAHs, respectively, than the corresponding values estimated for carbonaceous chondrites.

Next, the characteristics of the present-day fluxes of meteorites and micrometeorites have to be extrapolated over a time window of about 300 millions years, bracketed between 4.2 and 3.9 billions years ago. This extrapolation constitutes the major speculation in our scenario.

But we are guided by studies where the Moon was used as a unique "fossil" detector to estimate the flux of the large bodies (size ≥1 km) that impacted its surface during the accretionary tail, a period of much enhanced bombardment, when the Solar System was still heavily "congested" with material that did not get accreted into planets.

These earlier bodies did produce large impact craters, indicating that the collisions between the small bodies of the Solar System and the Earth/Moon system were much more frequent in the distant past. This is measured by the considerable increase in the crater density (i.e., number per unit area) with the age of the impacted lunar surface, as determined by dating rocks collected by the astronauts; see Chyba (1987) and Anders (1989), and references therein, for a short review of these studies.

But this direct method cannot be applied to the much smaller impacting bodies that generated micrometeorites and micrometeorites on the early Earth. Indeed, the relative abundance (in number) of such bodies drastically increases with decreasing size. Thus, their small craters overlap many times, and their density reaches a saturation value that cannot be used to determine their flux.

We assume that the big meteoroids of the accretionary tail were similar to the comets and asteroids that populate the contemporary Solar System and that they did produce meteorites and micrometeorites by the same processes we observe today, including the sublimation of cometary ices loaded with C-rich dust grains, and collisions between asteroids. Consequently, if the flux of these parent bodies was about ≈1,000 times higher than it is now, we assume that the transmitted fluxes of micrometeorites and meteorites were increased by the same factor. In this case, the total amount of carbon delivered to the early Earth by unmelted and partially dehydrated micrometeorites would have been about *150 times larger than that trapped in the present-day biosphere*, thus satisfying the first critical constraint imposed on any scenario.

We next discuss the role of this ancient micrometeorite flux in the delivery and/or synthesis of complex organics on the early Earth.

III.5. The "early micrometeorites" scenario

All these experimental measurements concerning the carbon chemistry of micrometeorites support a new scenario (Maurette et al. 1990, 1991, 1995), in which each micrometeorite could have functioned individually as a microscopic "chondritic" chemical reactor to generate complex organics (at least amino acids), as soon as they were in contact with water, by some catalyzed hydrolysis of their carbonaceous components.

Our model derives from the one already proposed by organic geochemists in the early 1970s to explain the formation of the long list of organic molecules (including up to 77 amino acids) found in the Murchison meteorite, which bears similarities with the carbonaceous micrometeorites. In the meteorite model the catalyzed hydrolysis was effective in the damp regolith of the parent bodies of meteorites, very early in the history of the Solar System (Bunch and Chang 1980).

For several years, the only experimental evidence supporting this micrometeorite scenario was TEM and SEM observations (see Section II.3.2.D) showing that most micrometeorites are made of complex aggregates of grains, in which a single micrometeorite with a size of ≈100 μm is made of millions of constituent grains embedded into an abundant carbonaceous material. Even the partially dehydrated scoriaceous AMM reported in Figure 7.8 still contains an abundant carbonaceous component.

Some of the minerals in contact with the organics, such as oxides and sulfides of metals, and clays showing different degrees of dehydration and/or disordering, are presently used as catalysts to enhance the rate of industrial chemical reactions. The thin shell of magnetite observed around all types of micrometeorites (Figure 7.5), while protecting them like the shell of an egg, was probably also acting as a kind of inorganic vesicle to confine reactants within the small volume of the grains; this would answer one major objection facing all other "primitive soup" scenarios, the one concerning the fast dilution of the reactants in water and the concomitant sharp decrease in reaction rates. The COPS phase, a dirty variety of ferrihydrite, also seems to be a very effective "sink" for concentrating complex organics, including amino acids (see Section II.3.2.D).

The recent detection of complex organics in micrometeorites by the groups of Jeffrey Bada (Brinton et al. in press) and Richard Zare (Clemett et al. in press) gave further support to this scenario, indicating in particular that the synthesis of amino acids in micrometeorites was already effective in some environment, and that some mineral constituent phases of micrometeorites behave as amazingly efficient cosmochromatographs to trap complex organics.

It was already pointed out (see Section II.3.2.C) that the high AIB-to-isovaline ratio observed in AMMs suggests a process of synthesis of amino acids involving HCN precursors. A major group of comets showing CN^- ions in their tails would provide the right environment for this process, which could have involved the percolation of hot water vapor (generated during the sublimation of cometary ice) through the dark cometary crust, in which micrometeorites could have resided temporarily. It would be much more efficient in the evolved crust of periodic comets, which were subjected many times to this reprocessing by hot water vapor (and other cometary gases), than in nonperiodic comets arriving for the first time in the inner Solar System with an immature surface.

III.6. Micrometeorites and/or hydrothermal sources?

An alternative scenario, also based on the study of real natural systems, postulates that prebiotic synthesis was occurring in hydrothermal sources on the ocean floors (Hennet, Holm and Engel 1992; Yanagawa and Kobayashi 1992). It is thus assumed that such sources were already functioning about 4 billion years ago, with characteristics similar to those observed today.

In this model, a mixture of CH_4, N_2, and CO_2 is released by the source. Furthermore, it is assumed that hydrogen would be generated during the decomposition of water by minerals precipitating from the hot and highly mineralized sources. One of the major difficulties of the model is the high temperature of the source, where complex organics should be quickly destroyed, coupled to the fast dilution of the reactants outside the source cores, which produces a sharp decrease in reaction rates.

This scenario was considered to present the unique advantage of shielding the chain of processes involved in the origin of life from the sterilization induced on the Earth's surface and/or shallow waters by the cataclysmic impacts of meteoroids of the accretionary tail (Sleep et al. 1990). But the "early micrometeorites" scenario would also benefit from the same advantage. In fact, about 500,000 micrometeorites were deposited over each square meter of the early Earth over the expected lifetime (about 10 years today) of a hydrothermal source, including the specific area where the sources were operating.

III.7. Extrapolation to the Martian environment

Micrometeorites deliver to the Earth about 50,000 times more PAHs than carbonaceous chondrites, and such a delivery contains a much richer variety of molecules, including highly alkylated PAHs and vinyl-PAHs (considered as being very reactive).

Even today, the Martian atmosphere is sufficiently thick (≈1 mbar) to fully decelerate all micrometeorites from the dominant mass range investigated in this work (Brack 1996) and to trigger their frictional heating. Thus, the micrometeorite flux would have also delivered large amounts of PAHs to the early Martian regolith.

It could be argued that such PAHs could have ended up being trapped in those Martian meteorites showing carbonates veins, during the percolation of Martian water that emplaced these carbonates in the rocks. In fact, it is known that many other organics, such as amino acids and carboxylic acids, can be preferentially adsorbed by carbonates (Carter 1978; Thomas, Clouse and Longo 1993).

It has been recently argued that the PAHs distribution observed in the meteorite ALH 84001, supposedly from Mars, is unique by its simplicity, which is defined by the small number of peaks appearing in the PAHs mass distribution, the very low degree of alkylation of the core PAHs, and the existence of a high mass envelope at $m/z \approx 300$. This in turn has been used as evidence supporting the existence of extinct microorganisms in this meteorite (McKay et al. 1996). But could this PAHs distribution just be inherited from the delivery of micrometeorites to the Martian regolith?

A comparison of the PAHs distribution observed in ALH 84001 (major peak at $m/z = 228$; a dominant peak at $m/z = 252$; and a low index of alkylation of these PAHs) and in 15 micrometeorites recently investigated at Stanford (Clemett et al. in press) reveals that the major peak is observed at $m/z = 228$ in six micrometeorites. But the micrometeorite PAH distributions differ from that observed in the Martian meteorite, for example, with regard to the absence and/or very low intensity of the 252 peak, and a much higher degree of alkylation. Consequently, the PAHs distribution observed in this

Martian meteorite can hardly derive even from the most similar ones observed in micrometeorites, except if carbonates adsorb preferentially a few core PAHs, or if the composition of the micrometeorite flux varies with time.

IV. Conclusions

Today, carbonaceous micrometeorites with sizes ranging from 50 to 500 μm, which survive well upon atmospheric entry, represent by far the dominant source of extraterrestrial organics on the Earth. We reported evidence suggesting that these micrometeorites could function individually as microscopic "chondritic" chemical reactors as soon as they are in contact with gases and water, either in their parent bodies or in their final host planetary environments, such as the Earth.

Their interactions with the terrestrial environment, which can be defined as a specific type of shooting star chemistry, greatly enhance their diversity. For example, their aerodynamical braking in the upper layers of the atmosphere might trigger both a polymerization of the initial constituent PAHs (leading to the formation of very reactive vinyl-PAHs) and the formation of a thin shell of magnetite encapsulating the grains, which mechanically shields them, increases their lifetime in sediments, and possibly acts as some sort of very primitive inorganic vesicle during chemical reactions.

The amazing diversity of both catalysts and organics trapped in such vesicles – when associated with their massive delivery to the early Earth by the accretionary tail, which allowed them to reach the most favorable (and possibly very rare) spots on the early Earth – supports a theory of the contribution of micrometeorites to the synthesis of prebiotic organic molecules on the Earth.

In this context, it would be tantalizing if future work further reveals that such a synthesis resulted from synergetic effects between the appropriate zones of the best hydrothermal sources on the ocean floor, and the most efficient micrometeorites, loaded with the best catalyzers and/or precursor organics.

This specific type of "shooting star" chemistry might also be effective in planetary bodies such as Mars, equipped with an atmosphere of at least 1 mbar, which can successfully slow down micrometeorites of the dominant 50–500-μm size range. It might also trigger exotic molecules in the atmosphere of the giant planets and their satellites.

We believe that micrometeorites collected on the Earth's surface are unique chemical reactors, which cannot be duplicated in the laboratory. Indeed, they were synthesized in a microgravity environment in the early Solar System, and subsequently modified both in their parent bodies and in the terrestrial environment. Thanks to collaborations with Sydney Leach,

André Brack, Jeffrey Bada, and Akira Shimoyama, we are developing methods to measure their catalytic activity for a variety of chemical reactions in the gaseous and liquid phases. These studies will improve our understanding of the "shooting star" chemistry that was induced by micrometeorites on the early Earth and which possibly contributed to the origin of life.

There was no witness to describe what the early Earth was like when the accretionary tail was still effective, somewhere between 4.2 and 3.9 billion years ago. Thus, all scenarios so far proposed to account for the origin of life, including ours, are based on speculations, such as the composition of the primitive Earth's atmosphere, the partial survival of cometary nuclei upon their cataclysmic impact with the Earth, the characteristics of hydrothermal sources on the deep sea floor, a 1,000–fold increase in the micrometeorite flux during the accretionary tail, and so forth.

The hydrothermal sources and the early micrometeorites scenarios represent two distinct attempts to reduce the number of such speculations. They are both based on experimental studies of real natural systems (e.g., measurements of the characteristics of the present-day micrometeorite flux and chemistry of hydrothermal vents), and not on laboratory investigations of questionable terrestrial "analogues."

We had still to make one major speculation, extrapolating the present-day characteristics of meteorites and micrometeorites at the time of the accretionary tail (see Section III.3). This speculation has been partially justified from studies of large impact craters on the Moon. We feel rather confident that comets were major contributors to the accretionary tail. Consequently, they produced carbonaceous micrometeorites by a process observed almost every year in the contemporary interplanetary medium, the sublimation of dirty cometary ices that replenishes the zodiacal cloud.

Acknowledgments

This work was made possible by grants from the French Space Agency, CNES (Programme d'Exobiologie). The recovery of micrometeorites from Antarctica was funded by the Institut Français de Recherches et de Techniques Polaires. We acknowledge the help of IN2P3 and INSU in France. Studies in Austria and in the United States were supported by grants from Fonds zur Förderung der wissenschaftlichen Forschung and NASA, respectively. The contributions of our colleagues Cécile Engrand and Gero Kurat were decisive in the development of this work, and the enthusiastic support and interest of André Brack and Sydney Leach were of great value.

References

Anders, E. 1989. Pre-biotic organic matter from comets and asteroids. *Nature* 342:255–256.

Becker, L., Glavin, D. P., and Bada, J. L. 1997. Polycyclic Aromatic Hydrocarbons (PAHs) in the Martian meteorite EETA 79001 and other Antarctic meteorites. *Geochim. Cosmochim. Acta* 61:475–481.

Becker, L., Poreda, R. J., and Bada, J. L. 1996. Extraterrestrial helium trapped in fullrenes in the Subdury impact structure. *Science* 272:249–252.

Bland, P. A., Smith, T. B., Jull, A. J., Berry, F. J., Bevan, A. W., Cloudt, S., and Pillinger, C. T. 1996. The flux of meteorites to the Earth over the last 50,000 years. *Mon. Not. R. Astron. Soc.* 233:551–565.

Bonny, P. 1989. Nouvelles collections de micrométéorites groenlandaises et antarctiques. *L'Astronomie (Spécial Exobiologie)* 498–504.

Bonny, Ph. 1990. Entrée atmosphérique de micrométéorites pierreuses chargées en matière organique. *ONERA, Rapport TP 110,* pp. 1–110.

Bonny, Ph., Balageas, D., Devezeaux, D., and Maurette, M. 1988. Atmospheric entry of micrometeorites containing organic material. *Lunar Planet. Sci.* XIX:118–119.

Brack, A. 1996. Why exobiology on Mars? *Planet. Space Sci.* 44:1435–1440.

Brearley, A. J. 1989. Nature and origin of matrix in the unique type 3 chondrite, Kakangari. *Geochim. Cosmochim. Acta* 53:2395–2411.

Brinton, K., Engrand, C., Glavin, D. P., Bada, J., and Maurette, M. In press. A search for extraterrestrial amino acids in carbonaceous Antarctic micrometeorites. *Origins Life Evol. Biosphere.*

Brownlee, D. E. 1981. Interplanetary dust: its physical nature and entry into the atmosphere of terrestrial planets. In *Comets and the Origin of Life*, pp. 63–70. Dordrecht: D. Reidel Publishing Co.

Brownlee, D. E. 1985. Cosmic dust: collection and research. *Ann. Rev. Earth Plan. Sci.* 13:147–173.

Bunch, T. E., and Chang, S. 1980. Carbonaceous chondrites. II. Carbonaceous chondrite phyllosilicates and light element geochemistry as indicators of parent bodies processes and surface conditions. *Geochim. Cosmochim. Acta* 44:1543–1577.

Carter, P. W. 1978. Adsorption of amino acid–containing matter by calcite and quartz. *Geochim. Cosmochim. Acta* 42:1239–1242.

Chyba, C. F. 1987. The cometary contribution to the oceans of primitive Earth. *Nature* 330:632–635.

Chyba, C. F., and Sagan, C. 1992. Endogenous production, exogenous delivery and impact–shock synthesis of organic molecules: an inventory for the origin of life. *Nature* 355:125–132.

Chyba, C. F., Thomas, P. J., Brookshaw, L., and Sagan, C. 1990. Cometary delivery of organic molecules to the early Earth. *Science* 249:366–373.

Clemet, S. J., Chillier, X. D., Gillette, S., Zare, R. N., Maurette, M., Engrand, C., and Kurat, G. In press. Search for polycyclic aromatic hydrocarbons in "giant" carbonaceous micrometeorites from Antarctica. *Origins Life Evol. Biosphere.*

Clemett, S. J., Maechling, C. R., Zare, R. N., and Alexander C. M. O. D. 1992. Analysis of polycyclic aromatic hydrocarbons in seventeen ordinary and carbonaceous chondrites. *Lunar Planet. Sci.* XXIII: 233–234.

Clemett, S. J., Maechling, C. R.,. Zare, R. N., Swan, P. D., and Walker, R. M. 1993. Identification of complex aromatic molecules in individual interplanetary dust particles. *Science* 262:721–723.

Cronin, J. R., and Chang, S. 1993. Organic matter in meteorites: molecular and isotopic analyses of the Murchison meteorite. In *The Chemistry of Life's Origin*,

ed. J. M. Greenberg et al., pp. 209–258. Dordrecht: Kluwer Academic Publishers.
Engrand, C. 1995. Micrométéorites antarctiques: vers l'exobiologie et la mission cométaire "Rosetta." Ph.D. thesis, Université Paris-Sud, Centre d'Orsay.
Engrand, C., Deloule, E., Hoppe, P., Kurat, G., Maurette, M., and Robert, F. 1996a. Water contents of micrometeorites from Antarctica. *Lunar Planet. Sci.* XXVII: 337–338.
Engrand, C., Deloule, E., Maurette, M., Kurat, G., and Robert, F. 1996b. Water contents of COPS-rich cosmic spherules from Antarctica (abstract). *Meteoritics Planet. Sci. Supp.* 31:A43.
Engrand, C., Maurette, M., Kurat, G., Brandstatter, F., and Perreau, M. 1993. A new carbon-rich phase ("COPS") in Antarctic micrometeorites. *Lunar Planet. Sci.* XXIV:441–442.
Graham, G. A., Wright, I. P., Grady, M. M., Perreau, M., Maurette, M., and Pillinger, C. T. 1996. Carbon stable isotope analyses of Antarctic micrometeorites. *Lunar Planet. Sci.* XXVII: 441–442.
Greenberg, J. M., and Hage, J. I. 1990. From interstellar dust to comets: a unification of observational constraints. *Astrophysical Journal* 361(1):260–274.
Halliday, I., Blackwell, A. T., and Griffin, A. A. 1989. The flux of meteorites on the Earth's surface. *Meteoritics* 24:173–178.
Hammer, C., and Maurette, M. 1996. Micrometeorite flux on the melt zone of the West Greenland ice sheet [abstract]. *Meteoritics Planet. Sci.* 31:A56.
Heide, F., and Wlotzka, F. 1995. *Meteorites, Messengers from Space.* Berlin-Heidelberg: Springer-Verlag.
Hennet, R. J. C., Holm, N. G., and Engel, M. H. 1992. Abiotic synthesis of amino acids under hydrothermal conditions and the origin of life: A perpetual phenomenon. *Naturwissenschaften* 79:361–365.
Hoppe, P., Kurat, G., Walter, J., Maurette, M. 1995. Trace elements and oxygen isotopes in a CAI-bearing micrometeorite from Antarctica. *Lunar Plan. Sci.* XXVI:623–624.
Jeanroy, E. 1990. Diagnostic des formes du fer dans les pédogénèses tempérées. Evaluation par les réactifs chimiques d'extraction et apport de la spectrométrie Mossbauer. Ph.D. Thesis, Université Nancy 1.
Jessberger, E. K., Bohsung, J., Chakaveh, S., and Traxel, K. 1992. The volatile element enrichment of chondritic interplanetary dust particles. *Earth Planet. Sci. Lett.* 112:91–99.
Krueger, F. R., and Kissel, J. 1987. The organic component in dust from comet Halley as measured by the PUMA mass spectrometer on board Vega 1. *Nature* 326:755–760.
Krueger, F. R., and Kissel, J. 1989. Biogenesis by cometary grains: Thermodynamic aspects of self-organization. *Origins Life Evol. Biosphere* 19:87–93.
Kurat, G., Hoppe, P., Walter, J., Engrand, C., and Maurette, M. 1994. Oxygen isotopes in spinels from Antarctic micrometeorites. *Meteoritics* 29:487.
Kurat, G., Koeberl, Ch., Presper, Th., Brandstatter, F., and Maurette, M. 1994. Petrology and geochemistry of Antarctic micrometeorites. *Geochim. Cosmochim. Acta* 58:3879–3904.
Lerner, N. R., Peterson, E., and Chang, S. 1993. The Strecker synthesis as a source of amino acids in carbonaceous chondrites: deuterium retention during synthesis. *Geochim. Cosmochim. Acta* 57:4713–4723.
Love, S. G., and Brownlee, D. E. 1993. A direct measurement of the terrestrial mass accretion rate of cosmic dust. *Science* 262:550–553.
Maurette, M., Beccard, B., Bonny, Ph., Brack, A., Christophe, M., and Veyssiere, P.

1990. C-rich micrometeorites on the early Earth and icy planetary bodies. *Proc. 24th ESLAB Symp. on the Formation of Stars and Planets, and the Evolution of the Solar System*. ESA, SP-315, 167–172.
Maurette, M., Bonny, Ph., Brack, A., Jouret, C., Pourchet, M., and Siry, P. 1991. Carbon-rich micrometeorites and prebiotic synthesis. *Lecture Notes in Physics* 390:124–132.
Maurette, M., Brack, A., Kurat, G,. Perreau, M., and Engrand, C. 1995. Were micrometeorites a source of prebiotic molecules on the early Earth? *Adv. Space Res.* 15(3):113–126.
Maurette, M., Immel, G., Hammer, C., Harvey, R., Kurat, G., and Taylor, S. 1994. Collection and curation of IDPs from the Greenland and Antarctic ice sheets. In *Analysis of Interplanetary Dust*, ed. M. E. Zolensky, T. L. Wilson, F. J. M. Rietmeijer, and G. Flynn, pp. 277–289. New York: Amer. Inst. Physics.
Maurette, M., Jehanno, C., Robin, E., and Hammer, C. 1987. Characteristics and mass distribution of extraterrestrial dust from the Greenland ice cap. *Nature* 301:473–477.
Maurette, M., Kurat, G., Presper, Th., Brandstatter, F., and Perreau, M. 1992. Possible causes of depletion and enrichment of minor elements in Antarctic micrometeorites. *Lunar Planet. Sci.* XXIII:861–862.
McDonald, G. E., and Bada, J. L. 1995. A search for endogenous amino acids in the Martian meteorite EETA 79001. *Geochim. Cosmochim. Acta* 59:1179–1184.
McKay, D. S., Gibson, E. K., Jr., Thomas-Keprta, K. L., Vali, H., Romanek, C. S., Clemett, S. J., Chillier, X. D., Maechling, C. R., and Zare, R. N. 1996. Search for past life on Mars: Possible relic biogenic activity in Martian meteorite ALH 84001. *Science* 273:924–930.
McKinnon, I. D. R., and Zolensky, M. E. 1984. Proposed structures for poorly characterized phases in CM2 carbonaceous chondrite meteorites. *Nature* 309:240–242.
Oberbeck, V. R., and Aggarwal, H. 1992. Comet impacts and chemical evolution on the bombarded Earth. *Origins Life Evol. Biosphere* 21:317–338.
Oró, J. 1961. Comets and the formation of biochemical compounds on the primitive Earth. *Nature* 190:389–390.
Oró, J., and Mills, T. 1989. Chemical evolution of primitive Solar System bodies. *Adv. Space Res.* 9:105–120.
Ponchelet, H. 1989. Astrophysique, tous fils du ciel. *Le Point* 860:121–122.
Rasmussen, K. L., Clausen, H. B., and Kallemeyn, W. 1995. No iridium anomaly after the 1908 Tungunska impact: Evidence from a Greenland ice core. *Meteoritics* 30:634–638.
Schidlowski, M. 1988. A 3,800-million-year isotopic record of life from carbon in sedimentary rocks. *Nature* 333:313–318.
Shock, B. L., and Schulte, M. D. 1990. Amino-acids synthesis in carbonaceous meteorites by aqueous alteration of polycyclic aromatic hydrocarbons. *Nature* 343:728–731.
Shu, F. H., Adams, F. C., and Lizano S. 1987. Star formation in molecular clouds: observation and theory. *Ann. Rev. Astron. Astrophys.* 25:23–81.
Sleep, N. H., Zahne, K. J., Kasting, J. F., and Morowitz, H. J. 1989. Annihilation of ecosystems by large asteroid impacts on the early Earth. *Nature* 342:139–142.
Taylor, S., Lever, J. H., Harvey, R. P. 1996. Terrestrial flux rates of micrometeorites determined from the South Pole water well [abstract]. *Meteoritics Planet. Sci.* 31:A140.
Thomas, M. M, Clouse, J. A., and Longo J. M. 1993. Adsorption of organic com-

pounds on carbonate minerals. 3. Influence on dissolution rates. *Chemical Geology* 109:227–237.

Tomeoka, K., and Buseck, P. R. 1985. Indicators of aqueous alteration in CM carbonaceous chondrites: microstructure of a layered mineral containing Fe, S, O, and Ni. *Geochim. Cosmochim. Acta* 49:2149–2163.

Tomeoka, K., and Buseck, P. R. 1988. Matrix mineralogy of the Orgueil CI carbonaceous chondrite. *Geochim. Cosmochim. Acta* 52:1627–1640.

Whipple, F. L. 1951. The theory of micrometeorites. 2. In a heterothermal atmosphere. *Proc. Nat. Acad. Sci. USA*. 37:19–31.

Whipple, F. L., and Hugues, R. F. 1955. On the velocity and orbits of meteors, fireballs and meteorites. In *Meteors*, ed. T. R. Kaiser, *Jour. Atmospheric Terrest. Phys.*, Spec. Suppl.2, 149–156.

Wolman, Y., Haverland, W. J., and Miller, S. L. 1972. Nonprotein amino acids from spark discharges and their comparison with the Murchison meteorite amino acids. *Proc. Nat. Acad. Sci. USA* 69:809–811.

Yanagawa, H., and Kobayashi, K. 1992. An experimental approach to chemical evolution in submarine hydrothermal systems. *Origins Life Evol. Biosphere* 22:147–160.

Yuasa, S., Flory, D., Basile, B., and Oró, J. 1984. On the abiotic formation of amino acids. I. HCN as a precursor of amino acids detected in extract of lunar samples. II. Formation of HCN and amino acids from simulated mixtures of gases released from lunar samples. *J. Mol. Evol.* 20:52–58.

Zolensky, M. E., Reindell, H. M., Wilson, I., and Weels, G. L. 1992. The age of the meteorite recovery surfaces of Roosevelt County, New Mexico, USA. *Meteoritics* 27:460–462.

Zolensky, M. E., Wilson, T. L., Rietmeijer, F. J. M., and Flynn, G. (eds.). 1994. *Analysis of Interplanetary Dust.* New York: Amer. Inst. Physics.

PART III

Possible starts for primitive life

8

Membrane compartments in prebiotic evolution

D. W. DEAMER
Department of Chemistry
University of California, Santa Cruz

This chapter addresses a question that must be answered if we are to understand life's origin fully: Did the living state arise a priori from preexisting cellular structures, or did cellular life develop only at a later evolutionary stage? Considerable insight into this basic question can be obtained from our knowledge of the self-assembly processes that are fundamental to contemporary cell structure and function. Before going on, it is useful to outline the biophysical principles that guide research in this area and indicate how they can be used to investigate cellular origins. We will particularly emphasize certain properties of lipidlike molecules that are well known to membrane biophysicists but may not be obvious to the general scientific community.

1. *Bilayers assemble from a variety of amphiphilic compounds.*

   Although contemporary cell membranes incorporate phospholipids as the major component of the lipid bilayer, it is not necessary to think that they were required for early cellular life. In fact, simpler amphiphilic molecules can also assemble into bilayer membranes. These include single-chain amphiphiles such as soap molecules, glycerol monooleate, oxidized cholesterol, and even detergents like dodecyl sulfate mixed with dodecyl alcohol. It seems likely that primitive cells incorporated lipidlike molecules from the environment, almost as a nutrient, rather than undertaking the much more complex process of synthesizing complex lipids *de novo*.

2. *Bilayer permeability strongly depends on chain length of the component amphiphilic molecules.*

   We tend to think of the lipid bilayer as being a nearly impenetrable barrier to ionic solutes and other large, polar molecules. This leads to a conundrum when we try to imagine how early forms of cellular life could have functioned in the absence of highly evolved transport enzymes that translocate ionic nutrients and metabolites across the bilayer barrier. It is true that lipid

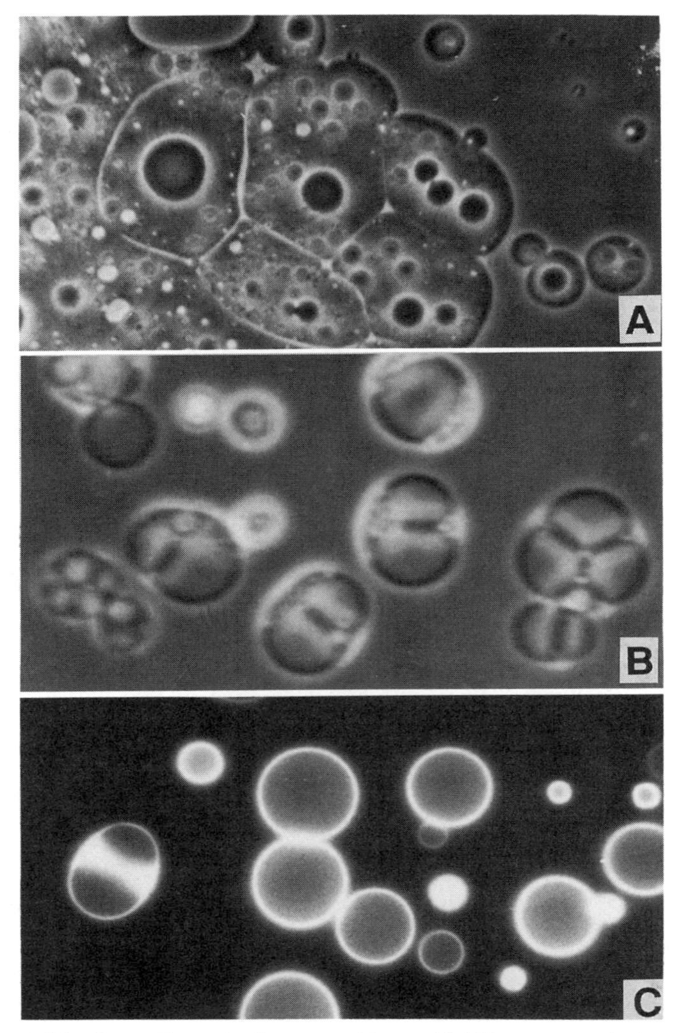

Figure 8.1. Membrane formation by meteoritic amphiphilic compounds. A sample of the Murchison meteorite was extracted with the chloroform–methanol–water solvent described by Deamer and Pashley (1989). Amphiphilic compounds were isolated chromatographically on TLC plates (fraction 1), and a small aliquot (~1 microgram) was dried on a glass microscope slide. Alkaline carbonate buffer (15 microliters, 10 mM, ph 9.0) was added to the dried sample, followed by a cover slip, and the interaction of the aqueous phase with the sample was followed by phase and fluorescence microscopy. **A.** The sample–buffer interface was 1 minute. The aqueous phase penetrated the viscous sample, causing spherical structures to appear at the interface and fall away into the medium. **B.** After 30 minutes, large numbers of vesicular structures are produced as the buffer further penetrates the sample. **C.** The vesicular nature of the structures in B are clearly demonstrated by fluorescence microscopy. A: Original magnification 160 ×. B and C: Original magnification 400 ×.

bilayers of contemporary cell membranes present a significant permeability barrier that is necessary for normal cell functions, particularly those related to bioenergetics of ion transport and chemiosmotic ATP synthesis. However, recent results to be described here show that shortening lipid chain length from 18 to 14 carbons increases the permeation of ionic solutes by several orders of magnitude. This level of permeability is sufficient to encapsulate large molecules such as proteins and polynucleotides, yet still to allow external substrate to reach an encapsulated enzyme. It follows that early cell membranes could have been composed of shorter chain lipids that provided access to nutrients for macromolecules undergoing growth and replication in an encapsulated microenvironment.

3. *Macromolecules can be encapsulated in bilayer vesicles under simulated prebiotic conditions.*

   A conceptual problem has been to imagine how lipid bilayers could capture macromolecules in the first place, given that the bilayer must present a nearly impenetrable barrier if the macromolecules are to be maintained within the membrane-bounded volume. We will describe how a mixture of lipid and protein or nucleic acids can undergo drying and wetting cycles that simulate tide pools. Under these conditions, the macromolecules are readily captured in membrane-bounded vesicles.

4. *Lipid bilayers grow by addition of amphiphilic compounds present in the bulk-phase medium.*

   It is not sufficient for a primitive cell to replicate its macromolecular components unless the boundary membrane can increase in area to accommodate the internal growth. We will discuss recent experimental results from liposome model systems that demonstrate a form of growth.

Sources of organic compounds on the early Earth

Is there a plausible source for amphiphilic molecules in the prebiotic environment? The current consensus is that the Archaean atmosphere was volcanic in origin, consisting of carbon dioxide and smaller amounts of CO, with nitrogen about at current levels. Miller–Urey reactions are not productive in carbon dioxide atmospheres (Stribling and Miller 1987), so that other sources of organics are being considered. The possibility that extraterrestrial infall provided significant amounts of organic carbon to the Earth's surface was first proposed by Oró (1961) and more recently elaborated by Anders (1989), Chyba et al. (1990), and Chyba and Sagan (1992). Figure 8.1 shows estimates of accumulated organic compounds during a 300-million-year bombardment period from about 4.2 to 3.9 bya. The carbon added by extraterrestrial infall is in the range of 10^{16}–10^{18} kg. This is less that the total organic carbon stored as oil shales, coal, and other fossil deposits on the Earth (10^{21} kg), which represents carbon dioxide reduced to organic compounds by photosynthetic processes. On the other hand, it is several orders of magnitude greater than the organic carbon now in the biosphere, estimated to be 6×10^{14} kg. It therefore seems reasonable to conclude that extraterrestrial infall was a significant source of organic carbon in the prebiotic environment, as long as it could be released from its mineral matrix.

Much of the cometary and meteoritic infall surviving atmospheric entry would presumably fall into oceans. A major fraction of the organic content would be buried as sediment, and a smaller fraction would be released into the marine environment over long time intervals. Mechanisms for releasing organic components from extraterrestrial infall have not been considered in detail. Recent observations from oil chemistry suggest that hydropyrolytic conditions may contribute to the extraction process (Siskin and Katritzky 1991, for review). In hydropyrolysis, organic compounds in a mineral matrix are simply heated in the presence of water and high pressure to temperatures of 200–350 °C for periods up to several days. Under these conditions, water takes on unusual characteristics. For instance, the dielectric constant is reduced to a value near that of acetone, permitting water to behave like an organic solvent. The dissociation constant of water also increases from 10^{-14} to $10^{-11.3}$ so that it is more effective as an acid/base catalyst.

Such conditions would presumably be relatively common on a prebiotic Earth with extensive vulcanism. We envisage extraterrestrial infall, largely in the form of micrometeorites and cometary debris, sinking in early oceans to regions of volcanic activity where water temperatures and pressures are in ranges suitable for hydropyrolysis. Hydropyrolytic reactions would disperse the organic components associated with the sediments. Water-soluble compounds would simply dissolve to form a highly dilute solution of organic solutes, whereas longer-chain hydrocarbons and their derivatives would accumulate at the ocean surface to form a thin film at the air–water interface. Such films would likely become concentrated at intertidal zones by the same mechanism that forms sea-foam from monolayers of surface-active organic compounds today. It follows that a probable site for the physical and chemical processes leading to the origin of living cells is a tide pool in which amphiphilic hydrocarbon derivatives accumulated and were mixed with water-soluble organics during cyclic drying and rehydration processes.

To test this idea, we have exposed samples of the Murchison carbonaceous meteorite to hydropyrolytic conditions and analyzed both the extracted components and the remaining powder by mass spectrometry (Mautner, Leonard and Deamer 1995). Hydropyrolysis was carried out at temperatures in the range of 100–300 °C over periods of hours to days. Gas chromatograms of products isolated from hydropyrolyzed Murchison samples showed an abundant array of organic compounds. When the carbon content of the powder was determined by mass spectral analysis before and after hydropyrolysis, over half of the carbon had been released as soluble organics. Significant amounts of amphiphilic compounds were present, including octanoic (C8) and nonanoic (C9) monocarboxylic acids, as well as a variety of polar aromatic hydrocarbon derivatives. Pure samples of these compounds are not sufficiently amphiphilic to produce stable membranes over a wide range of pH and concentration, but as we will see later, mixtures have self-stabilizing properties and can form long-lived membranous vesicles.

If extraterrestrial delivery followed by hydropyrolysis is a plausible source of organics, it seems likely that such compounds were among the most common components of the prebiotic organic inventory. This would have significant implications for our understanding of life's origin. Our aim now is to make quantitative estimates of individual components in order to understand their relative abundance in the prebiotic environment.

Sources of prebiotic hydrocarbon derivatives

All lipidlike molecules are polar derivatives of hydrocarbons. We can now ask whether it is likely that extraterrestrial infall would be expected to contain significant amounts of hydrocarbon derivatives that could become incorporated into primitive membranes. The best-studied examples of extraterrestrial organics are those in carbonaceous meteorites. Until recently, it was believed that long-chain hydrocarbons were present in such meteorites, and it seemed reasonable to think that normal and branched-chain hydrocarbons and their derivatives were available in the prebiotic environment. The hydrocarbons would have been either delivered directly as extraterrestrial infall, or perhaps synthesized by processes similar to those in the meteorite parent bodies. However, Cronin, and Pizzarello (1990) found that the long-chain hydrocarbons of the Murchison meteorite are in fact terrestrial contaminants, and that only cyclic aliphatic hydrocarbons occur in significant quantities. Mono- and dicarboxylic acids occur in the range of 100–200 ppm. This is approximately 10 times greater than the amino acid content, but the amount of a given molecular species decreases nearly exponentially with chain length (Lawless and Yuen 1979) so that only vanishingly small amounts of longer-chain compounds are present.

If hydrocarbon derivatives are not abundant in meteorites, perhaps lipidlike hydrocarbon derivatives were produced by chemical processes on the early Earth, rather than being delivered by infall. For instance, Fischer–Tropsch pathways can provide long-chain hydrocarbons and their derivatives (Hayatsu and Anders 1981), but such conditions require carbon monoxide and water vapor to be passed over a hot iron catalyst. It is not easy to imagine how such specialized conditions would be established on the prebiotic Earth. Another approach was taken by Ourisson and Nakatani (1994), who noted that molecular fossils commonly contain a variety of terpenoid derivatives. In particular, terpenoids dominate the lipid composition of Archaea, generally considered to be closely related to a universal ancestor present some 3.5 billion years ago. From this, it was proposed that acyclic polyprenols and their phosphate derivatives were the original lipids of membranes, with hydrocarbon chains synthesized by condensation reactions of isopentenols derived from formaldehyde and isobutene.

The evidence provided by molecular fossils and the structural role of terpenoids in membranes of Archaea is persuasive that terpenoids were a lipid component of early biological membranes. The case for terpenoids' playing a role in prebiotic systems is less convincing. As is true for most scenarios of chemical evolution, an abundant source of precursors is problematic. Formaldehyde is reasonably high on the list of potential prebiotic reactants, but isobutene is not. The requirement for phosphate further compounds the difficulty. Prebiotic synthesis of a 10-carbon compound such as geranyl phosphate would require a source of relatively concentrated isoprenol, phosphate, and a catalytic surface, none of which have been demonstrated experimentally under simulated prebiotic conditions.

To summarize, an abundant source of long-chain hydrocarbon components of prebiotic membranes is not obvious. On the other hand, one might argue that because the origin of cellular life absolutely requires lipidlike hydrocarbon derivatives, such molecules must have been available on the early Earth from a yet unknown source. Given that amphiphilic molecules were present (Deamer 1985; Deamer, Harang-Mahon and Bosco 1994), we can go on to discuss self-assembly processes that were likely to have played a role in primitive membranes.

Self-assembly processes in prebiotic organic mixtures

The first suggestion that membranes played a role in the origin of life was in J. B. S. Haldane's prescient note in *The Rationalist Annual* (1926). Haldane wrote that "the cell consists of numerous half-living chemical molecules suspended in water and enclosed in an oily film. When the whole sea was a vast chemical laboratory the conditions for the formation of such films must have been relatively favorable." Goldacre (1958) proposed that the first membranes could have been produced by wave action disturbing films of lipidlike surfactants. The first experimental approaches to this question were undertaken by Hargreaves et al. (1977) and Oró and co-workers (1978).

What physical properties are required if a molecule is to become incorporated into a stable bilayer? All bilayer-forming molecules are amphiphiles, with a hydrophilic "head" and a hydrophobic "tail" on the same molecule. Although we tend to think of membrane lipids as essentially phospholipids, in fact a surprising variety of amphiphiles take part in membrane structure, including phospholipids, sphingolipids, cerebrosides, sterols, and pigments like chlorophyll. Earlier studies (Hargreaves and Deamer 1978) showed that single-chain amphiphiles such as alkyl phosphates, alkyl sulfates, and even fatty acids assemble into bilayer membranes if they contain 10 or more carbons in their hydrocarbon chains. More recently, Roberts (1992) made similar observations with phospholipids. For instance, phosphatidylcholine (PC) with 12 carbon chains produces stable lipid bilayer membranes, whereas PC with 8 carbon chains assembles into micelles. PC with 10 carbon chains forms lipid vesicles that can be visualized by phase microscopy but are unable to provide a significant permeability barrier to flux of ionic solutes.

The physical state of membrane-forming lipids must also be taken into account. Fluid lipid chains are required for normal functions of cell membranes, and the fluid state is controlled by chain length, unsaturation, and, in certain prokaryotic cells, chain branching. Early cells presumably had the same requirement, so models of primitive cell membranes must be able to maintain a fluid state at ambient temperatures.

If amphiphilic molecules were present in the mixture of organic compounds available on the early Earth, it is not difficult to imagine that their self-assembly into molecular aggregates was a common process. Is this a plausible premise? To address this question experimentally, we might first investigate the mixture of organic compounds in carbonaceous meteorites, looking for evidence that certain components have amphiphilic properties. Most meteorites are composed of silicon-based minerals, and a small fraction (~5%) of these stony meteorites contain up to several percent of their mass in the form of organic carbon. These are referred to as carbonaceous meteorites, and the carbon compounds represent well-preserved samples of the organic material present in the early Solar System. A kerogen-like polymer composed of covalently linked polycyclic aromatic hydrocarbons is the most abundant material, whereas a series of organic acids (including 10–20 ppm of amino acids) represents the most abundant water-soluble fraction. Aliphatic and aromatic hydrocarbons, ureas, ketones, alcohols, aldehydes, and purines are also present. (See Cronin, Pizzarello and Cruickshank 1988 for review.)

A variety of amphiphilic molecules could also be present in the form of

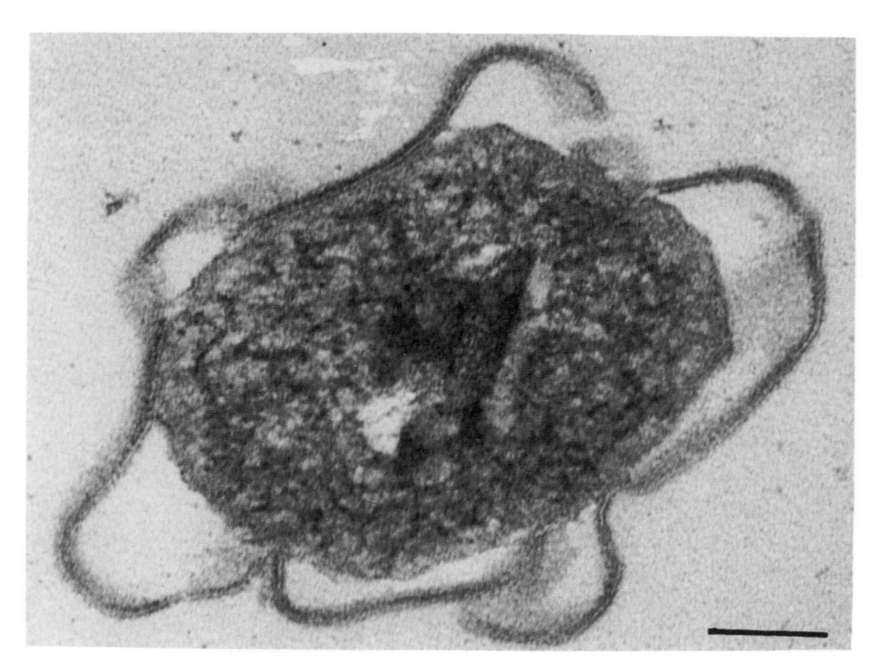

Figure 8.2. TEM of Murchison membranous vesicle. A sample equivalent to that described in Figure 8.1B was sonicated for 10 seconds to produce smaller fragments, and these were fixed in osmium tetroxide and ferricyanide, embedd in epoxy resin, and sectioned for transmission microscopy. Virtually all the particles had a trilaminar membrane surrounding a central core, as shown in the electron micrograph. A trilaminar feature is characteristic of bilayer lipid structures in cell membranes. Bar = 1.0 micron.

polar hydrocarbon derivatives. We therefore surveyed several carbonaceous meteorites, with a particular emphasis on the Murchison carbonaceous chondrite. It immediately became apparent that surface-active material was present in Murchison organics. In fact, if a few milligrams of fresh meteoritic powder is heated briefly in water, the surface tension decreases markedly, indicating formation of a monolayer of surface-active components. These could be solubilized in a chloroform–methanol–water system commonly used to extract membrane lipids from biological material. Two-dimensional thin layer chromatography showed that a complex mixture of oxidized aliphatic and aromatic hydrocarbons was present, which appeared as fluorescent fractions on the TLC plate. When this material was allowed to interact with aqueous phases, one class of compounds with acidic properties was clearly capable of forming membrane-bounded vesicles (Figures 8.1 and 8.2). The vesicles responded osmotically to sodium chloride or sucrose additions, and could maintain gradients of a negatively charged fluorescent dye (pyranine). This,

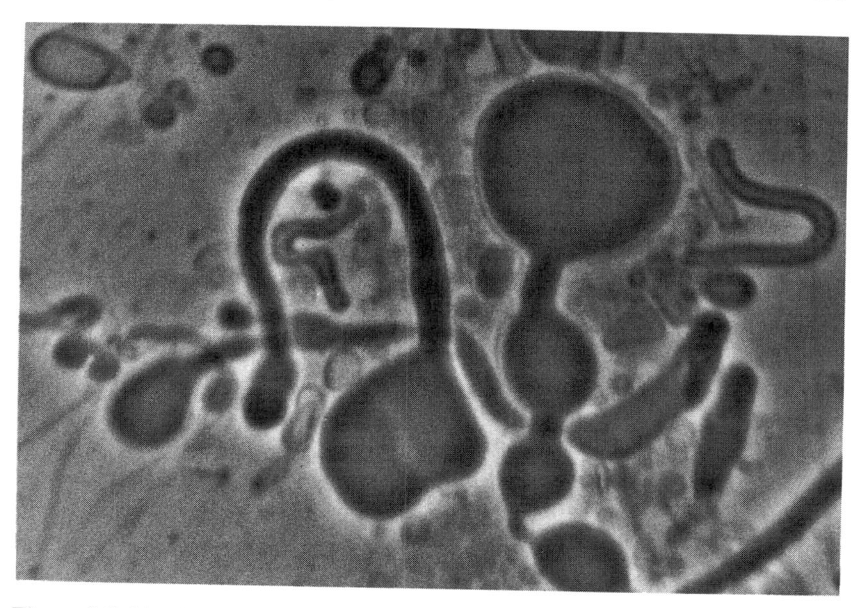

Figure 8.3. Membranes self-assembled from a monocarboxylic acid. A 9-carbon monocarboxylic acid (nonanoic acid) has been demonstrated by mass spectrometry to be present in the amphiphilic fraction of Murchison organic components (Mautner et al. 1995). This phase micrograph shows that pure nonanoic acid at high concentrations (0.1M) and neutral pH can self-assemble into membrane structures resembling those observed in Murchison extracts. Original magnification 160 ×.

together with the light and electron microscopic evidence, provides strong evidence that a mixture of abiotic organic compounds isolated from a carbonaceous meteorite contains amphiphiles capable of forming membranes.

Using mass spectrometry and infrared spectrophotometry (FTIR), we determined that one of the components of the mixture was nonanoic acid, a 9-carbon carboxylic acid. Nonanoic acid has too short a chain to form stable bilayers in dilute solutions, but at neutral pH and high concentrations of the amphiphile (>100 mM), membrane structures were readily observed by light microscopy (Figure 8.3). This evidence suggests that the meteoritic amphiphiles are a mixture of monocarboxylic acids, such as nonanoic acid, and polar polycyclic aromatic compounds that produce the characteristic fluorescence of the vesicle structures. Because only microscopic quantities of the membrane-forming components are available, we have not been able to directly analyze the membranes themselves.

To summarize these results, surface-active compounds capable of membrane formation are present in carbonaceous meteorites and are able to self-assemble into bilayer membranes. The membrane structures can be visualized

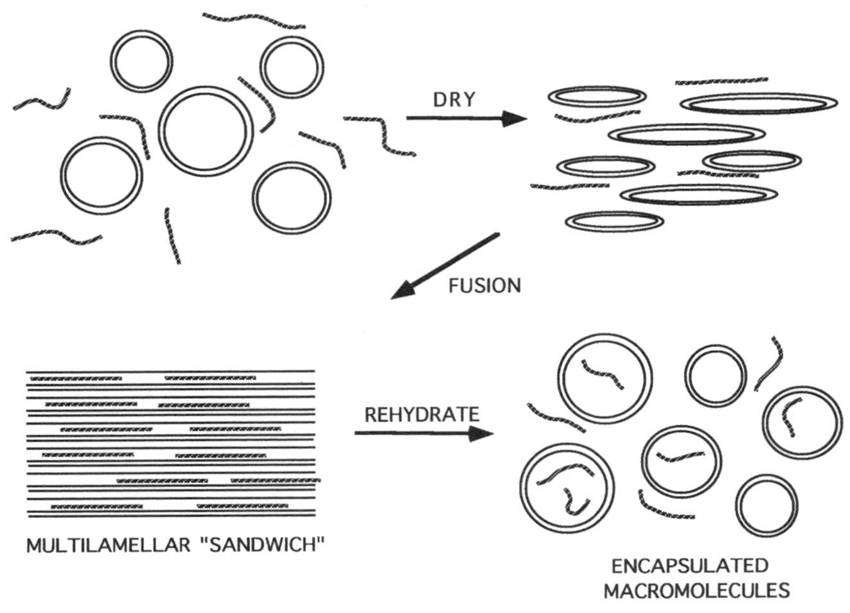

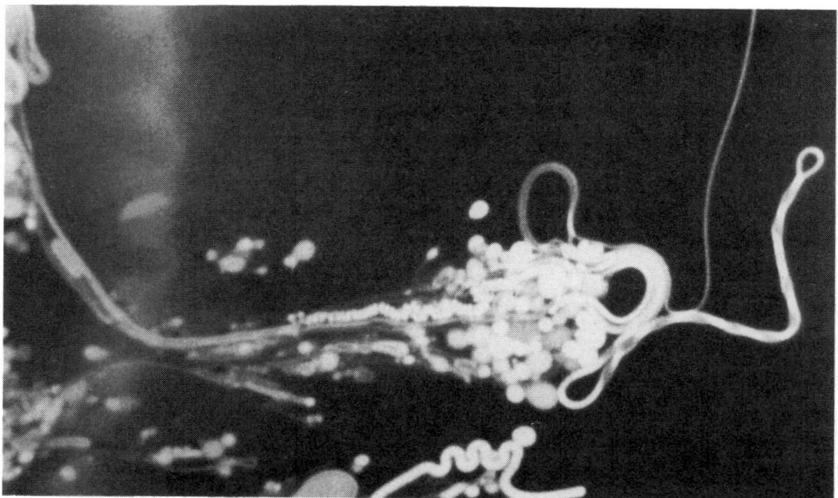

Figure 8.4. Macromolecules as large as proteins and nucleic acids are readily encapsulated in liposomes. When a mixture of liposomes and macromolecules is dried, the liposomes dry and fuse to form multilamellar structures with the macromolecules trapped between alternating layers. Upon rehydration, the vesicles capture approximately half of the large molecular solutes that were originally external to the liposomes. The fluorescence micrograph shows DNA (stained with the fluorescent dye acridine orange) encapsulated in lipid membranes following a dehydration–rehydration cycle.

by light microscopy, by thin sectioning of osmium-fixed specimens for electron microscopy, and by freeze-fracture methods. The membranes are approximately 10 nm thick and show the characteristic trilaminar staining pattern expected for a bilayer of amphiphilic material. The amount of such compounds in carbonaceous meteorites is probably too small to have provided an abundant source of lipidlike material on the early Earth. However, the fact that membranes can self-assemble from the amphiphilic components at least makes it more plausible that membrane-bounded structures were present at the time of life's origin. We will now go on to describe how lipid bilayer membranes could have been involved in the evolutionary process leading to primitive cells and bioenergetic functions.

Models of primitive cellular systems

We will first ask how prebiotic lipid bilayer vesicles could encapsulate macromolecules involved in catalysis and information processing. If this encapsulation is to occur, there must be a reversible process by which the bilayer barrier first can be broken, allowing entry of large molecules, then resealed. Laboratory models of membrane-bounded environments can be produced when small vesicles composed of lipid bilayers (liposomes) encapsulate solute molecules. Most liposome encapsulation processes are highly technical and involve dissolving the lipid in organic solvents, followed by processes like detergent solubilization, sonication, and extrusion of liposomes through polycarbonate filters. However, there is one encapsulation process simple enough to function in the prebiotic environment. This process depends on the fact that when liposomes are dried in the presence of other solute molecules, they tend to fuse into multilayered structures that "sandwich" the solutes (Deamer and Barchfeld 1982; Shew and Deamer 1983). As shown in Figure 8.4, solutes as large as proteins and nucleic acids are captured upon rehydration when the lipid layers reseal into vesicles. It is not difficult to imagine hydration–dehydration cycles occurring in intertidal zones, and it follows that encapsulated systems of solute molecules may have been reasonably common if amphiphilic molecules were in fact present.

Although membrane boundaries define all living cells, the membrane also limits access to nutrients and energy sources. The first living systems were unlikely to have evolved specialized membrane transport systems, and it is interesting to consider how an early cell could deal with the intrinsic barrier properties of its limiting membrane. To give a perspective on the magnitude of this barrier, we can compare the rates at which relatively permeable and relatively impermeable solutes cross contemporary lipid bilayers. For example, small, uncharged molecules such as water, oxygen, and carbon dioxide have permeabilities approximately 10^9 times greater than ions. This means that half the water in a small lipid vesicle exchanges in a few milliseconds, whereas sodium ions have half-times of exchange measured in days. The primary reason for the remarkable permeability barrier to ionic diffusion arises from the Born energy that must be overcome when an ion moves from a high dielectric medium (water) to a low dielectric medium represented by the hydrocarbon chains of the lipid bilayer. (See Parsegian 1969 and Deamer and Volkov 1995 for reviews.)

Typical nutrient solutes such as amino acids and phosphate are ionized, which means that they would not readily cross the permeability barrier. Permeability coefficients of liposome membranes to phosphate and amino acids were recently determined (Chakrabarti and Deamer 1994), and it was found that they were in the range of 10^{-11}–10^{-12} cm s^{-1}, similar to monovalent ions like sodium and potassium. From these figures one can estimate that if a primitive microorganism depended on passive transport of phosphate across a lipid bilayer composed of 18-carbon phospholipids, it would require several years to accumulate enough phosphate to double its DNA content, or to pass through one cell cycle.

One solution to this conundrum is simply to shorten the chain length of the lipids composing the bilayer. For instance, shortening phospholipid chains from 18 to 14 carbons increases permeability to ions nearly a thousandfold (Paula et al. 1996). This results from transient transmembrane defects that become increasingly common as the bilayer thins. In fact, bilayers composed of 10-carbon lipids are so permeable that ion gradients cannot be maintained for more than a few seconds. Intermediate-chain-length lipids therefore provide a barrier sufficient to encapsulate macromolecular catalysts, while allowing smaller ionic nutrients access to the interior.

In recent work, we have taken advantage of this effect to demonstrate that a polymer can be synthesized by an encapsulated catalyst using transmembrane transport of substrate and energy (Chakrabarti et al. 1994). At some point in the origin and evolution of early cellular life, such processes must have been involved. That is, the catalyzed polymerization reaction is captured in a membranous vesicle, but the sources of energy and nutrients are external. Polynucleotide phosphorylase (PNPase) is a useful enzyme for initial tests of this concept. PNPase can use ADP as a substrate to synthesize polyribonucleotides, in this case polyadenylic acid. The enzyme system does not require primers or templates, and the energy source is the pyrophosphate bond in the substrate, ADP. Some of Oparin's later research on the origin of life attempted to show that PNPase associated with coacervates could synthesize RNA (Oparin et al. 1976).

In the investigation to be described here, encapsulation of the enzyme was

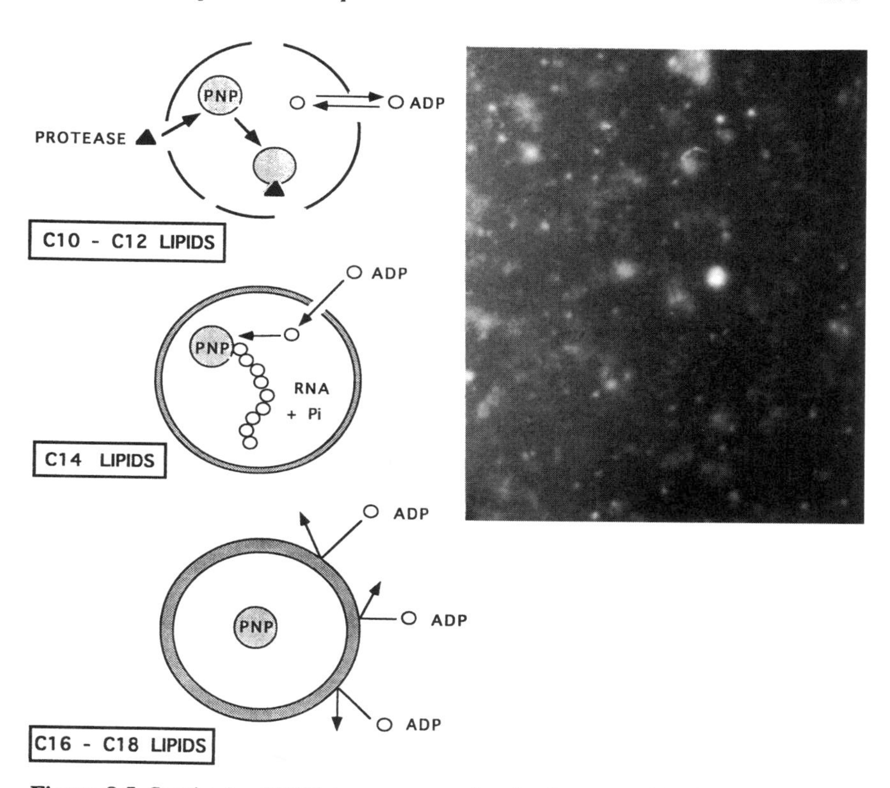

Figure 8.5. Synthesis of RNA by an encapsulated polymerase. By choosing an appropriate lipid chain length, liposomes can be made sufficiently permeable to ionic solutes so that even anions as large as ADP can permeate the bilayer barrier and supply substrate to enzymes in the liposomes. If the lipid chains are too short, as shown at the top of the figure, the bilayers are so unstable that the enzyme is not in a protected environment and is labile to protease digestion, for instance. If the lipid chains are too long, the bilayer is not permeable to ionic substrates. The fluorescence micrograph shows ethidium bromide–stained RNA synthesized in liposomes by polynucleotide.

carried out by dehydration of C14 phosphatidylcholine (DMPC) vesicles in the presence of the enzyme, followed by rehydration, a process that produces liposomes containing the enzyme (Figure 8.5). Any PNPase remaining outside of the liposomes was removed by gel filtration. To show that ADP could cross the bilayer, the release of encapsulated ADP from DMPC vesicles was measured at three different temperatures. We found the permeability coefficient to be 2.6×10^{-10} cm s^{-1} at 23 °C, sufficient to supply the enzyme with ADP. RNA synthesis was followed in two ways. In the first, ethidium bromide fluorescence increased as RNA filled the vesicles. We found that the reaction rate was only about 20% that of the free PNPase, suggesting that the DMPC bilayer, even though a thousand times more permeable than a C18 lipid bilayer, still represented a significant barrier to substrate permeation. Poly(A) synthesis was also monitored by polyacrylamide gel electrophoresis. In this experiment, to control for any PNPase bound to the outer surface of the liposomes, a protease was added. The protease completely inhibited RNA synthesis by the free enzyme, but the encapsulated PNPase was protected, with significant RNA synthesis observed. RNA could be visualized in the liposomes by fluorescence microscopy of the ethidium bromide. Luisi and co-workers have obtained similar results in a simpler membrane system to be discussed later.

These results, taken together, provide a helpful perspective on substrate transport by primitive forms of life. In the early Earth, there must have been a variety of amphiphilic hydrocarbon derivatives that could self-assemble into bilayer boundary structures. However, it is not necessary to think that early lipid bilayers had the same thickness and permeability properties of contemporary membranes. Instead, membrane-forming amphiphiles with 12–14 carbon chains, modeled here by DMPC, would produce bilayers that are permeable enough to allow passage of ionic substrates required for polymerization of macromolecules such as RNA, yet maintain those macromolecules within a boundary. Encapsulated catalysts and information-bearing molecules would thus have access to nutrients required for growth processes. Furthermore, specific groupings of macromolecules would be maintained. This would allow selection to occur at the cellular level, an important advantage not available in random mixtures of unencapsulated molecules. Last, there may have been some mechanism by which membranes could capture chemical energy for the polymerization reactions, in which case the energy would be immediately available to the encapsulated catalytic system.

In another experimental approach, Luisi and co-workers have established experimental systems in which membranes grow by addition of new components at the same time that polynucleotides are synthesized inside the vesicles. The term *self-reproduction* has been applied to such systems, in part because the membranes themselves catalyze the reaction by which the components are produced. As noted earlier, single-chain amphiphiles readily assemble into bilayer vesicles, and the Luisi group took advantage of this fact to develop a model membrane that undergoes growth and a kind of reproduction. The system depends on the hydrolysis of nonmembranous precursor molecules (such as fatty acid esters and anhydrides) to fatty acid–soap mixtures that can form bilayer vesicles at certain pH ranges (Bachman, Luisi and Lang 1992; Walde, Wick et al. 1994). In a system of giant liposomes, Wick, Walde and Luisi (1995) used light microscopy to follow the growth process and observed that vesicles grew over a 6-hour period from an average diameter of approximately 1 micrometer to diameters in the range of 4–6 micrometers.

This basic idea was then extended to the synthesis of RNA within the vesicles. In one such system, polynucleotide phosphorylase was first trapped in liposomes composed of oleic acid, and ADP was then added (Walde, Goto et al. 1994). Because the oleic acid/oleate membranes are considerably more permeable than phospholipid bilayers, that ADP readily permeates the membrane and can be used as a substrate by the enzyme. After a lag period of 4 days, characteristic of PNPase, polyadenylic acid began to be synthesized, at which point oleic anhydride was added. For the next 30 hours, the oleic anhydride was hydrolyzed to oleic acid added to the vesicles, and simultaneously polyadenylic acid was synthesized.

The next step was to determine whether a template-driven reaction can function within the vesicle environment. Oberholzer, Albrizio and Luisi (1995) approached this question by capturing a PCR reaction within liposomes, and showed that DNA could be produced in liposomes composed of a phospholipid. This result had several surprising aspects. First, it became clear that ordinary phospholipid is sufficiently stable to maintain membrane integrity even at PCR temperatures as high as 95 °C. Second, the result demonstrated the limitations of a membrane-encapsulated reaction that must be overcome before we can take further steps toward a model system with properties of the living state. The liposomes were prepared with phosphatidylcholine and were therefore relatively impermeable to ionic substrates. This means that eight different kinds of molecules must be captured in any one liposome if a successful reaction is to take place. These include the polymerase itself, a DNA template, two specific primers, and four nucleotide triphosphates, as well as magnesium ions. By calculation, the authors determined that only a very small fraction of liposomes would happen to capture all eight molecular species. Furthermore, a typical liposome in this preparation had a measured volume of 3.3×10^{-18} liters. At 1.0 mM NTP substrate concentrations, only 2,000 molecules of each substrate would be captured in a given liposome, barely sufficient to produce 10 or so double-stranded DNA product molecules (369 base pairs).

What would be required to reach the next step, to a true replicating system of encapsulated polynucleotides that can undergo cell-like growth? In concept, the answer is straightforward. Assuming PCR-type conditions, externally added NTP substrate must be supplied to the captured enzyme, and a mechanism for adding membrane components to the vesicles at the same time must also be in place. Furthermore, there should be a regulatory coupling between growth of the internal molecules and growth of the membrane, or one will outstrip the other. Last, protein enzymes cannot be used as catalysts, because they are not reproduced in the defined system. Instead, something like RNA should incorporate both the genetic information and the polymerase activity. In practice, there is still no way to deal simultaneously with all of these requirements. However, a few years ago it would have been inconceivable that we would soon reach a point at which intravesicular replication would be possible, so it seems likely that a laboratory version of primitive cellular life can be achieved in our scientific era.

References

Anders, E. 1989. Pre-biotic organic matter from comets and asteroids. *Nature* 342:255–257.

Bachman, P. A., Luisi, P. L., and Lang, J. 1992. Autocatalytic self-replicating micelles as models for prebiotic structures. *Nature* 357:57–59.

Chakrabarti, A., Breaker, R. R., Joyce, G. F., and Deamer, D. W. 1994. Production of RNA by a polymerase protein encapsulated within phospholipid vesicles. *J. Mol. Evol.* 39:555–559.

Chakrabarti, A., and Deamer, D. W. 1994. Permeation of membranes by the neutral form of amino acids and peptides: Relevance to the origin of peptide translocation. *J. Mol. Evol.* 39:1–5.

Chyba, C. F., and Sagan, C. 1992. Endogenous production, exogenous delivery and impact-shock synthesis of organic molecules: An inventory for the origin of life. *Nature* 355:125–13.

Chyba, C. F., Thomas, P. J., Brookshaw, L., and Sagan, C. 1990. Cometary delivery of organic molecules to the early Earth. *Science* 249:366–373.

Cronin, J. R., and Pizzarello, S. 1990. Aliphatic hydrocarbons of the Murchison meteorite. *Geochim. Geophys.* Acta 54:2859.

Cronin, J. R., Pizzarello, S., and Cruickshank, D. P. 1988. Organic matter in carbonaceous chondrites, planetary satellites, asteroids and comets. In *Meteorites and the Early Solar System*, ed. J. F. Kerridge and M. S. Matthews, pp. 819–857. Tucson: University of Arizona Press.

Deamer, D. W. 1985. Boundary structures are formed by organic components of the Murchison carbonaceous chondrite. *Nature* 317:792–794.

Deamer, D. W., and Barchfeld, G. L. 1982. Encapsulation of macromolecules by lipid vesicles under simulated prebiotic conditions. *J. Mol. Evol.* 18:203–206.

Deamer, D. W., Harang-Mahon, E., and Bosco, G. 1994. Self-assembly and function of primitive membrane structures. In *Early Life on Earth. Nobel Symposium No. 84.*, ed. S. Bengston, pp. 107–123. New York: Columbia University Press.

Deamer, D. W., and Pashley, R. M. 1989. Amphiphilic components of carbonaceous meteorites. *Origins Life Evol. Biosphere* 19:21–33.

Deamer, D. W., and Volkov, A. G. 1995. Proton permeation of lipid bilayers. In *Permeability and Stability of Lipid Bilayers*, ed. E. A. Disalvo and S. A. Simon, pp. 161–177. Boca Raton, Fla.: CRC Press.

Goldacre, R. J. 1958. Surface films; the collapse on compression, the sizes and shapes of cells, and the origin of life. In *Surface Phenomena in Biology and Chemistry*, ed. J. F. Danielli, K. G. A. Pankhurst, and A. C. Riddiford, pp. 12–27. New York: Pergamon Press.

Haldane, J. B. S. 1929. The origin of life. *The Rationalist Annual* 148:3–10.

Hargreaves, W. R., and Deamer, D. W. 1978. Liposomes from ionic, single-chain amphiphiles. *Biochemistry* 17:3759–3768.

Hargreaves, W. R., Mulvihill, S., and Deamer, D. W. 1977. Synthesis of phospholipids and membranes in prebiotic conditions. *Nature* 266:78–80.

Hayatsu, R., and Anders, E. 1981. Organic compounds in meteorites and their origins. *Top. Curr. Chem.* 99:1–37.

Lawless, J. G., and Yuen, G. U. 1979. Quantitation of monocarboxylic acids in the Murchison carbonaceous meteorite. *Nature* 282:396–398.

Mautner, M., Leonard, R., and Deamer, D. W. 1995. Meteoritic organics in planetary environments: Hydrothermal release and processing, surface activity and microbial utilization. *Planet. Space Sci.* 43:139–147.

Oberholzer, T., Albrizio, M., and Luisi, P. L. 1995. Polymerase chain reaction in liposomes. *Current Biology* 2:677–682.

Oparin, A. I., Orlovskii, A. F., Bukhlaeva, V. Ya., and Gladilin, K. L. 1976. Influence of the enzymatic synthesis of polyadenylic acid on a coacervate system. *Dokl. Akad. Nauk SSSR* 226:972–74.

Oró, J. 1961. Comets and the formation of biochemical compounds on the primitive Earth. *Nature* 190:389–390.

Oró, J., Sherwood, E., Eichberg, J., and Epps, D. 1978. Formation of phospholipids under primitive Earth conditions and the role of membranes in prebiological evolution. In *Light-Transducing Membranes: Structure, Function and Evolution*, ed. D. W. Deamer, pp. 1–21. New York: Academic Press.

Ourisson, G., and Nakatani, T. 1994. The terpenoid theory of the origin of cellular life: The evolution of terpenoids to cholesterol. *Chemistry and Biology* 1:11–23.

Parsegian, A. 1969. Energy of an ion crossing a low dielectric membrane: Solutions to four relevant electrostatic problems. *Nature* 221:844–846.

Paula, S., Volkov, A. G., Van Hoek, A. N., Haines, T. H., and D. W. Deamer. 1996. Permeation of protons, potassium ions and small polar molecules through phospholipid bilayers as a function of membrane thickness. *Biophys. J.* 70:339–348.

Roberts, M. 1992. Short-chain phospholipids: Useful aggregates in understanding phospholipase activity. In *Biomembrane Structure and Function: The State of the Art*, ed. B. P. Gaber and K. R. K. Eswaren, pp. 273–289. Schenectady, N.Y.: Adenine Press.

Shew, R., and Deamer, D. 1983. A novel method for encapsulating macromolecules in liposomes. *Biochim. Biophys. Acta.* 816:1–8.

Siskin, M., and Katritzky, A. R. 1991. Reactivity of organic compounds in hot water: geochemical and technological applications. *Science* 254:231–237.

Stribling, R., and Miller, S. L. 1987. Energy yields for hydrogen cyanide and formaldehyde syntheses: The HCN and amino acid concentrations in the primitive ocean. *Origins of Life* 17:261–273.

Walde, P., Goto, A., Monnard, P.-A., Wessicken, M., and Luisi, P. L. 1994. Oparin's reactions revisited: Enzymatic synthesis of poly(adenylic acid) in micelles and self-reproducing vesicles. *J. Am. Chem. Soc.* 116:7541–7547.

Walde, P., Wick, R., Fresta, M., Mangone, A., and Luisi, P. L. 1994. Autopoietic self-reproduction of fatty acid vesicles. *J. Am. Chem. Soc.* 116:11649–11654.

Wick, R., Walde, P., and Luisi, P. L. 1995. Light microscopic investigations of the autocatalytic self-reproduction of giant vesicles. *J. Am. Chem. Soc.* 117:1435–36.

9

Origin of life in an iron–sulfur world

GÜNTER WÄCHTERSHÄUSER

Introduction: the principle of common precursors

Higher organisms consist of a plurality of different cells. They can reproduce only by an orderly reproduction of all their individual component cells. Therefore, we hold that unicellular organisms are logically and phylogenetically prior to multicellular organisms.

Unicellular organisms are known to exist. They consist of a plurality of different components: a cell envelope, a genetic machinery, a machinery for chemiosmosis, a set of enzymes, and a host of metabolites. The organisms reproduce by reproducing all these components. Therefore, it is reasonable to hold that a subset of these components is logically and phylogenetically prior to unicellular organisms and that it is in fact the primordial form of life.

The cell envelope is an important component of cellular organisms. It is the physical basis for organismic individuation. Therefore, some have conjectured that a lipid vesicle (endowed with light-driven chemiosmosis) self-organized and self-reproduced in a primordial broth of lipids and later came to "invent" a genetic machinery and a metabolism (Morowitz 1992).

The genetic machinery is another important component. Nucleic acids are the physical basis for inheritable information. Therefore, others have conjectured that an RNA-like molecule (endowed with the capability of a replicase) self-organized and self-replicated in a broth of activated nucleotides and later came to "invent" a cell envelope and a metabolism ("RNA world") (James and Ellington 1995).

In contrast to these two proposals, which focus on complex machineries, a third proposal focuses on the underlying process of the cell, the metabolism that is more fundamental than any of the cell machineries (Wächtershäuser 1988a, 1990, 1992, 1994, 1997). In fact, the cell envelope and the genetic machinery are both derivatives of, and ancillary for, the metabolism, with the first having the main function of keeping all the constituents of the metabolism together, and the second having the main function of controlling all the metabolic reactions. Therefore, we may hold that the metabolism is logically

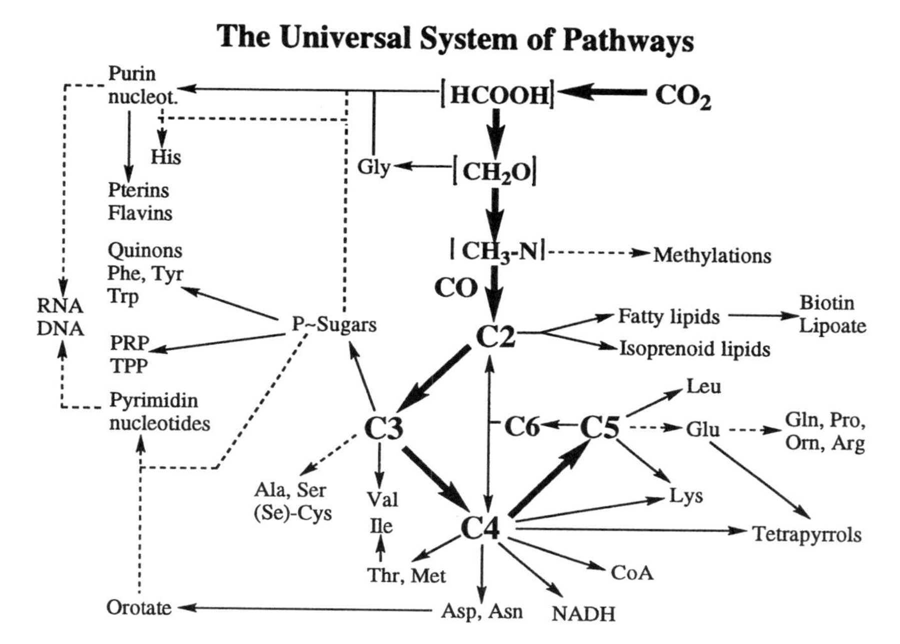

Figure 9.1. Idealized and unified pattern of the central metabolism. Full arrows show change of the carbon skeleton (bold arrows show central synthetic reactions). Dotted arrows show synthetic reactions not affecting the carbon skeleton.

and phylogenetically prior to cellular organization and genetic control; and we may conjecture that a rudimentary metabolism was the primordial life process and that the corresponding primordial organism ("metabolist") later came to "invent" cellular organization and genetic control.

The notions of a primordial vesicle or of a primordial RNA both assume a heterotrophic origin in a "prebiotic broth." Although the notion of an atmosphere-born broth is chemically supported by silent discharges in the gas phase that produce carboxylic acids from CH_4/H_2O (demonstrated) (Löb 1908), or glycin from $CO/NH_3/H_2O$ (demonstrated) (Löb 1913) or from $CO_2/NH_3/H_2O$ (conjectured) (Löb 1913), or glycin and higher amino acids and carboxylic acids from $CH_4/NH_3/H_2O$ (demonstrated) (Miller 1953), the notion of such primordial atmospheres is geochemically obscure (Walker 1977; Holland 1984). The metabolist proposal has the advantage that it is not dependent on the notion of a prebiotic broth. This makes room for the possibility that the earliest organisms were not heterotrophs but rather autotrophic metabolists.

A metabolism is a highly integrated, complex reaction network. An idealized reaction pattern, combining the salient central aspects of the various known metabolisms, is shown in Figure 9.1. Any rational methodology of

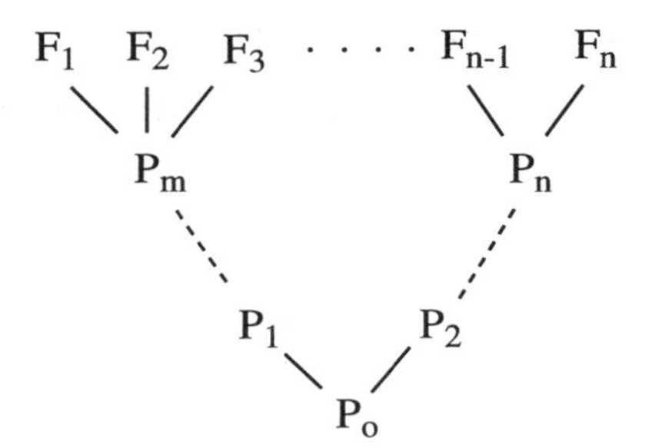

Figure 9.2. Schematic representation of the evolutionary tree of biological or biochemical functions. F_1 to F_n = extant functions; P_1 to P_n = precursor functions; P_0 = origin of functionality = origin of life.

locating within this pattern the remnants of the primordial metabolism must be based on Popper's principle of explanatory power (Popper 1963, 1972, 1983). By this principle, we have to guide the construction of a theory on early evolution so as to explain the greatest number of facts of extant organisms with the smallest number of evolutionary assumptions. This means that we have to search for "common precursor functions." A *common precursor function* is defined as a precursor function having several evolutionary successor functions. By a recursive application of this principle we may hope to arrive at the origin of life as the ultimate precursor function. This "methodology of retrodiction" (Wächtershäuser 1992) is illustrated schematically in Figure 9.2.

Surface metabolism and the fundamental polarity of life

We begin our inquiry with the problem of finding a functional precursor for cellular organization. The cell structure constitutes a flow-through reactor for the cellular metabolism, with a closed envelope for keeping the constituents of the metabolism together for interaction and with transport units for transporting food into the cell and waste products out of the cell. The only possibility for a functional precursor of this cellular organization seems to be a surface metabolist, whereby the constituents of the metabolism (metabolites) are bonded to a mineral surface, which constitutes a two-dimensional flow-through reactor (Wächtershäuser 1988a). Transport units are not required. The food material has free access and the waste products can leave unobstructed. The metabolites participate in the surface metabolism as long as they stay bonded to the mineral surface (Wächtershäuser 1988a).

It is a necessary condition of a surface metabolism that the surface bonding of the metabolites is strong enough (akin to chemisorption) to assure a sufficiently long residence time, but weak enough to allow a two-dimensional surface diffusion. This condition can be satisfied only by an ionic bonding, which is characterized by a dual selection rule: (a) The mineral surface must exhibit cationic (Lewis acid) binding centers; and (b) the constituents of the surface metabolism must be anionic (Lewis bases) (Wächtershäuser 1988a).

The conjectured ionic bonding in a surface metabolism explains why most constituents in the central metabolism are anionic, why the coenzymes are anionic, why the central sugar metabolism involves phosphorylated sugars, and why the nucleic acids are polyanionic (Wächtershäuser 1988a). And it assigns a precursor function to phosphate groups: the function of surface-bonding mediation (Wächtershäuser 1988a). Finally, it explains the polarity of lipid membranes, with anionic groups on the inside, by their origin in a surface metabolism (Wächtershäuser 1988a). In summary, the conjectured primordial surface metabolism may be considered the basis for a principle of polarity that permeates all extant organisms.

The functional precursor of enzyme surfaces

Nearly all metabolic reactions are catalyzed by enzymes. Enzymes must be accurately folded if they are to present properly shaped surfaces that provide a plurality of weak bondings to the metabolites to properly orient the metabolites for selective reaction. The only possible candidate for a functional precursor of enzyme surfaces seems to be a mineral surface. But mineral surfaces provide only singular bonds to metabolites and not the plurality of bonds of enzymes. As a consequence, these singular bonds cannot be weak, as in enzymes. They must instead be very strong, to give a definite orientation to the metabolites (Wächtershäuser 1988a). This leads us again to the principle of polarity, previously derived.

As a first example of surface-metabolic reaction, we consider the formation of polypeptides. The polarity principle means that the earliest polypeptides were polyanionic and made up exclusively of anionic amino acids (Wächtershäuser 1988a, 1990, 1992) (e.g., Asp, Glu, α-amino adipate, Cys, homo-Cys, phospho-Ser). On positively charged mineral surfaces, these will have the proper definite orientation for producing polyanionic polypeptides (Wächtershäuser 1988a). They become strongly surface-bonded in their nascent state and can undergo recursive chain extensions (Wächtershäuser 1988a). This prediction of the theory of surface metabolism has recently been corroborated experimentally by selecting Asp and Glu as anionic amino acids (Ferris et al. 1996).

As a second example, we turn to the surface-metabolic formation of nucleic acids. They are polyanionic, and nucleotides are at least dianionic. Therefore, nucleic acids satisfy the polarity principle, which speaks for their origin in a surface metabolism (Wächtershäuser 1988a). Polynucleotides will become strongly surface-bonded to cationic mineral surfaces in their nascent state and will be capable of undergoing recursive chain extensions (Wächtershäuser 1988a). This prediction of the theory of surface metabolism has recently been corroborated experimentally with a clay mineral that had been cationized on its surfaces by magnesium ions (Ferris et al. 1996).

As a third example, we turn to surface-metabolic conversions of phosphorylated sugars. The phosphorylated sugars are at least dianionic. Therefore, they satisfy the principle of polarity, which speaks for the origin of the sugar metabolism as a strictly phosphorylated sugar metabolism, whereby notably extant glycol aldehyde is the evolutionary successor of phosphoglycol aldehyde (Wächtershäuser 1988a). On cationic mineral surfaces, phosphorylated sugars have been assumed to adopt definite orientations lending specificity to their interconversions (Wächtershäuser 1988a). Phosphoglyceric acid, the biosynthetic precursor of phosphoglyceraldehyde, has been suggested to undergo surface-metabolic isomerization between the 2-phospho and 3-phospho isomers (Wächtershäuser 1988a). Phosphoglyceraldehyde and phosphoglycolaldehyde have been suggested to undergo surface-metabolic condensation to phosphorylated pentose (Wächtershäuser 1988a); this proposal has been recently corroborated experimentally (Pitsch et al. 1995).

A generalized mechanism of evolution

The RNA world theory assumes that the mechanism of evolution was always based on copying mistakes of nucleic acids. By definition, a theory that assumes a metabolist as the first organism must postulate a generalized mechanism of evolution that comprises the nucleic acid–implemented mechanism as a special case. Metaphorically speaking, such a theory must assume an early evolution of analog information, which later gave rise to the digital information of nucleic acids and to the digital-to-analog conversion in ribosomes (Wächtershäuser 1994).

Chemically speaking, the template effect of nucleic acids is a special case of chemical catalysis. DNA, therefore, is a catalyst that derives from the metabolism and exerts two kinds of catalytic feedback: altruistic feedback into the metabolism whence it derives, and egotistic feedback into its own pathway of derivation. (Viral DNA/RNA shows only an egotistic feedback.) From this vantage point we may generalize that evolution proceeds by virtue of by-products of metabolism – "vitalizers" – that exhibit a dual altruistic/egotistic feedback. Every new dual feedback constitutes an expansion loop to the metabolism, and the process of evolution can be characterized as a concatenation of such expansion loops (Wächtershäuser 1992).

If we trace the process of evolution backwards, we come to simpler and simpler metabolic networks. Ultimately we arrive at the simplest of all metabolic networks, which must be an autocatalytic cycle, whereby a food acceptor takes up food and produces a product that is catalytic for this very process. Two limiting cases can be distinguished:

a. A pathway of initiation is facile enough for continuously feeding a food acceptor into the autocatalytic cycle. In this case, the autocatalytic cycle does not have to reproduce (multiply) the food acceptor.
b. The pathway of initiation is extremely slow and has only a triggering function. In this case, the autocatalytic cycle must be a reproduction cycle that multiplies the food acceptor.

It may be assumed that the process of evolution begins in a homestead of life where case (a) holds. From this vantage point, a somewhat advanced form of life may spread into a chemical space where case (a) and case (b) hold and, finally, may spread into a chemical space where only case (b) holds. Alternatively, it may be assumed that the process of evolution begins in a homestead of life where only case (b) holds, and the process then spreads into a chemical space where cases (a) and (b) both hold (Wächtershäuser 1990, 1992).

The aforementioned generalized mechanism of evolution is a universal mechanism that holds independently of special chemical conditions like nucleic acids. It has the important consequence that, anywhere in the universe, any early stage of life will be based on an archaic pattern of metabolism that consists of

a. a central autocatalytic cycle (alpha cycle);
b. a pathway of initiation that feeds a constituent into the alpha cycle; and
c. biosynthetic pathways radiating from (a) and (b) and giving rise to ever more complex vitalizers.

The known extant patterns of metabolism are all highly evolved. But they still satisfy the archaic pattern to a high degree. They all have a complete (or interrupted and/or partially reversed) central autocatalytic cycle: the reductive citric acid cycle. They all have a complete (or interrupted and/or reversed) pathway of initiation: the reductive acetyl-CoA pathway. And all their biosynthetic pathways start from constituents of the reductive citric acid cycle and/or the reductive acetyl-CoA pathway (Wächtershäuser 1990, 1992, 1994; Huber and Wächtershäuser 1997).

Chemistry of the alpha cycle and its initiation pathway

The reductive acetyl-CoA pathway and the reductive citric acid cycle build up carbon skeletons by carbon fixation. It has therefore been concluded that the archaic metabolism is an autotrophic metabolism feeding on CO_2, CO (Wächtershäuser 1988a, 1990, 1992), and COS (Wächtershäuser 1992). This assumption has an enormous advantage. It allows us to do away with the geochemically obscure notion of a "prebiotic broth." CO_2, CO, and COS are found in magmatic exhalations of volcanos and hydrothermal vents and must always have been available at such sites.

Thioesters play a crucial role in the central metabolism. The reductive citric acid cycle is dependent on thioesters as activated intermediates, and the reductive acetyl-CoA pathway injects the thioester CH_3–CO–S–CoA into the reductive citric acid cycle. De Duve has postulated the formation of thioesters from mercaptans and carboxylic acids in a prebiotic broth at the very onset of a heterotrophic origin of life (de Duve 1991). By contrast, the theory of an autotrophic origin of life entails that thioesters must have functional precursors that can only be thioacids (R–CO–SH). This means that hydrogen sulfide is the functional precursor of all extant biochemical constituents with sulfhydryl groups, such as Cys and CoA (Wächtershäuser 1988a, 1990, 1992).

Several of the enzymes in the reductive citric acid cycle and in the reductive acetyl-CoA pathway have Fe/S-clusters as active components. These must have inorganic precursors, which can only be like the polynuclear clusters in amorphous iron sulfide (Wächtershäuser 1988b). Similarly, the Ni/Fe/S-clusters in acetyl-CoA synthase and carbon monoxide dehydrogenase must have a functional precursor, which can only be like a polynuclear cluster in nickel-iron sulfide. The suggestion of an involvement of catalytic nickel in the origin of life (Wächtershäuser 1988a, 1990, 1992) does not introduce a further postulate regarding the geochemical conditions, since natural iron sulfides always contain nickel. Similarly, inorganic Mo(W)-sulfur compounds may have been involved in the origin of life (Wächtershäuser 1992) as functional precursors of extant molybdo(tungsto) pterins.

The reductive acetyl-CoA pathway and the reductive citric acid cycle require a reducing agent. It has been suggested that the oxidative formation of pyrite from FeS/H_2S (or FeS alone) is the functional precursor of all extant biochemical reducing agents and that it is the primordial energy source of life that drives the primordial metabolists (Wächtershäuser 1988b):

$$FeS + H_2S \rightarrow FeS_2 + 2e^- + 2H^+.$$

Again, nothing is postulated for this energy source that is not available at volcanic or hydrothermal sites.

The extant reductive acetyl-CoA pathway comprises a multistep enzymatic reduction of CO_2 to a methyl-pterin that is subsequently converted by a

Ni/Fe/S-enzyme with coenzyme A and CO to acetyl-CoA. It has been postulated that in the archaic pathway, CH_3–SH is the functional precursor of methyl-pterin and that CH_3–SH is formed reductively with FeS/H_2S and subsequently converted to CH_3–CO–SH with CO under the sulfidic conditions of volcanic or hydrothermal sites (Wächtershäuser 1990, p. 203):

$$CO_2 \text{ (or CO)} + FeS/H_2S \rightarrow CH_3\text{–}SH + FeS_2.$$
$$CH_3\text{–}SH + CO \rightarrow CH_3\text{–}CO\text{–}SH \rightarrow CH_3\text{–}CO\text{–}OH.$$

Both predictions have been confirmed (Heinen and Lauwers 1996; Huber and Wächtershäuser 1997).

The extant reductive citric acid cycle as shown in Figure 9.3. comprises two types of carbon fixation. It has been suggested that both are driven by the overall reductive energy flow under the chemical potential of FeS/H_2S (Wächtershäuser 1990). The conversion of a thioester into 2-keto acids with CO_2 and $Na_2S_2O_4$ in the presence of Fe/S-clusters has been demonstrated (Nakajima, Yabushita and Tabushi 1975). It is remarkable that the aforementioned archaic reductive acetyl-CoA pathway produces a thioacid (CH_3–COSH), which has precisely the kind of activation that is required for the subsequent CO_2-fixation to pyruvate:

$$CH_3\text{–}COSH + CO_2 + FeS \rightarrow CH_3\text{–}CO\text{–}COOH + FeS_2.$$

This formulation makes no mechanistic presumption. It includes the possibility that CO_2 is first reductively converted to CO (or HCOOH), which subsequently enters the carbon skeleton.

In the reductive citric acid cycle, oxaloacetate is converted in five steps into succinoyl-CoA. Experiments have shown that in the presence of aqueous FeS/H_2S (Blöchl et al. 1992) a direct conversion of the 2-keto acid Ph–CH_2–CO–COOH to Ph–CH_2–CH_2–COOH takes place. It is assumed that the reaction proceeds through a succession of 2-mercapto intermediates: Ph–CH_2–C$(SH)_2$–COOH and Ph–CH_2–CHSH–COOH. This assumption is supported by the finding that HS–CH_2–COOH is converted to CH_3–COOH by FeS/H_2S. For the catalysis by H_2S, the following mechanism is suggested (Blöchl et al. 1992):

$$HS\text{–}CH_2\text{–}COOH + HS^- \rightarrow {}^-S\text{–}CH_2\text{–}CO\text{–}SH + H_2O,$$
$${}^-S\text{–}CH_2\text{–}CO\text{–}SH + FeS \rightarrow {}^-CH_2\text{–}CO\text{–}SH + FeS_2,$$
$${}^-CH_2\text{–}CO\text{–}SH + H_2O \rightarrow CH_3\text{–}COO^- + H_2S.$$

This proposal is supported by the formation of acetanilide if aniline is

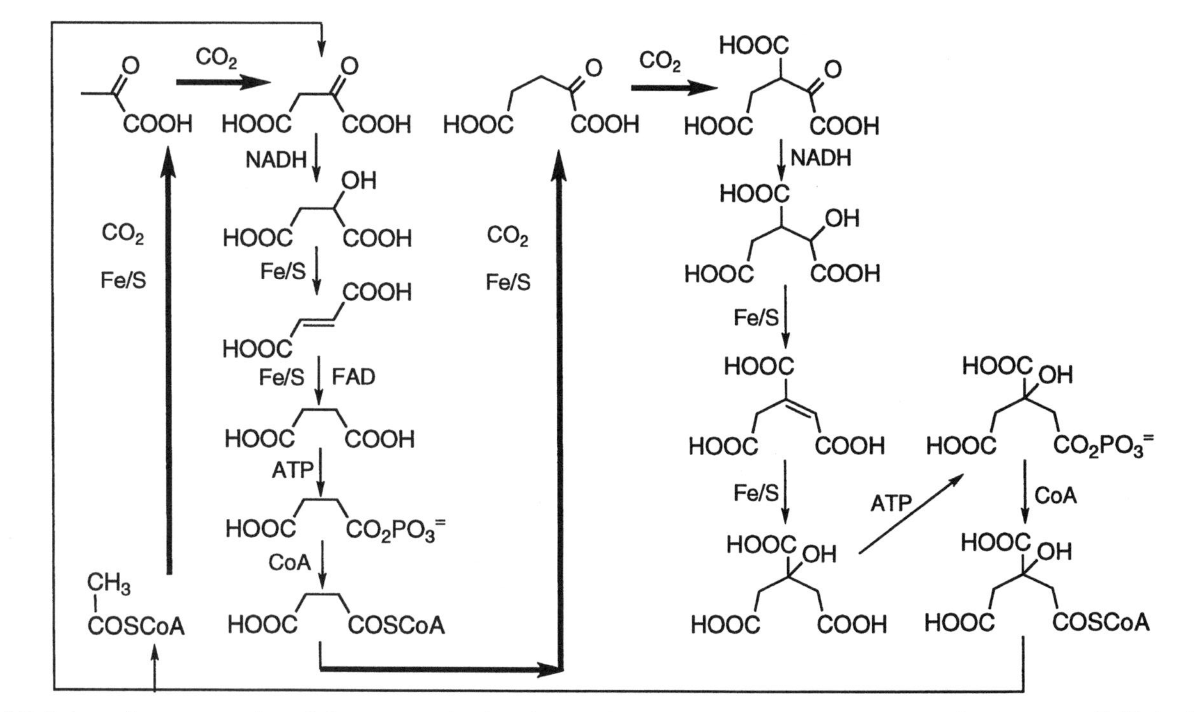

Figure 9.3. Schematic representation of the extant reductive citric acid cycle. Bold arrows signify carbon fixation steps. Fe/S signifies an enzyme with iron–sulfur cluster.

added as a trapping agent (Keller et al. 1994). It satisfies the principle of explanatory power, since

a. the five steps from oxaloacetate to succinoyl-CoA are replaced by a single one-pot reaction in the presence of FeS/H²S;
b. the system FeS/H_2S is the source for (1) redox energy, (2) carbanion energy (in water without a base), and (3) group activation energy (in water without ATP); and
c. thioacids are again the precursors of thioesters.

The earliest vitalizers

The lipids of Archaea are formed by an isoprenoid phosphate biosynthesis and the lipids of bacteria are formed by fatty acid biosynthesis. The principle of explanatory power is satisfied by the proposal that isoprenoid acids are the functional precursors of isoprenoid phosphates and that their synthesis is the sibling of the fatty acid biosynthesis (Wächtershäuser 1992).

The common first step of all extant fatty and isoprenoid lipid biosyntheses is dependent on acetyl-CoA (requiring deprotonation with an enzymatic base group):

$$2CH_3\text{–}COCoA \rightarrow CH_3\text{–}CO\text{–}CH_2\text{–}CO\text{–}CoA + HCoA.$$

The thioacid analogue of this product will form by the proposed FeS-induced anionic Claissen-type condensation (Wächtershäuser 1992, 1993) of $HS\text{–}CH_2\text{–}COOH$. It is suggested that further Claissen-type condensations of this compound give rise to both fatty acid and isoprenoid acid lipids. This is again an example of explanatory power. The lipids will accumulate on the mineral surface and change the reaction conditions for the surface metabolism, including lipid biosynthesis. This means that the lipids are vitalizers with a dual feedback. As source for $HS\text{–}CH_2\text{–}COOH$, the reduction of OHC–COOH is suggested, which may arise as a by-product of either the reductive citric acid cycle or the reductive acetyl-CoA pathway.

The amino acids are biosynthesized by reductive amination of 2-keto acids. It has been suggested, and confirmed experimentally, that 2-keto acids can be aminated by NH_3/FeS to amino acids like Gly, Ala, and Glu in high yields (Hafenbradl et al. 1995; Huber unpublished results). From the point of view of the preexisting metabolism, the formation of these amino acids results from detrimental reactions of decay. Therefore, it is a consequence of the present theory that the origin of life will be first situated in a chemical space wherein amino acids do not form, and that this primordial organism somewhat later colonizes another chemical space wherein certain amino acids that exhibit a dual feedback are formed. A similar pattern must hold for all other vitalizers. It is proposed that these earliest vitalizers will be ligands for the transition metal centers with the effect of increasing rate and/or selectivity. An assumption that any of these later vitalizers like amino acids must form from the start (from CO_2/NH_3/FeS, etc.) (and experimental failure to demonstrate it) is "soup inspired" and puts the cart before the horse (Keefe et al. 1995).

Chemically speaking, nucleic acids are catalysts that operate by base-pairing and that involve two 9-bonded purines and two pyrimidines. It has been suggested that the pyrimidines are functional successors to 3-bonded purines, which are isoelectronic to the pyrimidines (Wächtershäuser 1988c). A functional precursor of purine base-pairing is seen in acid–base catalysis by imidazole units. These arise in the early steps of purine biosynthesis. Histidine is a derivative of the purine pathway and it operates by imidazole catalysis. Therefore, it is suggested that both – the purines as well as histidine – are the functional successors of a primitive imidazole-type vitalizer and that the extant purine/histidine pathways are successors to an early pathway producing surface-bonded imidazoles.

Nucleic acids have a poly(ribose phosphate) backbone. Its formation requires advanced activated nucleotides, like ATP. Therefore, the nucleic acid backbone must have a functional precursor that does not require group activation. The extant biosynthesis of ribose phosphate involves a triose phosphate as a biosynthetic precursor. These observations, together with the principle of surface metabolism, led to the notion of a "tribonucleic acid" (Wächtershäuser 1988a), which comprises interconnected surface-bonded phospho-trioses. The phosphate groups are not highly energetic and have merely a surface-bonding function. The backbone is a poly(hemiacetal) backbone of the phospho-trioses, which is stabilized by surface bonding. The bases are attached to this backbone by glycosidic bonds, as in extant RNA/DNA.

The transition metal sulfides, notably pyrite, have surfaces with cationic surface charges. Therefore, they provide the surfaces required for the surface metabolism. This is another case of explanatory power. Thioacids (in contrast to thioesters) are excellent anionic surface bonders or ligands (Wächtershäuser 1988a), notably functioning as ligands for bonding to the free valences of transition metal sulfides and pyrites (or to Fe/Ni/S-clusters bonded to their surfaces). This is yet another case of explanatory power and of the principle of common functional precursors. It means that the thioacids and other similarly activated constituents are formed in the ligand sphere of an iron–sulfur mineral surface or a metal–sulfur cluster attached to said surface and that they subsequently react further without leaving the ligand sphere.

Conclusion

We have seen how a systematic application of Popper's principle of explanatory power requires the construction of common functional precursors and how a recursive application of this principle promises to lead us deeper and deeper into the earliest phases of biochemical evolution. The immediate benefit of such an enterprise is the establishment of an explanatory theory of biochemistry. It is hoped that we may be able to go far enough into the past to uncover the very origin of metabolic networks, that is, the origin of life, as a by-product of sorts.

But it should not be forgotten that any theory of evolution has the main and primary purpose of explaining biology, including biochemistry. A theory on the origin of life is scientific and capable of progress to the extent that it satisfies this requirement. A rational comparison of the relative scientific merits of competing theories on the origin of life is possible only by the comparison of their relative explanatory power. A theory on the origin of life that fails to have explanatory power is a myth. This is the challenge of the field.

Acknowledgment

I acknowledge support of experimental work (C. Huber) by the Deutsche Forschungsgemeinschaft.

References

Blöchl, E., Keller, M., Wächtershäuser, G., and Stetter, K. O. 1992. Reactions depending on iron sulfide and linking geochemistry with biochemistry. *Proc. Natl. Acad. Sci. USA* 89:8117–8120.

De Duve, C. 1991. *Blueprint for a Cell.* Burlington, N.C.: Neil Patterson Publ.

Ferris, J. P., Hill, A. R., Jr., Liu, R., and Orgel, L. E. 1996. Synthesis of long prebiotic oligomers on mineral surfaces. *Nature* 381:59–61.

Hafenbradl, D., Keller, M., Wächtershäuser, G., and Stetter, K. O. 1995. Primordial amino acids by reductive amination of α-oxo acids in conjunction with the oxidative formation of pyrite. *Tetrahedron Lett.* 36:5179–5182.

Heinen, W., and Lauwers, A. M. 1996. Organic sulfur compounds resulting from the interaction of iron sulfide, hydrogen sulfide and carbon dioxide in an anaerobic aqueous environment. *Origins Life Evol. Biosphere* 26:131–150.

Holland, H. D. 1984. *The Chemical Evolution of the Atmosphere and Oceans.* Princeton: Princeton University Press.

Huber, C., and Wächtershäuser, G. 1997. Activated acetic acid by carbon fixation on (Fe, Ni)S under primordial conditions. *Science* 276:245–247.

James, K. D., and Ellington, A. D. 1995. The search for missing links between self-replicating nucleic acids and the RNA world. *Origins Life Evol. Biosphere* 25:515–530.

Keefe, A. D., Miller, S. L., Mc Donald, G., and Bada, J. 1995. Investigation of the prebiotic synthesis of amino acids and RNA bases from CO_2 using FeS/H_2S as a reducing agent. *Proc. Natl. Acad. Sci. USA* 92:11904–11906.

Keller, M., Blöchl, E., Wächtershäuser, G., and Stetter, K.O. 1994. Formation of amide bonds without a condensation agent and implications for origin of life. *Nature* 368:836–838.

Löb, W. 1908. Die Einwirkung der stillen elektrischen Entladung auf feuchtes Methan. *Ber. Dtsch. Chem. Ges.* 41:87–90.

Löb, W. 1913. Über das Verhalten des Formamids unter der Wirkung der stillen Entladung. Ein Beitrag zur Stickstoff-Assimilation. *Ber. Dtsch. Chem. Ges.* 46:684–697.

Miller, S. L. 1953. A production of amino acids under possible primitive earth conditions. *Science* 117:528–529.

Morowitz, H. J. 1992. *Beginnings of Cellular Life.* New Haven: Yale University Press.

Nakajima, T., Yabushita, Y., and Tabushi, I. 1975. Amino acid synthesis through biogenic CO_2 fixation. *Nature* 256:60–61.

Pitsch, S., Eschenmoser, A., Gedulin, B., Hui, S., and Arrhenius, G. 1995. Mineral-induced formation of sugar phosphates. *Origins Life Evol. Biosphere* 25:297–334.

Popper, K. R. 1963. *Conjectures and Refutations.* London: Routledge & Kegan Paul.

Popper, K. R. 1972. *Objective Knowledge.* London: Oxford University Press.

Popper, K. R. 1983. *Realism and the Aim of Science*, ed. W. W. Bartley III. London: Hutchinson.

Wächtershäuser, G. 1988a. Before enzymes and templates: theory of surface metabolism. *Microbiol. Rev.* 52:452–484.

Wächtershäuser, G. 1988b. Pyrite formation, the first energy source for life: a hypothesis. *Syst. Appl. Microbiol.* 10:207–210.

Wächtershäuser, G. 1988c. An all-purine precursor of nucleic acids. *Proc. Natl. Acad. Sci. USA* 85:1134–1135.

Wächtershäuser, G. 1990. Evolution of the first metabolic cycles. *Proc. Natl. Acad. Sci. USA* 87:200–204.

Wächtershäuser, G. 1992. Groundworks for an evolutionary biochemistry: the iron–sulfur world. *Prog. Bioph. Mol. Biol.* 58:85–201.

Wächtershäuser, G. 1993. The cradle chemistry of life: on the origin of natural products in a pyrite-pulled chemo-autotrophic origin of life. *Pure and Applied Chemistry* 65(6):1343–1348.

Wächtershäuser, G. 1994. Life in a ligand sphere. *Proc. Natl. Acad. Sci. USA* 91:4283–4287.

Wächtershäuser, G. 1997. The origin of life and its methodological challenge. *J. Theor. Biol.* 187:483–494.

Walker, J. C. G. 1977. *Evolution of the Atmosphere.* New York: Macmillan.

10

Clues from present-day biology: the thioester world

CHRISTIAN DE DUVE
Institute of Cellular Pathology
Brussels, Belgium

Organic chemistry no longer deserves the name it was given by the founders of chemistry. There is now abundant evidence that many so-called organic molecules can arise without the help of living organisms (including chemists) and that they do so on a very large scale throughout the cosmos (see Chapter 5). It seems likely that such abiotically formed compounds provided the first building blocks of life. Little is known, however, of the manner in which these compounds interacted further to generate increasingly complex molecules and molecular assemblages, up to the first living cells. To date, in spite of much experimental and theoretical work, a striking discontinuity still separates the most successful attempts at reproducing biogenic processes in the laboratory from the manner in which these processes take place in living organisms. Yet, an uninterrupted sequence of events must link abiotic chemistry historically to biochemistry.

Many suggestions have been made concerning the nature of this link. The most radical suggestion, inspired by the genetic-takeover concept of Cairns-Smith (1982), postulates that the early chemistry was totally unrelated to biochemistry and served only as a temporary framework, later to be dismantled without leaving traces, for the development of the new chemistry. At the other end of the spectrum, the proposal made in this chapter rests on the premise that abiotic chemistry merged into present-day biochemistry already at a very early stage. This contention is supported by the perceived need for *congruence* between the two chemistries (de Duve 1991, 1993, 1995).

The case for congruence

It is widely believed, on the strength of cogent arguments, that RNA preceded protein (and probably also DNA) in the development of life (see Chapters 11, 12, 13). I shall adopt this "RNA world" hypothesis, but with two provisos. First, I reserve the term *protein* to polypeptides made by an RNA-based

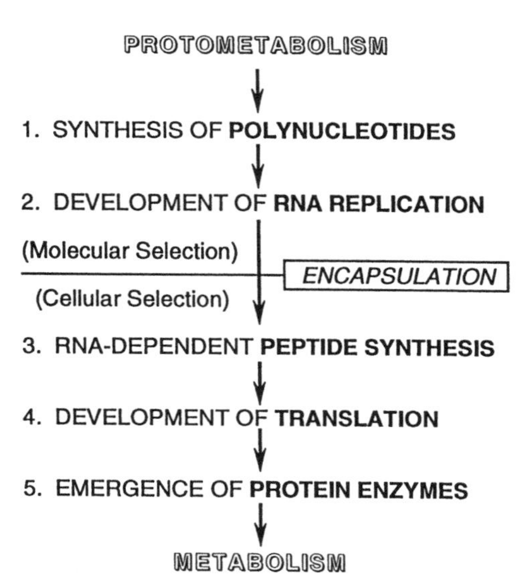

Figure 10.1. Five stages in the transition from protometabolism to metabolism. For details, see text.

machinery from the 19 L–α-amino acids plus glycine found in proteins today. I do not exclude – in fact, I consider likely – the possibility that RNA was preceded by differently made peptidelike substances of more heterogeneous composition (multimers, see later). Next, I leave out Gilbert's (1986) qualification that "RNA molecules and cofactors (were) a sufficient set of enzymes to carry out all the chemical reactions necessary for the first cellular structures." What is known of present-day ribozymes does not support such a wide spectrum of catalytic activities. In addition, RNA molecules could obviously not have played a role in the chemical reactions that led to their own genesis. To suppose that ribozymes later became involved as catalysts in these reactions is a gratuitous – and unnecessary – assumption without factual foundation. The simpler hypothesis is that the chemistry that did the job in the first place continued to do so until protein enzymes took over.

For the purpose of this discussion, I shall use the term *protometabolism* to designate the set of chemical reactions that generated the RNA world and sustained it during all the time it took for the RNA-protein world to emerge and to produce the protein enzymes that inaugurated *metabolism*. The question, then, is, how did protometabolism evolve into metabolism? Whatever the nature of protometabolism, this transition must almost obligatorily have gone through the five successive stages shown schematically in Figure 10.1.

The first stage is obviously the chemical synthesis of polynucleotides. Although all attempts at reproducing this stage in the laboratory under plausibly prebiotic conditions have so far failed, it must clearly have depended on sturdy processes, which went on supporting the RNA world throughout its evolution. The sometimes invoked alternative possibility that the first RNA molecules somehow appeared by a chance accident and were then perpetuated thanks to their capacity for self-replication is not chemically realistic.

Once the synthesis of polynucleotides was initiated as a chemical process, the next logical stage was the development of their replication, presumably by the inclusion of a directing template within the synthetic machinery. It seems likely, in this connection, that the first polynucleotides were more heterogeneous than present-day RNA molecules. Sugars other than D-ribose may have been involved, linkages other than 3'–5' phosphodiester bonds may have joined mononucleotides together, and, especially, bases other than the canonical four may have participated in the formation of the mononucleotides. The early chemistry that gave rise to adenine, guanine, cytosine, and uracil can hardly have been so selective as to produce only those four molecules. It presumably generated other purines and pyrimidines, as well as a number of related heterocyclic compounds, some of which, like flavins, exist in nucleotide-like combinations today. If such was the case, RNA may have emerged from the primitive polynucleotide mixture by a process of *molecular selection* dependent on the ability of molecules capable of base-pairing and possessing the appropriate topology to act as templates and thus be amplified. Further selection would then have proceeded along the lines first reproduced in the laboratory by the late Sol Spiegelman (1967) and since developed to a fine art by a number of workers.

The next stage in our reconstructed transition from protometabolism to metabolism is the development of RNA-dependent peptide synthesis. How this key process arose is still highly mysterious, but a basic sequence of events may be inferred from present knowledge. The initiating steps must almost necessarily have consisted in interactions between certain RNA molecules, which became today's transfer RNAs, and amino acids, leading to the first aminoacyl–RNA complexes. Presumably, these complexes reacted with other RNA molecules, possible precursors of ribosomal and messenger RNAs, resulting in the linking of the RNA-bound amino acids by peptide bonds.

Whatever the details, the assumption that RNA molecules interacted with amino acid molecules (which may have been activated) seems almost mandatory if we wish to account for the involvement of RNA in protein synthesis. These interactions could possibly have been mediated by some catalyst acting in the manner of today's aminoacyl–tRNA synthases. Or, more likely, they took place directly between the two kinds of molecules. This assumption, in turn, implies as an important corollary the occurrence of a mutual selection of the reacting partners on the basis of chemical affinities. This may be how certain specific amino acids first were singled out for protein synthesis among those that were available. Such a selection phenomenon may have a bearing on the chirality problem. It is, indeed, noteworthy that the exclusion against D-amino acids concerns only protein synthesis. There are plenty of D-amino acids in nature today, especially as components of bacterial compounds; and specific enzymes exist for their metabolism, even in mammalian cells. The proposed explanation does not, however, entirely solve the chirality problem, because the RNA molecules involved were themselves chiral. Would RNA molecules made from L-ribose have selected D-amino acids? If so, did something favor D-ribose over the L enantiomer as a building block of RNA? Or did pure chance decide between the D-ribose/L-amino acid combination and its enantiomeric counterpart? Molecular modeling could possibly help clarify this issue.

Another question raised by the development of RNA-dependent protein synthesis concerns the nature of the selective forces that favored this key event. Association with amino acids may conceivably have given the RNA molecules involved a selective edge by rendering them more stable, or more readily replicatable, or both. It is, however, more difficult to explain the selection of the other RNA components of the machinery on the basis of molecular qualities. In any case, a stage must eventually have been reached where the protein product provided the advantage. Since protein molecules are not directly replicatable and presumably never were (see, however, the recent work of Lee et al. 1996, referred to later), the advantage could no longer have been molecular. It must have been indirect, a condition that requires the system to be partitioned into a large number of units, or *protocells,* capable of multiplying and of competing with one another for their ability to survive and reproduce under prevailing conditions. The development of RNA-dependent protein synthesis thus sets the latest limit for the encapsulation of life. Cellularization could, of course, have taken place any time earlier, including at the very beginning of the life-generating process, as is postulated in a number of theories. The need for cellularization implies that protometabolism must have included a pathway for the formation of the necessary boundaries, whatever their chemical nature.

At first, the protein-synthesizing machinery probably used only a small number of amino acids, which it assembled more or less at random. The resulting peptides nevertheless must have been of some benefit to the protocells concerned, to account for the evolutionary selection of the machinery. Soon, however, information must have entered the system, leading to the progressive development of translation and the genetic code, associated perhaps with the coevolutionary recruitment of additional amino acids. Biochemistry suggests a possible explanation for this intriguing occurrence. Perhaps the precursors of messenger RNAs started by playing a conformational role in the protein-synthesizing machinery – a role messenger RNAs still play today in the joining of the two ribosomal subunits. If, in addition, a phenomenon analogous to codon–anticodon recognition were involved in the interaction of the pre–messenger RNAs with aminoacyl-bearing RNAs, the chemical seeds of coding would have been present. The advantages linked to increasing fidelity in the making of certain useful peptides could then have driven the evolutionary development of translation. Several models of this sort have been proposed (see de Duve 1991, 1995).

With the emergence of a population of protocells possessing an RNA genome and appropriate machineries for replicating the genes and expressing them into the corresponding polypeptides with a reasonable degree of accuracy, the last stage, leading to the formation of enzymes, was finally reached. Presumably, this development took place in stepwise fashion, along familiar Darwinian lines. A mutation in a given protocell was expressed into a modified polypeptide that turned out to display a new or improved catalytic activity from which the protocell concerned derived a selective advantage. As a result, the mutant protocell grew and multiplied faster than the others, eventually replacing the original population with its own progeny. The same sequence of events must have taken place a number of times in succession, until a complete set of genetically encoded enzymes had arisen.

Here is where congruence comes in as a necessary condition. In order to be useful to the mutant protocell concerned, each new enzyme must have found one or more substrates in the existing chemistry. For what is the good of a catalyst, however efficient, if it is not given the possibility of exerting its activity? The existing chemistry must also have provided an outlet for the product or products of the reaction catalyzed by the new enzyme. Otherwise, the mutation would more likely have been harmful than useful. Thus, by acting as a screen for the selection of enzymes, protometabolic intermediates to some extent set the pathways metabolism was to follow.

This conclusion is at variance with the often-voiced opinion that the two chemistries must have been very different. The divergence, however, is only a matter of degree. It is not disputed that life probably derived its first building blocks from the kind of reactions that fill the cosmos with a variety of carbon compounds and that are being successfully reproduced in many laboratories. Neither is it doubted by anyone that abiotic chemistry must have, at some stage, spawned reactions that are part of present-day biochemistry. The only point at issue is one of timing. According to the congruence argument, the transition occurred earlier than many believe, which implies that biochemistry may contain more clues to the origin of life than is often assumed.

The anatomy of protometabolism

From the analysis offered in the preceding section, it is clear that protometabolism must have depended on a rich, coherent, and dynamically stable network of reactions capable, at the very least, of supporting the continuing synthesis of polynucleotides, of polypeptides, and of whatever materials made up the boundaries of the primitive protocells. Even reduced to their starkest requirements, such elaborate constructions must have depended on many supporting reactions, attended, most likely, by a number of side reactions and ancillary processes.

How long this network functioned is difficult to estimate. Chemically, the reactions involved could have been quite fast. From the evolutionary point of view, however, the numerous selective episodes involved in the successive emergence of RNA replication, RNA-dependent peptide synthesis, translation and the genetic code, and, finally, protein enzymes, call for more than a fleeting moment transiently created by some chancy combination of events. It seems reasonable to think in terms at least of years, more likely of centuries or more. We are obviously dealing with a set of robust, *deterministic* processes causally linked to the prevailing physical–chemical conditions.

Such a network of reactions could almost certainly not have operated without the help of catalysts. It no doubt required a source of utilizable energy, and probably also involved the participation of electron-transfer reactions. Finally, it must have provided the chemical basis for encapsulation into protocells. I shall briefly discuss these features in the light of the congruence condition, with special reference to a possible role of thioesters in protometabolism (de Duve 1988, 1991, 1995, in press). As an introduction to this discussion, the role played by thioesters in present-day metabolism will first briefly be surveyed. Any biochemistry textbook can be consulted for additional details.

The biological role of thioesters

Thioesters are formed by the dehydrating condensation of carboxylic acids with thiols:

$$R{-}COOH + R'{-}SH \Longleftrightarrow R'{-}S{-}CO{-}R + H_2O \qquad (1)$$

Under physiological conditions, the equilibrium of this reaction is shifted very far to the left. The physiological free energy of hydrolysis of the thioester bond is of the order of -12 kcal/mole (about -50 kJ/mole), which makes this bond energetically equivalent to the pyrophosphate bonds of ATP.

In living organisms, thioesters have the uniquely important property that they can arise by two distinct mechanisms, of which one depends on group transfer and the other on electron transfer. The first one is supported, via two consecutive group-transfer reactions, by the hydrolysis of one or the other of the pyrophosphate bonds of ATP, as shown schematically by the following equations (in which P stands for a phosphoryl group, P_i for inorganic phosphate, and PP_i for inorganic pyrophosphate):

$$\begin{aligned} &\text{R–COO}^- + \text{ATP} \Longleftrightarrow \text{R–CO–P} + \text{ADP} \\ &\text{R–CO–P} + \text{R'–SH} \Longleftrightarrow \text{R'–S–CO–R} + \text{P}_i \end{aligned} \qquad (2)$$

$$\begin{aligned} &\text{R–COO}^- + \text{ATP} \Longleftrightarrow \text{R–CO–AMP} + \text{PP}_i \\ &\text{R–CO–AMP} + \text{R'–SH} \Longleftrightarrow \text{R'–S–CO–R} + \text{AMP} \end{aligned} \qquad (3)$$

ATP is split into ADP and inorganic phosphate in reaction (2), into AMP and inorganic pyrophosphate in reaction (3). The key intermediate is the acyl-phosphate in reaction (2), the acyl-adenylate in reaction (3).

The second mechanism for the formation of thioesters involves the oxidation of aldehydes or α-ketoacids in the presence of a thiol, most often with NAD^+ as electron acceptor:

$$\text{R–CHO} + \text{R'–SH} + \text{NAD}^+ \Longleftrightarrow \text{R'–S–CO–R} + \text{NADH} + \text{H}^+ \qquad (4)$$

$$\text{R–CO–COO}^- + \text{R'–SH} + \text{NAD}^+ \Longleftrightarrow \text{R'–S–CO–R} + \text{NADH} + \text{CO}_2 \qquad (5)$$

Thanks to the two kinds of mechanisms, thioesters can serve as central intermediates in the all-important interconversion of group-linked and electron-linked energy. With reactions (2) or (3) running from left to right, and reactions (4) or (5) running in the opposite direction, ATP can support the endergonic reduction by NADH (or NADPH) of carboxylic acids to aldehydes or, in the presence of CO_2, their carboxylating reduction to α-ketoacids. Running in the reverse direction, the coupled reactions can harness the assembly of ATP to exergonic electron transfers. The key processes called substrate-level phosphorylations depend on this sort of mechanism.

A typical example of such a coupled process involving an aldehyde (reactions [4] + [2]) is the central reaction for energy retrieval in glycolysis, namely the reversible, NAD-dependent oxidation of glyceraldehyde-3-phosphate to 3-phosphoglycerate, with a cysteinyl residue of the enzyme glyceraldehyde-phosphate dehydrogenase acting as the acyl-carrying thiol. The coupled

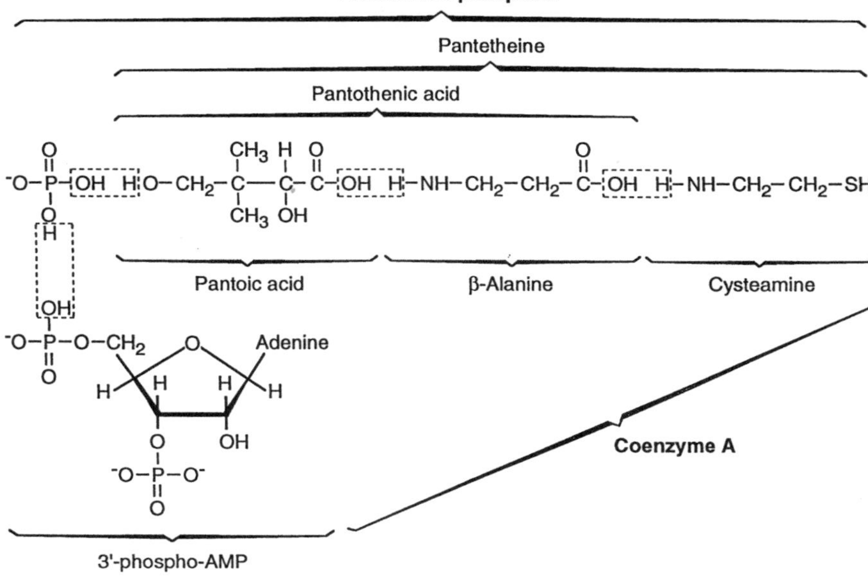

Figure 10.2. The structure of coenzyme A.

process serves catabolically in the assembly of ATP at the expense of the oxidative conversion of half a glucose molecule to pyruvate (which, in anaerobic fermentations, serves as the final electron acceptor, to be reduced either to lactate or to ethanol and CO_2). Anabolically, the coupled process allows ATP to support the reductive step of the Calvin cycle in photosynthesis and of gluconeogenesis in animal cells.

Important examples of reactions involving α-ketoacids (reactions [5] + [2]) are the phosphorylation processes coupled to the oxidative decarboxylation of pyruvate or of α-ketoglutarate in the Krebs cycle. In these cases, there is involvement of lipoate, a dithioacid that can exist in reduced dithiol form and in oxidized disulfide form. The final thiol acceptor of the acyl group is coenzyme A, a complex molecule formed from the joining by a pyrophosphate bond of 3'-phospho-AMP with 4-phospho-pantetheine. The latter molecule is itself made of pantoic acid (a dimethylated, dihydroxy derivative of butyric acid), β-alanine, and cysteamine (Figure 10.2).

In addition to their fundamental role in the interconversion of the two main forms of biological energy, thioesters also serve important functions in biosyntheses. With either coenzyme A or enzyme-bound phospho-pantetheine as the acyl carrier, thioesters act as universal donors of acyl groups in the assembly of lipids and other esters, and of aminoacyl groups in the formation of a number of bacterial and fungal peptides; they also provide activated building blocks for the synthesis of citrate in the Krebs cycle, citrate and malate in the glyoxylate cycle, fatty acids, porphyrins, and terpene derivatives, including sterols.

Thioesters and the origin of life

Considering the unique, all-pervasive role of thioesters in present-day metabolism and the likely requirement for early congruence between protometabolism and metabolism, it seems highly plausible that thioesters carried out key functions in protometabolism. This hypothesis is all the more attractive because the necessary ingredients were probably present in adequate quantities. Organic acids are known to be among the main products of abiotic chemistry, and thiols may be expected to have been formed also if, as seems probable, the prebiotic setting was rich in hydrogen sulfide (see Kaschke, Russell and Cole 1994; Keller et al. 1995; Heinen and Lauwers 1996). The main problem concerns the joining of the two, which, as mentioned earlier, is opposed by a substantial thermodynamic barrier. Leaving aside this problem for the moment, let us examine briefly the functions thioesters could have fulfilled in protometabolism.

CATALYSIS. Clays and other minerals, metal ions, organic coenzymes, ribozymes, proteins or proteinlike substances, even amino acids have all been considered as potential prebiotic catalysts. The possibility that the early chemistry may have generated its own catalysts in autocatalytic fashion has also received attention. Space limitations do not permit an extensive discussion of this active field of research. I shall confine myself to an examination of the possible involvement of peptidelike compounds formed from the thioesters of amino acids and other difunctional acids.

Discovered by Lipmann and co-workers in the course of an investigation of the biosynthesis of gramicidin S (Gevers, Kleinkauf and Lipmann 1969), the assembly of peptides from aminoacyl thioesters has now been identified as an important mechanism, responsible for the formation of more than two hundred compounds in bacteria and also in some eukaryotic cells (Kleinkauf and von Döhren 1990, 1996). The building blocks include D- as well as L-amino acids, some not used in protein synthesis, and sometimes also hydroxy acids. The reactions involved in these syntheses are catalyzed by complex multienzyme systems using protein-bound 4-phospho-pantetheine (Figure 10.2) as a general carrier. The same cofactor is used by fatty acid–synthesizing systems, with which the peptide-synthesizing systems have a number of features in common, except that they carry in their specificities informational blueprints of much greater complexity.

The basic chemistry underlying peptide synthesis is, however, very simple. As shown by the late Theodor Wieland more than 15 years before the enzyme-catalyzed reactions were discovered, peptides assemble spontaneously from aminoacyl thioesters in slightly alkaline aqueous solution, without the help of any catalyst (Wieland 1988). Recently, a thioester-dependent mechanism for the ligation of unprotected peptides has been described by Dawson et al. (1994). It is thus not implausible that, if thioesters of amino acids and other difunctional acids had been present in the prebiotic setting, they would have reacted to form *multimers*, which is the name I have given to the resulting heterogeneous assemblages to distinguish them from true peptides (de Duve 1988, 1991). Also suggestive with respect to this possibility, and possibly relevant, are recent experiments by the Miller group showing that the components of pantetheine, the key cofactor of thioester-dependent syntheses, can form spontaneously under plausibly prebiotic conditions and assemble correctly merely upon concentration by evaporation (Keefe, Newton and Miller 1995).

In Figure 10.3 are illustrated three steps in the assembly of a hypothetical multimer from thioesters. The composition of such molecules would have been ruled exclusively by physical–chemical factors, as no information would have guided their assembly. (This statement may possibly have to be revised some day, in view of recent results by Lee et al. [1996], who have observed the template-directed ligation of peptides, using the thioester-dependent mechanism devised by Dawson et al. [1994].) As long as conditions remained the same, however, a stable steady-state mixture of multimers would have been maintained, determined, on one hand, by the concentrations of the various thioesters and by the steric and other kinetic factors that directed their interactions and, on the other hand, by the relative stabilities of the resulting multimers. Would such a mixture have included the kind of enzymelike catalysts needed to support a protometabolism congruent with metabolism?

There is no direct experimental answer to this question. In theory, the possibility is at least plausible, if not likely (de Duve 1991). There are plenty of examples of short peptides exhibiting catalytic activity (for literature, see Barbier and Brack 1992; Brack 1993b). Unfortunately, the activities tested for are mostly hydrolytic, which would not be very helpful for protometabolism. But there is no reason to assume that other reactions, such as group transfers or, in the presence of appropriate redox substrates, electron transfers, would not be likewise facilitated by some short peptides. Indeed, if the first RNA genes were no more than 70–100 nucleotides long, as demanded by the expected rate of error of their replication (Eigen and Schuster 1977), then the

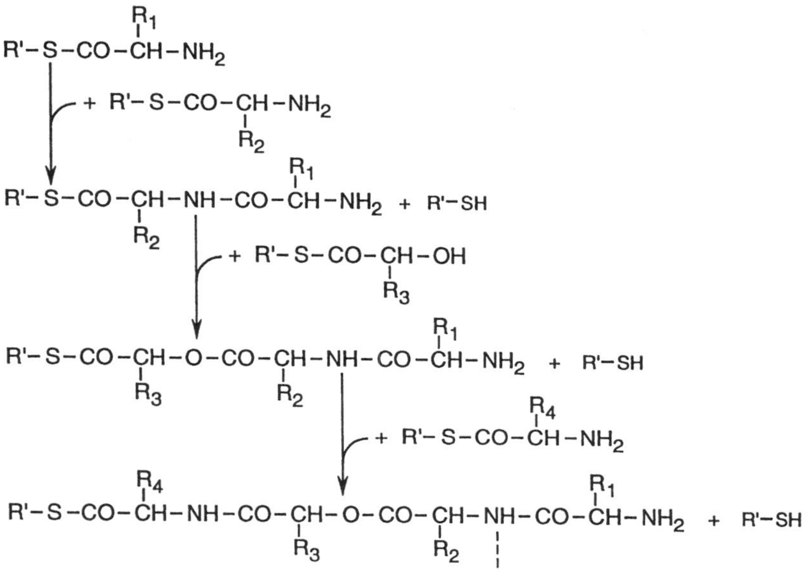

$$R'{-}S{-}CO{-}CH(R_1){-}NH_2$$

$$\downarrow + R'{-}S{-}CO{-}CH(R_2){-}NH_2$$

$$R'{-}S{-}CO{-}CH(R_2){-}NH{-}CO{-}CH(R_1){-}NH_2 + R'{-}SH$$

$$\downarrow + R'{-}S{-}CO{-}CH(R_3){-}OH$$

$$R'{-}S{-}CO{-}CH(R_3){-}O{-}CO{-}CH(R_2){-}NH{-}CO{-}CH(R_1){-}NH_2 + R'{-}SH$$

$$\downarrow + R'{-}S{-}CO{-}CH(R_4){-}NH_2$$

$$R'{-}S{-}CO{-}CH(R_4){-}NH{-}CO{-}CH(R_3){-}O{-}CO{-}CH(R_2){-}NH{-}CO{-}CH(R_1){-}NH_2 + R'{-}SH$$

Figure 10.3. Three steps in the synthesis of a hypothetical multimer from thioesters. The scheme shows the successive addition to an N-terminal L-amino acid (R_1) of a D-amino acid (R_2), a D-hydroxy acid (R_3), and another L-amino acid (R_4).

first RNA-encoded enzymes that presumably supported incipient metabolism could not have been more than 20–30 residues long. Some, as suggested by the reconstructed phylogeny of the *Clostridium pasteurianum* ferredoxin (Eck and Dayhoff 1966), could have been as short as 4 residues long. It should be noted further that, of all possible prebiotic catalysts, the postulated multimers are chemically closest to proteins and, therefore, most likely to mimic catalytic activities now carried out by protein enzymes, an important point in relation to congruence.

Granting that certain short peptides can exhibit the kind of activities needed for protometabolism, it remains to be seen whether such peptides could have been present in the postulated multimer mixture. This is indeed likely, to the extent that increased stability would be associated with structural properties (such as greater length, a cyclic or compact conformation, perhaps even homochirality) likely also to favor the creation of catalytic active centers. Work by Brack and colleagues (reviewed by Barbier and Brack 1992; Brack 1993b) has demonstrated such a correlation experimentally.

It appears from these considerations that peptides and other multimers formed from thioesters – or otherwise formed (see Brack 1993a) – could have provided significant enzymelike activities for the early development of protometabolism. This possibility does not preclude the additional participation of metal ions (which would have benefited from the support provided by multimers), mineral catalysts, autocatalytic cycles, and, at later stages, organic coenzymes and ribozymes.

ENERGY. That protometabolism could have derived energy from thioesters is obvious. According to what is known of their biological role, thioesters could have fueled a large number of synthetic mechanisms dependent on group transfer, including the assembly of multimers (Figure 10.3), the assembly of lipid esters (possibly important in the formation of membranes; see later section, "Encapsulation"), and many others.

Thioesters could also, in the presence of inorganic phosphate, have supported the formation of inorganic pyrophosphate, a compound that is generally believed to have preceded ATP as purveyor of energy in early protometabolism (Baltscheffsky and Baltscheffsky 1992; Baltscheffsky 1996). The reaction involved would have been similar to reaction (2) running in reverse, with ADP replaced by inorganic phosphate:

$$\begin{aligned} &R'\text{–}S\text{–}CO\text{–}R + P_i \Longleftrightarrow R\text{–}CO\text{–}P + R'\text{–}SH \\ &R\text{–}CO\text{–}P + P_i \Longleftrightarrow R\text{–}COO^- + PP_i \end{aligned} \tag{6}$$

It has been found by Weber (1981, 1982) that the two steps of this process can take place readily in the absence of catalysts. This finding is all the more important because the prebiotic availability of mineral pyrophosphate (or polyphosphates) is being seriously questioned (see de Duve in press).

Interestingly, one could even account for the first appearance of ATP, assuming AMP to be present, by a thioester-dependent mechanism following reaction (3) in reverse:

$$\begin{aligned} &R'\text{–}S\text{–}CO\text{–}R + AMP \Longleftrightarrow R\text{–}CO\text{–}AMP + R'\text{–}SH \\ &R\text{–}CO\text{–}AMP + PP_i \Longleftrightarrow R\text{–}COO^- + ATP \end{aligned} \tag{7}$$

This intriguing possibility would make inorganic pyrophosphate the direct precursor of the terminal pyrophosphate group of ATP. In living organisms, reaction (7) no longer serves in the assembly of ATP. It runs normally in the reverse direction as the main mechanism of activation of organic acids, with coenzyme A as the thiol acceptor. In the activation of amino acids for protein synthesis, only the ATP-dependent step is used, and the resulting aminoacyladenylate then donates its aminoacyl group to a transfer RNA molecule (refer to Figure 10.4).

OXIDATION–REDUCTIONS. Thioesters could also have played an important role in prebiotic electron transfers. In particular, thioesterification of carboxylic acids could, as it does in living organisms today, have helped overcome the energy barrier that opposes the reduction of carboxyl groups to the carbonyl groups of aldehydes or α-ketoacids. Such processes would have taken place by the reversal of reactions (4) or (5):

$$\text{R'–S–CO–R} + 2\text{ H} \Longleftrightarrow \text{R–CHO} + \text{R'–SH} \tag{8}$$

$$\text{R'–S–CO–R} + CO_2 + 2\text{H} \Longleftrightarrow \text{R–CO–COOH} + \text{R'–SH} \tag{9}$$

These reactions could also have taken place in the opposite direction (oxidation of aldehydes or α-ketoacids), thanks to the ability of thiols to form addition complexes (hemithioacetals) with carbonyl groups:

$$\begin{aligned} &\text{R–CHO} + \text{R'–SH} \Longleftrightarrow \text{R'–S–CHOH–R} \\ &\text{R'–S–CHOH–R} \Longleftrightarrow \text{R'–S–CO–R} + 2\text{H} \end{aligned} \tag{10}$$

$$\begin{aligned} &\text{R–CO–COOH} + \text{R'–SH} \Longleftrightarrow \text{R'–S–COH(COOH)–R} \\ &\text{R'–S–COH(COOH)–R} \Longleftrightarrow \text{R'–S–CO–R} + CO_2 + 2\text{H} \end{aligned} \tag{11}$$

Interestingly, Weber (1984) has found that reaction (10) can occur spontaneously, without a catalyst and even without an added electron acceptor, with glyceraldehyde molecules serving both as the substrate for oxidation and as the acceptor for the electrons released, and N-acetylcysteine acting as the acyl-accepting thiol.

In general, the direction adopted by such processes would have depended on the availability of appropriate electron donors or acceptors, a question that falls outside the scope of this chapter. An intriguing possibility, in the postulated sulfide-rich context of the reactions, is that the reactions were catalyzed by iron–sulfur complexes, perhaps in association with some cysteine-containing multimer. There are, indeed, reasons to suspect that iron–sulfur proteins may be the most ancient electron-transfer catalysts operating in metabolism today (Cammack 1983).

To be noted is the key connecting link provided by thioesters between electron transfers (reactions [8], [9], [10], and [11]) and phosphorylations (reactions [6] and [7]). This link forms the basis of substrate-level phosphorylations, which are in all likelihood the most ancient coupled processes of this type. Carrier-level phosphorylations, which depend on complex membrane-embedded systems operating by way of protonmotive force, almost certainly arose later.

ENCAPSULATION. Much has been written on the problems raised by the notion of an unstructured prebiotic soup and on the need of some form of containment to ensure a sufficiently high concentration of reactants for protometabolism to take place. Irrespective of such a requirement, we have seen that the existence of discrete protocells at a fairly early stage in the RNA world must be postulated if the development of protein synthesis is to be explained on the basis of Darwinian selection (see earlier section, "The case for congruence"?). Protometabolism must therefore have provided the chemical constituents of the protocellular envelopes.

Phospholipids, the universal components of the lipid bilayers that make up the fabric of all biological membranes, can assemble spontaneously into vesicles (liposomes) and are often considered as the likely constituents of the first membranes. This is chemically conceivable within the necessary complexity of a protometabolism capable of generating RNA and other elaborate molecules. Thioesters could have been involved in the esterification steps of phospholipid synthesis and, perhaps, in the formation of fatty acids.

A problem with liposomes' making up the first protocellular membranes is represented by the relative impermeability of phospholipid bilayers to small molecules and ions. An alternative possibility is that the first membranes consisted of proteolipids, which could have built more porous boundaries. It appears, from what is known of amino acid formation under abiotic conditions, that hydrophobic amino acids could have been abundant on the prebiotic Earth. Thus, the prebiotic presence of proteolipid-like multimers derived from aminoacyl thioesters seems not inconceivable. Such multimers could have assembled to form the first membranes, which would later have been enriched by the insertion of phospholipids.

THE PRIMARY FORMATION OF THIOESTERS. The model of a thioester-based protometabolism presupposes the ready availability of thioesters on the prebiotic Earth. As mentioned earlier, the ingredients for their formation may have been there, but their assembly raises an energy problem.

In aqueous solution, only traces of thioesters can form from free acids and thiols. The yield would be better at low pH and high temperature (de Duve 1991), but still insufficient for practical purposes, unless very soluble reactants were made to condense to a less soluble product upon progressive concentration by evaporation, as in Miller's "drying lagoon" hypothesis (see Keefe, Newton, and Miller 1995).

Another possibility, not yet explored, involves synthesis in vapor phase from volatile reactants. It is also conceivable that thioesters first formed at the expense of pyrophosphates, as in reactions (6) or (7). But this implies that inorganic pyrophosphate or polyphosphates were present, a point on which there is much debate.

More likely, perhaps, is oxidative synthesis by reaction (10) or (11), provided an appropriate electron acceptor were available. For example, an acetyl thioester could readily be formed by reaction (11) from pyruvic acid, which could itself arise by the dehydrogenation of lactic acid, a likely prebiotic substance:

$$CH_3\text{–CHOH–COOH} \Longleftrightarrow CH_3\text{–CO–COOH} + 2H \qquad (12)$$

A difficulty with mechanisms restricted to one acid or a small number of acids concerns the synthesis of other thioesters, for example, that of an aminoacyl thioester from the acetyl thioester made oxidatively from pyruvate. A prebiotic analogue of transthiolation would have to be postulated:

$$R'\text{–S–CO–}R_1 + R_2\text{–COOH} \Longleftrightarrow R'\text{–S–CO–}R_2 + R_1\text{–COOH} \qquad (13)$$

The thioester world

According to the thioester world model, life started in a sulfide-rich, volcanic setting, amply supplied with abiotically formed thiols and organic acids of various kinds, including amino acids and hydroxy acids. Somehow, these components found local conditions that favored their condensation to thioesters. These, in turn, spontaneously assembled into a variety of multimers. Some of these compounds may have served to form the first protocellular membranes. Especially, the primitive multimers may have included a number of enzymelike catalysts, which, together with available minerals, set into motion a protometabolic network containing the seeds of present-day metabolism.

Initially, protometabolism would have been fueled by the free energy of hydrolysis of the thioester bond. Soon, however, thioesters would have reacted with inorganic phosphate to generate inorganic pyrophosphate, which, in turn, would, by transphosphorylation, have given rise to various phosphate esters, greatly enriching protometabolism. At some stage, AMP and other nucleoside monophosphates would have appeared. The thioester-fueled condensation of such molecules with inorganic pyrophosphate would then have led to the first nucleoside triphosphates and, through these, to the first polynucleotides, ushering in the RNA world – not, however, the self-sufficient RNA world often designated by this term, but an RNA world that continued to be

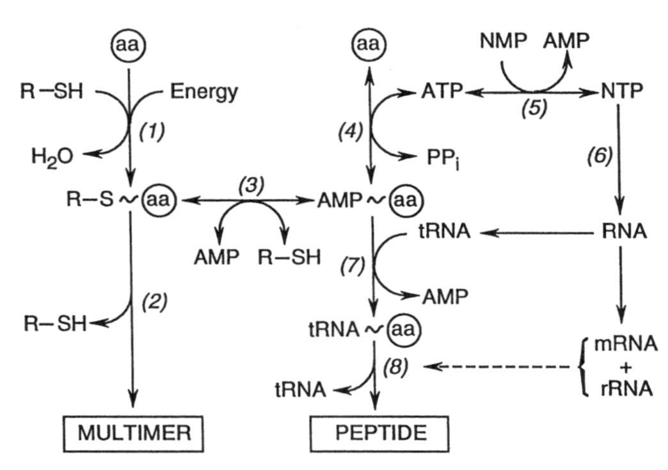

Figure 10.4. Certain key steps in the hypothetical transition from the thioester world to the RNA world (from multimers to peptides). Reaction **1** depicts the primary formation of thioesters from amino acids (aa), which condense into multimers by reaction **2**. With the appearance of AMP, the first aminoacyl adenylate is formed from a thioester by reaction **3**. ATP arises from the aminoacyl adenylate by reaction **4**, thereby opening the way to the formation of other nucleoside triphosphates (NTP) (reaction **5**) and, eventually, to RNA molecules (reaction **6**). With the availability of RNAs, the first aminoacyl–tRNA complexes can be formed (reaction **7**), leading to the assembly of proteins (reaction **8**) with the help of mRNAs and rRNAs.

supported by the thioester world until it had generated a sufficient set of protein enzymes to allow the transition from protometabolism to a congruent metabolism. Some key steps in this evolution are shown schematically in Figure 10.4.

At present, this scheme is purely hypothetical, resting on only presumptive evidence concerning the prebiotic environment and its supply in organic substances, and on key features of present-day metabolism believed, on the strength of the congruence argument, to have appeared very early in prebiotic times. Future experimental work could, however, help to bolster the model or to disprove it. First, there is a great need for more investigations into sulfur chemistry, such as have been conducted with remarkably intriguing results by Weber (1981, 1982, 1984) and, more recently, by Keefe, Newton and Miller (1995) and by Liu and Orgel (1997). The iron–sulfide models of Wächtershäuser and of Russell (for literature, see de Duve in press) also merit more attention. In another area, more effort should be devoted to the detection of enzymelike catalysts among short oligopeptides and other multimers prepared in more or less random fashion from the amino acids and hydroxy acids most likely to have been abundant on the prebiotic Earth, not omitting cysteine. In this search, some of the principal kinds of reactions that make up metabolism should be tested for under plausibly prebiotic conditions, in particular, in the absence of molecular oxygen.

References

Baltscheffsky, H. 1996. Energy conversion leading to the origin and early evolution of life: Did inorganic pyrophosphate precede adenosine triphosphate? In *Origin and Evolution of Biological Energy Conversion,* ed. H. Baltscheffsky, pp. 1–9. New York: VCH Publishers.

Baltscheffsky, M., and Baltscheffsky, H. 1992. Inorganic pyrophosphate and inorganic pyrophosphatases. In *Molecular Mechanisms in Bioenergetics,* ed. L. Ernster, pp. 331–348. Amsterdam: Elsevier Science Publishers B. V.

Barbier, B., and Brack, A. 1992. Conformation-controlled hydrolysis of polyribonucleotides by sequential basic polypeptides. *J. Am. Chem. Soc.* 114:3511–3515.

Brack, A. 1993a. Early proteins. In *The Chemistry of Life's Origins,* ed. J. M. Greenberg et al., pp. 357–388. Amsterdam: Kluwer Academic Publishers.

Brack, A. 1993b. From amino acids to prebiotic active peptides. *Pure Appl. Chem.* 65:1143–1151.

Cairns-Smith, A. G. 1982. *Genetic Takeover and the Mineral Origins of Life.* Cambridge: Cambridge University Press.

Cammack, R. 1983. Evolution and diversity in the iron–sulfur proteins. *Chem. Scr.* 21:87–95.

Dawson, P. E., Muir, T. W., Clark-Lewis, I., and Kent, S. B. H. 1994. Synthesis of proteins by native chemical ligation. *Science* 266:776–779.

de Duve, C. 1988. Prebiotic syntheses and the mechanism of early chemical evolution. In *The Roots of Modern Biochemistry*, ed. H. Kleinkauf, H. von Döhren, and L. Jeanicke, pp. 881–894. Berlin-New York: Walter de Gruyter.

de Duve, C. 1991. *Blueprint for a Cell.* Burlington, N.C.: Neil Patterson Publishers, Carolina Biological Supply Company.

de Duve, C. 1993. The RNA world: Before and after? *Gene* 135:29–31.

de Duve, C. 1995. *Vital Dust.* New York: Basic Books.

de Duve, C. In press. The origin of life: Energy. In *Frontiers of Biology, Vol. 1: From Atoms to Mind*, ed. W. Gilbert and G. Tocchini-Valentini. Rome: Istituto della Enciclopedia Italiana.

Eck, R. V., and Dayhoff, M. O. 1966. Evolution of the structure of ferredoxin based on living relics of primitive amino acid sequences. *Science* 152:363–366.

Eigen, M., and Schuster, P. 1977. The hypercycle: a principle of self-organization. Part A: Emergence of the hypercycle. *Naturwissenschaften* 65:7–41.

Gevers, W., Kleinkauf, H., and Lipmann, F. 1969. Peptidyl transfers in gramicidin S biosynthesis from enzyme-bound thioester intermediates. *Proc. Natl. Acad. Sci. USA* 63:1335–1342.

Gilbert, W. 1986. The RNA world. *Nature* 319:618.

Heinen, W., and Lauwers, A. M. 1966. Organic sulfur compounds resulting from the interaction of iron sulfide, hydrogen sulfide and carbon dioxide in an anaerobic aqueous environment. *Origins Life Evol. Biosphere* 26:131–150.

Kaschke, M., Russell, M., and Cole, W. J. 1994. (FeS/Fes_2). A redox system for the origin of life. *Origins Life Evol. Biosphere* 24:43–56.

Keefe, A. D., Newton, G. L., and Miller, S. L. 1995. A possible prebiotic synthesis of pantetheine, a precursor of coenzyme A. *Nature* 373:683–685.

Keller, M., Hafenbradl, D., Stetter, K. O., Teller, G., Natakani, Y., and Ourisson, G. 1995. Einstufige Synthese von Squalen aus Farnesol unter präbiotische Bedingungen. *Angew. Chem.* 107:2015–2017.

Kleinkauf, H., and von Döhren, H. 1990. Nonribosomal biosynthesis of peptide antibiotics. *Eur. J. Biochem.* 192:1–15.

Kleinkauf, H., and von Döhren, H. 1996. A nonribosomal system of peptide biosynthesis. *Eur. J. Biochem.* 236:335–351.

Lee, D. H., Granja, J. R., Martinez, J. A., Severin, K., and Ghadiri, M. R. 1996. A self-replicating peptide. *Nature* 382:525–528.

Liu, R. E., and Orgel, L. E. 1997. Oxidative acylation using thioacids. *Nature* 389:52–54.

Spiegelman, S. 1967. An in vitro analysis of a replicating molecule. *Amer. Scient.* 55:221–264.

Weber, A. L. 1981. Formation of pyrophosphate, tripolyphosphate, and phosphorylimidazole with the thioester, N,S-diacetylcysteamine, as the condensing agent. *J. Mol. Evol.* 18:24–29.

Weber, A. L. 1982. Formation of pyrophosphate on hydroxyapatite with thioesters as condensing agents. *BioSystems* 15:183–189.

Weber, A. L. 1984. Prebiotic formation of "energy-rich" thioesters from glyceraldehyde and N-acetylcysteine. *Origins of Life* 15:17–27.

Wieland, T. 1988. Sulfur in biomimetic peptide syntheses. In *The Roots of Modern Biochemistry*, ed. H. Kleinkauf, H. von Döhren, and L. Jaenicke, pp. 212–221. Berlin-New York: Walter de Gruyter.

11

Origins of the RNA world

ALAN W. SCHWARTZ

Evolutionary Biology Research Group, Faculty of Science
University of Nijmegen, Toernooiveld, The Netherlands

1. A paradigm for the origin of life

The theory that life began in an "RNA world" postulates that the first chemical system to appear on Earth with the capability to replicate itself was a set of RNA molecules. Replication of sequence information is an implicit and critical concept here, rather than mere "reproduction," since only a system that quite faithfully reproduces the properties of its parents is capable of Darwinian evolution. The arguments in favor of the theory that life originated with RNA are both chemical and biological in nature (Joyce 1989). The chemical properties that support such a potential will be considered briefly later in connection with the template-directed oligomerization of mononucleotides. The central role of RNA in the expression of genetic information is probably the most persuasive of the biological lines of evidence. The most impressive demonstration of the evolutionary potential of RNA is the fact that selection experiments reveal a growing catalogue of new catalytic functions of RNA molecules (see Chapter 13). Once self-replicating RNA molecules appear on the scene, it becomes possible to understand how evolution could produce increasingly efficient self-replicating systems, although the origin of encoded protein synthesis still presents serious theoretical difficulties. However, the task set for this chapter is to look backward from RNA and ask how the RNA world got its start. Because RNA molecules are not only self-organizing, but self-organizing in an information-transferring sense, speculation has often produced a scenario of the following kind:

1. Relatively short RNA molecules (oligonucleotides) were first formed by a random oligomerization of mononucleotides (Figure 11.1).
2. Some oligonucleotides began to catalyze the synthesis of complementary copies of themselves simply by functioning as templates, with chemical energy for the synthesis of daughter oligonucleotides being provided in the form of small reactive organic molecules,combined with a supply of mononucleotides.

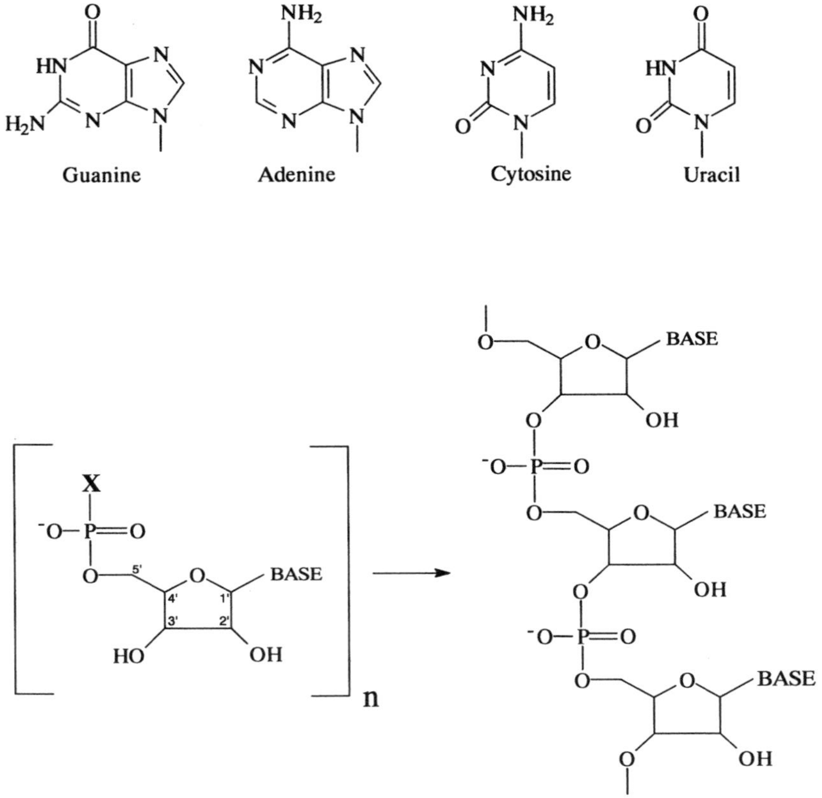

Figure 11.1. The structure of RNA and its bases. In a hypothetical synthesis, a mononucleotide with an activated 5'-phosphate (X is the activating group) oligomerizes to produce a 3'–5' phosphodiester-linked oligonucleotide. Other possibilities include the production of 2'–5' phosphodiester and 5'–5' pyrophosphate linkages.

3. Chemical replication of oligonucleotides led ultimately to the development of RNA molecules that could catalyze their own synthesis. At this point, evolution by natural selection would have become possible.

Once stage 3 was reached, evolution could have overcome many bottlenecks that existed prior to that point, for example, by leading to the development of synthetic pathways that liberated the first RNA molecules from dependency upon an external supply of reactive intermediates. Implicit in this extended scenario is the development of a primitive cell-like membrane, which would have been necessary to isolate individual RNA "organisms" from the environment (see Chapter 8).

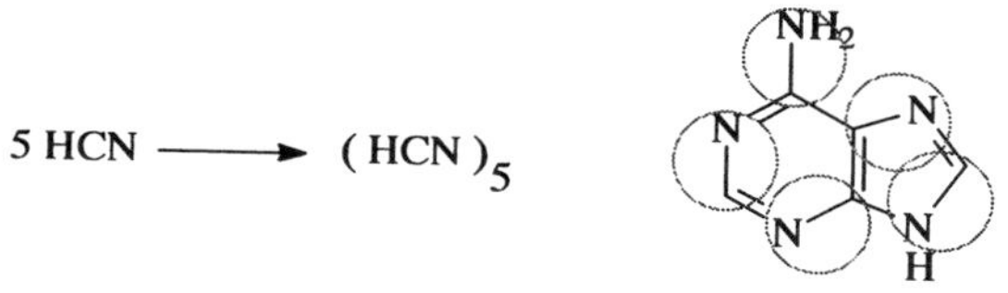

$$5\ HCN \longrightarrow (HCN)_5$$

Figure 11.2. The oligomerization of HCN to form adenine. In an idealized scheme, adenine is regarded as a pentamer of HCN. The encircled C–N bonds show a hypothetical arrangement of the monomer. In fact, the reaction in dilute aqueous solution is very much more complex, probably involving the formation and rearrangement of at least hexamers and heptamers of HCN. (Voet and Schwartz 1983.)

Our problem really is that there are many difficulties inherent in stages 1 and 2. This chapter may therefore be a little frustrating for readers hoping to find a neat summary of accomplishments, as it will be necessary to show why some new thinking is necessary. The chapter will end on a positive note, however, due to some exciting new developments in the field.

2. Pieces of the puzzle: purines and pyrimidines

Although studies of possible prebiotic syntheses are highly model-dependent and can only suggest possibilities concerning the historical course of evolution, meteorites provide an indisputable source of information concerning the prebiotic synthesis of organic compounds. All of the biologically important purines are present in quite reasonable amounts in carbonaceous meteorites, but only one pyrimidine (uracil) has been detected, and that in much lower concentration than any of the purines (Stoks and Schwartz 1979, 1981). Can we explain this distribution as the product of a known synthetic pathway? Recent data from model prebiotic syntheses seem to predict a different distribution. Since the discovery by Oró (1960) that adenine is produced from the oligomerization of hydrogen cyanide (Figure 11.2), this synthesis has become established as "the rock of the faith" of prebiotic chemistry (Joyce and Orgel 1993). Actually, significant yields of adenine have been demonstrated only under rather extreme conditions (Schwartz, Voet and Van Der Veen 1984). Even when a strong acid hydrolysis step is included to increase the yield of adenine, it does not exceed 0.1% (Voet and Schwartz 1983). Nevertheless, the synthesis is remarkable, if only because of the large number of steps involved (at least nine) and the numerous other possible reactions. The prebiotic synthesis of guanine has been modeled by Sanchez, Ferris and Orgel (1967), but only by utilizing high concentrations of unstable precursors, including ammonia. It is now generally recognized that such conditions are unlikely to have been prebiotic.

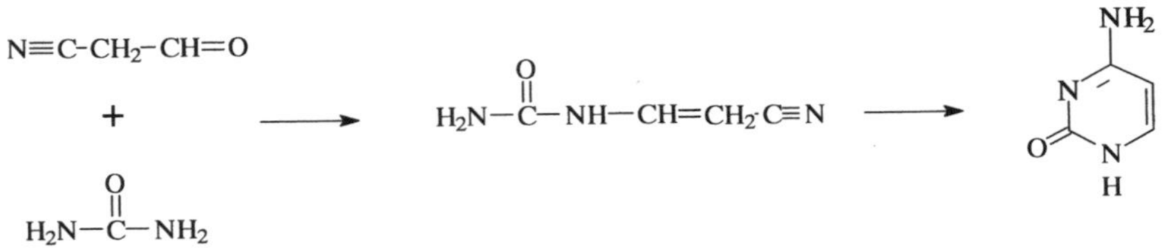

$$N{\equiv}C{-}CH_2{-}CH{=}O \; + \; H_2N{-}\overset{O}{\overset{\|}{C}}{-}NH_2 \longrightarrow H_2N{-}\overset{O}{\overset{\|}{C}}{-}NH{-}CH{=}CH_2{-}C{\equiv}N \longrightarrow$$

NH2 N O N H

Figure 11.3. A relatively simple pathway for pyrimidine synthesis. The reactants (cyanoacetaldehyde and urea) are both produced in good yield under reasonable, although differing, prebiotic conditions.

In contrast to adenine synthesis, recent results now suggest, the pyrimidines could have been produced in quite high yields. The starting point for this pathway is the demonstration by Ferris and co-workers (Ferris, Sanchez and Orgel 1968; Ferris et al. 1974) that cytosine and uracil are produced by the reaction of cyanoacetylene or cyanoacetaldehyde (the hydrolysis product of cyanoacetylene) with urea (Figure 11.3). Cyanoacetylene is the second most abundant product (after HCN) produced by sparking mixtures of CH_4 and N_2, and urea is one of the major products of the oligomerization of HCN. In "evaporating pond" experiments, in which very high concentrations of urea can be produced, yields of cytosine in excess of 50% can be obtained (Robertson and Miller 1995). Both purines and pyrimidines have been synthesized in a model solar nebula process, a Fischer–Tropsch type reaction, in which CO, H_2, NH_3, and an iron catalyst are ingredients. This pathway, however, has been criticized on several grounds (for a summary of this controversy, see Stoks and Schwartz 1981). We clearly do not yet completely understand the processes that led to the presence of purines in meteorites. A complicating factor is the fact that carbonaceous chondrites are as old as the Earth, and therefore questions of stability also arise. Preferential degradation of pyrimidines, which are inherently less stable than purines, is a possible explanation for this apparent discrepancy. In comparison with other problems yet to be introduced, however, trying to formulate a complete theory for the prebiotic production of both purines and pyrimidines seems like quibbling. It is reasonable to assume that, as a first approximation, their presence would not have been a major problem in chemical evolution.

3. Nucleosides, nucleotides, and sources of phosphorus

Deferring for the time being the question of ribose synthesis, it is possible to be quite brief in reviewing prebiotic synthesis of nucleosides. Purine nucleosides have been synthesized in modest yield by heating adenine, hypoxanthine, and guanosine in the presence of ribose and magnesium salts (Fuller, Sanchez and Orgel 1972a and 1972b). However, this procedure is ineffective

$Ca_{10}(PO_4)_6F_2$ $(Fe, Ni)_3P$ $R{-}P(=O)(OH){-}OH$

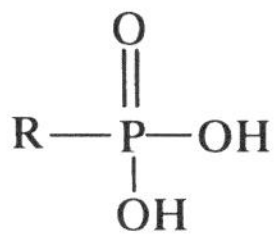

A B C

Figure 11.4. Sources of phosphorus on the Earth and in meteorites. **A.** Fluorapatite is the only geologically significant source of phosphorus in the crust of the Earth, although the core may contain phosphorus in the form of dissolved phosphide. **B.** The mineral schreibersite is commonly found in iron meteorites, where it is of secondary origin, and in small amounts in stony meteorites. **C.** The only organic compounds containing phosphorus that have been identified in meteorites are alkyl phosphonic acids. (Cooper, Onwo and Cronin 1992.)

with pyrimidines, nor is any other prebiotic synthesis known for the pyrimidine nucleosides. This clearly represents a gap that must be bridged in any pathway leading to the synthesis of RNA.

The synthesis of nucleotides requires both a source of phosphate and a mechanism for activating phosphate for the synthesis of phosphate esters. Several pathways have been suggested for solubilizing and activating rock phosphate (the mineral apatite; Figure 11.4). These include the effects of evaporating solutions of ammonium chloride and urea (Lohrmann and Orgel 1971), evaporation of ammonium oxalate in the presence of urea or other condensing agents (Schwartz et al. 1975), and the possibility of carrying out phosphorylations with the mineral struvite ($MgNH_4PO_4{\cdot}6H_2O$), which could precipitate under conditions of elevated magnesium and ammonium ion concentrations (Handschuh and Orgel 1973). Other mechanisms have also been suggested, such as precipitation and thermal activation of brushite (Gedulin and Arrhenius 1994) and volcanic volatilization of P_4O_{10} (Yamagata et al. 1991). A recent criticism of all of these pathways led Keefe and Miller (1995) to question the participation of phosphate in early biogenesis.

A rather different source of phosphorus has been discovered recently in the Murchison meteorite, in the form of simple alkyl phosphonic acids (Cooper, Onwo and Cronin 1992). Phosphonic acids differ from phosphate esters such as nucleotides in having a C–P bond, rather than a C–O–P bond, that links the PO_3 grouping with the organic part of the molecule (Figure 11.4). The synthesis of phosphonic acids in a meteorite parent body or in the early atmosphere of the Earth during ablation of meteorites or comets has been modeled photochemically (De Graaf, Visscher and Schwartz 1995). Exogenous sources of organic molecules, such as comets and meteorites, could have sup-

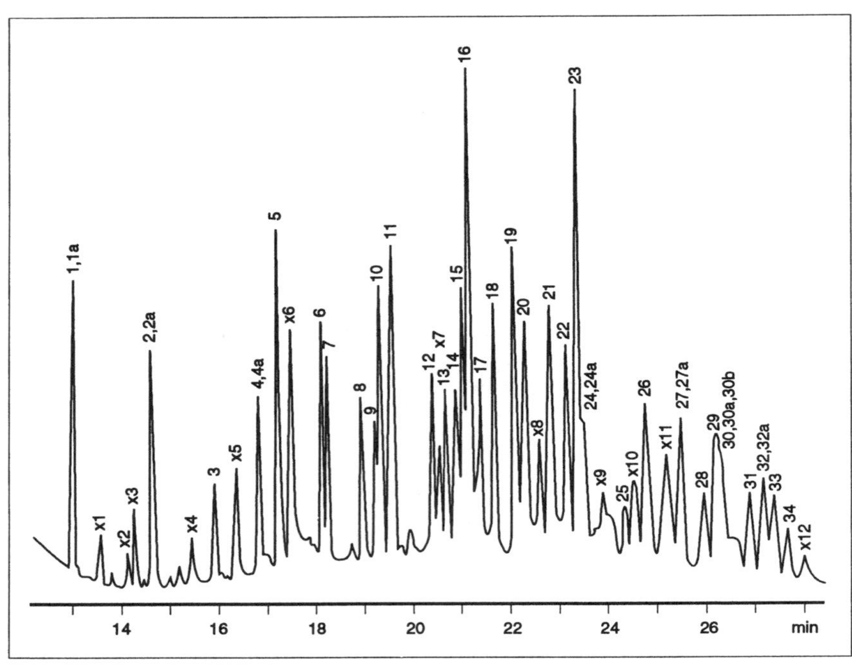

Figure 11.5. The complexity of the formose reaction. The number of peaks (derivatives of carbohydrates) on this gas chromatogram gives a rough idea of the nature of the problem. (Reproduced with kind permission of Elsevier Science from Decker, Schweer and Pohlmann 1982.)

plied the early hydrosphere of the Earth with analogues of phosphate esters, if not actually with organic phosphates. We will return to the question of nucleotide analogues later.

The logistics involved in arranging for a supply of purines and pyrimidines for nucleoside synthesis and for the necessary phosphorylation to produce nucleotides may be complex, but are not intrinsically insoluble. The problem of a supply of ribose, however, is one that has seemed, until recently, to be vastly more difficult.

4. Ribose: a critical ingredient

The problem of ribose synthesis has attracted the attention of a number of workers. Until fairly recently, the only synthesis of ribose that was thought to be even remotely prebiotic was the base-catalyzed condensation of formaldehyde known as the formose reaction. This autocatalytic reaction produces a highly complex mixture of sugars, of which ribose is merely a minor component. Shapiro (1988) has summarized most of the accumulated experience with the formose reaction, and he concluded that only low concentrations of ribose, as part of highly complex mixtures of other compounds, can reasonably have been present on the primitive Earth (Figure 11.5). Previous claims of more selective formose reactions' being catalyzed by minerals have also been reexamined and not found to produce such selectivity (Schwartz and De Graaf 1993). As a source of ribose, the formose reaction is one of the few examples in prebiotic chemistry where "negative results" have become fairly widely known. Because all sugars have closely similar chemical properties, it is clear to any chemist that only complex separation mechanisms could result in enrichment of ribose. A simple process such as absorption on a mineral surface would therefore seem to be incapable of achieving this end. It has been found possible, in photochemical reactions, to produce selective aldol condensations of formaldehyde, but ribose is not one of the products (Shigemasa et al. 1977; Schwartz and De Graaf 1993).

An intriguing new approach to this problem has been introduced by Albert Eschenmoser, who decided to look closely at the most important intermediate in the formose reaction, glycolaldehyde. Realizing that in biochemistry, sugar phosphates rather than free sugars are formed, Eschenmoser studied the condensation of glycolaldehyde phosphate in the presence of a limited amount of formaldehyde and base. He and his colleagues (Müller et al. 1990) were able to show that a set of products was formed with strikingly limited complexity compared to the usual course of events (Figure 11.6). In more recent work, a mineral-catalyzed version of this reaction has been reported, which can take place in neutral solution and low concentration (Pitsch et al. 1995). We will return to these observations and some interesting consequences somewhat later. Let us now jump ahead to the next problem on the way to RNA and consider the second stage in our ideal scenario. (The synthesis of oligonucleotides from mononucleotides will be covered in Chapter 12.)

5. Chemical replication of oligonucleotides

In DNA and RNA, the combination of Watson–Crick base-pairing and the tendency of the purines and (to a lesser extent) pyrimidines to form stacks in aqueous solution, accounts for the stability of the helix formed between two chains with complimentary nucleotide sequences. It has long been understood that under certain conditions double helix–like complexes can be formed with one chain consisting of a sequence of pyrimidines, such as poly(C), and a second "chain" consisting of only monomers such as the mononucleotide guanosine-5' monophosphate (pG). In theory, this phenomenon could form the basis

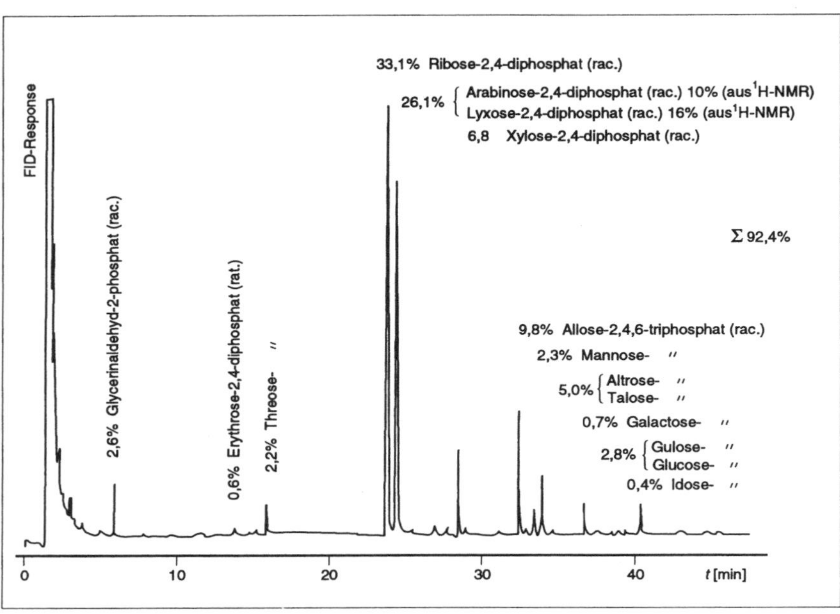

Figure 11.6. The relatively simple gas chromatogram showing the carbohydrate products produced when glycolaldehyde phosphate is condensed in the presence of formaldehyde. (Analysis after dephosphorylation of the products. Reproduced with permission from Müller et al. 1990.)

for a chemically driven replication mechanism for oligonucleotides if the aligned mononucleotides could be linked together to form a daughter chain, and if this chain could in turn catalyze the synthesis of a copy of the original template. Although in practice the template-directed oligomerization of monomers has proven to be somewhat more difficult than theory suggests, considerable success has been achieved. Using a simple organic chemical group[1] to activate the phosphate of a pG, Orgel has demonstrated that poly(C) will catalyze the synthesis of long chains of oligoguanylic acid, which forms a very stable double helix with a poly(C) template (Figure 11.7). The details of this and related oligomerizations have been worked out in exquisite detail in Orgel's laboratory (Inoue and Orgel 1982, 1983). The reaction is highly dependent on the nature of the activating group that is used, as well as on the chemical nature of the individual nucleotides. Recently, Orgel's laboratory has completed an extensive evaluation of the reaction by making use of so-called hairpin RNA structures in which both template and a primer sequence of nucleotides are incorporated into one molecule. Under these conditions, the

[1] The activating group is 2-methylimidazole; thus the abbreviated name of the activated pG is 2-MeImpG.

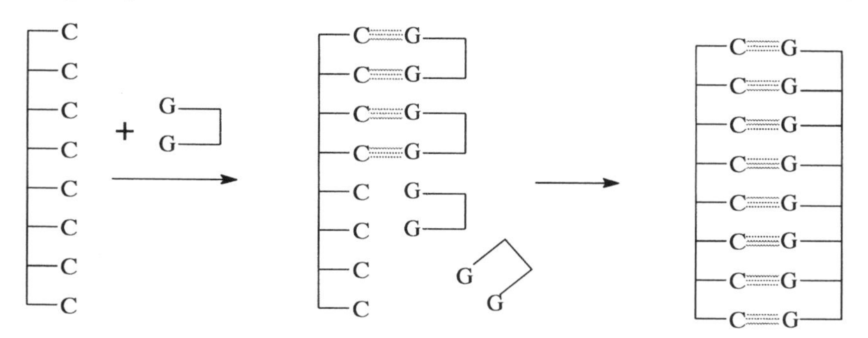

Figure 11.7. Template-directed oligomerization. The principle process in template-directed oligomerization of mononucleotides actually occurs at the dimer or trimer level. When an activated mononucleotide is allowed to react in the presence of a complementary polynucleotide template, dimers are first formed in a noncatalyzed phase of the reaction; they then bind to the template and are ligated and extended.

template-directed oligomerization can be studied under highly dilute conditions, reducing problems caused by the tendency of chains of oligo(G) to associate with each other. The rather sobering results can be summarized as follows: Templates consisting primarily of C residues, but including isolated A,T, G, or short oligo(G) sequences, present no serious obstacle to copying. However, the complementary situation – an oligonucleotide rich in G and incorporating isolated A or T residues – cannot be copied, and isolated C is at best copied inefficiently. This means that a replication cycle is not possible by means of template-catalysis alone. It is suggested that in order to increase the efficiency of the process, a "sophisticated catalyst" would be needed (Hill, Orgel and Wu 1993).

These studies have all been conducted with synthetic mononucleotides that are homochirally pure, that is, based on D-ribose only. It is now necessary to introduce a set of problems that are due to chirality.

6. Confronting chirality

The separation of optical isomers of organic compounds is a problem that has intrigued chemists since 1848, when Pasteur, who regarded chirality as the exclusive product of living organisms, discovered the phenomenon. There has been considerable theoretical interest in the possibility that the prebiological world may have been homochiral. Among other mechanisms that have been considered, a "parity violating energy difference" between enantiomers has been proposed as an explanation for the emergence of such a world prior to the origin of life. This effect, which has been estimated to produce an energy difference of the order of 10^{-17} to 10^{-15} between chiral forms, would require

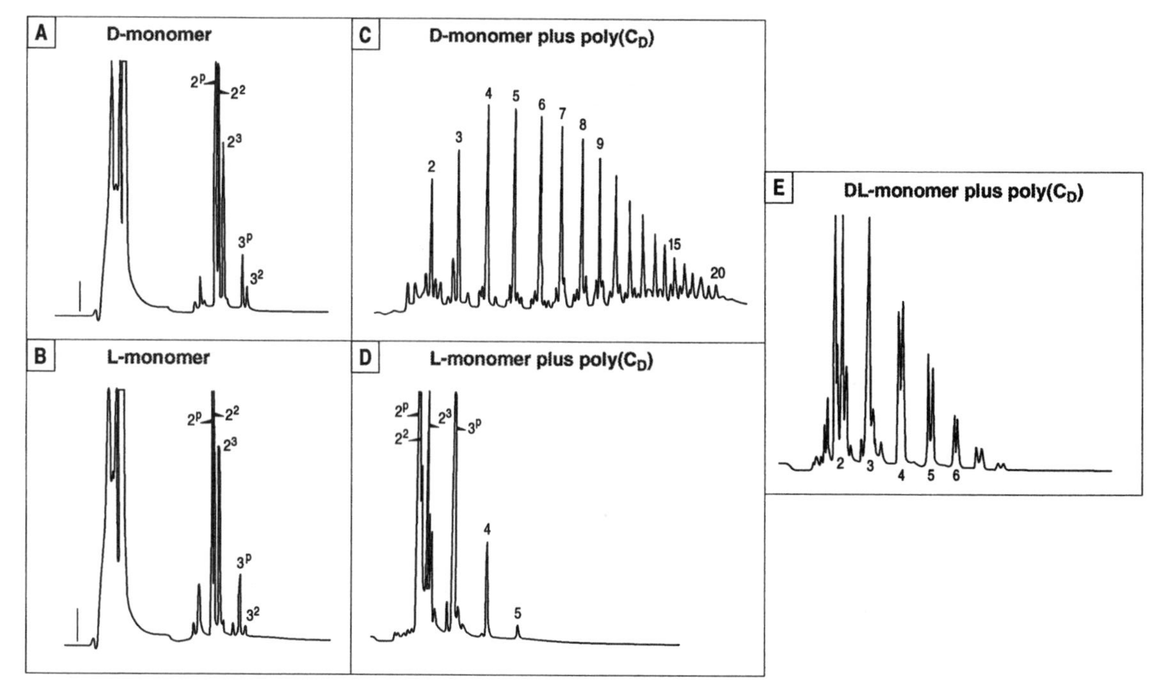

Figure 11.8. The phenomenon of enantiomeric cross-inhibition. **A** and **B** show the products (mostly dimers) of the oligomerization of the D- and L-forms of 2-MeImpG (an activated derivative of guanosine monophosphate), respectively, in the absence of a template. **C** demonstrates the oligomerization of the D-monomer on a (D-)poly(C) template, and **D** shows the virtual lack of oligomerization of the L-monomer on the same template. When a racemic mixture of D- and L-isomers was oligomerized on poly(C), the products shown in **E** were observed (compare **E** with **C**). Multiple peaks result from the fact that a large proportion of the oligomers are terminated by pyrophosphate linkages. (Adapted from Joyce et al. 1984.)

an amplification mechanism in order to produce homochirality, and several potential mechanisms have been considered and discounted (see Bonner 1991). It has been argued (Bada and Miller 1987) that racemization of amino acids is too rapid a process to permit amplification (by as yet unspecified reactions) to give rise to homochirality. Carbohydrates are more resistant to racemization than are amino acids. However, they are also chemically less stable, a factor that could equally defeat an amplification cycle. The seemingly inevitable presence of both enantiomers of ribose in any prebiotically synthesized nucleotide has consequences that now have to be considered.

The aforementioned studies of the selectivities and limitations of template-directed oligomerization of oligonucleotides have been conducted under carefully controlled conditions using templates, as well as monomers, whose structural and chiral purity were rigorously controlled. In an experiment designed to test the requirement for chiral purity (Figure 11.8), it was demonstrated that incorporation of even a single mononucleotide of opposite chirality into the end of a growing chain in template-directed oligomerization is sufficient to terminate the reaction (Joyce et al. 1984). This observation (the phenomenon is referred to as enantiomeric cross-inhibition) has had serious consequences for theories of the origin of life, since it seemingly put an end to speculation that template-directed oligomerization might be capable of acting as an amplification mechanism to produce or maintain homochirality, but has led, instead, to a rather different conclusion. If the structural requirements necessary to permit chemical replication of RNA are so restrictive, perhaps the process of replication began in an achiral system and was taken over later in evolution by RNA (Joyce et al. 1987). A proposal of a similar kind was originally made by Cairns-Smith (1966), who postulated that the first genetic material was a clay mineral. Although it is not clear how information could be transferred from such a mineral to RNA, it does seem possible to imagine a "nucleic acid–like" polymer that incorporated purines and pyrimidines, but was attached to an achiral backbone, serving as a precursor of RNA.

7. Templating experiments with nucleotide analogues

In one of the first experiments indirectly bearing on this possibility, Pitha and Pitha (1972) showed that a polynucleotide analogue, incorporating uracil on an atactic (disordered) polyvinyl backbone, could catalyze the formation of dinucleoside monophosphate from the reaction of adenosine-2',3'-cyclic phosphate with adenosine. In our own laboratory, the first analogue to be synthesized and tested in this context was the guanosine analogue shown in Figure 11.9A. This structure, in which ribose has been replaced by a diphosphorylated

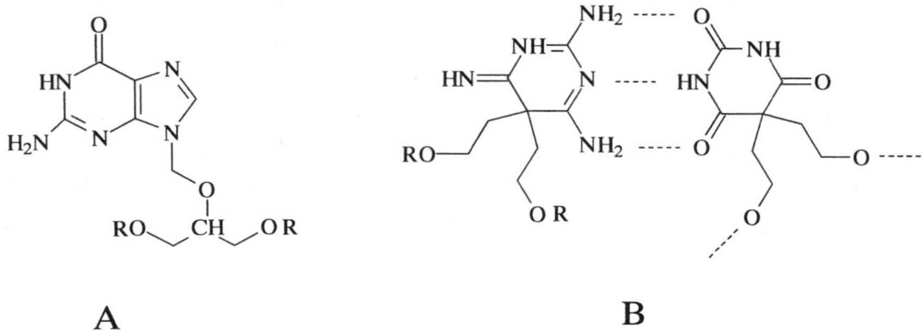

A B

Figure 11.9. Nonchiral nucleotide analogues that oligomerize by formation of pyrophosphate linkages. **A** is a prochiral monomer based on guanine, glycerol, and formaldehyde. **B** shows a pair of achiral analogues that have complementary structures. R indicates that the monomers in each case are phosphorylated and activated in the form of phosphoimidazolides.

glycerol-formaldehyde adduct, is properly described as being prochiral rather than achiral, since chiral centers are created when the monomers are linked together (by pyrophosphate bond formation), creating a random sequence of configurations (an atactic chain). In spite of this drawback, however, a chain of this kind incorporating the cytidine analogue was shown to have a limited catalytic effect in the oligomerization of monomer 11.9A (Visscher and Schwartz 1990). A further exploration of this possibility was conducted with the truly achiral pair of analogues shown in Figure 11.9B. The oligomerization of the rather strange looking monomer on the left was shown be catalyzed to a small degree by the "complementary" oligomer indicated on the right, but only in a solvent mixture consisting of 50% dimethylformamide in water (Van Vliet, Visscher and Schwartz 1995). Although these conditions favor hydrogen bonding and therefore improve the formation of a complex between template and monomer, they clearly cannot be considered to be prebiotic.

For reasons having nothing to do with the question of the origin of life, a large number of other nucleic acid analogues have been synthesized and studied during the past several years. The stimulus has been the possibility of using "antisense" molecules therapeutically as weapons against a variety of diseases and organisms. Most of these analogues are based on ribonucleotides, the modifications being mostly to the mode of linkage between the residues. In nearly every case, the properties of the analogue, in terms of the formation of stable Watson–Crick–based complexes, have been less favorable than those of native nucleic acid. Quite recently, however, some striking exceptions to the usual experience with analogues have been described.

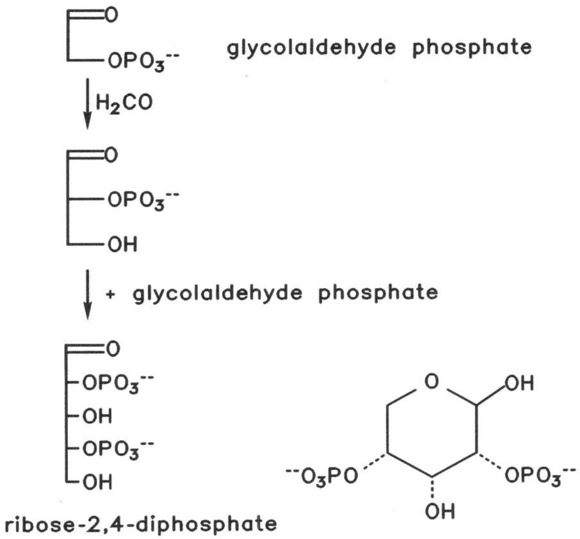

glycolaldehyde phosphate

H_2CO

+ glycolaldehyde phosphate

ribose-2,4-diphosphate

Figure 11.10. The formation of ribose-2,4-diphosphate by the base-catalyzed condensation of glycolaldehyde phosphate in the presence of formaldehyde. (Müller et al. 1990.)

8. Pyranose-RNA

Let us return now to the observation that the condensation of glycolaldehyde phosphate differs from the comparable reaction of free glycolaldehyde and results in a rather selective synthesis. Note that the presence of a phosphate group at the 4- position of ribose prevents the formation of a five-membered, or furanose ring, but permits formation of a six-membered, or pyranose ring (Figure 11.10). Eschenmoser (1994) has carried out an extensive theoretical analysis of the possibility of constructing nucleic acid–like polymers based on such six-membered rings, as well as the actual synthesis and analysis of the most promising candidates. Of all the analogues studied,the most interesting properties were displayed by what was named pyranosyl-RNA (p-RNA), a 2',4'-linked isomer of RNA (Figure 11.11). The characteristics of this molecule in forming Watson–Crick paired duplexes are striking indeed. Complementary p-RNA chains interact in a way that is both stronger and more selective than either DNA or RNA.

This success immediately raises the question of whether this analogue could have been synthesized prebiotically. It is not at all clear if a logical sequence of steps can be found to explain the prebiotic formation of p-RNA,

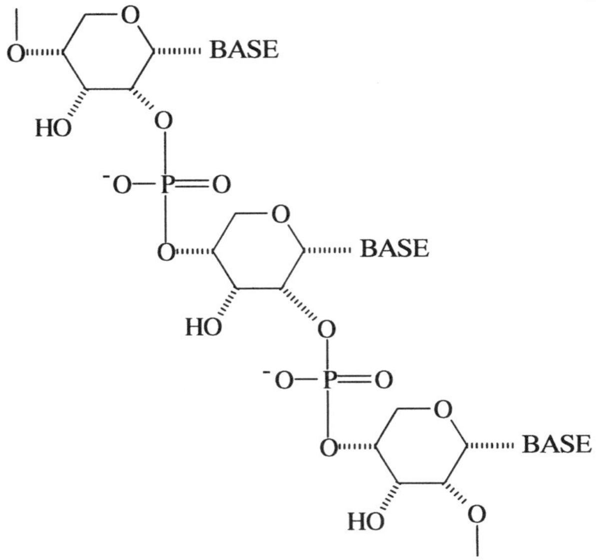

Figure 11.11. The structure of pyranosyl-RNA, a nucleic acid analogue based on ribose-2,4-diphosphate. (Eschenmoser 1994.)

although Eschenmoser has presented a most elegant, but thus far theoretical, scheme to this purpose. Elsewhere I have suggested a possible pathway for the prebiotic synthesis of a related analogue structure, starting with phosphonic acids rather than phosphates (Schwartz 1997). Only future work will be able to address the question of the prebiotic plausibility of p-RNA or related analogues.

9. "Achiral" peptide nucleic acid

The "Peptide Nucleic Acid" (PNA) of Nielsen and co-workers was designed for the purpose of antisense interaction with DNA (Wittung et al. 1994). It is a remarkable exception in all respects to the other analogues that have been studied recently. The backbone is based on a repeating peptide unit, although of an unusual kind, the most striking feature being the attachment of purines and pyrimidines to the backbone via a trigonal nitrogen atom, producing a truly nonchiral structure (Figure 11.12). Due to the absence of charge on the backbone, double helical complexes between complementary chains are exceptionally stable. These flexible analogues can assume either right-handed or left-handed double helices, which, of course, are opposite chiral forms. Recently it has been shown that template-directed oligomerization of 2- MeImpG can be conducted using a PNA oligomer containing cytosine as template ($PNAC_{10}$). PNA-directed oligomerization of dimers[2] of PNA has

[2] The monomer is subject to internal cyclization.

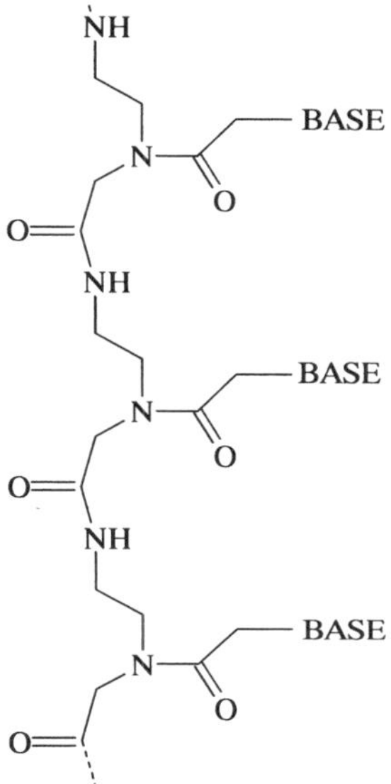

Figure 11.12. "Peptide nucleic acid" (PNA). (Wittung et al. 1994.)

also been achieved, although less efficiently (Böhler, Nielsen and Orgel 1995). These results confirm, at least in principle, the possibility that switching from a hypothetical precursor self-replicating system to an RNA system is possible. It is conceivable that an achiral molecule such as PNA, which can assume either right-handed or left-handed helices, might be capable of being part of a chiral amplification mechanism. However, recent results from Orgel's laboratory show that this system is as subject to enantiomeric cross-inhibition as is RNA or DNA (Schmidt, Nielsen and Orgel 1997).

10. Self-assembly and chiral amplification

Very recently, Bolli, Micura and Eschenmoser (1997) have demonstrated that p-RNA is capable of undergoing a self-assembly process that is chirally selective. They have achieved this by making use of tetramers with sequences chosen to be partially self-complementary and overlapping, such as ATCG. If two such tetramers combine to form the sequence ATCGATCG, this product is capable of acting as a template for linking further tetramers in the following manner:

```
ATCG + ATCG → ATCGATCG

ATCGATCG + ATCG → ATCGATCG
                      GCTA     (Duplex)
ATCGATCG + ATCG → ATCGATCG
```

GCTA GCTAGCTA

ATCGATCG + ATCG → ATCGATCGATCG
GCTAGCTA GCTAGCTA etc.

Although the repeating sequence formed in the example shown is relatively devoid of information, the oligomerization of a mixture of such sequences is capable of producing a great diversity of product sequences. When the authors introduced a mixture of enantiomers into the starting sequences, they were able to demonstrate that there was a strong selectivity for homochirality in the product sequences. In other words, the rate of the template-directed reaction of the homochiral oligomers with each other was many times faster than the rate of reaction of heterochiral oligomers.

11. Concluding comments and future prospects

RNA has clearly been designed by evolution to do the job it does. It is obvious that the ancestral RNA molecules were not assembled by accident as part of a population of randomly synthesized RNA molecules, since the spontaneous chemical synthesis of even a short structurally *and* chirally homogeneous oligonucleotide is implausible in the extreme. Furthermore, experience to date with the template-directed oligomerization of mononucleotides has led to the conclusion that RNA-directed nonezymatic replication may have required help from a "sophisticated catalyst" to permit a replicative cycle to occur (Hill, Orgel and Wu 1993). However, results with analogue systems such as p-RNA, for which complementary duplex formation is both stronger and more specific than in RNA, reinforce the suggestion that residue-by-residue replication of sequence information, which is a requirement for evolution, could have arisen in an RNA precursor system in which the replicative specificity was supplied by Watson–Crick base-pairing alone. As we have seen, just such a possibility has been demonstrated with the help of p-RNA. It is perhaps unlikely, however, that p-RNA was actually the historical system in which self-replication arose. Nevertheless, one more roadblock has been removed on the path to understanding the origin of the RNA world and consequently, if the current models prove to be correct, of the origin of life itself.

References

Bada, J. L., and Miller, S. L. 1987. Racemization and the origin of optically active organic compounds in living organisms. *BioSystems* 20:21–26.
Böhler, C., Nielsen, P. E., and Orgel, L. E. 1995. Template switching between PNA and RNA oligonucleotides. *Nature* 376:578–581.
Bolli, M., Micura, R., and Eschenmoser, A. 1997. Pyranosyl-RNA: chiroselective self-assembly of base sequences by ligative oligomerization of tetranu-

cleotide-2',3'-cyclophosphates (with a commentary concerning the origin of biomolecular homochirality). *Chem. Biol.* 4:309–320.

Bonner, W. A. 1991. The origin and amplification of biomolecular chirality. *Origins Life Evol. Biosphere* 21:59–111.

Cairns-Smith, A. G. 1966. The origin of life and the nature of the primitive gene. *J. Theoret. Biol.* 10:53–88.

Cooper, G. W., Onwo, W. M., and Cronin, J. R. 1992. Alkyl phosphonic acids and sulfonic acids in the Murchison meteorite. *Geochim. Cosmochim. Acta* 56:4109–4115.

Decker, P., Schweer, H., and Pohlmann, R. 1982. Bioids. X. Identification of formose sugars, presumable prebiotic metabolites, using capillary gas chromatography/ gas chromatography–mass spectrometry of *n*-butoxime trifluoroacetates on OV-225.J. *Chromat.* 244:281–291.

De Graaf, R. M., Visscher, J., and Schwartz, A. W. 1995. A plausibly prebiotic synthesis of phosphonic acids. *Nature* 378:474–477.

Eschenmoser, A. 1994. Chemistry of potentially prebiological natural products. *Origins Life Evol. Biosphere* 24:389–423.

Ferris, J. P., Sanchez, R. A., and Orgel, L. E. 1968. Studies in prebiotic synthesis. III. Synthesis of pyrimidines from cyanoacetylene and cyanate. *J. Mol. Biol.* 33:693–704.

Ferris, J. P., Zamek, O. S., Altbuch, A. M., and Freiman, H. 1974. Chemical evolution. XVIII. Synthesis of pyrimidines from guanidine and cyanoacetaldehyde. *J. Mol. Evol.* 3:301–309.

Fuller, W. D., Sanchez, R. A., and Orgel, L. E. 1972a. Studies in prebiotic synthesis. VI. Synthesis of purine nucleosides. *J. Mol. Biol.* 67:25–33.

Fuller, W. D., Sanchez, R. A., and Orgel, L .E. 1972b. Studies in prebiotic synthesis. VII. Solid-state synthesis of purine nucleosides. *J. Mol. Evol.* 1:249–257.

Gedulin, B., and Arrhenius, G. 1994. Sources and geochemical evolution of RNA precursor molecules: The role of phosphate. In *Early Life on Earth,* ed. S. Bengston, pp. 91–106. Nobel Symposium 84. New York: Columbia University Press.

Handschuh, G. J., and Orgel, L. E. 1973. Struvite and prebiotic phosphorylation. *Science* 179:483–484.

Hill, A. R., Jr., Orgel, L. E., and Wu, T. 1993. The limits of template-directed synthesis with nucleoside-5'-Phosphoro(2-methyl)imidazolides. *Origins Life Evol. Biosphere* 23:285–290.

Inoue, T., and Orgel, L. E. 1982. Oligomerization of (guanosine 5'-Phosphor)-2-methylimidazolideon poly(C). An RNA polymerase model. *J. Mol. Biol.* 162:201–217.

Inoue, T., and Orgel, L. E. 1983. A nonenzymatic RNA polymerase model. *Science* 219:859–862.

Joyce, G. F. 1989. RNA evolution and the origin of life. *Nature* 338:217–224.

Joyce, G. F., and Orgel, L. E. 1993. Prospects for understanding the origin of the RNA world. In *The RNA World,* ed. R. F. Gesteland and J. F. Atkins, pp. 1–25. Plainview, N.Y.: Cold Spring Harbor Laboratory Press.

Joyce, G. F., Schwartz, A. W., Miller, S. L., and Orgel, L. E. 1987. The case for an ancestral genetic system involving simple analogues of the nucleotides. *Proc. Natl. Acad. Sci. USA* 84:4398–4402.

Joyce, G. F., Visser, G. M., Boeckel, C. A. A. van, Boom, J. H. van, Orgel, L. E., and Westrenen, J. van. 1984. Chiral selection in poly(C)-directed synthesis of oligo(G). *Nature* 310:602–604.

Keefe, A. D., and Miller, S. L. 1995. Are polyphosphates or phosphate esters prebiotic reagents? *J. Mol. Evol.* 41:693–702.

Lohrmann, R., and Orgel, L. E. 1971. Urea–inorganic phosphate mixtures as prebiotic phosphorylating agents. *Science* 171:490–494.

Müller, D., Pitsch, S., Kittaka, A., Wagner, E.,Wintner, C. E., and Eschenmoser, A. 1990. Chemie von alpha-Aminonitrilen (135). *Helv. Chim. Acta* 73:1410–1468.

Oró, J. 1960. Synthesis of adenine from ammonium cyanide. *Biochem. Biophys. Res. Com.* 2:407–412.

Pitha, P. M., and Pitha, J. 1972. Nonenzymatic synthesis of oligoadenylates on template lacking steric regularity. *Nature New Biology* 240:78–80.

Pitsch, S., Krishnamurthy, R., Bolli, M., Wendeborn, S., Holzner, A., Minton, M., Lesueur, C., Schlönvogt, I., Jaun, B., and Eschenmoser, A. 1995. Pyranosyl-RNA (p-RNA): Base-pairing selectivity and potential to replicate. *Helv. Chim. Acta* 78:1621–1635.

Robertson, M. P., and Miller, S. L. 1995. An efficient prebiotic synthesis of cytosine and uracil. *Nature* 375:772–774.

Sanchez, R. A., Ferris, J. P., and Orgel, L. E. 1967. Studies in prebiotic synthesis. II. Synthesis of purine precursors and amino acids from aqueous hydrogen cyanide. *J. Mol. Biol.* 30:223–253.

Schmidt, J. G., Nielsen, P. E., and Orgel, L. E. 1997. Enantiomeric cross-inhibition in the synthesis of oligonucleotides on a nonchiral template. *J. Am. Chem. Soc.* 119:1494–1495.

Schwartz, A. W. 1997. Speculation on the RNA precursor problem. *J. Theoret. Biol.* 187:523–527.

Schwartz, A. W., and De Graaf, R. M. 1993. The prebiotic synthesis of carbohydrates: A reassessment. *J. Mol. Evol.* 36:101–106.

Schwartz, A. W., Van der Veen, M., Bisseling, T., and Chittenden, G. J. F. 1975. Prebiotic nucleotide synthesis: Demonstration of a geologically plausible pathway. *Origins of Life* 6:163–168.

Schwartz, A. W., Voet, A. B., and Van Der Veen, M. 1984. Recent progress in the prebiotic chemistry of HCN. *Origins of Life* 14:91–98.

Shapiro, R. 1988. Prebiotic ribose synthesis: A critical analysis. *Origins Life Evol. Biosphere* 18:71–85.

Shigemasa, Y., Sakazawa, C., Nakashima, R., and Matsuura, T. 1978. Selectivities in the formose reaction. In *Origin of Life*, ed. H. Noda, pp. 211–216. Tokyo: Japan Scientific Societies Press.

Stoks, P. G., and Schwartz, A. W. 1979. Uracil in carbonaceous meteorites. *Nature* 282:709–710.

Stoks, P. G., and Schwartz, A. W. 1981. Nitrogen-heterocyclic compounds in meteorites: Significance and mechanisms of formation. *Geochim. Cosmochim. Acta* 45:563–569.

Van Vliet, M. J., Visscher, J., and Schwartz, A. W. 1995. Hydrogen-bonding in the template-directed oligomerization of a pyrimidine nucleotide analogue. *J. Mol. Evol.* 41:257–261.

Visscher, J., and Schwartz, A. W. 1990. Template-catalyzed oligomerization with an atactic, glycerol-based, polynucleotide analog. *J. Mol. Evol.* 31:163–166.

Voet, A. B., and Schwartz, A. W. 1983. Prebiotic adenine synthesis from HCN: Evidence for a newly discovered major pathway. *Bioorganic Chem.* 12:8–17.

Wittung, P., Nielsen, P. E., Buchardt, O., Edholm, M., and Norden, B. 1994. DNA-like double helix formed by peptide nucleic acid. *Nature* 368:561–563.

Yamagata, Y., Watanabe, H., Saitoh, M., and Namba, T. 1991. Volcanic production of polyphosphates and its relevance to prebiotic evolution. *Nature* 352:516–519.

12

Catalyzed RNA synthesis for the RNA world

JAMES P. FERRIS
Department of Chemistry
Rensselaer Polytechnic Institute
Troy, New York

I. Introduction

The RNA world and the research on the forming RNA monomers was described in Chapter 11. Progress in the polymerization of monomers to RNA or RNA-like structures will be considered in this chapter. One of the arguments central to the discussion will be that catalysis was essential for the conversion of the complex mixtures of small organic molecules present on the primitive Earth to the biopolymers that led to the origin of life. This catalysis will be illustrated by examples from studies of the synthesis of RNA (Figure 12.1) from its monomeric units.

II. Catalysis was essential

Life as we know it on the Earth today is built on a complicated array of protein- and RNA-catalyzed reactions. These chemical processes, which occur on the surfaces of organic catalysts, result in, among other things, the formation of a complex array of biopolymers. None of the reactions proceed rapidly enough in the absence of these catalysts to sustain life. It is likely that catalysis was also essential for the prebiotic synthesis of biopolymers on the primitive Earth (Ferris 1993).

No prebiotic simulation experiments have been reported in which biopolymers are formed directly from simple inorganic and organic starting materials. It appears likely that the biopolymers that initiated life were formed by the reactions of more complicated monomers formed in previous prebiotic reactions. Scenarios proposed previously for biopolymer formation have involved the accumulation of solutions of monomers of the same type in one location, evaporation of the water in which they were dissolved, and then the heating of the monomers to the appropriate temperature to form polymers (but not so hot that the polymer decomposes) (Fox 1988). It is assumed in these scenarios that on the primitive Earth the polymers were then washed off the

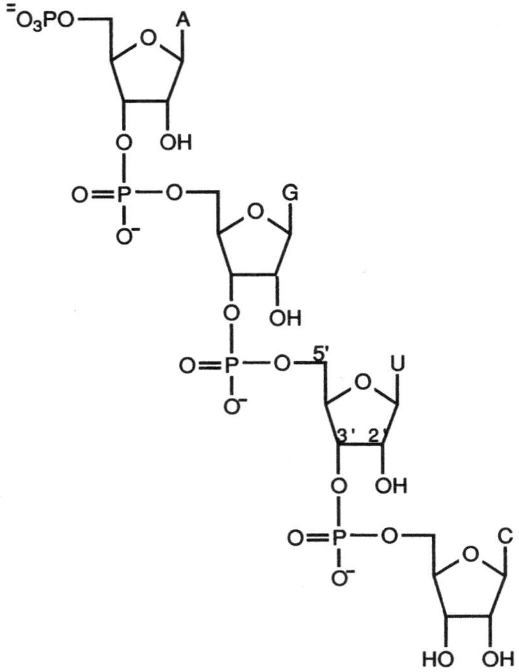

Figure 12.1. An RNA composed of four monomer units. A = adenine, G = guanine, U = uracil, and C = cytosine.

surface into solution where they reacted further. One can imagine that, in special instances, one stage of this process may have occurred, but it is difficult to see how a series of these processes essential for the origin of the first life took place. For example, if amino acids polymerized to small catalytic peptides in the solid phase and were then washed into a lake or ocean, how were they concentrated in another locale to catalyze the formation of additional polypeptides or polynucleotides?

In the new scenario, it is proposed that mineral surfaces catalyzed polymer formation and that water was the medium that carried the organics that were essential for life to the surfaces where reactions took place. Since the reactions central to the origin of life had to proceed faster than decomposition processes such as hydrolysis by water, catalysis was essential. Once the central synthetic reactions had occurred, the first life must have originated with the polymers bound at the catalytic surface or at a nearby surface "downstream" from the reaction center. The specificity essential for the binding of one type of monomer from the prebiotic soup was provided by specific binding interactions between the monomers and the mineral surface. Adsorption from solution provided a high local concentration of one class of monomers, a situation favorable for polymer formation. Specific sites of the mineral, such as surface defects, may have been where polymer formation was catalyzed.

A possible site for this scenario may have been near a hydrothermal system where there was extensive circulation of seawater through reefs of crustal minerals (Ferris 1992; Holm 1992). An assemblage of interacting polymers bound to a rock may have been the first living system. This system may have been similar to that of contemporary life that is bound to a rock, for example, abalone and mussels. They separate out the nutrients for life from the ocean water flowing around them, much in the same way that the proposed prebiotic polymers may have grown by reacting with the activated monomers flowing past them.

III. Why investigate the prebiotic synthesis of RNA?

Problems exist with the proposed prebiotic synthesis of ribose from formaldehyde, as well as with other aspects of the prebiotic synthesis of ribonucleotides. These problems have prompted the search for simpler structures, similar to ribonucleotides, which may have been the basis for the first life on Earth (see Chapter 11). The principal problem associated with the simpler structures proposed to date is their propensity to cyclize rather than form oligomers. Ribose provides the essential molecular architecture for nucleotides where the oligomerization via a 5'-phosphate group is favored over cyclization with the 3'-hydroxyl grouping. These data suggest that alternatives to ribose must have been sufficiently complex so that cyclization was not a principal reaction pathway (Hill et al. 1988; Ferris and Kamaluddin 1989). At present, it appears that a prebiotically plausible nucleotide analogue, the oligomerization of which is not susceptible to cyclization and enantiomeric cross-inhibition (see Chapter 11), may have been as complicated as contemporary ribonucleotides. Consequently, the discussion in this chapter focuses on the proposed prebiotic formation of RNA from activated ribonucleotides. It is possible that a pre-RNA will be devised that overcomes the problems noted for the pre-RNAs postulated so far. It is our belief that this derivative will have many of the characteristics of ribonucleotides, so that our studies on the mineral-catalyzed formation of RNA will be applicable to it as well.

IV. Prebiotic catalysts

The need for catalysis of prebiotic reactions was first recognized by Bernal in a lecture given in 1947 (Bernal 1949). Bernal was mainly concerned with the need for the concentration of organics, with protecting them from destruction by ultraviolet light, and with the polymerization of the adsorbed organics, and

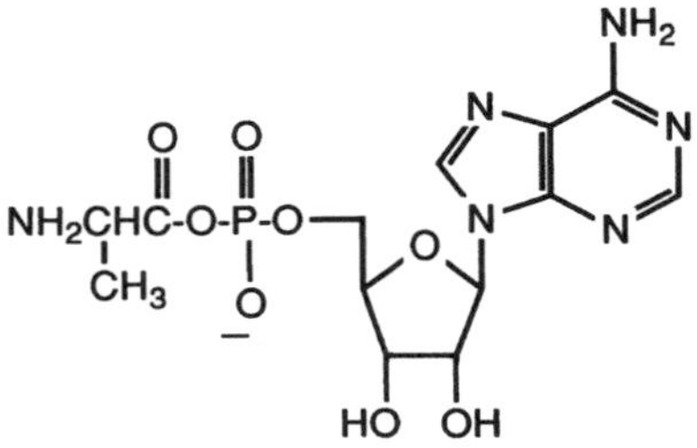

Figure 12.2. The amino acid adenylate of alanine with 5'-AMP.

he proposed that adsorption of prebiotic organics on clays could have accomplished these objectives. In subsequent proposals, Cairns-Smith suggested not only that clay minerals catalyzed organic reactions but that their surfaces contained the genetic material for life, and he proposed that clay minerals were the first life on Earth (Cairns-Smith 1982). Much later, Wächtershäuser (1988) proposed that the metabolic processes leading to the origin of life occurred on minerals.

There have been a number of reports of experiments designed to identify mineral catalysts that may have catalyzed prebiotic reactions (Rao, Odom and Oró 1980; Ponnamperuma, Shimoyama and Friebele 1982). The most successful early experiment was the use of a montmorillonite clay to catalyze formation of polypeptides from amino acid adenylates (Figure 12.2) (Paecht-Horowitz and Eirich 1988). This approach to polypeptide synthesis has been criticized because of the low probability of the formation of such a high-energy intermediate, and there is a certain amount of unease because others have not been able to repeat the reported formation of the aminoacyladenylate by the reaction of an amino acid with ATP in a process catalyzed by a zeolite clay (Warden et al. 1974). A recent paper reports the synthesis of polypeptides containing more than 55 monomer units on illite clay and hydroxylapatite (Ferris et al. 1996).

Although there have been few successful attempts to find minerals that catalyze polymer formation, there has been much greater success in demonstrating metal ion catalysis. For example, mainly 3',5'-linked oligo(G)s are formed in the template-directed reaction of the phosphorimidazolide of guanine (ImpG, N = G, R = H) (Figure 12.3) on oligo(C)s when Zn^{2+} is the catalyst; 2',5'-linked oligomers are formed when the reaction is catalyzed by Pb^{2+} (Lohrmann, Bridson and Orgel 1980). The template-directed synthesis of oligo(A)s is catalyzed by Pb^{2+} but not Zn^{2+} (Sleeper, Lohrmann and Orgel 1979). UO_2^{2+} catalyzes the formation of oligonucleotides from ImpN where N is A, U, and C (Sawai, Kuroda and Hojo 1988; Sawai, Higa and Kuroda 1992).

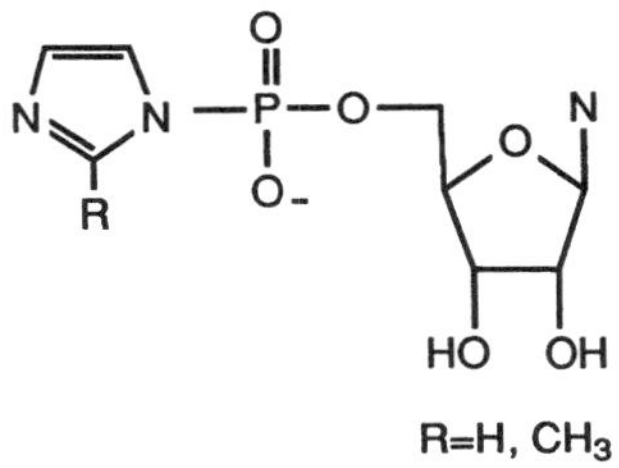

Figure 12.3. The phosphorimidazolide of a nucleotide where N = A, G, U, C. A = adenine, G = guanine, U = uracil, and C = cytosine.

V. Montmorillonite catalysis of RNA formation: a search for catalysts

Initial studies on the search for the catalysis centered on studies of the oligomerization of HCN. Unexpectedly, montmorillonite inhibited oligomer formation (Ferris et al. 1979). Since a HCN tetramer (diaminomaleonitrile, DAMN) had been shown to be the precursor of these oligomers, the reaction of DAMN with montmorillonite was investigated. It was observed that DAMN underwent a rapid oxidation to DISN (diiminosuccinonitrile) by the Fe^{3+} in the montmorillonite, and this oxidation was the cause of the inhibition of oligomer formation (Figure 12.4) (Ferris et al. 1982).

This negative result had a positive outcome, since the very reactive DISN served as an agent for the formation of phosphodiester bonds as shown by the cyclization of 3'-AMP to 2',3'-cyclic AMP. It was observed that the cyclization reaction proceeded more efficiently when a mixture of DISN and montmorillonite was used as the condensing agent instead of DISN alone (Ferris, Huang and Hagan 1988). This suggested that the clay was catalyzing the reaction, and led to the proposed reaction scheme shown in Figure 12.5.

Since cyclization with the proximate 2'-hydroxyl is strongly favored, no oligomers are formed when the 3'-phosphate group of 3'-AMP is activated. No oligomers were formed when the 5'-phosphate group of 5'-AMP was activated with DISN in the presence of montmorillonite, but dimers were observed

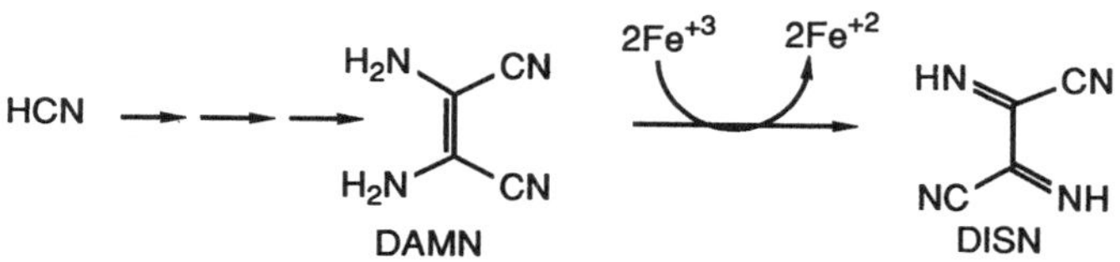

Figure 12.4. The condensation of HCN to diaminomaleonitrile (DAMN) and the oxidation of DAMN to diiminosuccinonitrile (DISN).

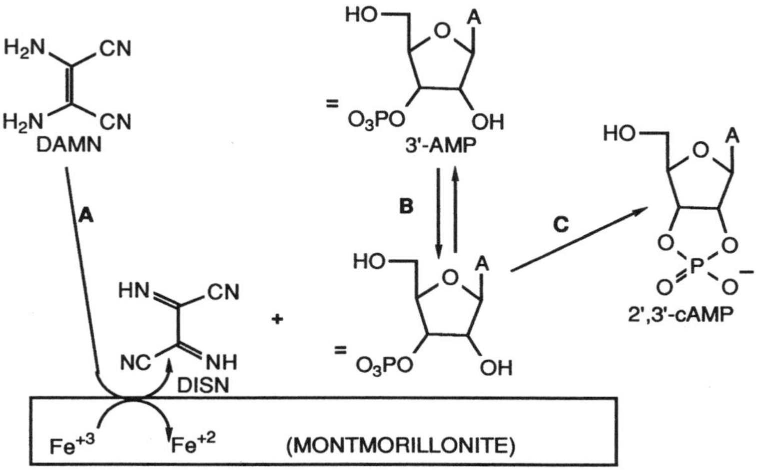

Figure 12.5. Montmorillonite catalysis of the formation of 2', 3'-cyclic AMP from 3'-AMP. A. The reversible binding of 3'-AMP to montmorillonite. B. DAMN reacts with ferric iron in the montmorillonite to generate DISN proximate to 3'-AMP. C. Montmorillonite catalysis of the reaction of DISN with 3'-AMP to generate 2', 3-cyclic AMP, which is released from the surface of the montmorillonite. (Ferris, Huang and Hagan 1988; with kind permission from Kluwer Academic Publishers.)

when a water-soluble carbodiimide was the activating agent (Ferris, Ertem and Agarwal 1989). In addition, the montmorillonite changed the regioselectivity of phosphodiester bond formation from mainly 2',5'-links in the absence of clay (Lohrmann and Orgel 1978) to a 0.6:1 ratio of 3',5'/2',5' in its presence.

Although it is encouraging that montmorillonite catalyzed the formation of RNA oligomers, the dimers formed are too short to have initiated the RNA world. Studies on the condensation of RNA dimers in an attempt to build the oligomers two units at a time led to tetramers joined by the pyrophosphate grouping, but none were formed that were linked by phosphodiester bonds (Ferris and Peyser 1994).

Longer RNA oligomers were formed when an activated nucleotide monomer was reacted in the presence of montmorillonite. This approach was prompted by the observation that use of the preformed, activated nucleotide greatly enhanced the yields of longer oligomers in template-directed synthesis (Weimann et al. 1968). Reaction of the ImpA in the presence of montmorillonite at room temperature in pH 8 aqueous solution containing Mg^{2+} and Na^{+} resulted in the formation of dimers to decamers (Figure 12.6) (Ferris and Ertem 1992; Ferris and Ertem 1993). The compositions of the dimer, trimer, and tetramer fractions were determined by selective enzymatic hydrolysis after separation of the fractions by HPLC. Extension of the reaction to the

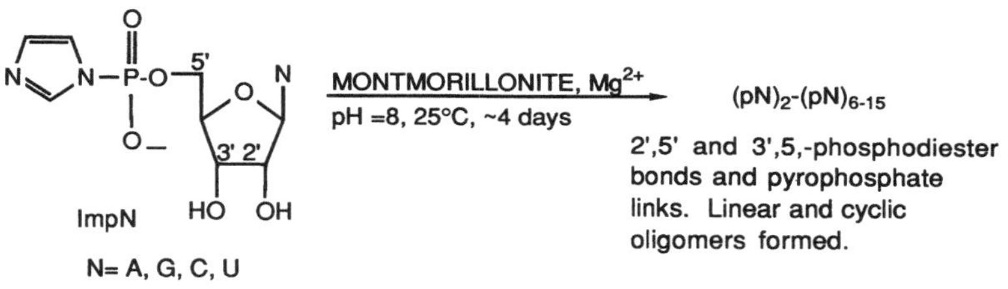

ImpN

N= A, G, C, U

MONTMORILLONITE, Mg^{2+}

pH =8, 25°C, ~4 days

$(pN)_2$-$(pN)_{6-15}$

2',5' and 3',5,-phosphodiester bonds and pyrophosphate links. Linear and cyclic oligomers formed.

Figure 12.6. The condensation of the phosphorimidazolides of nucleotides on montmorillonite in aqueous solution. N = A, G, U, C. A = adenine, G = guanine, U = uracil, and C = cytosine.

phosphorimidazolides of the nucleotides of U, C, I, and G established that the longest oligomers are obtained with the C and the shortest with G. (Ding, Kawamura and Ferris 1996; Ertem and Ferris 1996; Kawamura and Ferris unpublished results). The average regioselectivity for the formation of 3',5'-links was greatest for A and I (67% and 94% [dimer fraction only] respectively) and lowest for U and C (24% and 35% respectively). Since only 6 mers are formed in the reaction of ImpG, the regioselectivity of its phosphodiester bond formation was not investigated.

The promising results obtained using a preformed, activated phosphate prompted a search for other activating groups that would give longer oligomers and a greater regioselectivity for the formation of 3',5'-linked phosphodiester bonds. An investigation of a variety of potential activating groups resulted in the discovery that 4-aminopyridine and its alkylated derivatives are excellent activating groups (Prabahar, Cole and Ferris 1994). Reaction of 5'-AMP in which the phosphate grouping is activated by 4-dimethylaminopyridine (DMAP) (Figure 12.7a) resulted in the formation of 12 mers, with an average regioselectivity for 3',5'-phosphodiester bond formation of 88%. The structural similarity between 4-aminopyridines with purines and 4-aminopyrimidines prompted their consideration as activating groups. The use of purines and pyrimidines prebiotically is more plausible than the use of imidazole or DMAP as activating groups because purines and pyrimidines must have been available for the formation of RNA, so they would therefore also have been available as activating groups for phosphate condensation reactions. When 1-methyladenine is used to activate the 5'-phosphate of adenosine (Figure 12.7b), oligomers as long as 11 mers are formed, with an average 3',5'-regioselectivity of 84% (Prabahar and Ferris unpublished results). 2-Methyladenine and adenine initiated the formation of 7 mers and 5 mers respectively, with a regioselectivity for forming 3',5'-links of 78% and 72% respectively.

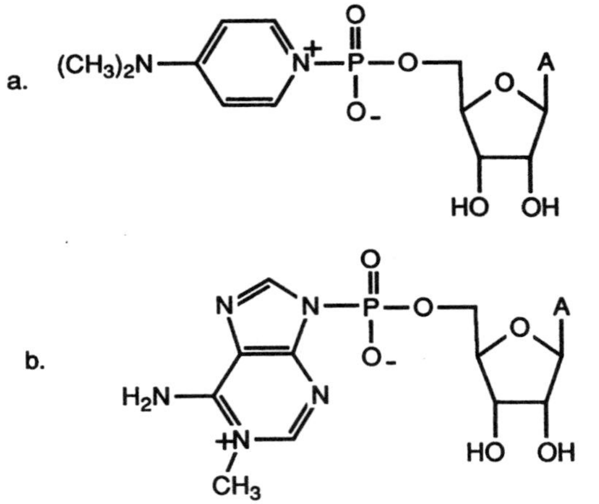

Figure 12.7. Activated nucleotides. a. Adenosine 5'-phosphoro-4-(dimethylamino) pyridinium. b. Adenosine 5'-phosphoro-1-methyladeninium.

VI. Synthesis of 20–50 mers of RNA

Although the synthesis of 10–15 mers was a major incremental increase in the length of RNA over the 2–4 mers formed using carbodiimide, the 10–15 mers formed are not sufficiently long enough to expect that they will have significant catalytic activity (Joyce and Orgel 1993; Szostak and Ellington 1993). Elongation of primers was investigated since it is likely that some of the 10–15 mers formed by mineral-surface catalysis remain bound to the surface. These primers would have been elongated by the addition of activated monomers. The alternative possibility, the self-condensation of the activated monomers to form dimers, is less likely since the rate constant for dimer formation is 10 times smaller than the rate constant for primer elongation (Kawamura and Ferris 1994). Experimental studies in which fresh activated monomer was added daily to a 10-mer primer over a 14-day period resulted in the incremental growth of the decamer to a mixture of 20–50 mers (Ferris et al. 1996). Oligomers in the 30–50 mer size range are considered to be sufficiently long to exhibit catalytic activity (Joyce and Orgel 1993; Szostak and Ellington 1993). These studies suggest that it may have been possible to form copolymers of RNA that were sufficiently long to have exhibited catalytic activity. Formation of catalytic RNA would have been a crucial step in the origin of the RNA world.

VII. Template-directed synthesis

RNAs are believed to have been central to the origin of the first life on Earth not only because of their catalytic properties but also because they can pre-

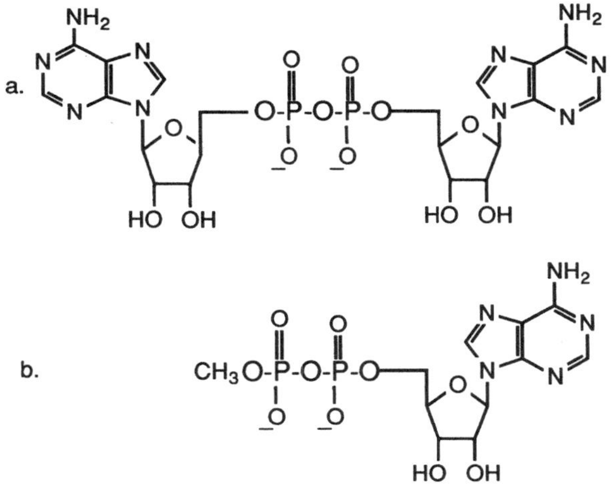

Figure 12.8. a. P[1],P[2]-diadenosine pyrophosphate. b. P[1]- adenosine P[2]-methylpyrophosphate.

serve sequence information by template-directed synthesis (see Chapter 11). The extensive investigations of template-directed synthesis by Leslie Orgel and co-workers utilized linear templates containing only 3',5'-phosphodiester bonds (Lohrmann, Bridson and Orgel 1980). RNA formed by processes that occurred on the primitive Earth probably also contained 2',5'- and pyrophosphate-linked oligomers (Figure 12.8a) along with cyclic oligomers. It is not clear whether templates containing a complex mixture of linkages would have been capable of catalyzing the formation of the templates' complementary oligomers.

The effect of nonhomogeneous templates on template-directed synthesis was investigated by using the oligo(C)s formed by the reaction of ImpC on montmorillonite (Ertem and Ferris 1996). As noted previously, a high proportion of 2',5'-linked oligomers are formed in the reaction of ImpC on montmorillonite. The hexamers and higher oligomers were isolated from the reaction mixture and were shown to catalyze the conversion of 2-methylImpG (Figure 12.3, N = G, R = CH_3) to 3-6 mers. A 2',5'-linked oligo(C) template also catalyzed the synthesis of oligomers of oligo(G). Thus heterogeneous templates of the type likely to have been formed on the prebiotic Earth were capable of forming their complementary oligomers and thus preserving their information content. Catalysis other than that provided by a template is required for the effective template-directed synthesis of oligomers containing A, U, and C nucleotides (Chapter 11).

VIII. The nature of clay catalysis

A. The formation and structure of montmorillonite clays

Several lines of evidence indicate the presence of clays on the primitive Earth. First, clays are formed by the weathering of volcanic rock. Volcanos are believed to have been prevalent on the prebiotic Earth because this would have been a way to release the heat produced in the accretion process. Reaction of volcanic rock with water results in the dissolution of aluminum and silicon compounds, which then react with each other to form the aluminosilicates that constitute clays. Today, clays are one of the more abundant constituents of the crust and are often found in large deposits all over the Earth. There is evidence for the presence of clay minerals in the 3.8 bya Isua Supercrustal formation in Greenland and they have also been found in meteorites. Montmorillonite clay consists of aluminosilicate sheets (platelets) (Figure 12.9) in which Fe(II), Fe(III), and Mg(II) are substituted for some of the Al(III), and some of the Si(IV) is substituted by Al(III). Because the oxygen content of the sheets does not change, the substitution for a higher valent metal ion by a lower valent one, for example, Al(III) by Mg(II), generates a negatively charged sheet. This negative charge is balanced by cations; in contemporary clays, these are mainly Na^+ and Ca^{2+}. Montmorillonite clays tend to associate like a deck of cards, but they also have edge-to-face interactions in which the acidic edges bind to the negative faces.

Cationic organic compounds bind to the surfaces of montmorillonite by displacement of the exchangeable metal ions. Divalent metal ions are especially helpful in enhancing the binding of anionic organics because they can bind simultaneously to the organic compound and the clay surface (Figure 12.9).

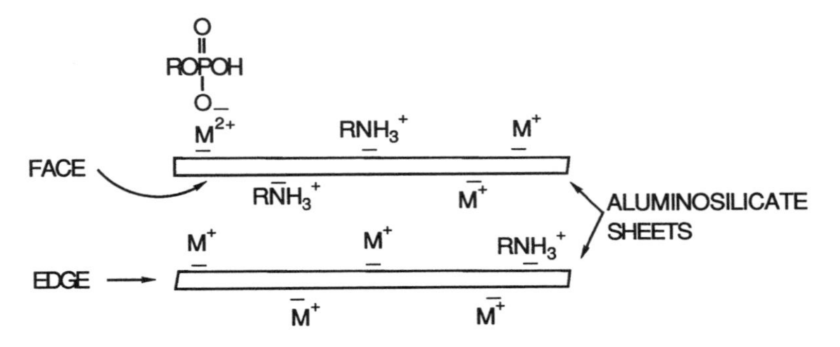

Figure 12.9. A simplified diagram of a montmorillonite clay with bound anionic and cationic organics.

B. Demonstration of catalysis

1. Kinetic studies

It was determined that the rate constants for the formation of 2 mers from ImpA were enhanced by a factor of about 1.4×10^3 in the presence of montmorillonite as compared to the rate constant in its absence (Kawamura and Ferris 1994). The rate constant for the hydrolysis of ImpA was shown to increase by a factor of about 35 in the presence of montmorillonite. The 400-fold difference between the hydrolytic rate and oligomerization results explains the enhanced formation of oligo(A)s in the clay-catalyzed reaction. Longer oligomers are formed in this clay-catalyzed reaction because the rate constants increase in the order 2 mer < 3 mer < 4 mer for the formation of dimers to tetramers before leveling off. The 10-fold difference between the rate of formation of dimers versus the rate of formation of tetramer and longer oligomers means that chain elongation is favored over dimerization. This accounts for the observation of longer oligomers; only a more rapid formation of dimers and trimers would be observed if the rate constants did not increase.

2. Inhibition studies

Inhibition studies provided insight into the site of catalysis on the platelets of montmorillonite (Ertem and Ferris in press). Binding of alkyltrimethylammonium salts to the clay interlayers, where the alkyl group is a bulky dodecyl grouping that fills the interlayer, prevents the binding of activated nucleotides and inhibits oligomer formation. Similar inhibition was observed when $Al_{13}O_4(OH)_{28}{}^{3+}$ cations were bound between the clay platelets (Pinnavia 1983). Reaction of montmorillonite with trimethylsilyl chloride results in the chemical binding of bulky trimethylsilyl groups to the silanol and aluminol groupings at the edge sites on the montmorillonite (Sindorf and Maciel 1983). These trimethylsilyl-substituted clays have catalytic activity that is comparable to that of the nonsilylated clay, a result consistent with the observation that the oligomer formation does not occur at the edges of the montmorillonite.

Organic compounds that don't react with ImpU have been observed to inhibit its condensation to oligomers on montmorillonite. For example, the pyrophosphates CH_3ppA (Figure 12.8b) and $dA^{5'}$ppA strongly inhibit oligomer formation (Ding, Kawamura and Ferris 1996). The deoxypyrophosphate $dA^{5'}$ppA does not react with ImpU because its 3'-hydroxyl group is much less reactive than the vicinal 2',3'-dihydroxy grouping present in the ribonucleotides. This inhibition suggests that these pyrophosphates prevent the binding and reaction of ImpU at the catalytic site. This result is consistent with the finding that mixtures that contain ImpU and the ribopyrophosphate $A^{5'}ppA$ react mainly to form oligomers in which pU units are added to the $A^{5'}ppA$. In these reactions, the $A^{5'}ppA$ binds preferentially to the site where its reaction with ImpU is catalyzed.

IX. Conclusions

Catalyzed reactions were essential for the formation of the biopolymers that initiated life on Earth. Experimental studies have shown that it may have been possible to make the transition from the prebiotic world to the RNA world by the montmorillonite-catalyzed formation of RNA. Ten to 15 mers are formed in the one-step reaction of the phosphorimidazolide of a nucleoside in the presence of montmorillonite clay, whereas up to 50 mers are formed in feeding reactions that proceed for 14 days. The heterogeneous mixture of oligomers formed in the montmorillonite-catalyzed reaction serve as templates for the formation of the complementary oligomers. This finding demonstrates that information storage is possible even though heterogeneous oligomers are formed by mineral catalysis. What remains to be done is the demonstration of catalytic activity in the random mixture of oligomers formed by clay mineral catalysis. Since information storage and catalysis are essential for the origin of the RNA world, the observation of catalysis will establish that RNA-based life may have evolved from prebiotic monomers via polymerization reactions that were catalyzed by minerals or other catalysts.

References

Bernal, J. D. 1949. The physical basis of life. *Proc. Roy. Soc. London* 357A:537–558.

Cairns-Smith, A. G. 1982. *Genetic Takeover and the Mineral Origins of Life.* Cambridge: Cambridge University Press.

Ding, P. Z., Kawamura, K., and Ferris, J. P. 1996. Oligomerization of uridine phosphorimidazolides on montmorillonite: A model for the prebiotic synthesis of RNA on minerals. *Origins Life Evol. Biosphere* 26:151–171.

Ertem, G., and Ferris, J. P. 1996. Synthesis of RNA oligomers on heterogeneous templates. *Nature* 379:238–240.

Ertem, G., and Ferris, J. P. In press. Formation of RNA oligomers on montmorillonite: Site of catalysis. *Origins of Life Evol. Biosphere* 28.

Ferris, J. P. 1992. Marine hydrothermal systems and the origin of life: Chemical markers of prebiotic chemistry in hydrothermal systems. *Origins Life Evol. Biosphere* 22:109–134.

Ferris, J. P. 1993. Catalysis and prebiotic RNA synthesis. *Origins Life Evol. Biosphere* 23:307–315.

Ferris, J. P., Edelson, E .H., Mount, N. M., and Sullivan, A. E. 1979. The effect of clays on the oligomerization of HCN. *J. Mol. Evol.* 13:317–30.

Ferris, J. P., and Ertem, G. 1992. Oligomerization reactions of ribonucleotides on montmorillonite: Reaction of the 5'-phosphorimidazolide of adenosine. *Science* 257:1387–1389.

Ferris, J. P., and Ertem, G. 1993. Montmorillonite catalysis of RNA oligomer formation in aqueous solution: A model for the prebiotic formation of RNA. *J. Am. Chem. Soc.* 115:12270–12275.

Ferris, J. P., Ertem, G., and Agarwal, V. K. 1989. Mineral catalysis of the formation of dimers of 5'-AMP in aqueous solution: The possible role of montmorillonite clays in the prebiotic synthesis of RNA. *Origins Life Evol. Biosphere* 19:165–178.

Ferris, J. P., Hagan, W. J., Jr., Alwis, K. W., and McCrea, J. 1982. Chemical evolution. 40. Clay-mediated oxidation of diaminomaleonitrile. *J. Mol. Evol.* 18:304–309.

Ferris, J. P., Hill, A. R., Jr, Liu, R., and Orgel, L. E .1996. Synthesis of long prebiotic oligomers on mineral surfaces. *Nature* 381:59–61.

Ferris, J. P., Huang, C.-H., and Hagan, W. J., Jr. 1988. Montmorillonite: A multifunctional mineral catalyst for the prebiological formation of phosphate esters. *Origins Life Evol. Biosphere* 18:121–133.

Ferris, J. P., and Kamaluddin, 1989. Oligomerization reactions of deoxyribonucleotides on montmorillonite clay: The effect of mononucleotide structure on phosphodiester bond formation. *Origins Life Evol. Biosphere* 19:609–619.

Ferris, J. P., and Peyser, J. R. 1994. Rapid and efficient syntheses of phosphorylated dinucleotides. *Nucleosides and Nucleotides* 13:1087–1111.

Fox, S. 1988. *The Emergence of Life*. New York: Basic Books Inc.

Hill, A. R., Jr., Dee Nord, L., Orgel, L. E., and Robins, R. K. 1988. Cyclization of nucleotide analogs as an obstacle to polymerization. *J. Mol. Evol.* 28:170–171.

Holm, N. G. 1992. Why are hydrothermal systems proposed as plausible environments for the origin of life? *Origins Life Evol. Biosphere* 22:5–14.

Joyce, G. F., and Orgel, L. E. 1993. Prospects for understanding the origin of the RNA world. In *The RNA World*, ed. R. F. Gesteland and J. F. Atkins, pp. 1–25. Cold Spring Harbor, N.Y.: Cold Spring Harbor Laboratory Press.

Kawamura, K., and Ferris, J. P. 1994. Kinetic and mechanistic analysis of dinucleotide and oligonucleotide formation from the 5'-phosphorimidazolide of adenosine on Na^+-montmorillonite. *J. Am. Chem. Soc.* 116:7564–7572.

Lohrmann, R., Bridson, P. K., and Orgel, L. E. 1980. Efficient metal-ion catalyzed template-directed oligonucleotide synthesis. *Science* 208:1464–1465.

Lohrmann, R., and Orgel, L. E. 1978. Preferential formation of (2'-5')-linked internucleotide bonds in non-enzymatic reactions. *Tetrahedron* 34:853–855.

Paecht-Horowitz, M., and Eirich, F. R. 1988. The polymerization of amino acid adenylates on sodium-montmorillonite with preadsorbed peptides. *Origins Life Evol. Biosphere* 18:359–387.

Pinnavia, T. J. 1983. Intercalated clay catalysis. *Science* 220:365–371.

Ponnamperuma, C., Shimoyama, A., and Friebele, E., 1982. Clay and the origin of life. *Origins Life Evol. Biosphere* 12:9–40.

Prabahar, K. J., Cole, T. D., and Ferris, J. P. 1994. Effect of phosphate activating group on oligonucleotide formation on montmorillonite: The regioselective formation of 3',5'-linked oligoadenylates. *J. Am. Chem. Soc.* 116:10914–10920.

Prabahar, K. J., and Ferris, J. P. 1997. Adenosine derivatives as phosphate-activating groups for the regioselective formation of 3',5'-linked oligoadenylates on montmorillonite: possible phosphate-activating groups for the prebiotic synthesis of RNA. *J. Am. Chem. Soc.* 119:4330–4337.

Rao, M., Odom, D. G., and Oró, J. 1980. Clays in prebiological chemistry. *J. Mol. Evol.* 15:317–331.

Sawai, H., Higa, K., and Kuroda, K. 1992. Synthesis of cyclic and acyclic oligocytidylates by uranyl ion catalyst in aqueous solution. *J .Chem. Soc. Perkin* I:505–508.

Sawai, H., Kuroda, K., and Hojo, T. 1988. Efficient oligoadenylate synthesis catalyzed by uranyl ion complex in aqueous solution. In *Nucleic Acids Research, Symposium Series No. 19*, ed. H. Hayatsu, pp. 5–7. Oxford: IRL Press Limited.

Sindorf, D. W., and Maciel, G. E. 1983. Silicon-29 nuclear magnetic resonance study of hydroxyl sites on dehydrated silica gel surfaces, using silylation as a probe. *J. Phys. Chem.* 87:5516–5521.

Sleeper, H. L., Lohrmann, R., and Orgel, L. E. 1979. Temple-directed synthesis of oligoadenylates catalyzed by Pb^{2+} ions. *J. Mol. Evol.* 13:203–214.

Szostak, J. W., and Ellington, A. D. 1993. In vitro selection of functional RNA sequences. In *The RNA World*, ed. R. F. Gesteland and J. F. Atkins, pp. 511–533. Cold Spring Harbor, N.Y.: Cold Spring Harbor Laboratory Press.

Wächtershäuser, G. 1988. Before enzymes and templates: theory of surface metabolism. *Microbiol. Rev.* 52(4):452–484.

Warden, J. T., McCullough, J. J., Lemmon, R. M., and Calvin, M. 1974. A re-examination of the zeolite-promoted, clay-mediated peptide synthesis. *J. Mol. Evol.* 4:189–194.

Weimann, B. J., Lohrmann, R., Orgel, L. E., Schneider-Bernloehr, H., and Sulston, J. E. 1968. Template-directed synthesis with adenosine-5'-phosphorimidazolide. *Science* 161:387–388.

13

Catalysis in the RNA world

KENNETH D. JAMES and ANDREW D. ELLINGTON
Department of Chemistry
Indiana University, Bloomington, Indiana

Introduction

There are three routes to the discovery of the history of life. First, since modern biology and biochemistry are the result of genetic and phenotypic divergence, descriptions of past life can be derived from comparative biology and biochemistry; this route extrapolates backwards from the secure present. Second, since all biology is of necessity constrained by the laws of chemistry and physics, descriptions of past life can be derived by attempting to understand what living systems may have been a priori possible, rather than what living systems have a posteriori evolved; this route looks forward from an imagined past (for a perspective, see James and Ellington 1995). Finally, since evolutionary changes in form and function (either molecular or organismal) have frequently been shown to be based on preexisting form and function, the biochemistry of ancient and modern organisms should be linked by a continuous (but not necessarily gradual) path. This route is an amalgam of the first two, and it attempts to connect the present and the past. For example, if this last proposition is true, it is unlikely that complex assemblages such as ribosomes arose fully formed, but rather they were likely derived from molecules of similar form or function that preceded them. By demonstrating that molecules with properties similar to the ribosome could have existed, we provide credence for this route; the more intermediates that can be demonstrated or extrapolated, the more sure is the link between present and past. We fully realize that this proposition is less subject to evidential constraints and relies on a purely naturalistic interpretation of the world. However, it is useful to see what the utility of these assumptions is; that is, if we can construct a self-consistent and coherent story of the origins and evolution of life by assuming the existence of molecular intermediates, then this is in itself a validation that such a process may have occurred in fact. Similar "proofs from self-consistency" have previously been used by Weiner and Maizels to argue for the antiquity of molecular biological idiosyncrasies (Weiner and Maizels 1987,

1990). In the limit, we might expect that the boundary conditions that determined the nature of ancient replicators would eventually be played out in the nature of contemporary organisms and that the biochemical mechanisms of contemporary organisms would have counterparts in the first organic replicators. For example, the fact that chemical replicators both in the lab and in modern organisms in the field pass DNA between generations via the back-and-forth process of semiconservative replication implies that this mechanism may have been continuously operative from the first living systems onwards.

These three explanatory routes can be used to assess the so-called RNA world hypothesis (Gilbert 1986). In brief, this hypothesis holds that the earliest self-replicating molecules were likely either nucleic acids or compounds that could have templated the synthesis of nucleic acids. Self-replicating nucleic acids acquired additional metabolic functions and formed symbiotic associations with other functional molecules and macromolecular assemblies, such as amino acids or peptides, and lipids or lipidlike membranes. The elaboration of nascent nucleic acid catalysts led to the development of a wide variety of catalytic activities and the invention of cofactors that were derived from (and therefore appropriate for) nucleic acids (Hager, Pollard and Szostak 1996). A self-consistent set of metabolic reactions arose based on the nucleic acid catalysts and nucleotide cofactors (Benner, Ellington and Tauer 1989; Benner et al. 1993). The ribozymatic metabolic machinery eventually sowed the seeds of its own destruction when some form of translation was invented. RNA catalysts were supplanted by protein catalysts, and only a few vestiges of the RNA world can still be charted in modern metabolism.

Three lines of evidence that support each of the three explanatory routes will be examined in turn. First, we will look at modern RNA molecules and try to discern what their likely antecedents were. Second, we will look at what RNA molecules could have existed and thereby try to discern what types of organismal biochemistry and metabolism may have been possible. And, finally, we will attempt to tie these two lines of evidence together in the form of a narrative for how life arose and evolved. To the extent that this narrative is self-consistent, it further validates the facts used in its construction.

Modern natural history of the ribozyme

Since the discovery that RNA molecules can catalyze reactions in vivo (Kruger et al. 1982), seven general classes of catalytic RNAs (ribozymes) have been found. Each of these ribozymes utilizes metal cations as cofactors, and each manipulates RNA phosphodiester linkages. Although the natural ribozymes all cleave RNA with a high degree of site specificity, several of them also perform ligation reactions. The size range of extant ribozymes is large: the smallest functional ribozymes are less than 20 nucleotides in length, and the largest are over 300 nucleotides. We will first review the ribozyme classes in greater detail, focusing on attributes that may tie them to the past. We will then examine what their mechanisms may reveal about catalysis in a putative RNA world.

Description and origins

Several small (<100 nucleotide) ribozymes are found in viruses and carry out site-specific cleavage reactions that facilitate viral replication. With a functional domain of less than 20 nucleotides, the hammerhead ribozyme is the smallest known ribozyme (for a review, see Sigurdsson and Eckstein 1995). The hammerhead was initially identified in the avocado sunblotch viroid (Hutchins et al. 1986) but has since been found in other viruses. A high-resolution structure has been obtained by X-ray crystallography and reveals that the three helices are arranged in a simple Y (Figure 13.1; Pley, Flaherty and McKay 1994; Scott, Finch and Klug 1995). The hairpin ribozyme (for a review, see Chowira and Burke 1991) was found in a satellite RNA of the tobacco ringspot virus (Prody et al. 1986). The 50 or so nucleotides in the catalytic core are arranged into four short helices (Butcher and Burke 1994). The small size and structural simplicity of the hammerhead and hairpin ribozymes and their presence in selfish genetic elements are facts that are consistent with a recent, chance genesis.

Several larger (>100 nucleotide) ribozymes have been found in organisms and carry out both cleavage and ligation reactions. The Group I and Group II introns (for reviews, see Saldanha et al. 1993; Michel and Ferat 1995) are complex catalysts with cores that are well over 200 nucleotides in length. The Group I self-splicing intron was discovered when Tom Cech and co-workers observed that the intervening sequence of preribosomal RNA underwent cleavage in the absence of any enzyme or protein (Kruger et al. 1982). Hundreds of different Group I introns have been found in nuclear and endosymbiont genomes in a wide variety of lineages, and have even been discovered in organisms not normally known to have introns, such as bacteria and bacteriophages (Reinhold-Hurek and Shub 1992). The structure of the Group I intron was predicted by a combination of comparative sequence analysis and molecular modeling (Michel and Westhof 1990). Whereas the helices of the hammerhead ribozyme are sparsely interconnected, the helices of the Group I ribozyme form long coaxial stacks and participate in intricate, sequence-specific, tertiary structural interactions. The Group II intron was discovered in chloroplasts and in fungal mitochondria (Michel and Dujon

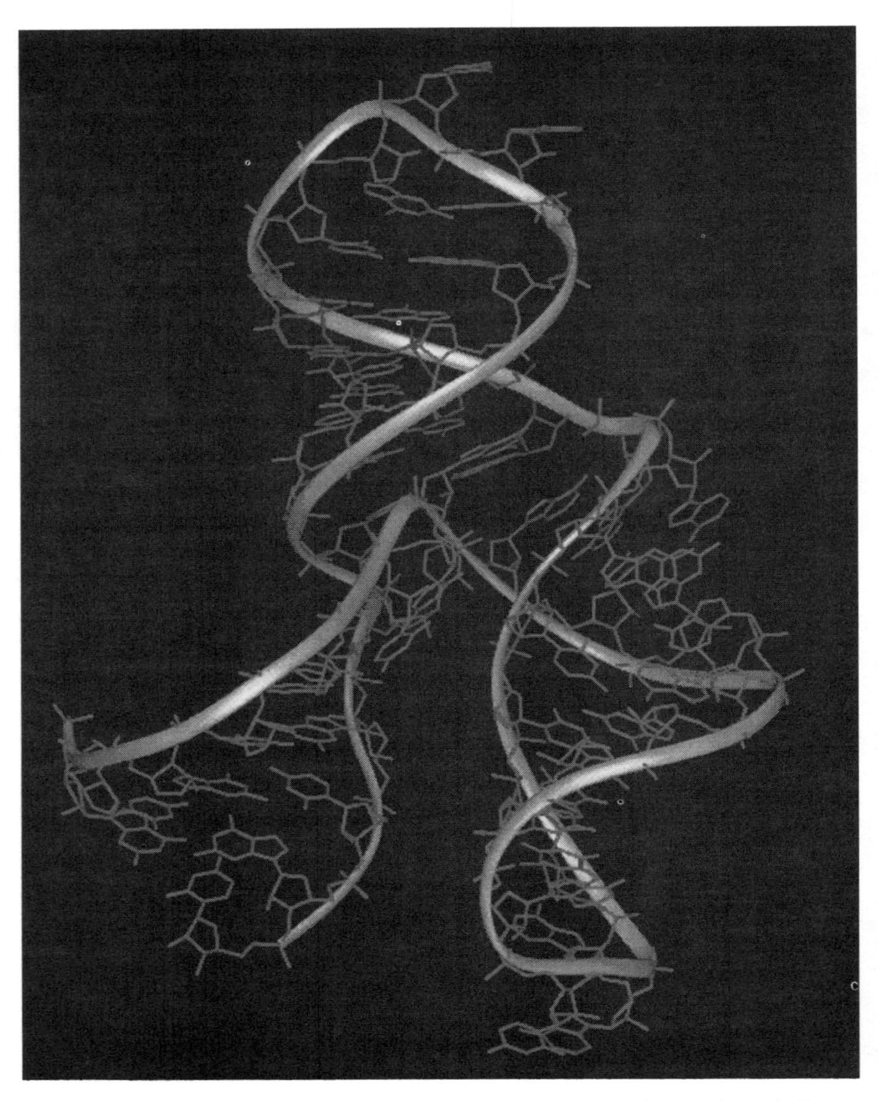

Figure 13.1. The hammerhead ribozyme. The hammerhead is the simplest of all known naturally occurring ribozymes, yet its complex tertiary structure shows how even small RNAs may assist in biochemical transformations. (Pley, Flaherty and McKay 1994.)

1983; Peebles et al. 1986; Van der Veen et al. 1986), also self-splices, and has a secondary structure that is similarly complex. Group II introns are phylogenetically well distributed and have been found in cyanobacteria and in a subdivision of proteobacteria (Ferat and Michel 1993). The large size, structural complexity, and wide (albeit uneven) phylogenetic distribution of self-splicing introns suggest that they could be remnants of the RNA world. As an example, it was originally argued that the extreme sequence conservation of some Group I introns found in tRNAs in cyanobacteria and plant chloroplasts proved their existence at least predated this ancient endosymbiosis (Xu et al. 1990). However, the lineages of both types of introns are obscured by the fact that they are mobile genetic elements. The Group I intron has been shown to "reverse splice" into mRNAs (Woodson and Cech 1989), which can then potentially be reverse transcribed and inserted back into an organism's genome with their newfound information. In fact, it has been recently demonstrated that the supposedly ancient cyanobacterial introns show an uneven distribution and are more likely the result of recent lateral transfer rather than long divergence (Biniszkiewicz, Cesnaviciene and Shub 1994).

Ribonuclease P (for a review, see Brown and Pace 1991) is a tRNA processing enzyme that is found in all domains of life, and is thus the only ribozyme whose ancestry can be securely extrapolated at least to the progenote, the last common ancestor of modern life. Although the enzyme is always a nucleoprotein complex in vivo (Gardiner and Pace 1980), naked RNAs from some eubacterial species have been shown to be catalytically active in vitro (Guerrier-Takada et al. 1983). RNase P was one of the first ribozymes discovered, and it is the only ribozyme that carries out a multiple turnover function as part of its natural role. Although high-resolution structural studies of RNase P have not yet been completed, modeling experiments have revealed a complex three-dimensional structure that rivals that of the Group I and Group II self-splicing introns.

For the five ribozymes we have so far considered, it appears as though the simple hammerhead and hairpin ribozymes could have arisen de novo in the organisms in which they are found; the complex Group I and Group II introns likely arose at some undetermined but ancient date and spread throughout phylogeny; and RNase P was present at least as far back as the last common ancestor. This analysis is both bolstered and contradicted by the stories of the last two natural ribozymes, the hepatitis delta virus (HDV) ribozyme and the *Neurospora* VS ribozyme. Hepatitis delta virus has a circular RNA genome that is distinct from known animal viruses. However, it was suggested by Hoyer et al. (1983) and Riesner and Gross (1985) that the structure and replication of this virus are similar to those of known plant viruses. Since plant viruses can undergo site-specific cleavage, Sharmeen et al. (1988) hypothesized (and experimentally confirmed) that HDV could do the same. Deletion and mutational analysis have been used to delimit the functional residues in the HDV ribozyme (Suh et al. 1992; Kumar et al. 1993) and led Kawakami et al. (1993) to propose minimal sequence requirements for activity (around 86 residues). The pseudoknotted secondary structures of the genomic and the antigenomic HDV ribozymes are similar to one another (Rosenstein and Been 1991), but differ significantly from the structures of the hammerhead and hairpin ribozymes. In fact, the sequence and structure of the HDV ribozyme are so far unique in phylogeny. The relative simplicity of the ribozyme and its phylogenetic singularity lend credence to the notion that the HDV ribozyme could have arisen relatively recently. In contrast, the VS RNA is nearly twice the size of the HDV ribozyme (164 nucleotides) but also enjoys an extremely limited phylogenetic distribution. Unlike the similarly sized Group I and Group II introns, the VS ribozyme has so far been found only in plasmids in *Neurospora* species (Collins and Olive 1993). Like the HDV ribozyme, it has a primary and a secondary structure distinct from those of the other ribozymes (Beatty, Olive and Collins 1995), and it is further unique in that it cleaves double-stranded RNA (Guo and Collins 1995). Thus, it appears as though the VS ribozyme, despite its sequence complexity, could have arisen at a relatively recent juncture. Overall, only RNase P can be confidently extrapolated to a time period that is remotely close to the RNA world.

Diversity and sophistication

Whether or not their ancestry can be traced to the RNA world, the natural ribozymes provide examples of what may have been possible in a nascent metabolism. Although these ribozymes all carry out the same basic reaction (phosphodiester bond rearrangement), the diversity and sophistication of the chemistries that they employ augurs well for the range and efficiency of catalysts that may have been present in the RNA world.

The chemical diversity of natural ribozymes is apparent from the range of nucleophiles that are used to attack phosphodiester bonds. The simplest mechanism for RNA cleavage has appropriately been adopted by the small hairpin, hammerhead, hepatitis delta virus, and larger *Neurospora* VS ribozymes (Figure 13.2). A 2'-hydroxyl is activated by a bound metal and cleaves the adjacent phosphodiester bond, forming a 2',3'-cyclic phosphate. Whereas the hammerhead is catholic in its choice of divalent metal cations, the hairpin and other ribozymes are generally dependent on magnesium or a much more limited set of divalent cations for activity (Chowira, Berzal-Herranz and Burke 1993). The Group I and Group II ribozymes (Figures 13.3 and 13.4, respectively) utilize two-step mechanisms for their excision from longer transcripts.

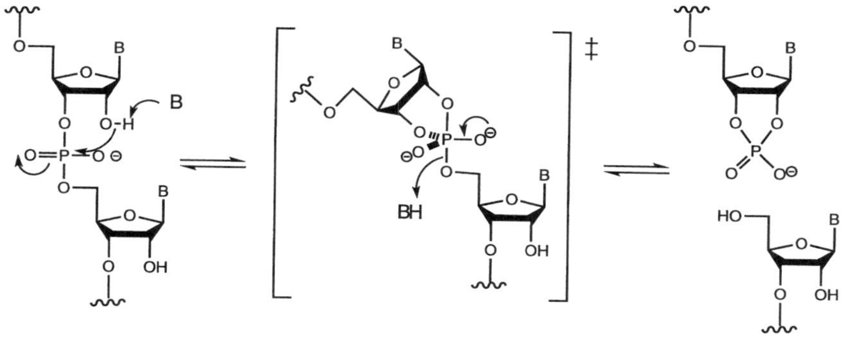

Figure 13.2. Mechanism of self-cleavage used by the hairpin, hammerhead, *Neurospora* VS, and HDV ribozymes. With the assistance of a Brønsted base, the 2'-hydroxyl attacks the adjacent phosphodiester linkage to generate a trigonal bipyrimidal transition state with both the attacking nucleophile and the subsequent leaving group in the apical positions. The RNA is cleaved upon regeneration of the tetrahedral geometry about the phosphorus, resulting in the release of one oligonucleotide with a free 5'-hydroxyl and another oligonucleotide with a 2', 3'-cyclic phosphate.

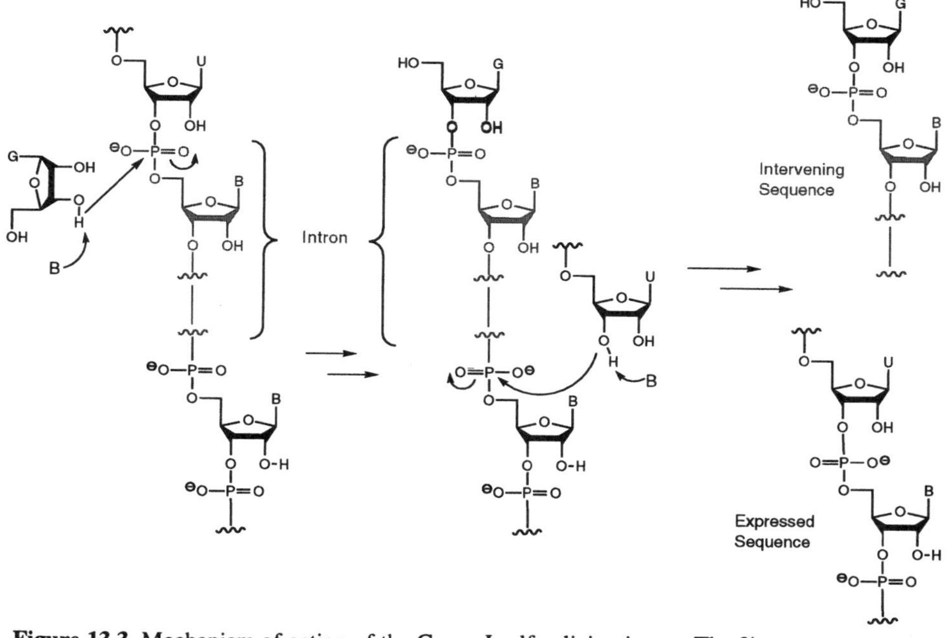

Figure 13.3. Mechanism of action of the Group I self-splicing intron. The 3'-hydroxyl of a guanosine nucleoside attacks the phosphate at the 5' end of the intron, leading to release of an oligomer that has a uridine with free vicinal hydroxyls as the terminal residue. This same oligomer then attacks the phosphodiester linkage at the 3' end of the intron, thus releasing the intervening sequence and generating the sequence that is to be expressed.

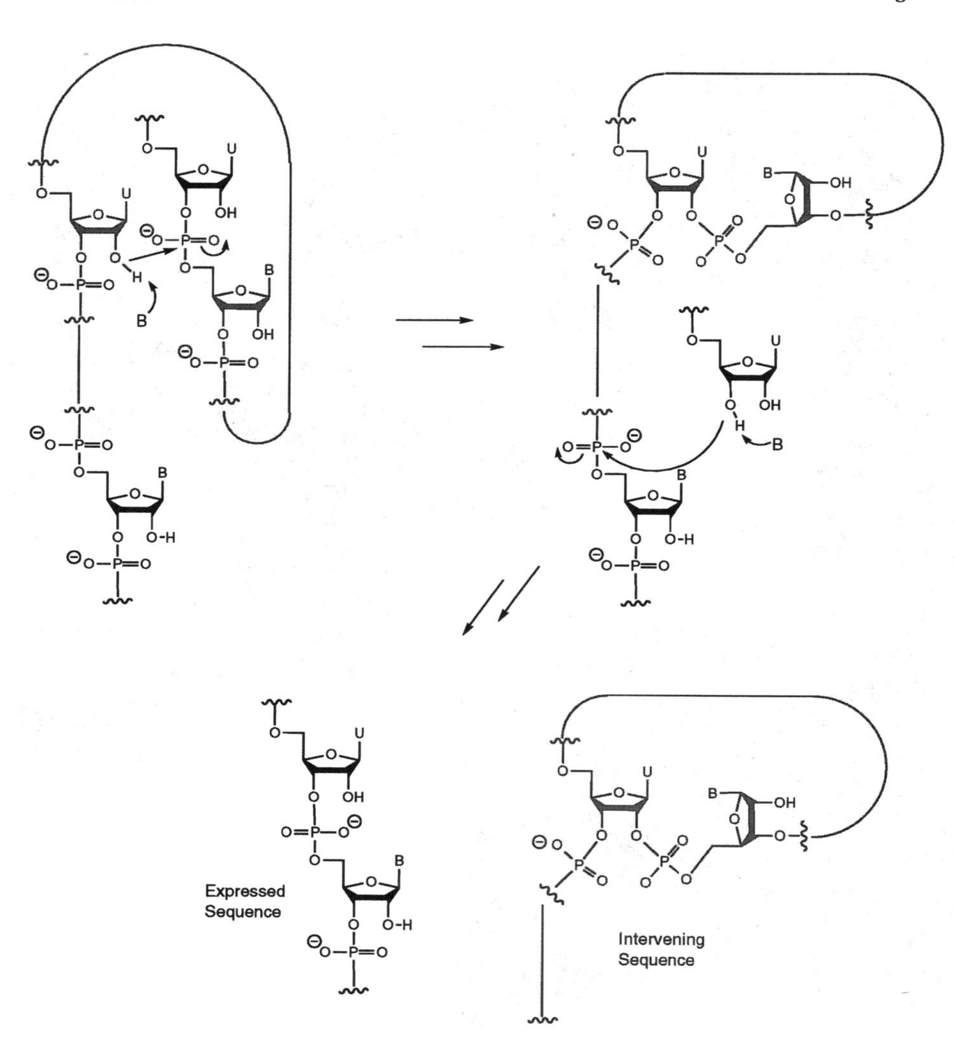

Figure 13.4. Mechanism of the Group II self-splicing intron. A 2'-hydroxyl of a uridine that is within the intron attacks the phosphodiester linkage at the 5' end of the intron, thus leading to release of an oligomer that has a uridine with free vicinal hydroxyls as the terminal residue. The splicing is completed in a manner identical to that of Group I. The excised oligomer bearing the terminal uridine then attacks the phosphodiester linkage at the 3' end of the intron, leading to release of the intervening sequence as a lariat and generation of the sequence that is to be expressed.

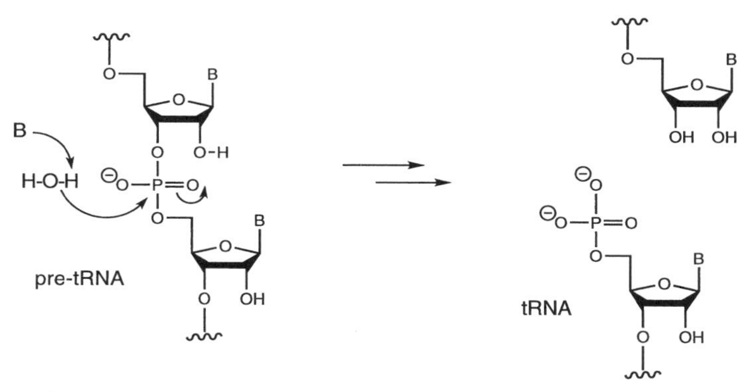

Figure 13.5. Mechanism of RNase P in the generation of mature tRNA. Unlike the previously mentioned ribozymes, naturally occurring RNase P cleaves a molecule other than itself. Within the catalytic domain of the ribozyme, a water molecule is activated to become the nucleophile that attacks pre-tRNA. The mature tRNA with a phosphate at the 5' terminus is released and the unchanged RNase P is able to catalyze more reactions.

The Group I intron uses the 3'-hydroxyl of an external guanosine as the nucleophile for the first step, whereas the Group II intron uses the 2'-hydroxyl of an internal uridine. The second step of splicing is chemically identical for both introns and is initiated by the 3'-hydroxyl of the newly released 5' exon. Finally, RNase P relies on a relatively simple mechanism for cleaving pre-tRNAs (Figure 13.5): It activates a water molecule proximal to the cleavage site.

The mechanisms by which natural ribozymes specify their substrates is also indicative of their diversity. Many of the ribozymes form base pairs with target RNAs in order to bring a particular phosphodiester bond into apposition with the catalytic site. Because of the simplicity of the Watson–Crick "code," substrate specificities can be changed at will. For example, both the hammerhead and hairpin ribozymes have been engineered in the laboratory to function in trans and their sequence specificities altered (Haseloff and Gerlach 1989; Hampel et al. 1990; Perrotta and Been 1990; Branch and Robertson 1991). The internal guide sequence of the Group I self-splicing intron has been manipulated to recognize different RNA substrates (Been and Cech 1986; Zaug, Been and Cech 1986; Zaug, Grosshans and Cech 1988). In addition, though, the Group I intron forms tertiary structural contacts with 2'-hydroxyls in the P1 stem (Herschlag, Eckstein and Cech 1993; Strobel and Cech 1993) and specifically recognizes a G:U wobble pairing at the 5' intron–exon junction (Doudna, Cormack and Szostak 1989) by forming tertiary structural contacts with the exocyclic 2 amino position of the G (Strobel and Cech 1996). In contrast, RNase P must recognize a wide variety of tRNA sequences that all fold into the same basic structure, and therefore it discriminates between substrates primarily on the basis of their shape, as dictated by the arrangement of phosphate and sugar moieties along the acceptor stem and T stem-loop (Harris et al. 1994; Pan, Loria and Zhong 1995). Interestingly, RNase P can also recognize substrates other than tRNAs; some of these substrates, such as 4.5S RNA, lack a structural motif similar to the T stem-loop and may bind to different subsites on RNase P (Talbot and Altman 1994). Building on this observation, Pan has selected RNase P substrates that lack structures corresponding the T stem-loop (Pan 1995), and has selected and characterized RNase P substrates that bind at a different site than tRNA (Pan and Jakacka 1996).

The diversity of ribozyme chemistries and interactions with substrates implies that nucleic acids are structurally and functionally plastic, and that therefore nucleic acid strings with nascent or novel functions could have readily evolved. To the extent that this proposition is true, we might hypothesize that it should be possible to engineer ribozymes to carry out new functions. This hypothesis has been experimentally validated: The Group I intron has proven extremely amenable to manipulation and has been engineered to carry out a variety of reactions in trans, including RNA cleavage, RNA ligation, hydrolysis of phosphate monoesters, and oligonucleotide disproportionation (Cech 1987, 1990).

The mechanisms utilized by RNA catalysts are not only diverse but, in many cases, as sophisticated as those employed by protein enzymes. Detailed kinetic studies of the hammerhead ribozyme (Fedor and Uhlenbeck 1992; Hertel, Herschlag and Uhlenbeck 1994) and the Group I self-splicing intron (Mei and Herschlag 1996) reveal that these catalysts appear to promote catalysis in part through substrate destabilization. In addition, the hammerhead ribozyme may drive the cleavage reaction by setting two entropic "traps": first, the hammerhead is rigid prior to cleavage but can access numerous conformational states following cleavage; second, the catalytic magnesium ion dissociates from the hammerhead following cleavage. The Group I intron not merely positions a nucleophile adjacent to a sissile bond, but activates the nucleophile by electrostatic destabilization and further stabilizes the developing charge on the product (Narlikar et al. 1995). The complex cascade of phosphodiester transesterifications carried out by the Group I intron (5' exon cleavage followed by 5' and 3' exon ligation) is abetted by thermodynamic coupling between different sets of substrates: Exogenous guanosine stabilizes interactions between the substrate P1 and the ribozyme (McConnell, Cech and Herschlag 1993), and the presence of the 3' exon appears to suppress the dissociation of the cleaved 5' exon (and thus helps to ensure the completion of the splicing cascade) (Mei and Herschlag 1996). These results strongly argue that the kinetic parameters of natural ribozymes have been evolutionarily optimized and that therefore any natural ribozymes, including those in the RNA world, would have enjoyed some of the same mechanistic advantages as protein enzymes.

Modern unnatural history of the ribozyme

Directed evolution experiments can be performed in the laboratory much faster than selections that occur in the wild. Artificial (or in vitro) selection methods have been used both to modify the functions of natural ribozymes and to generate novel ribozymes *de novo* (Szostak and Ellington 1993; Kumar and Ellington 1995; Hager, Pollard and Szostak 1996). Although selected ribozymes oftentimes have no biotic counterparts in terms of sequence or mechanism, they further reveal the potential diversity and sophistication of the RNA world.

Modifying natural ribozymes

Natural selection sieves organismal variants for those that are most fit or fecund. The phenotypic diversity of organisms is in turn the expression of a mutable genotype. Since the phenotype and genotype of ribozymes are held in the same molecule, ribozymes can be selected in vitro in much the same way that organisms are selected in the wild: a variety of genotypes are generated; the most fit or fecund individuals are selected ("fitness" being defined by the experimenter rather than the environment); and the selected individuals are preferentially amplified, generally by reverse transcription (RT) and the polymerase chain reaction (PCR).

In a typical in vitro selection experiment, ribozyme populations of varied sequence are generated by chemical synthesis of randomized oligonucleotides followed by in vitro transcription. The randomization strategy employed depends on the nature of the selection experiment. For example, a population based on relatively infrequent sequence substitutions (0.5%–5% per position) peppered throughout a ribozyme serves as a starting point for the identification of particular sites that inactivate a ribozyme. Similarly, complete randomization of short sequence tracts (5–30 residues) can be employed to identify residues or substructures that are essential for function. Finally, more extensive sequence substitution (5%–30% throughout a ribozyme, or complete randomization of tracts >30 residues in length) can be used to scan sequence space for more novel structures or functions.

Two general schemata have been used for ribozyme selections: cleavage and ligation (Figure 13.6). Since all of the natural ribozymes are known to

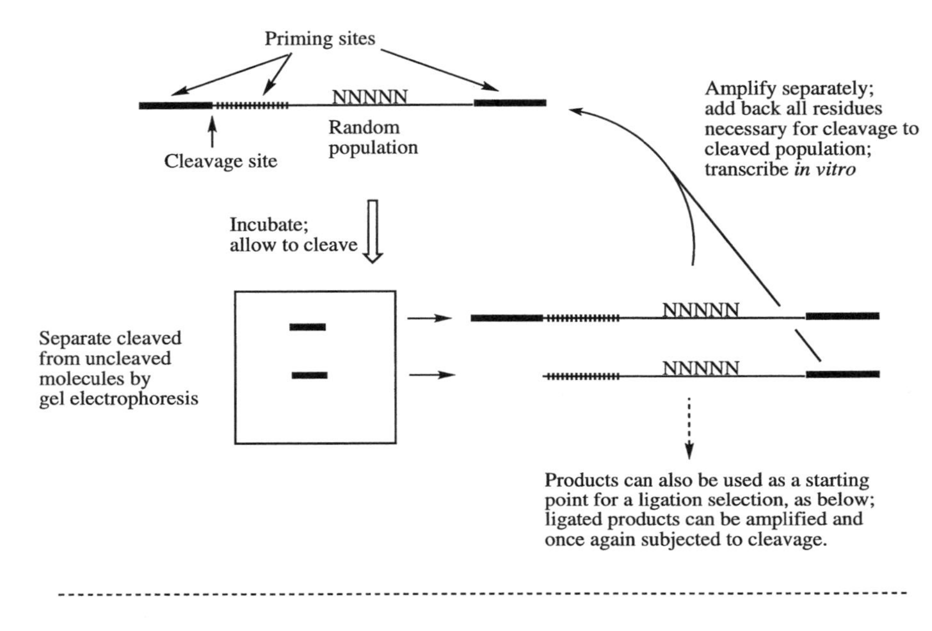

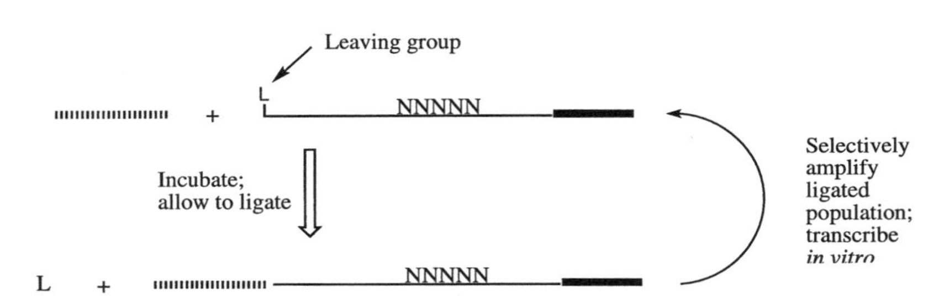

Figure 13.6. Selection schemata. (a) Cleavage selection. A randomized ribozyme population is flanked by constant sequences necessary for amplification. Different constant sequences flank the cleavage site. After incubation, cleaved molecules are separated from uncleaved (for example, by gel electrophoresis) and then separately amplified. Uncleaved molecules are amplified via primers that can anneal to the bolded lines at either end of the randomized region; cleaved molecules are amplified via a combination of primers that can anneal to the 3' bolded region and the 5' hatched region. During amplification, constant sequences necessary for a subsequent cycle of cleavage are added back to the cleaved population. (b) Ligation selection. A randomized ribozyme population is flanked at its 3' end by a constant sequence (bold line) necessary for amplification. A separate constant sequence (dashed line) is incubated with the population; those ribozymes that can add the S' hatched region can be selectively and exponentially amplified. During the ligation step some leaving group

cleave phosphodiester bonds, cleavage selections find universal application. The ribozyme population is placed in a buffer that should facilitate catalysis. Those variants that carry out the cleavage reaction within a given time are separated from uncleaved variants by gel electrophoresis (Figure 13.6a). The cleaved (or uncleaved) variants are isolated and amplified, and can undergo further rounds of selection. Unfortunately, separation for RNA substrates and products is generally inefficient and both can be amplified by RT/PCR. Thus cleavage selections can be plagued with high backgrounds of unselected molecules. Ligation selection can best be applied to ribozymes that already catalyze ligation reactions, such as the Group I self-splicing intron and the hairpin ribozyme. In a ligation selection protocol (Figure 13.6b), ribozyme variants in a population add new sequence information to themselves, and the product can be preferentially amplified even in the presence of the substrate. The two selections are sometimes carried out in tandem: for instance, hairpin ribozyme variants have been selected that first cleave themselves from a particular primer sequence and then ligate themselves to a different primer sequence (legend, Figure 13.6).

Some of the most instructive experiments in ribozyme selection and evolution have been carried out with the Group I self-splicing intron. For example, Green, Ellington and Szostak (1990) randomized two small segments of the *Tetrahymena* Group I intron, comprising nine residues total. Ribozymes that could add a 5′ exon mimic to themselves (a reversal of the first step in the Group I splicing cascade) were selected from the population of about 256,000 variants. After several rounds of selection and amplification, the selected ribozymes were sequenced. Although the basic secondary structure of the ribozyme remained unchanged, novel base-pairings that could support catalytic activity were identified. Similarly, Green and Szostak (1992) selected an efficient ribozyme variant from an inefficient, minimal version of the sunY Group I intron. The foreshortened variant was partially mutagenized at a rate

---

(L) must be displaced to make the reaction thermodynamically feasible. For example, during group I intron selections a guanosine residue or a short oligonucleotide sequence can be released; during hairpin ribozyme selections, a 2'-hydroxyl can be displaced from a terminal 2', 3'-cyclic phosphate (in the figure, this latter example would be represented by a reversal of polarity: the leaving group, L, would be on the 3' end of a randomized core). As detailed in (a), cleavage and ligation selections can be used in tandem. For example, in the hairpin selections, the cleaved population was subsequently allowed to ligate to a different constant sequence than was originally present at the 3' end of the ribozyme. This combination of steps in effect switched constant regions at one end of the ribozyme and thus allowed selective amplification of those sequences that could perform sequential cleavage and ligation steps. (Kumar and Ellington 1995.)

of 5%, and ribozymes that could efficiently ligate themselves to a substrate supplied in trans were again selected. After four rounds of selection and amplification, the population not only had regained the activity that had been lost due to the structural deletion, but also displayed enhanced catalytic activity at the low magnesium concentrations employed during the selection. The improvement in activity was largely due to four sequence substitutions scattered throughout the core. A second selection, which started from a population that contained these four sequence substitutions, resulted in the identification of an additional functional substitution. The final variant was 350-fold more active than the original, deleted ribozyme and four times more active than the undeleted wild type; at reduced magnesium concentrations, the difference was far greater. In fact, the selected variant was so stable that it could be divided into three separate pieces and could still reassemble to form a catalytically active complex more efficiently than a similar, tripartite wild-type ribozyme. These experiments were amongst the first to show that the basic sequence, structure, and properties of ribozymes could be readily manipulated by artificial selection. The fact that relatively few selected sequence substitutions resulted in significant changes in structure or function implies that the ribozymes in the RNA world may also have been able to evolve quite quickly.

More impressively, Joyce and co-workers have evolved the *Tetrahymena* Group I ribozyme to carry out novel catalytic tasks for which it initially seemed unsuited. These workers have developed a different ligation selection schema in which a noncovalently bound oligonucleotide substrate is attacked at the equivalent of the ligated exon junction by the 3' terminal guanosine of the ribozyme (a reversal of the second step in the splicing cascade); the rearrangement of phosphodiester bonds results in the transfer of a primer sequence from the oligonucleotide substrate to the ribozyme. This scheme was first applied to altering the substrate specificity of the ribozyme reaction. Starting from a small pool of deletion variants, Robertson and Joyce (1990) were able to identify a ribozyme that could better utilize a DNA oligonucleotide as a substrate. However, the rate enhancement that was achieved was quite small. In an effort to further improve catalysis with DNA substrates, Beaudry and Joyce (1992) partially (5%) randomized 140 positions in the catalytic core of the *Tetrahymena* intron. After 10 rounds of selection and amplification, the best selected variants could cleave DNA substrates up to 100-fold more efficiently than the wild-type parent could. Although sequence substitutions at a number of positions contributed to the altered activities, most of the catalytic improvement was due to changes in only four residues. Interestingly, different evolutionary strategies were represented in the selected population. For example, some ribozymes had sacrificed fidelity for activity and could splice themselves into positions on the DNA primer that did not correspond to the wild-type ligated exon junction. Alternatively, the most active clones freed themselves from the fetters of fidelity via a double mutation that disrupted base-pairings that normally position the 3' terminal guanosine in the ribozyme active site.

The activities of artificially evolved ribozymes can progressively improve over time, just as the activities of naturally evolved proteins do. The DNA cleavage activity of the selected Group I population looked as though it was still improving after 10 rounds, so Tsang and Joyce (1994) continued the selection for an additional 17 generations. The concentration of the DNA substrate was decreased by 50-fold relative to the early rounds in order to favor those catalysts that either could bind the substrate most tightly or could efficiently react following substrate binding. By the 18th (total) round, the selected variants were saturated with substrate even at the reduced concentration. In order to encourage further improvements, the time allowed for the splicing reaction to occur was reduced 12-fold and the selection was continued for additional cycles. The ribozymes that were selected after the full 27 rounds of varied selection pressure could cleave DNA substrates 100,000-fold better (in terms of k_{cat}/K_m) than the wild-type parent.

Although it is apparent that the catalytic parameters of RNAs can be substantially altered by mutation, the chemistries available to the four canonical nucleotides are limited. Therefore, in order to demonstrate that ribozymes in the RNA world may have been able to carry out a variety of reactions, it is necessary to show that they could have readily acquired new cofactors. For example, a likely reason why all natural ribozymes are metalloenzymes may be that there are few good nucleophiles in RNA. The wild-type Group I intron uses one or two magnesium ions to polarize the vicinal hydroxyls of both endogenous and exogenous guanosine residues, but cannot normally use calcium for RNA cleavage. Starting from the same partially randomized sequence population that was used to isolate DNA-cleaving ribozymes, Lehman and Joyce (1993a) selected variants that could function in the presence of calcium with only trace amounts of magnesium. After 12 rounds of selection (Lehman and Joyce 1993b), the calcium-utilizing ribozymes could cleave RNA to the same extent as could the wild-type enzyme with magnesium. As was the case with ribozymes selected for DNA cleavge, the activity of the selected variants was due largely to specific changes at relatively few (six) positions. However, although the evolutionary path to these six substitutions was progressive, it was not direct. Lehman and Joyce (1993b) examined the sequences and activities of ribozymes at different points during the selection. Substitutions at position 260 are observed early during the evolutionary process but are later displaced by substitutions at position 270. No individuals contained both substitutions; site-directed mutational analysis revealed that, together, the two substitutions would render the ribozyme inactive. The substitution at position 270 appears relatively late because it is dependent on the preexistence of a substitution at position 271. Additionally, the 270 substitution attains its full potential only when accompanied by an additional substitution at position 103. It is notable that the sequence substitutions selected to optimize calcium utilization generally did not overlap with those that optimized DNA cleavage (Kumar and Ellington 1995).

Taken together, these studies suggest that the sequence, structure, and function of the Group I ribozyme can evolve quickly along a large number of divergent mutational pathways. By implication, multiple pathways may also have been available to ribozymes in the RNA world. However, there are two caveats to this conclusion. First, mutagenesis rates used in artificial evolution experiments have been far higher than those observed in natural organisms, even viruses. If the evolution of altered or novel function requires multiple, simultaneous substitutions, then ribozymes in the natural world may not be able to move via the same grand leaps as those in artificial selection experiments. Thus, the old saw that states that "you can't get there from here" may (in some cases) be an evolutionary reality. On the other hand, the evolved variants that have been observed to date required relatively few substitutions to increase latent activities by several orders of magnitude. Similarly, it may be that ribozymes can evolve first along simple paths (i.e., the substitutions at position 260 observed in the selection for calcium utilization), accumulate comparatively neutral mutations (i.e., substitutions at 271 and 103), and then move to new, greater, previously unaccessible fitness peaks. The role of neutral mutations in ribozyme (and protein) evolution remains an unresolved question, but theoretical treatments by Schuster (Schuster et al. 1994) suggest that there are myriad neutral pathways by which RNA secondary structure can move. Second, although different structures and functions may be accessible to variants of the Group I ribozyme, the same does not necessarily appear to be true of smaller ribozymes, such as the hammerhead and hairpin ribozymes. Selection experiments that attempted to alter the substrate specificity of the hammerhead ribozyme returned only wild-type sequences, despite the fact that the most of the core was randomized (Nakamaye and Eckstein 1994). Selection experiments that targeted the stems of the hairpin ribozyme confirmed predicted or known secondary structural features, but did not return sequences or structural variants that differed significantly from the wild type (Berzal-Herranz, Joseph and Burke 1992). It may be that ribozymes must reach a threshold of sequence and structural complexity in order to evolve new functions. If so, this may in turn imply that aboriginal self-replicating nucleic acids may have acquired catalytic functionality slowly until they reached a certain size (ca. 100 residues?), at which point they could have explosively diversified.

Generating novel ribozymes

Although the experiments described here demonstrate that extant ribozymes might evolve quickly and are functionally quite plastic, they do not tell us whence these extant ribozymes may have come. Although small ribozymes, such as the hammerhead, could (and perhaps still do) arise at random in the universe of genomes, the larger ribozymes could not have easily arisen *de novo* and have an unknown heritage. It is tempting to speculate that at least some simple catalytic functions did arise *de novo* from random sequence nucleic acid strings and then were further elaborated in the context of an early RNA world. Indeed, there are relatively few steps from observing that in vitro selection can alter the function of a natural ribozyme, to extrapolating that at least some modern ribozymes may have been derived from ancestral sequences that could catalyze related reactions, to hypothesizing that there were once many ribozymes that could catalyze novel reactions. Experimental support for this scenario is now at hand.

In vitro genetic selections have now been used to isolate novel ribozymes from completely random sequence populations. In a classic experiment, Bartel and Szostak (1993) used a ligation selection scheme similar to the one devised by Green, Ellington, and Szostak (1990) for the Group I intron to select ribozymes from a random sequence RNA library that spanned 220 positions and contained $>10^{15}$ members. The selected ligases were mechanistically different from their natural counterparts in that they utilized a 5' pyrophosphate (rather than a mono- or oligonucleotide) as the leaving group. Multiple, different classes of ribozyme ligases were identified. Although several classes utilized a terminal 3'-hydroxyl as the attacking nucleophile, others instead preferentially utilized a 2'-hydroxyl, suggesting that both mechanistic solutions may have been available during the course of natural evolution (and thus further clouding the choice of the 3',5' phosphodiester linkage in natural nucleic acids). The rate enhancements achieved by the selected ribozyme ligases were a stunning 7×10^6-fold over background and within a few orders of magnitude of those observed for natural ribozymes. One of the ribozyme ligases was partially randomized, and functional variants were reselected (Ekland, Szostak and Bartel 1995). A consensus sequence was constructed based on the results of this selection and was found to have a turnover rate of about 100 per minute, as fast as natural ribozymes and within orders of magnitude of protein ligases. Interestingly, when the conserved positions in this ribozyme (the so-called class I ligase) were tallied up, it became apparent that the ribozyme should have been found on average once every 3×10^{18} sequences – or roughly once every 2,000 times the experiment was performed (based on a pool size of ca. 1.4×10^{15}). Although it is possible that the researchers were extraordinarily lucky, these estimates are also amenable to a different explanation: Although a particular functional RNA sequence may be vanishingly rare, functional RNA sequences as a whole may be relatively common. In other words, if Bartel and Szostak's experiment were repeated, the class I ligase might not be found, but another catalyst of equal sequence and structural complexity would be. This analysis accords nicely with our previous supposition that above a certain size, RNA catalysts may have readily diversified.

Other selection schemes have yielded even more novel activities. For example, Lorsch and Szostak (1994) have selected a completely new type of ribozyme, a ribo-kinase, from a random sequence pool. They built a previously discovered ATP-binding domain (Sassanfar and Szostak 1993) into the original pool in order to encourage the selected species to use ATP as a cofactor. They then randomized 100 surrounding residues and incubated the pool with an ATP analogue containing a gamma-thiophosphate. Those sequences that catalyzed the transfer of the thiophosphate to themselves were isolated from the population by reversible immobilization on an affinity column bearing thiol groups. During the course of the selection, stringency was controlled by modulating the incubation time; the number of variants was expanded by mutagenic PCR. After 13 cycles of selection, amplification, and mutagenesis, several major catalytic species had been resolved. The best selected catalyst had a rate acceleration of 10^9-fold over background. Its catalytic efficiency (k_{cat}/K_m) approached that of the mechanistically unrelated *Tetrahymena* ribozyme (although it was still five orders of magnitude less efficient than a mechanistically related protein enzyme). The ribo-kinases that survived the selection were not necessarily those that would have been predicted in advance: Most of the classes had significantly altered ATP-binding domains and only one class actually retained the ability to bind ATP by the criterion of affinity chromatography. In addition, different classes of catalysts used either the terminal 5'- or internal 2'-hydroxyls as reaction substrates.

Novel nucleic acid catalysts are now being derived from random sequence pools at a frenetic pace. Using an in vitro selection similar to that used to identify catalytic antibodies, Prudent, Uno and Schultz (1994) have generated an RNA catalyst that can carry out the isomerization of a bridged biphenyl compound. Aptamers that could bind to a transition state analogue of the isomerization reaction were selected from a pool of oligonucleotides with 128 randomized positions. One individual isolated from this selection had a K_i of 7 micromolar and could catalyze the isomerization at a rate 88-fold greater than background. Although the isomerization reaction was not of metabolic importance, this was one of the first demonstrations that ribozymes could catalyze reactions other than phosphodiester bond rearrangements or phosphotransfers. Similarly, Wilson and Szostak (1995) have selected a ribozyme that can catalyze its own alkylation using a reactive version of biotin as a substrate. This ribo-alkyl transferase can form a carbon–nitrogen bond and thereby makes the existence of ancient ribo-methylases more plausible.

Even DNA, long considered to be merely an informational macromolecule even after the discovery of ribozymes, can catalyze reactions. Breaker and Joyce (1995) have derived lead-dependent deoxyribozymes that cleave an RNA linkage from a random sequence population. Cuenoud and Szostak (1995) have shown that selected DNAs can catalyze ligation reactions. These experiments further expanded the diversity of nucleic acid ligase mechanisms, since the leaving group was imidazole rather than a mononucleotide or pyrophosphate. Both ribozymes and deoxyribozymes have been selected that can catalyze the metallation of porphyrins (Conn, Prudent and Schultz 1996; Li and Sen 1996). These selections have been especially instructive in that they demonstrated that nucleic acids can augment and direct the natural chemistry of cofactors. In addition, the rates of the selected nucleic acids can be compared with the rates of both natural protein enzymes and artificially selected catalytic antibodies that catalyze porphyrin metallation. Astoundingly, these very different catalysts appear to be more or less equally efficient. This apparent equivalence may reflect the catalytic power of nucleic acids, or possibly the inherent facility of the uncatalyzed porphyrin metallation reaction. Overall, the discovery of deoxyribozymes is much more than a novelty and has implications for the RNA world. Because the prebiotic synthesis of ribose and ribotides has proven to be difficult, it has been postulated that some other, more synthetically accessible polymer may have been the forebear of nucleic acids (Joyce 1989). If so, then the fact that DNA catalysts can be selected from random pools may imply that other candidates for life's origin, such as nucleic acids with pyrophosphate linkages (Visscher, Bakker and Schwartz 1990) or with acyclic sugars (Visscher and Schwartz 1988), may also be capable of catalysis.

New catalysts are sometimes discovered fortuitously, rather than by design. For example, although it normally rearranges phosphodiester bonds, the Group I self-splicing intron has also been shown to slowly hydrolyze ester bonds (Piccirilli et al. 1992). Similarly, an evolved variant of the Group I intron that hydrolyzes DNA has been shown to also cleave an amide bond (Dai, De Mesmaeker and Joyce 1995). Finally, the class I ribo-ligase can polymerize short stretches of RNA using mononucleoside triphosphates as substrates (Ekland and Bartel 1996). This last observation has sparked

Table 13.1. *Examples of the variety of artificially evolved ribozymes and their sources. Although the list of all evolved ribozymes is much longer, these selections illustrate the diversity of function that has already been achieved. The remarks about specificity are qualitative and subjective.*

Nucleic Acid	Function	From Scratch/ Wild Type	Specificity High 1 2 3 low	Reference
RNA	Cleavage of DNA/ RNA Hybrid	Wild Type Group I	1	Robertson and Joyce 1990
RNA	DNA Cleavage	Wild Type Group I	1	Beaudry and Joyce 1992
RNA	RNA Ligation	Scratch	1	Green and Szostak 1992
RNA	Transesterification	Wild Type Group I	2	Piccirilli et al. 1992
RNA	Peptidyl Transfer	Wild Type Ribosome	2	Noller, Hoffarth and Zimniak 1992
DNA	DNA Ligation	Scratch	1	Cuenoud and Szostak 1995
RNA	N-Alkylation	Scratch	2-3	Wilson and Szostak 1995
RNA	Amide Cleavage	Wild Type Group I	2	Dai, De Mesmaeker and Joyce 1995
RNA	Aminoacylation	Scratch	2	Illangasekare et al. 1995
RNA	Acyl Transfer	Scratch	3	Lohse and Szostak 1996
DNA	Porphyrin Metallation	Scratch	3	Li and Sen 1996

speculation that an RNA "self-replicase" may soon be engineered (Hager, Pollard and Szostak 1996; Joyce 1996). Taken together, these examples further support a leitmotif of this review: that the discovery of new functionalities by complex nucleic acid catalysts may have been inevitable. A sampling of the wide variety of reactions to date that have been catalyzed by artificially evolved ribozymes is presented in Table 13.1.

Evolutionary significance

At the outset, we suggested that there were three routes to viewing abiogenesis. First, our view of life's origin is constrained by history. The discovery of sophisticated RNA catalysts that may not have arisen in the modern world by chance suggests that the molecular precursors of modern biopolymers could have been nucleic acids. Second, our view is constrained by chemistry. Any organism in a putative RNA world should have been capable of sustaining itself in much the same way that modern organisms do: by transforming materials and energy from the environment into a complex series of metabolic reactions that can replicate and sustain themselves. Although the discovery of ribozymes that can cleave viruses into unit-length pieces or splice themselves out of an mRNA is immensely important, it does not on its own provide a basis for constructing such a complete and interwoven metabolism (Benner, Ellington and Tauer 1989; Benner et al. 1993). Although in vitro selection experiments are sometimes chided as researcher's toys with no relevance to the origin of life (could even Darwin's warm pond have accommodated 10^{15} 100-mers?), they forcefully demonstrate that the ribozymes predicted by the RNA world hypothesis (ribo-kinases, ribo-alkyl transferases, ribo-polymerases) can be experimentally derived and therefore could have existed. These experiments also give us an idea of the information content of ribozymes of varying degrees of catalytic efficiency and sophistication, and thus allow us to begin to set probabilities for abiogenetic scenarios. Finally, our view should be a logical narrative that interrelates history and necessity. Though the links between natural and selected ribozymes may not have always been apparent in this review, there is at least one example where the two routes converge: the ribosome.

One of the key facts in support of the RNA world hypothesis is that an RNA molecule is in charge of making cellular proteins. Although it can be argued that nucleic acids are inherently suitable for replication and for a genetic code, there is no a priori reason why translation should be supported by an RNA scaffold. If necessity does not suffice, then the best explanation may be history: Proteins are made by RNA because RNA preceded proteins. In this view, the fact that ribosomal RNA was central to translation would have made it one of the last ribozymes to have been supplanted by protein catalysts. In confirmation of this bold hypothesis, it has been shown that ribosomal RNA stripped of almost all of its attendant proteins can still carry out peptide bond formation (Noller, Hoffarth and Zimniak 1992). However, even if the ribosome was (and is) a ribozyme, translation is far too complex to have arisen *de novo* in an RNA world of almost any conceivable size. Rather, the machinery for translation would have to have accumulated piecemeal, used for other metabolic tasks, until eventually the stage was set for the template-directed synthesis of the first peptides (Weiner and Maizels 1987); given what we have learned about the catalytic abilities that lie hidden within known ribozymes, the first peptides may even have been made fortuitously. Nonetheless, if this scenario is accurate, then it should be possible to derive ribozymes that individually comprise many of the central functions of the translation apparatus. Again, this hypothesis is now backed by fact. Yarus and co-workers have selected RNA molecules that can aminoacylate themselves from random sequence pools (Illangasekare et al. 1995), and Szostak's group has shown that selected RNA molecules can catalyze an acyl transfer reaction that results in amide bond formation (Lohse and Szostak 1996). In other words, two simple ribozymes can carry out the core reactions of protein biosynthesis: esterifying a RNA molecule, and transferring the acyl group to a primary amine. Although these reactions have not yet been joined to one another and do not in and of themselves suggest how templating and the genetic code may have arisen, they nonetheless provide credence to the view that translation could have evolved in an RNA world (Hager, Pollard and Szostak 1996). That such a self-consistent narrative can be constructed based on the few experiments done to date encourages that the RNA world hypothesis will be found to be increasingly viable in the years to come.

References

Bartel, D. P., and Szostak, J. W. 1993. Isolation of new ribozymes from a large pool of random sequences. *Science* 261:1411–1418.

Beatty, T. L., Olive, J. E., and Collins, R. A. 1995. A secondary-structure model for the self-cleaving region of *Neurospora* VS RNA. *Proc. Natl. Acad. Sci. USA* 92:4686–4690.

Beaudry, A. A., and Joyce, G. F. 1992. Directed evolution of an RNA enzyme. *Science* 257:635–641.

Been, M. D., and Cech, T. R. 1986. One binding site determines sequence specificity of *Tetrahymena* pre-rRNA self-splicing, trans-splicing, and RNA enzyme activity. *Cell* 47:207–216.

Benner, S. A., Cohen, M. A., Gonnet, G. H., Berkowitz, D. B., and Johnson, K. P. 1993. Reading the palimpsest: contemporary biochemical data and the RNA world. In *The RNA World*, ed. R. F. Gesteland and J. F. Atkins, pp. 27–70. Cold Springs Harbor, N.Y.: Cold Spring Harbor Press.

Benner, S. A., Ellington, A. D., and Tauer, A. 1989. Modern metabolism as a palimpsest of the RNA world. *Proc. Natl. Acad. Sci. USA* 86:7054–7058.

Berzal-Herranz, A., Joseph, S., Burke, J. M. 1992. In vitro selection of active hairpin ribozymes by sequential RNA-catalyzed cleavage and ligation reactions. *Genes Dev.* 6:129–134.

Biniszkiewicz, D., Cesnaviciene, E., and Shub, D.A. 1994. Self-splicing group I intron in cyanobacterial initiator methionine tRNA: evidence for lateral transfer of introns in bacteria. *EMBO J.* 13:4629–4635.

Branch, A. D., and Robertson, H. D. 1991. Efficient trans cleavage and a common structural motif for the ribozymes of the human hepatitis delta agent. *Proc. Natl. Acad. Sci. USA* 88:10163–10167.

Breaker, R., and Joyce, G. F. 1995. Self-incorporation of coenzymes by ribozymes. *J. Mol. Evol.* 40:551–558.

Brown, J. W., and Pace, N. 1991. Structure and evolution of ribonuclease P RNA. *Biochemie* 73:689–697.

Butcher, S. E., and Burke, J. M. 1994. Structure mapping of the hairpin ribozyme. *J. Mol. Biol.* 244:52–63.

Cech, T. R. 1987. The chemistry of self-splicing RNA and RNA enzymes. *Science* 236:1532–1539.

Cech, T. R. 1990. Self-splicing of group I introns. *Annu. Rev. Biochem.* 59:543–568.

Chowira, B. M., Berza-Herranz, A., and Burke, J. M. 1993. Ionic requirements for RNA binding, cleavage, and ligation by the hairpin ribozyme. *Biochemistry* 32:1088–1095.

Chowira, B. M., and Burke, J. M. 1991. Binding and cleavage of nucleic acids by the hairpin ribozyme. *Biochemistry* 30:8518–8522.
Collins, R. A., and Olive, J. E. 1993. Reaction conditions and kinetics of self-cleavage of a ribozyme derived from *Neurospora* VS RNA. *Biochemistry* 32:2795–2799.
Conn, M. M., Prudent, J. R., and Schultz, P. G. 1996. Porphyrin metalation matalyzed by a small RNA molecule. *J. Am. Chem. Soc.* 118:7012–7013.
Cuenoud, B., and Szostak, J. W. 1995. A DNA metalloenzyme with DNA ligase activity. *Nature* 375:611–614.
Dai, X., De Mesmaeker, A., and Joyce, G. F. 1995. Cleavage of an amide bond by a ribozyme. *Science* 267:237–240.
Doudna, J. A., Cormack, B. P., and Szostak, J. W. 1989. RNA structure, not sequence, determines the 5' splice-site specificity of the Group I intron. *Proc. Natl. Acad. Sci. USA* 86:7402–7406.
Ekland, E. H., and Bartel, D. P. 1996. RNA-catalysed RNA polymerization using nucleoside triphosphates. *Nature* 382 373–376.
Ekland, E. H., Szostak, J. W., and Bartel, D. P. 1995. Structurally complex and highly active RNA ligases derived from random RNA sequences. *Science* 269:364–370.
Fedor, M. J., and Uhlenbeck, O. C. 1992. Kinetics of intermolecular cleavage by hammerhead ribozymes. *Biochemistry* 31:12042–12054.
Ferat, J. L., and Michel, F. 1993. Group II self-splicing introns in bacteria. *Nature* 364:358–361.
Gardiner, K., and Pace, N. 1980. RNase P of *Bacillus subtilis* has a RNA component. *J. Biol. Chem.* 255:7507–7509.
Gilbert, W. 1986. The RNA world. *Nature* 319:618.
Green, R., Ellington, A. D., and Szostak, J. W. 1990. In vitro genetic analysis of the *Tetrahymena* self-splicing intron. *Nature* 347:406–408.
Green, R., and Szostak, J. W. 1992. Selection of a ribozyme that functions as a superior template in a self-copying reaction. *Science* 258:1910–1915.
Guerrier-Takada, C., Gardiner, K., Marsh, T., Pace, N., and Altman, S. 1983. The RNA moiety of ribonuclease P is the catalytic subunit of the enzyme. *Cell* 35:849–857.
Guo, H. C., and Collins, R. A. 1995. Efficient trans-cleavage of a stem-loop RNA substrate by a ribozyme derived from *Neurospora* VS RNA. *EMBO J.* 14:368–376.
Hager, A. J., Pollard, J. D., and Szostak, J. W. 1996. Ribozymes: aiming at RNA replication and protein synthesis. *Chem. Biol.* 3:717–725.
Hampel, A., Tritz, R., Hicks, M., and Cruz, P. 1990. "Hairpin" catalytic RNA model: evidence for helices and sequence requirements for substrate RNA. *Nucleic Acids Res.* 18:299–304.
Harris, M. E., Nolan, J. M., Malhotra, A., Brown, J. W., Harvey, S. C., and Pace, N. R. 1994. Use of photoaffinity crosslinking and molecular modeling to analyze the global architecture of ribonuclease P RNA. *EMBO J.* 13:3953–3963.
Haseloff, J., and Gerlach, W. L. 1989. Sequences required for self-catalysed cleavage of the satellite RNA of tobacco ringspot virus. *Gene* 82:43–52.
Herschlag, D., Eckstein, F., and Cech, T. R. 1993. Contributions of 2'-hydroxyl groups of the RNA substrate to binding and catalysis by the *Tetrahymena* ribozyme: An energetic picture of an active site composed of RNA. *Biochemistry* 32:8299–8311.
Hertel, K. J., Herschlag, D., and Uhlenbeck, O. C. 1994. A kinetic and thermodynamic framework for the hammerhead ribozyme reaction. *Biochemistry* 33:3374–3385.
Hoyer, B., Bonino, F., Ponzetto, A., Denniston, K., Nelson, J., Purcell, R., and Gerin,

J. L. 1983. Properties of delta-associated ribonucleic acid. In *Viral Hepatitis and Delta Infection,* ed. G. Vermi, F. Bonino, and M. Rizzetto, pp. 91–97. New York: Alan R. Liss.

Hutchins, C. J., Rathjen, P. D., Forster, A. C., and Symons, R. H. 1986. Self-cleavage of plus and minus RNA transcripts of avocado sunblotch viroid. *Nucleic Acids Res.* 14:3627–3640.

Illangasekare, M., Sanchez, G., Nickles, T., and Yarus, M. 1995. Aminoacyl-RNA synthesis catalyzed by an RNA. *Science* 267:643–647.

James, K. D., and Ellington, A. D. 1995. The search for missing links between self-replicating nucleic acids and the RNA World. *Origins Life Evol. Biosphere* 25:515–530.

Joyce, G. F. 1989. Amplification, mutation and selection of catalytic RNA. *Gene* 82:83–87.

Joyce, G. F. 1996. Ribozymes: building the RNA world. *Curr. Biol.* 6:965–967.

Kawakami, J., Kumar, P. K., Suh, Y. A., Nishikawa, F., Kawakami, K., Taira, K., Ohtsuka, E., and Nishikawa, S. 1993. Identification of important bases in a single-stranded region (SSrC) of the hepatitis delta virus ribozyme. *Eur. J. Biochem.* 217:29–36.

Kruger, K., Grabowski, P. J., Zaug, A. J., Sands, J., Gottschling, D. F., and Cech, T. R. 1982. Autoexcision and autocyclization of the ribosomal RNA intervening sequence of *Tetrahymena. Cell* 31:147–157.

Kumar, P. K. R., and Ellington, A. D. 1995. Artificial evolution and natural ribozymes. *FASEB J.* 9:1183–1195.

Kumar, P. K., Suh, Y. A., Taira, K., and Nishikawa, S. 1993. Point and compensation mutations to evaluate essential stem structures of genomic HDV ribozyme. *FASEB J.* 7:124–129.

Lehman, N., and Joyce, G. F. 1993a. Evolution in vitro of an RNA enzyme with altered metal dependence. *Nature* 361:182–185.

Lehman, N., and Joyce, G. F. 1993b. Evolution in vitro: analysis of a lineage of ribozymes. *Curr. Biol.* 3:723–734.

Li, Y., and Sen, D. 1996. A catalytic DNA for porphyrin metallation. *Nat. Struct. Biol.* 3:743–747.

Lohse, P. A., and Szostak, J. W. 1996. Ribozyme-catalysed amino-acid transfer reactions. *Nature* 381:442–444.

Lorsch, J. R., and Szostak, J. W. 1994. In vitro evolution of new ribozymes with polynucleotide kinase activity. *Nature* 371:31–36.

McConnell, T. S., Cech, T. R., and Herschlag, D. 1993. Guanosine binding to the *Tetrahymena* ribozyme: thermodynamic coupling with oligonucleotide binding. *Proc. Natl. Acad. Sci. USA* 90:8362–8366.

Mei, R., and Herschlag, D. 1996. Mechanistic investigations of a ribozyme derived from the *Tetrahymena* group I intron: insights into catalysis and the second step of self-splicing. *Biochemistry* 35:5796–5809.

Michel, F., and Dujon, B. 1983. Conservation of RNA secondary structures in two intron families including mitochondrial-, chloroplast-, and nuclear-encoded members. *EMBO J.* 2:33–38.

Michel, F., and Ferat, J. L. 1995. Structure and activities of group II introns. *Annu. Rev. Biochem.* 64:435–461.

Michel, F., and Westhof, E. 1990. Modelling of the three-dimensional architecture of group I catalytic introns based on comparative sequence analysis. *J. Mol. Biol.* 216:585–610.

Nakamaye, K. L., and Eckstein, F. 1994. AUA-cleaving hammerhead ribozymes: attempted selection for improved cleavage. *Biochemistry* 33:1271–1277.

Narlikar, G. J., Gopalakrishnan, V., McConnell, T. S., Usman, N., and Herschlag, D. 1995. Use of binding energy by an RNA enzyme for catalysis by positioning and substrate destabilization. *Proc. Natl. Acad. Sci. USA* 92:3668–3672.

Noller, H. F., Hoffarth, V., and Zimniak, L. 1992. Unusual resistance of peptidyl transferase to protein extraction procedures. *Science* 256:1416–1419.

Pan, T. 1995. Novel RNA substrates for the ribozyme from *Bacillus subtilis* ribonuclease P identified by in vitro selection. *Biochemistry* 34:8458–8464.

Pan, T., and Jakacka, M. 1996. Multiple substrate binding sites in the ribozyme from *Bacillus subtilis* RNase P. *EMBO J.* 15:2249–2255.

Pan, T., Loria, A., and Zhong, K. 1995. Probing of tertiary interactions in RNA: 2'-hydroxyl-base contacts between the RNase P RNA and pre-tRNA. *Proc. Natl. Acad. Sci. USA* 92:12510–12514.

Peebles, C. L., Perlman, P. S., Mecklinburg, K. L., Petrillo, M. L., Tabor, J. H., Jarrell, K. A., and Cheng, H. -L. 1986. A self-splicing RNA excises an intron lariat. *Cell* 44:213–223.

Perrotta, A. T., and Been, M. D. 1990. The self-cleaving domain from the genomic RNA of hepatitis delta virus: sequence requirements and the effects of denaturant. *Nucleic Acids Res.* 18:6821–6827.

Piccirilli, J. A., McConnell, T. S., Zaug, A. J., Noller, H. F., and Cech, T. R. 1992. Aminoacyl esterase activity of the *Tetrahymena* ribozyme. *Science* 256:1420–1424.

Pley, H. W., Flaherty, K. M., and McKay, D. B. 1994. Three-dimensional structure of a hammerhead ribozyme. *Nature* 372:68–74.

Prody, G. A., Bakos, J. T., Buzayan, J. M., Schneider, I. R., and Bruening, G. 1986. Autolytic processing of dimeric plant virus satellite RNA. *Science* 231:1577–1580.

Prudent, J. R., Uno, T., and Schultz, P. G. 1994. Expanding the scope of RNA catalysis. *Science* 264:1924–1927.

Reinhold-Hurek, B., and Shub, D.A. 1992. Self-splicing introns in tRNA genes of widely divergent bacteria. *Nature* 357:173–176.

Riesner, D., and Gross, H. J. 1985. Viroids. *Annu. Rev. Biochem.* 54:531–564.

Robertson, D. L., and Joyce, G. F. 1990. Selection in vitro of an RNA enzyme that specifically cleaves single-stranded DNA. *Nature* 344:467–468.

Rosenstein, S. P., and Been, M. D. 1991. Evidence that genomic and antigenomic RNA self-cleaving elements from hepatitis delta virus have similar secondary structures. *Nucleic Acids Res.* 19:5409–5416.

Saldanha, R., Mohr, G., Belfort, M., and Lambowitz, A. M. 1993. Group I and Group II introns. *FASEB J.* 7:15–24.

Sassanfar, M., and Szostak, J. W. 1993. An RNA motif that binds ATP. *Nature* 364:550–553.

Schuster, P., Fontana, W., Stadler, P. F., and Hofacker, I. L. 1994. From sequences to shapes and back: a case study in RNA secondary structures. *Proc. R. Soc. Lond. B. Biol. Sci.* 255:279–284.

Scott, W. G., Finch, J. T., and Klug, A. 1995. The crystal structure of an all-RNA hammerhead ribozyme: a proposed mechanism for RNA catalytic cleavage. *Cell* 81: 991–1002.

Sharmeen, L., Kuo, M. Y. P., Dinter-Gottlieb, G., and Taylor, J. 1988. Antigenomic RNA of human hepatitis delta virus can undergo self-cleavage. *J. Virol.* 62:2674–2679.

Sigurdsson, S. T., and Eckstein, F. 1995. Structure–function relationships of hammerhead ribozymes: from understanding to applications. *Trends Biotechnol.* 13:286–289.

Strobel, S. A., and Cech, T. R. 1993. Tertiary interactions with the internal guide sequence mediate docking of the P1 helix into the catalytic core of the *Tetrahymena* ribozyme. *Biochemistry* 32:13593–13604.

Strobel, S. A., and Cech, T. R. 1996. Exocyclic amine of the conserved G:U pair at the cleavage site of the *Tetrahymena* ribozyme contributes to 5'-splice site selection and transition state stabilization. *Biochemistry* 35:1201–1211.

Suh, Y. A., Kumar, P. K., Nishikawa, F., Kayano, E., Nakai, S., Odai, O., Uesugi, S., Taira, K., and Nishikawa, S. 1992. Deletion of internal sequence on the HDV ribozyme: elucidation of functionally important single-stranded loop regions. *Nucleic Acids Res.* 20:747–753.

Szostak, J. W., and Ellington, A. D. 1993. In vitro selection of functional RNA sequences. In *The RNA World*, ed. R. F. Gesteland, and J. F. Atkins, pp. 511–534. Cold Spring Harbor, N.Y.: Cold Spring Harbor Laboratory Press.

Talbot, S. J., and Altman, S. 1994. Gel retardation analysis of the interaction between C5 protein and M1 RNA in the formation of the ribonuclease P holoenzyme from *Escherichia coli. Biochemistry* 33:1399–1405.

Tsang, J., and Joyce, G. F. 1994. Evolutionary optimization of the catalytic properties of a DNA cleaving ribozyme. *Biochemistry* 33:5966–5973.

Van der Veen, R. T., Arnberg, A. C., Van der Horst, G., Bonen, L., Tabak, H. F., and Grivell, L. A. 1986. Excised Group II introns in yeast mitochondria are lariats and can be formed by self-splicing in vitro. *Cell* 44:225–234.

Visscher, J., Bakker, C. G., and Schwartz, A. W. 1990. Oligomerizations of deoxyadenosine bis-phosphates and of their 3'-5', 3'-3', and 5'-5' dimers: effects of a pyrophosphate-linked, poly(T) analog. *Origins Life Evol. Biosphere* 20:369–375.

Visscher, J., and Schwartz, A. W. 1988. Template-directed synthesis of acyclic oligonucleotide analogues. *J. Mol. Evol.* 28:3–6.

Weiner, A. M., and Maizels, N. 1987. tRNA-like structures tag the 3' ends of genomic RNA molecules for replication: implications for the origin of protein synthesis. *Proc. Natl. Acad. Sci. USA* 84:7383–7387.

Weiner, A. M., and Maizels, N. 1990. RNA editing: guided but not templated? *Cell* 61:917–920.

Wilson, C., and Szostak, J. W. 1995. In vitro evolution of a self-alkylating ribozyme. *Nature* 374:777–782.

Woodson, S. A., and Cech, T. R. 1989. Reverse self-splicing of the *Tetrahymena* Group I intron: Implication for the directionality of splicing and for intron transposition. *Cell* 57:335–345.

Xu, M. Q., Kathe, S. D., Goodrich-Blair, H., Nierzwicki-Bauer, S. A., and Shub, D. A. 1990. Bacterial origin of a chloroplast intron: conserved self-splicing group I introns in cyanobacteria. *Science* 250:1566–1570.

Zaug, A. J., Been, M. D., and Cech, T. R. 1986. The *Tetrahymena* ribozyme acts like an RNA restriction endonuclease. *Nature* 324:429–433.

Zaug, A. J., Grosshans, C. A., and Cech, T. R. 1988. Sequence-specific endoribonuclease activity of the *Tetrahymena* ribozyme: enhanced cleavage of certain oligonucleotide substrates that form mismatched ribozyme–substrate complexes. *Biochemistry* 27:8924–8931.

14

Self-replication and autocatalysis

JENS BURMEISTER
Institute for Organic Chemistry
Ruhr-University
Bochum, Germany

1. Introduction

Long before science began, the question of the origin of life was answered by religion and mythology. With the work of Lamarck and Darwin, science started to try to give new answers to an old question. The theory of Darwinian evolution describes the origin of biological information. In general, an evolving system (i.e., an information-gaining system) is able to metabolize, to self-replicate, and to undergo mutations (as was stated by Oparin in 1924). Thus self-replication is one of the three criteria that enable us to distinguish non-living from living systems. Since nucleic acids carry the inherent ability for complementary base-pairing (and replication), they are very likely candidates to have been the first reproducing molecules. Kuhn and others (Crick 1968; Orgel 1968; Eigen and Schuster 1979; Kuhn and Waser 1981; for a review, see Joyce 1989) have drawn the picture of an RNA world that might have existed before translation was invented. This picture was supported by the finding that RNA (and also DNA) can act as an enzymelike catalyst (Sharp 1985; Cech 1987; Joyce 1989; Breaker and Joyce 1994). It is a challenge for organic and bioorganic chemists to trace back the original path of evolution by "reinventing" simple self-replicating systems in the lab and to learn more about the principles of replication on a molecular scale.

A simple three-step model can be used to conceptualize the process of molecular self-replication (Figure 14.1). In this model, the template molecule **T** is self-complementary and thus able to autocatalytically augment itself. In the first step, the template **T** reversibly binds its constituents **A** and **B** to yield a termolecular complex **M**. Within this complex, the reactive ends of the precursors are held in close proximity, which facilitates the formation of a covalent bond between them. In the following step, the termolecular complex **M** is irreversibly transformed into the duplex **D**. Reversible dissociation of **D** gives two template molecules, each of which can initiate a new replication

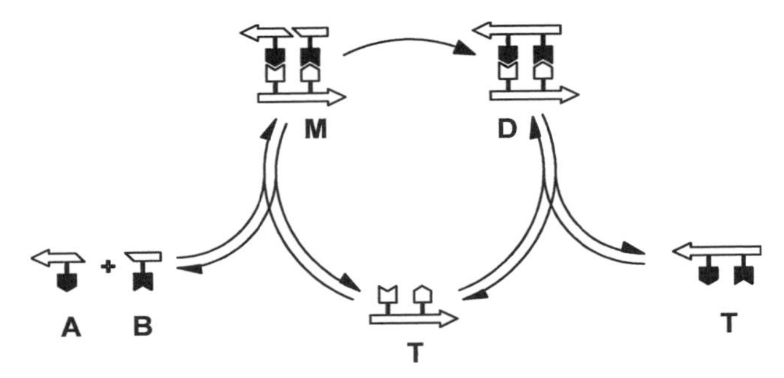

Figure 14.1. Schematic representation of a minimal self-replicating system.

cycle. The minimal representation given in Figure 14.1 has served as a successful aid for the development of nonenzymatic self-replicating systems based on nucleotidic and nonnucleotidic precursors, as will be shown.

2. Nucleic acid replicators

2.1. Polycondensations of activated mononucleotides directed by templates

The template-directed synthesis of oligonucleotides from activated mononucleotide precursors (such as monoribonucleoside-5'-phosphorimidazolides) has been studied extensively by Orgel and co-workers since 1968 (Orgel and Lohrmann 1974; Inoue and Orgel 1983). These studies revealed two basic principles:

1. The polycondensation yields two different regioisomeric products, one with a 2'-5'- and one with the natural 3'-5'-linkage. The regioselectivity depends on various parameters (e.g., the template, nature of the leaving group, presence of metal ions, etc.).
2. The polycondensations were found to be rather efficient with a pyrimidine-rich template. In contrast, a purine-rich oligo- or polynucleotide acts as a poor template.

The findings concerning the templates were envisioned as a major obstacle for further attempts to realize self-replicating systems based on mononucleotide precursors. Despite these difficulties, the Orgel group succeeded in demonstrating the nonenzymatic template-directed synthesis of fully complementary products: The pentamer pGGCGG, for example was obtained in 18% yield from a mixture of the 5'-(2-methyl)-phosphorimidazolide derivatives of guanosine and cytidine in the presence of CCGCC as a template (Inoue et al. 1984). Later on, templates as long as 14 mers were successfully transcribed

(Orgel 1992). More recently, Orgel and collaborators have studied the template-directed condensations of nucleoside 5'-phosphorimidazolides, using the long end (5'-end) of J-shaped oligonucleotide hairpins as "internal templates." Hairpin elongation reactions (forming U-shaped products) proved to be far less restrictive than reactions on external templates with respect to pyrimidine incorporation (for a review, see Orgel 1992; Orgel 1995).

2.2. *From chemical ligation to self-replication*

The work of Orgel and co-workers clearly demonstrated that transcription with information transfer can occur in the absence of enzymes. However, no complete replication cycle could be achieved when using mononucleotides as precursors, due to the problems previously described. It thus seemed worthwhile to use oligonucleotides as building material and to employ chemical ligations instead of polycondensations as coupling reactions. The first example of a chemical ligation, namely a template-directed condensation of activated oligonucleotides, was reported by Naylor and Gilham in 1966. They demonstrated that the condensation of pentathymidylic acid and hexathymidylic acid molecules could be catalyzed by a poly(A) template in the presence of the water-soluble carbodiimide EDC (1-ethyl-3-(3-dimethylaminopropyl)-carbodiimide). Further examples of chemical ligation reactions have been reported by the group of Shabarova in Moscow (Dolinnaya et al. 1988, 1991). A successful demonstration of enzyme-free nucleic acid replication based on an autocatalytic chemical system was reported in 1986 (von Kiedrowski 1986). A 5'–terminally protected trideoxynucleotide 3'-phosphate d(Me-CCG-p) (**1**) and a complementary 3'-protected trideoxynucleotide d(CGG-p') (**2**) were reacted in the presence of EDC to yield the self-complementary hexadeoxynucleotide d(Me-CCG-CGG-p') (**3**) with natural phosphodiester linkage as well as the 3'-3'-linked pyrophosphate of **1**. The sequences chosen were such that the product **3** could act as a template for its own production. The hexamer formation proceeds via the termolecular complex **C** formed from **1**, **2**, and **3**, in which the reactive ends are in close spatial proximity and thus ready to be ligated. During the course of reaction, the activated 3'-phosphate of **1** is attacked by the adjacent 5'-hydroxyl group of **2**, forming a 3'-5'-internucleotide bond between the trimers (see Figure 14.2).

The resulting template duplex then can dissociate to yield two free template molecules, each of which can initiate a new replication cycle. Two parallel pathways for the formation of hexameric templates exist, as can be shown by kinetic studies using HPLC. Both pathways, the template-dependent, autocatalytic pathway and the template-independent, non-autocatalytic pathway, contribute to product formation. The latter pathway has been found to be

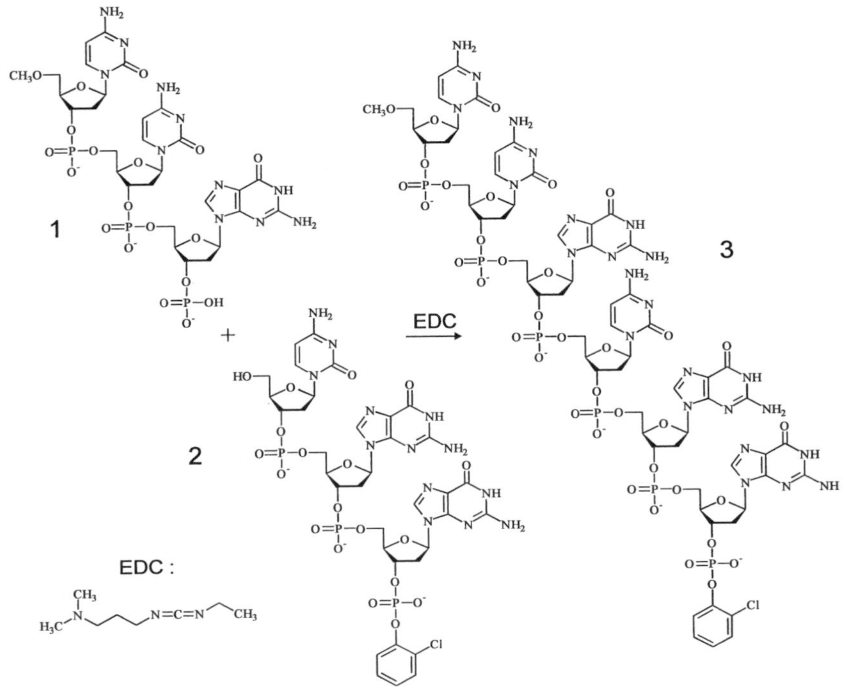

Figure 14.2. First nonenzymatic self-replicating system.

predominant. Moreover, the experiments revealed that the addition of template did not increase the rate of autocatalytic template formation in a linear sense. Instead, the initial rate of autocatalytic synthesis was found to be proportional to the square root of the template concentration (a finding that was termed the square root law of autocatalysis). Thus, the reaction order in this autocatalytic self-replicating system was found to be ½ rather than 1, a finding in contrast to most autocatalytic reactions known so far. According to theory, a square root law is expected in the previously described cases, in which most of the template molecules remain in their double-helical (duplex) form, which leaves them in an "inactive" state. In other words, a square root law reflects the influence both of autocatalysis and of product inhibition.

Another example of an autocatalytic system following the square root law was published by Zielinsky and Orgel in 1987. The diribonucleotide analogues ($G_{NHp}C_{NH_2}$) **(4)** and ($_pG_{NHp}C_{N_3}$) **(5)** were ligated in the presence of water-soluble carbodiimide (EDC) and self-complementary tetraribonucleotide triphosphoramidate ($G_{NHp}C_{NHp}G_{NHp}C_{N_3}$) **(6)**, serving as a template (see Figure 14.3). This was the first demonstration of self-replication of nucleic

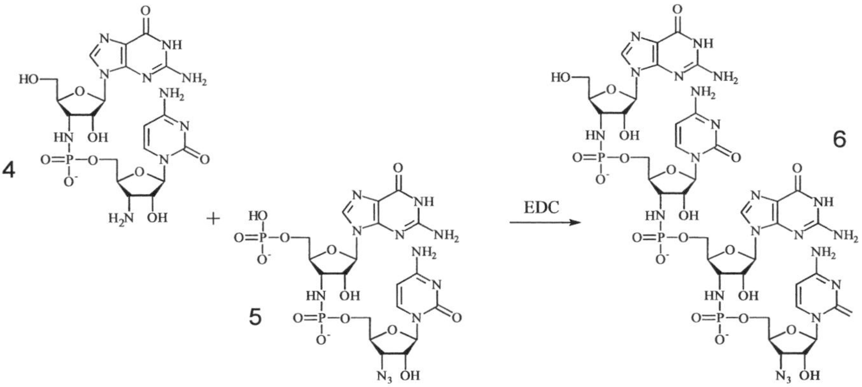

Figure 14.3. Self-replicating tetraribonucleotide with artificial backbone according to Zielinsky and Orgel (1987).

acid–like oligomers bearing an artificial backbone structure. In kinetic studies, the autocatalytic nature of template synthesis was obvious from the square root dependence of the initial reaction rate on the template concentration. In theory, every true autocatalytic system should show a sigmoidal concentration–time profile (von Kiedrowski 1993). Due to the predominance of the non-autocatalytic pathway, this was not true for either system just described. The autocatalytic nature of these systems was ascertained indirectly by observing the increase in the initial reaction rate when seeding the reaction mixtures with increasing amounts of template.

In the years of research following 1987, a major goal was to enhance the template-instructed autocatalytic synthesis while keeping the reaction rate of the noninstructed synthesis as low as possible. It can be shown that autocatalytic synthesis particularly benefits from an increased nucleophilicity of the attacking 5'-group. When trimer **2** was used in its 5'-phosphorylated form instead of the 5'-hydroxyl form, the carbodiimide-dependent condensation with **1** yielded hexamers bearing a central 3'-5'-pyrophosphate linkage. Due to this modification, the rate of the template-induced hexamer formation was increased by roughly two orders of magnitude (von Kiedrowski, Wlotzka and Helbing 1989). Replacing the 5'-phosphate by a 5'-amino group led to the formation of a 3'-5'-phosphoramidate bond and resulted in a rate enhancement of almost four orders of magnitude as compared to the phosphodiester system. In addition to that, the autocatalytic synthesis of template molecules was found to be more selective in the case of the faster replicators. The quantity *e* (a measure of the ratio of autocatalytic over background synthesis) could be increased from $16M^{-1/2}$ in the phosphodiester system to $430M^{-1/2}$ in the 3'-5'-phosphoramidate system. The first direct evidence for a sigmoidal increase in template concentration was found in the latter system (von Kiedrowski 1990; von Kiedrowski et al. 1991). Shortly after these observations, a second chemical self-replicating system was reported to exhibit sigmoidal growth (Rotello, Hong and Rebek 1991). A sigmoidal shape for template formation gives direct evidence of autocatalytic growth since this type of growth is a direct consequence of the square root law of autocatalysis. Following the square root law, the increase of template concentration at early reaction times is parabolic rather than exponential. For these early points in time, the integrated form of $dc/dt = a \times c^{1/2}$ can be approximated using a second order polynomial of time whose graph shows a parabola. Parabolic growth is a direct consequence of the square root law in cases where non-autocatalytic synthesis is negligible. A detailed analytical treatment revealed three types of autocatalytic growth as borderline cases (von Kiedrowski 1993). From theoretical considerations, it was also concluded that the autocatalytic growth order is solely dependent on the thermodynamics of the coupled equilibria, rather than on the energy on the transition state.

Further studies were devoted to the subject of sequence selectivity in the self-replication of hexadeoxynucleotides, giving further evidence for the autocatalytic nature of this reaction. Although the first data were obtained using the 3'-5'-pyrophosphate system (von Kiedrowski, Wlotzka and Helbing 1989) more recent studies were performed using the more efficient 3'-5'-phosphoramidates (Wlotzka 1992). A homologous set of trimer-3'-phosphates bearing the general sequence $^{PG}XYZp$ (PG = protective group) was synthesized, where X, Y, and Z could represent either C or G monomers. Each of the trimer 3'-phosphates was reacted with the aminotrimer $^{H_2N}CGGp^{PG}$ in the presence of EDC. The trimer 3'-phosphate bearing the sequence CCG was formed significantly faster than all the other trimers. Moreover, addition of template CCGCGG, complementary to the ligated trimers, stimulated only the synthesis of the proper phosphoramidate while having a negligible influence on the variant sequences. These studies demonstrated that autocatalysis can occur only if the sequences of both trimers match the sequence of the resulting hexamer according to the Watson–Crick base-pairing rules. They also showed that the condensation reactions are controlled predominantly by the stacking of nucleobases flanking the newly formed internucleotide linkage. Hexamers bearing a central G–C subsequence, for example, are formed one order of magnitude faster than hexamers with a central C–G subsequence. In general, the following reactivity order was found: G–G > G–C > C–G > CC, a finding that is in good correspondence with experiments with non-autocatalytic chemical ligations. Further studies revealed a remarkable temperature dependence in hexadeoxynucleotide self-replication. Each parabolic replicator shows a rate optimum at a certain temperature, which was found to be close to the measurable melting temperature of the respective hexamer duplex (Wlotzka 1992). Again, this is in good correspondence with minimal replicator theory. Generally, the autocatalytic reaction rate is given by k[C], where k equals the rate constant for irreversible internucleotide bond formation and [C] is the equilibrium concentration of the termolecular complex. Because the rate constant k increases with temperature (according to Arrhenius' law) and [C] decreases due to the melting of the termolecular complex, the reaction rate as a function of temperature is expected to pass a maximum at the temperature T_{opt}. T_{opt} itself depends on both the concentrations of template and its precursors as well as the thermodynamic stabilities of the termolecular complex and the template duplex.

2.3. Competition and cooperation in nonenzymatic self-replicating systems

Recent studies in our lab were devoted to the question of information transfer in a more complex system, in which a number of alternative templates can be produced from a set of common precursors. Such a system was realized in an experiment in which the sequence CCGCGG was synthesized from three fragments (Achilles and von Kiedrowski 1993). The trimer-3'-phosphate ^{PG}CCGp (**A**), the 5'-aminodimer3'-phosphate ^{H_2N}CGp (**B**) and the 5'-aminomonomer ^{H_2N}G (**C**) were allowed to react in the presence of 1-methylimidazole (MeIm) and EDC. The five products **AB**, **AC**, **BC**, **BB**, and **ABC**, all bearing central 3'-5'-phosphoramidate linkages, could be identified. To monitor the whole system using HPLC kinetic analysis, it was necessary to reduce its complexity. In order to detect possible catalytic, cross-catalytic, or autocatalytic pathways (couplings) induced by the different products, the whole reaction system was divided into less complex subsystems. For example, in order to analyze the formation of the pentamer **AB** separately, we employed a 5'-aminodimer **B'**, which was protected at its 3'-phosphate.

In a series of experiments, each subsystem was studied with respect to the effect of each reaction product. Standard oligodeoxynucleotides were employed as model templates. These experiments allowed us to decipher the dynamic structure of the whole system, which can be understood as a catalytic network, namely, an autocatalytic set with a total of six feedback couplings (see Figure 14.4). However, only those couplings with sufficient efficiency (denoted ⊕⊕ and ⊕⊕⊕) exert a notable influence. These strong couplings affect only the synthesis of the hexamer **ABC** and its pentameric precursor **AB**. Both products are coupled autocatalysts: They behave as autocatalytic "egoists" and, at the same time, as mutually catalytic altruists. On the other hand, the tetramer **AC**, which is formed as the main product via the non-auto-

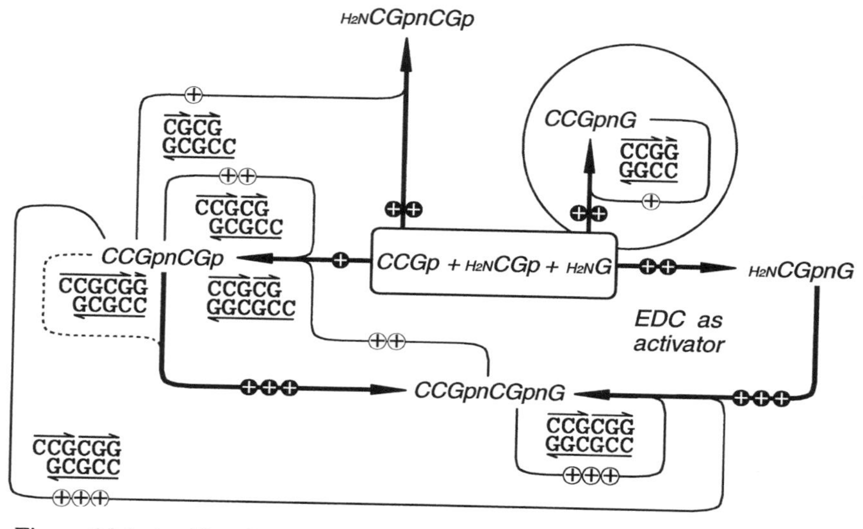

Figure 14.4. A self-replicating system from three starting materials: synthetic pathways and catalytic couplings.

catalytic channel (G–G-stack leads to fast condensation), can be described as an isolated autocatalyst. The molecules **AB** and **ABC** compete with the main product **AC** for the incorporation of their common precursors **A** and **C**. This competition was indeed observed: Upon the seeding of the reaction mixture with the hexameric template CCGCGG (**ABC**), the autocatalysts **AB** and **ABC** were formed more quickly, whereas the formation of the egoistic tetramer **AC** decreased. Competition of replicators for common resources is the prerequisite for selection. "Selection" in the biological sense usually means the takeover of resources by a species that reproduces more efficiently than its competitors (survival of the fittest). However, because selection also depends on the population level of a species (its concentration), a less efficient species may win if it starts at a higher population level. In any case, the population size of a species directs the flow of resource consumption. Although our experimental findings resemble selection, they represent only a rudimentary form of it: "True" Darwinian selection necessitates exponential, and not parabolic, growth.

All synthetic replicators described so far are based on the simplification of an autocatalytic, self-complementary system. However, the natural prototype of nucleic acid replication utilizes complementary, rather than self-complementary, strands (that is, replication via (+)- and (–)-strands). The underlying principle of this type of replication is a cross-catalytic reaction in which one strand acts as a catalyst for the formation of the other strand and vice versa. It

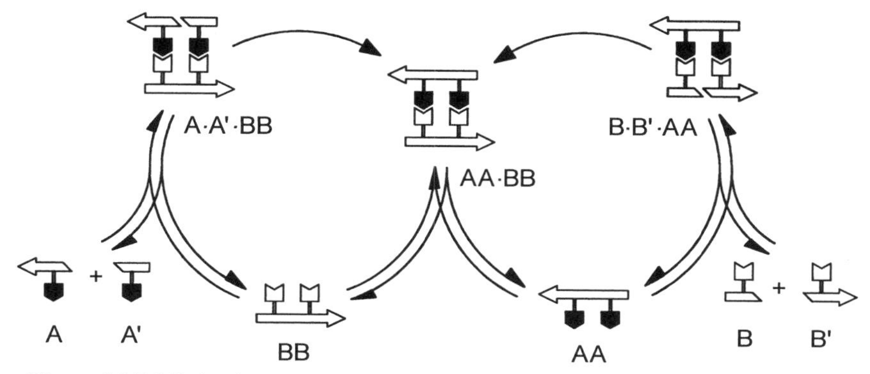

Figure 14.5. Minimal representation of a cross-catalytic self-replicating system.

seemed worthwhile to test whether or not a replication of complementary hexadeoxynucleotides in the absence of any enzyme could be achieved. A minimal implementation of such a cross-catalytic self-replicating system can be represented by a simple reaction scheme (see Figure 14.5) where **AA** and **BB** denote templates, and **A**, **A'**, **B**, and **B'** denote fragment molecules complementary to the templates.

In this scheme, two parallel pathways exist, both leading to the same template duplex. Since **AA** catalyzes the formation of **BB** and vice versa, one can speak of cross-catalysis. As long as both templates are formed using the same type of chemistry, the same conditions that enable cross-catalytic formation of complementary molecules **AA** and **BB** will also allow for the autocatalytic self-replication of both self-complementary products **AB** and **BA**. For best observing the cross-catalytically coupled reactions, all four possible condensations between the precursor molecules must be equally efficient. Earlier experiments in our lab had shown that the efficiency of template-directed condensation reactions between suitably protected trideoxynucleotides was predominantly determined by the stacking of nucleobases flanking the reaction site within the termolecular complex (Wlotzka 1992). Accordingly, the cross-catalytic system could be realized experimentally by Sievers and von Kiedrowski (Sievers and von Kiedrowski 1994; Sievers et al. 1994) using the trimer precursors **A** (for CCG) and **B** (for CGG) to form a central GC subsequence in all four possible hexamers (**AB**, **BB**, **AB**, **BA**). The formation of 3'-5'-phosphoramidate linkages was used to join the trimers. Kinetic analysis revealed that the two complementary sequences **AA** and **BB** are formed with similar efficiency, despite the difference in their pyrimidine content. When the experiment was performed in such a way that all products could form simultaneously, both the complementary and the self-complementary hexadeoxynucleotides exhibited the same time course of formation. In other words, cross-catalysis was as efficient as autocatalysis. On the contrary, in single experiments where only one hexamer could form, the autocatalysts were formed much faster than the complementary oligonucleotides. This is the expected result, since by definition only the autocatalyst is able to accelerate its own synthesis. Cross-catalysis, again by definition, necessitates the simultaneous formation of both complementary products and thus cannot be observed in single experiments in which only one product is formed. As expected, the cross-catalytic self-replication of complementary hexamers revealed parabolic growth characteristics due to product inhibition.

The general scheme of a minimal self-replicating system (Figure 14.1) has served as a successful aid for the design of various new replicators. In these systems, the rate of autocatalytic synthesis typically depends on the square root of the template concentration, and thus the template growth is parabolic rather than exponential. It was shown theoretically by Szathmáry and Gladkih that parabolic growth leads to the coexistence of self-replicating templates, which compete for common resources under stationary conditions (Szathmáry and Gladkih 1989). Coexistence in this context means that the faster replicator is not able to take over the common resources completely. If two non–self-replicating molecules, denoted C_1 and C_2, are competing for common precursors, the ratio of their concentrations is determined solely by their reactivity: $[C_1]/[C_2] = k_1/k_2$. The quotient $[C_1]/[C_2]$ describes selectivity, which is a metric for coexistence. For two parabolic replicators with the autocatalytic rate constants k_1 and k_2, it follows from Szathmáry's treatment that $[C_1]/[C_2] = (k_1/k_2)^2$. Hence, small differences in reactivity lead to a higher selectivity in parabolic replicators as compared to non–self-replicating molecules. This enhancement of selectivity is partly implicit in the results reported and clearly needs to be elaborated in further experiments.

2.4. Self-replication through triple helices: a different approach

A different approach to self-replication has been reported recently by Li and Nicolaou: The principles of triple strand (via Watson–Crick–Hoogsteen pairs) and double strand formation (via Watson–Crick pairs) are exploited to achieve a complete replication cycle (Li and Nicolaou 1994). Switching the pH of the aqueous medium leads from triple helices to double helices. The palindromic duplex DNA (**S1.S2**) serves as a template for the single-stranded DNA fragments **F1** and **F2*** (step 1) under slightly acidic conditions (pH 6). Now the reactive ends come in close proximity and are thus ready to be ligated. N-cyanoimidazole functions as the condensing agent leading to the triple helix (step 2). Raising the pH to 7 enables dissociation of the triple helix (**S1.S2.S1***) into one double strand (**S1.S2**) and a single-stranded molecule

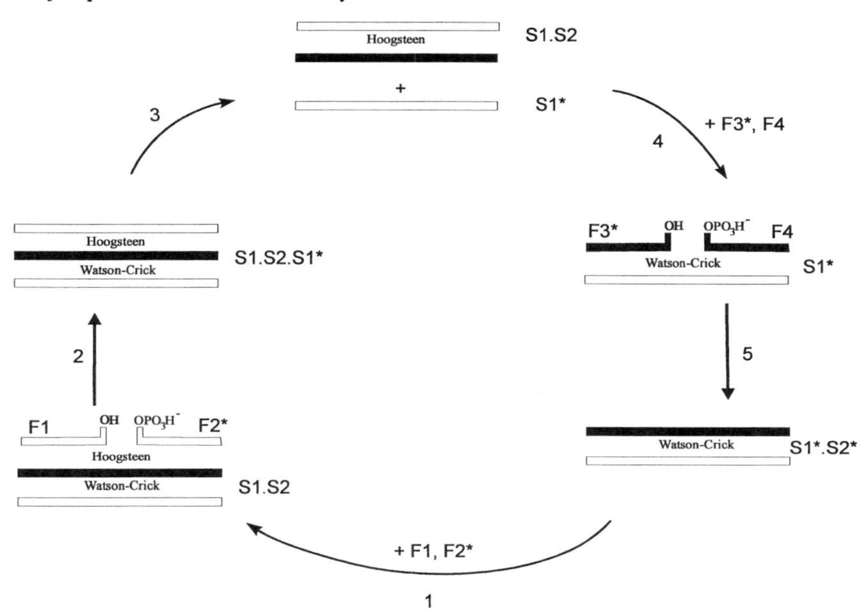

Figure 14.6. Self-replication through triple helix according to Li and Nicolaou (1994). (Purine strands are black; pyrimidine strands are white.)

(**S1***, step 3). The latter functions as a template for the complementary fragments **F3*** and **F4** (step 4) and leads to the duplex (**S1*.S2***) in the presence of N-cyanoimidazole. Turning the pH back to 6 closes the replication cycle. As a result, the original double helix has been replicated by using four different oligonucleotide precursors. This is an example of a stepwise replication scheme that clearly bears the potential of overcoming the problem of product inhibition. However, it is limited to palindromic sequences in which purine and pyrimidine bases are excluded from occurring simultaneously in the same template strand.

3. Replication in nonnucleotidic model systems

Another exciting field in bioorganic chemistry is the development of nonnucleotidic replicators. Since molecular recognition and catalysis are common features among various organic molecules, it seems feasible to develop new replicators based on truly artificial precursors. Rebek and co-workers designed a replicator consisting of an adenosine derivative as the natural component and a derivative of Kemp's acid as the artificial part (Tjivikua, Ballister and Rebek 1990; Nowick et al. 1991). As shown in Figure 14.7, the reactive ends of the reactants **7** and **8** come into spatial proximity when the

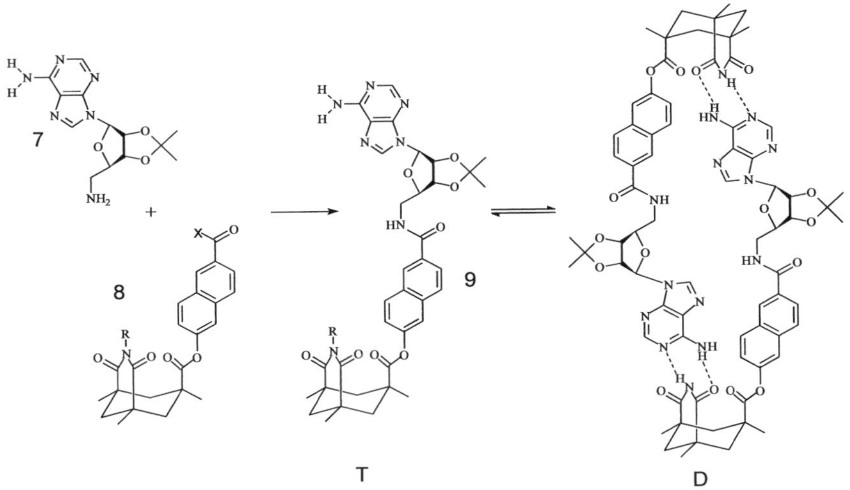

Figure 14.7. Synthetic self-replicating system according to Rebek (Tjivikua, Ballister and Rebek 1990).

reactants interact with the template **9**. Nucleophilic attack of the primary amine of **7** on the activated carboxyl ester of **8** leads to amide bond formation, giving a new template molecule of **9**. Dissociation of the self-complementary template duplex closes the replication cycle.

The source of autocatalysis observed in this system has been recently reinterpreted by Menger and co-workers. (Menger, Eliseev and Khanjin 1994; Menger et al. 1995). Evidence from their experiments showed that simple amides catalyze the acylation reactions between **7** and **8** leading to **9**. Because of the amide moiety in **7**, it was argued that the observed rate enhancement is a consequence of amide catalysis rather than template catalysis. It is not clear yet to what extent the autocatalysis in this system depends on template effects. However, it should be emphasized that the reaction kinetics of a later system, in which the template was able to form much stronger complexes, clearly revealed a square root law dependence (Rotello, Hong and Rebek, 1991; Rebek 1994). This finding strongly supports self-replication with product inhibition.

Rebek's replicator challenged us to think about an even simpler self-replicating system. Terfort and von Kiedrowski used the condensation of 3-aminobenzamidine (**10**) and (2-formylphenoxy)-acetic acid (**11**) to develop an artificial self-replicating system based on simple organic molecules (Terfort and von Kiedrowski 1992). As illustrated in Figure 14.8, autocatalytic condensation of **10** and **11** ($R^1 = R^3 = Me$; $R^2 = R^4 = NO_2$) giving the anil (**13**) was followed by ^1H NMR spectroscopy in dimethylsulfoxide. As expected, the

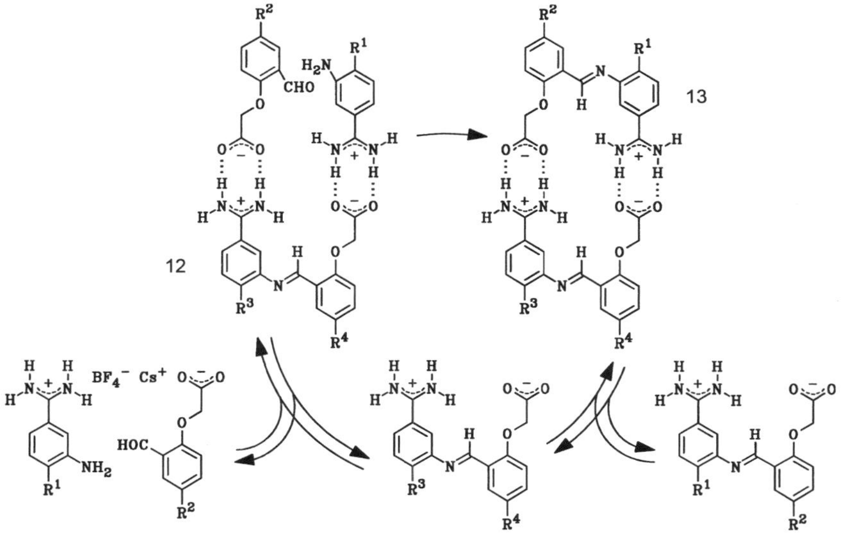

Figure 14.8. Self-replication of amidinium-carboxylate templates.

autocatalytic contribution of the condensation reaction shows a square root law. When R^3 was changed to t-Bu, and R^4 to H, the condensation yielded an unexpected result: The reaction rate was shown to depend linearly on the template concentration. For reasons that are still unknown, product inhibition is not operative in this case. The observation of first-order catalysis in a self-complementary template molecule suggests that exponential replication of nonnucleotidic replicators might be possible. A possible recipe might be what could be termed the modulation of molecular recognition: to search for a template in which the linkage moieties affect the recognition properties of the binding sites such that it forms less stable salt bridges with itself than with its bound precursors.

4. Outlook: toward exponential replication

It was pointed out by Szathmáry that the Darwinian kind of selection (survival of the fittest) necessitates exponential growth of competing replicators. For equilibrated self-replicating systems (in which the template-directed condensation is slow compared to internal equilibration), the autocatalytic reaction order (which is 1 in the case of exponential growth and ½ in the case of parabolic growth) is determined solely by the population of the complexes involved. A minimal self-replicating system as presented in Figure 14.1 is

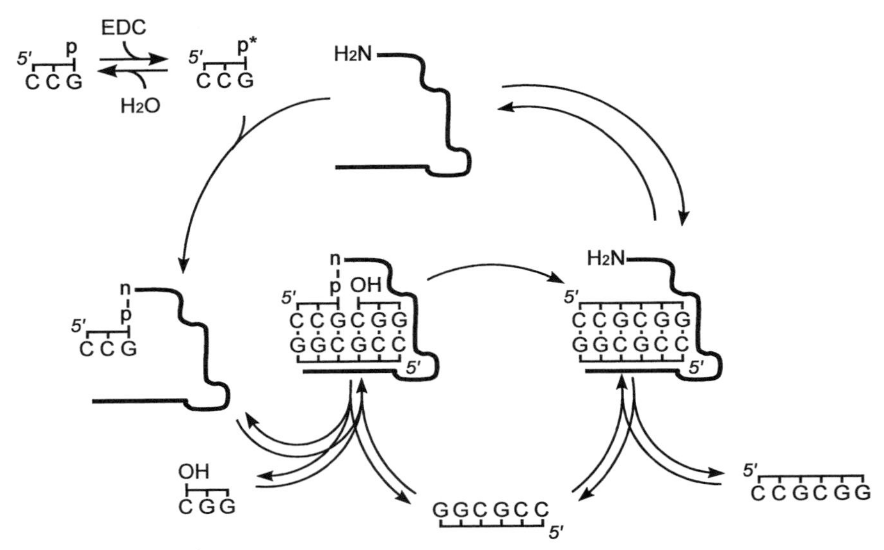

Figure 14.9. Principle of a minimal replicase.

expected to exhibit exponential growth if the termolecular complex is thermodynamically more stable than the template duplex (due to entropic reasons, the usual situation is just opposite to that).

These theoretical considerations lead us to a possible experimental approach toward achieving exponential replication by increasing the thermodynamic stability of the termolecular complex. Using the technique of directed molecular evolution (Ellington and Szostak 1990; Tuerk and Gold 1990; Beaudry and Joyce 1992), we are currently searching through artificial pools of random oligodeoxynucleotides to isolate molecules that are able to catalyze the EDC-driven self-replication of a self-complementary hexadeoxynucleotide. The goal is to find an oligodeoxynucleotide that would act as a nucleophilic catalyst of the phosphoryl transfer step occurring in the termolecular complex (covalent catalysis). Such a catalyst may be considered as a minimal replicase (see Figure 14.9). As long as the replicase is part of the termolecular complex, it stabilizes the latter, for example by wrapping itself around the complex. Stabilization occurs in an intramolecular sense. After phosphoryl transfer, the replicase can bind to the resulting duplex only in an intermolecular way. Thus, both before and after the phosphoryl transfer, we are dealing with termolecular complexes.

A scheme for the directed molecular evolution of a minimal replicase implies the search for a catalytically active leaving group. Our selection starts with a synthetic pool of random oligodeoxynucleotides bearing 5'- and 3'-constant regions as primer binding sites for PCR amplification. A modified primer containing a 5'-terminal biotin and a 3'-5'-phosphoramidate linkage following the leader sequence CCG (primer–trimer–conjugates) was synthesized. PCR amplification of the random pool using the modified 5'-primer, as well as a conventional 3'-primer, gives a pool of double-stranded DNA. Immobilization of the double-stranded pool on a streptavidin column and subsequent denaturation results in a single-stranded DNA pool containing one 3'-5'-phosphoramidate linkage. The in vitro selection can be performed on the column in the presence of the attacking trimer (CGG) and the hexameric template (CCGCGG). The 5'-amino oligomers that are released during selection can be eluted from the column. PCR amplification of the column eluate yields the pool for the next round of selection. Successive rounds of kinetic selection and PCR amplification should result in selective enrichment of active sequences. Negative selection steps in the absence of trimer and template should enable us to distinguish between those molecules that are released during phosphoryl transfer and those molecules that undergo simple hydrolysis of the phosphoramidate bond (preliminary experiments studying the uncatalyzed phosphoryl transfer indicate that hydrolysis is the most predominant reaction).

An alternative way to overcome product inhibition in minimal self-replicating systems might be the destabilization of template duplexes. Performing the template replication at surfaces could help to separate the template strands. This approach might ultimately lead to the nonenzymatic exponential replication of arbitrary sequences. Once exponential replication with information transfer has been achieved in the lab, molecular evolution in vitro mimicking Darwinian behavior seems feasible. This may also lead to interesting practical applications of a field that is currently driven by academic interest.

Acknowledgments

Financial support by Deutsche Forschungsgemeinschaft, Fonds der Chemischen Industrie, NATO, and German Israeli Foundation is gratefully acknowledged.

References

Achilles, T., and von Kiedrowski, G. 1993. A self-replicating system from three starting materials. *Angew. Chem. Int. Ed. Engl.* 32:1198–1201.

Bag, B. G., and von Kiedrowski, G. 1996. Templates, autocatalysis and molecular replication. *Pure Appl.* 68:2145–2152.

Beaudry, A. A., and Joyce, G. F. 1992. Directed evolution of an RNA enzyme. *Science* 257:635–641.

Breaker, R., and Joyce, G. F. 1994. A DNA enzyme that cleaves RNA. *Chemistry and Biology* 1:223–229.

Cech, T. R. 1986. A model for the RNA-catalyzed replication of RNA. *Proc. Natl. Acad. Sci. USA* 83:4360–4363.

Crick, F. H. C. 1968. The origin of the genetic code. *J. Mol. Biol.* 38:367–379.

Dolinnaya, N. G., Sokolova, N. I., Gryaznova O. I., and Shabarova Z. A. 1988. Site-directed modification of DNA duplexes by chemical ligation. *Nucleic Acids Res.* 16:3721–3738.

Dolinnaya, N. G., Tsytovich, A. V., Sergeev, V. N., Oretskaya, T. S., and Shabarova, Z. A. 1991. Structural and kinetic aspects of chemical reactions in DNA duplexes: Information of DNA local structure obtained from chemical ligation data. *Nucleic Acids Res.* 19:3073–3080.

Eigen, M., and Schuster, P. 1979. *The Hypercycle: A Principle of Natural Self-Organization.* Berlin: Springer.

Ellington, A. D., and Szostak, J. W. 1990. *In vitro* selection of RNA molecules that bind specific ligands. *Nature* 346:818–822.

Inoue, T., Joyce, G. F., Grzeskowiak, K., Orgel, L. E., Brown, J. M., and Reese, C. B. 1984. Template-directed synthesis on the pentanucleotide CpCpGpCpC. *J. Mol. Evol.* 178:669–676.

Inoue, T., and Orgel, L. E. 1983. A nonenzymatic RNA polymerase model. *Science* 219:859–862.

Joyce, G. F. 1989. RNA evolution and the origins of life. *Nature* 338:217–224.

Kuhn, H., and Waser, J. 1981. Molekulare Selbstorganisation und Ursprung des Lebens. *Angew. Chem.* 93:495–515.

Li, T., and Nicolaou, K. C. 1994. Chemical self-replication of palindromic duplex DNA. *Nature* 369:218–221.

Menger, F. M., Eliseev, A. V., and Khanjin, N. A. 1994. *J. Am. Chem. Soc.* 116:3613–3615.

Menger, F. M., Eliseev, A. V., Khanjin, N. A., and Sherrod, M. J. 1995. *J. Org. Chem.* 60:2870–2872.

Naylor, R, and Gilham, P. T. 1966. Studies on some interactions and reactions of oligonucleotides in aqueous solution. *Biochemistry* 5:2722–2728.

Nowick, J. S., Feng, Q., Tjivikua, T., Ballester, P., and Rebek, J. 1991. Kinetic studies and modeling of a self-replicating system. *J. Am. Chem. Soc.* 113:8831–8832.

Oparin, A. I. 1924. *The Origin of Life.* Moscow: Pabochii.

Orgel, L. E. 1968. Evolution of the genetic apparatus. *J. Mol. Biol.* 38:381–393.

Orgel, L. E. 1992. Molecular replication. *Nature* 358: 203–209.

Orgel, L. E. 1995. Unnatural selection in chemical systems. *Acc. Chem. Res.* 28:109–119.

Orgel, L. E., and Lohrmann, R. 1974. Prebiotic chemistry and nucleic acid replication. *Acc. Chem. Res.* 7:368–377.

Rebek, J. 1994. A template for life. *Chem. Brit.* 30: 286–290.

Rotello, V., Hong, J. I., and Rebek, J. 1991. Sigmoidal growth in a self-replicating system. *J. Am. Chem. Soc.* 113: 9422–9423.

Sharp, P. A. 1985. *Cell* 42:397–403.

Sievers, D., Achilles, T., Burmeister, J., Jordan, S., Terfort, A., and von Kiedrowski, G. 1994. Molecular replication: from minimal to complex systems. In *Self-Production of Supramolecular Structures*, ed. G. R. Fleischaker et al., pp. 45–64. Dordrecht: Kluwer Publishers.

Sievers, D., and von Kiedrowski, G. 1994. Self-replication of complementary nucleotide-based oligomers. *Nature* 369:221–224.

Szathmáry, E., and Gladkih, I. 1989. Subexponential growth and coexistence of nonenzymatically replicating templates. *J. Theor. Biol.* 138:55–58.
Terfort, A., and von Kiedrowski, G. 1992. Self-replication by condensation of 3-amino-benzamidines and 2-formylphenoxyacetic acids. *Angew. Chem. Int. Ed. Engl.* 31:654–656.
Tjivikua, T., Ballister, P., and Rebek, J. 1990. A self-replicating system. *J. Am. Chem. Soc.* 112:1249–1250.
Tuerk, C., and Gold, L. 1990. Systematic evolution of ligands by exponential enrichment: RNA ligands to bacteriophage T4 DNA polymerase. *Science* 249:505–510.
von Kiedrowski, G. 1986. A self-replicating hexadeoxynucleotide. *Angew. Chem. Int. Ed. Engl.* 25:932–935.
von Kiedrowski, G. 1990. *Selbstreplikation in chemischen Minimalsystemen. 40 Jahre Fonds der Chemischen Industrie 1950–1990.* VCI, Frankfurt, 197–218.
von Kiedrowski, G. 1993. Minimal replicator theory I: Parabolic versus exponential growth. *Bioorg. Chem. Front.* 3:113–146.
von Kiedrowski, G., Wlotzka, B., and Helbing, J. 1989. Sequence dependence of template-directed syntheses of hexadeoxynucleotide derivatives with 3'-5'-pyrophosphate linkage. *Angew. Chem. Int. Ed. Engl.* 28:1235–1237.
von Kiedrowski, G., Wlotzka, B., Helbing, J., Matzen, M., and Jordan, S. 1991. Parabolic growth of a self-replicating hexadeoxynucleotide bearing a 3'-5'-phosphoamidate linkage. *Angew. Chem. Int. Ed. Engl.* 30:423–426; and corrigendum p. 892.
Wlotzka, B. 1992. Selbstreplizierende Oligonucleotide. Untersuchungen zur Sequenz- und Temperaturabhängigkeit bei der Synthese hexamerer 3'-5'-Phosphoamidate. Dissertation thesis, University of Göttingen.
Zielinski, W. S., and Orgel, L. E. 1987. Autocatalytic synthesis of a tetranucleotide analogue. *Nature* 327: 346–347.figure and table captions

Part IV

Clues from the bacterial world

15

Hyperthermophiles and their possible role as ancestors of modern life

KARL O. STETTER
Lehrstuhl für Mikrobiologie, Universität Regensburg
Germany

1. Introduction

Various microorganisms grow fastest ("optimally") at temperatures above 40 °C. In an anthropocentric kind of view (the human body temperature is only 37 °C) these are designated as thermophiles (Brock 1986; Kristjansson and Stetter 1992). These "heat lovers" flourish within natural and anthropogenic biotopes like hot waters, sun-heated soils, geothermal areas, self-heated waste dumps, and industrial cooling waters. Depending on the isolates, most thermophiles exhibit an upper temperature limit of growth between 50 and 70 °C. On the other hand, these moderate thermophiles are still able to grow (with reduced propagation rates, however) at 25–40 °C within the mesophilic temperature range. Thermophiles are widespread within many different taxonomic groups of pro- and eukaryotic microorganisms like Bacilli, Clostridia, Streptomycetes, Cyanobacteria, fungi, algae, and protozoa, which harbor mainly mesophilic species. Due to their close phylogenetic relationship to mesophiles and their modest thermophily, moderate thermophiles may represent secondary adaptation to heat in the history of life.

The upper temperature border of life is represented by hyperthermophilic prokaryotes with optimal growth temperatures between 80 and 113 °C (under slight overpressure). In contrast to moderate thermophiles, as a rule, hyperthermophiles are unable to grow below 60 °C. The most extreme hyperthermophile, *Pyrolobus fumarii,* is unable to propagate even at 90 °C since this temperature is still too low to support its growth (Blöchl et al. 1997). So far, an unanticipated variety of distant groups of hyperthermophiles have been successfully cultivated (Table 15.1; Stetter 1996). This result is in contrast to earlier assumptions of a decreasing diversity of microorganisms with increasing temperatures of the biotope (e.g., Castenholz 1979). Hyperthermophiles belong to the *Bacteria* and *Archaea* (former: archaebacteria; Woese and Fox 1977; Woese, Kandler, and Wheelis 1990). Most likely, they represent very ancient adaptations to high temperatures. So far, 56 species of hyperther-

Table 15.1. *Taxonomy of hyperthermophilic prokaryotes*

Order	Genus	Species	T. max (°C)
I. BACTERIAL DOMAIN			
Thermotogales	*Thermotoga*	*T. maritima*	90
		T. neapolitana	90
		T. thermarum	84
		T. elfii	72*
	Thermosipho	*T. africanus*	77*
	Fervidobacterium	*F. nodosum*	80*
		F. islandicum	80*
	Geotoga	*G. petrea*	55*
		G. subterranea	60*
	Petroga	*P. miotherma*	65*
Aquificiales	*Aquifex*	*A. pyrophilus*	93
II. ARCHAEAL DOMAIN			
Sulfolobales	*Sulfolobus*	*S. acidocaldarius*	85
		S. solfataricus	87
		S. shibatae	86
		S. metallicus	75*
	Metallosphaera	*M. sedula*	80*
		M. prunae	80*
	Acidianus	*A. infernus*	95
		A. brierleyi	75*
		A. ambivalens	95
	Stygiolobus	*S. azoricus*	89
Thermoproteales	*Thermoproteus*	*T. tenax*	97
		T. neutrophilus	97
		T. uzoniensis	97
	Pyrobaculum	*P. islandicum*	103
		P. organotrophum	103
		P. aerophilum	104
	Thermofilum	*T. pendens*	95
		T. librum	95
Desulfurococcales	*Desulfurococcus*	*D. mobilis*	95
		D. mucosus	97
		D. saccharovo-rans	97
		D. amylolyticus	97
	Staphylothermus	*S. marinus*	98

Order	Genus	Species	T. max (°C)
Pyrodictiales	*Pyrodictium*	*P. occultum*	110
		P. brockii	110
		P. abyssi	110
	Pyrolobus	*P. fumarius*	113
	Hyperthermus	*H. butylicus*	108
	Thermodiscus	*T. maritimus*	98
Thermococcales	*Thermococcus*	*T. celer*	93
		T. litoralis	98
		T. stetteri	98
		T. profundus	90
		T. alcaliphilus	90
		T. chitonophagus	93
	Pyrococcus	*P. furiosus*	103
		P. woesii	103
Archaeoglobales	*Archaeoglobus*	*A. fulgidus*	92
		A. profundus	92
Methanobacteriales	*Methanothermus*	*M. fervidus*	97
Methanococcales		*M. sociabilis*	97
	Methanococcus	*M. thermolithotrophicus*	70*
		M. jannaschii	86
Methanopyrales		*M. igneus*	91
	Methanopyrus	*M. kandleri*	110

* Moderate thermophiles related to hyperthermophiles

mophilic *Bacteria* and *Archaea* are known (Table 15.1). They are divergent by their phylogeny and physiological properties and are grouped into 24 genera and 11 orders. Within the *Bacteria, Aquifex pyrophilus* and *Thermotoga maritima* exhibit the highest growth temperatures with 95 and 90 °C, respectively (Table 15.1). Within the *Archaea*, the organisms with the highest growth temperatures (between 103 and 113 °C) are members of the genera *Pyrobaculum, Pyrococcus, Pyrolobus, Pyrodictium,* and *Methanopyrus*. Direct microscopic inspection of hot biotopes indicates there are other morphotypes yet to be cultivated. Moreover, extraction and amplification of DNA corresponding to 16S rRNA from a hot pool in Yellowstone Park indicates the presence of many so far uncultivated members of *Archaea* with unknown properties (Barns et al. 1994).

In this chapter, the hyperthermophiles will be introduced and their biotopes, strategies of living, their phylogeny, and possible role as rather primitive descendants of a hyperthermophilic common ancestor of life will be discussed.

Table 15.2. *Biotopes of hyperthermophiles*

	TYPE OF THERMAL AREA	
Characteristics	Terrestrial	Marine
Locations	Solfataric fields; steam-heated soils; mud holes and surface waters; deep hot springs; geothermal power plants; deep terrestrial oil fields	Submarine hot springs and fumaroles; hot sediments and vents ("black smokers"); active seamounts; deep submarine oil fields
Temperatures	Surface: up to 100 °C* Depth: above 100 °C	Up to about 400 °C
Salinity	Usually low (0.1 to 0.5 % salt)	Usually about 3% salt
pH	0.5–9	5–8.5 (rarely: 3)
Major gases and sulfur compounds	CO_2, CO, CH_4, H_2, H_2S, S°, $S_2O_3^{2-}$, SO_3^{2-}, SO_4^{2-}, NH_4^+, N_2, NO_3^-	CO_2, CO, CH_4, H_2, H_2S, S°, $S_2O_3^{2-}$**, SO_3^{2-}, NH_4^+, N_2, NO_3^-

* At sea level, depending on the altitude
** Seawater contains about 30 mmol/l of sulfate

2. Biotopes of hyperthermophiles

Hyperthermophiles have so far been isolated from water-containing terrestrial and marine high-temperature areas (Table 15.2) where they form communities. The most common biotopes are volcanically and geothermally heated hydrothermal systems like solfataric fields, neutral hot springs, and submarine saline hot vents. As a rule, hot solfataric soils consist of two different zones: an upper acidic layer, which contains significant amounts of oxygen. Due to the presence of ferric iron, it exhibits an ochre color. From this layer, aerobic hyperthermophilic acidophiles like *Sulfolobus, Acidianus,* and *Metallosphaera* can be isolated. The area below shows a blackish-blue color due to the presence of ferrous iron. Its pH is neutral and its redox state is strongly reduced. From there, strictly anaerobic members of the *Thermoproteales* and *Methanothermus* can be isolated. Boiling solfataric mud holes in Iceland (Krafla area) contained up to 10^8 cells/ml that consisted mainly of the characteristic "golf club"–shaped *Thermoproteales* and the irregularly coccoid-shaped *Sulfolobales* (Stetter unpublished). Cultivation attempts from Javan solfataric fields revealed the presence of complex communities with species similar to those known already from Italian and Yellowstone Park solfataric fields (G. Huber et al. 1991). From a hot spring at the Dieng Plateau, Java, *Ammonifex degensii,* a novel, strictly autotrophic nitrate ammonifier was isolated (R. Huber et al. 1996). Artificial, man-made solfataric field–like biotopes are smoldering coal refuse piles and (coal-bearing) uranium mines (e.g., Wismut AG, Ronneburg, East Germany) that contain members of *Thermoplasma, Metallosphaera,* and *Sulfolobus* (Darland et al. 1970; Marsh and Norris 1985; Fuchs et al. 1995; H. Huber and Stetter unpublished). Further nonnatural environments for hyperthermophiles are boiling waste waters from geothermal power plants. From an overpressure valve at the Krafla geothermal power plant in Iceland, *Pyrobaculum islandicum* has been isolated (Isolate GEO3; R. Huber, Kristjansson and Stetter 1987).

Submarine hydrothermal systems are situated in shallow and abyssal depths. They consist of hot fumaroles, springs, sediments, and deep sea vents with temperatures up to about 400 °C (e.g., "black smokers"). Shallow marine hydrothermal systems are located at the beaches of Vulcano, Naples, and Ischia (all Italy), Sao Michel (Azores), and Djibouti (Africa). At the Kolbeinsey area at the Mid-Atlantic Ridge north of Iceland, a submarine hydrothermal system is situated at a depth of about 120 m (Fricke et al. 1989). Examples of deep sea hydrothermal systems are the Guaymas Basin (depth: 1,500 m) and the East Pacific Rise (21° N; depth about 2,500 m), both in Mexico, and the Mid-Atlantic Ridge ("Snake Pit" and "TAG" site; 26° N, 44° W; depth about 3,700 m). Further submarine high-temperature areas are active seamounts like Teahicya and Macdonald in the Tahiti area. Samples taken from the active crater and the cooled-down open ocean plume of erupting Macdonald seamount contained communities of hyperthermophiles with up to 10^6 viable cells per liter (R. Huber et al. 1990). Submarine hydrothermal fluids are usually high in NaCl and exhibit a slightly acidic to alkaline pH of 5 to 8.5 (Table 15.2). As an exception, alkaline fresh water hot springs are emerging at the sea bottom off the coast of Reykjanes in the Isafjardardjup Bay in northwest Iceland (Alfredsson et al. 1988). Shallow, as well as deep, marine hydrothermal systems harbor members of the *Pyrodictiales, Thermococcales, Methanococcales, Archaeoglobales,* and *Thermotogales.* So far, members of *Methanopyrus* were found only in greater depths, whereas *Aquifex* was isolated exclusively from shallow hydrothermal systems.

There is evidence for the presence of hyperthermophiles within cold seawater. Although unable to grow, the hyperthermophiles may survive for many years in a kind of dormant state. Even after storage for 10 years at 4 °C, samples of originally hot submarine environments gave rise to positive enrichment cultures of hyperthermophiles (Stetter unpublished). Water samples taken from the (cold) open sea around Easter Island contained about one viable cell of hyperthermophiles per cubic meter of seawater (Keller and Stetter unpublished). In agreement with this finding, sand filters passed by cold Arctic seawater at the Beaufort Sea, North Alaska, harbored viable

Table 15.3. *Examples of viable hyperthermophiles in Thistle production fluids (Stetter et al. 1993)*

Well no.	Original titer (cells/ml)	Identified species
A02	100	*Archaeoglobus fulgidus; Archaeoglobus lithotrophicus*
	10	*Thermococcus celer; Thermococcus litoralis*
	1,000	*Pyrococcus* sp. nov.
A05	1,000	*Archaeoglobus fulgidus; Archaeoglobos lithotrophicus*
	10	*Thermococcus litoralis*
	10	*Pyrococcus* sp. nov.
A08	10,000	*Archaeoglobus fulgidus; A. profundus; A. lithotrophicus*
	100	*Thermococcus litoralis*
	10	*Pyrococcus* sp. nov.
A25	100	*Archaeoglobos fulgidus; Archaeoglobus lithotrophicus*
	1,000	*Thermococcus celer; Thermococcus litoralis*
	10	*Pyrococcus* sp. nov.
A31	100	*Archaeoglobus fulgidus; Archaeoglobus lithotrophicus*
	10	*Thermococcus celer; Thermococcus litoralis*
	1,000	*Pyrococcus* sp. nov. (enrichment on crude oil)

hyperthermophiles that gave rise to positive enrichment cultures at 85 °C (Stetter et al. 1993). A novel, recently discovered biotope of hyperthermophiles is deep, geothermally heated oil reservoirs some 3,500 m below the bed of the North Sea and the permafrost soil at the North Slope, northern Alaska (Stetter et al. 1993). The upwelling production fluids contained up to 10^4 and 10^7 viable cells of hyperthermophiles per milliliter at Thistle platform, North Sea (Table 15.3) and Prudhoe Bay, Alaska, respectively. This indicates the presence of as yet unknown communities of hyperthermophiles within deep oil reservoirs. In the laboratory, some of the crude oil hyperthermophiles are able to grow anaerobically even on unknown hydrophobic crude oil components (R. Huber and Stetter unpublished).

3. Possible archaean hydrogen and ammonia sources created by anaerobic pyrite formation

Many hyperthermophiles require H_2 as an energy source. Within their biotopes, H_2 may be supplied by (a) volcanic gases, (b) fermentative organisms like *Thermotoga* and *Pyrococcus* present in hyperthermophilic microbial

Table 15.4. *Abiotic formation of ammonia from nitrate in the presence of FeS and H_2S (Böchl et al. 1992)*

Experiment no.	FeS (2 mmol)	H_2S (2mmol)	NH_3 formed (μmol)
1	+	+	200
1a	+	+	206
2	+	–	54
2a	+	–	53
3	–	+	0
4	–	–	0

(72 hours at 100 °C; pH 4; 480 μmol $NaNO_3$)

communities, or (c) abiotic anaerobic pyrite formation from FeS and H_2S by the following reaction (Drobner et al. 1990):

$$FeS + H_2S \rightarrow FeS_2 + H_2.$$

This reaction has been postulated as a possible direct source of energy for early life (Wächtershäuser 1988). Pyrite and H_2 are formed simultaneously from ferrous sulfides like pyrrhotite or amorphous FeS and H_2S. In our quest for phylogenetically deep-branching hyperthermophilic *Bacteria* and *Archaea,* we tried very hard, though unsuccessfully, to enrich for organisms possibly using such a proposed primitive type of metabolism. However, we were able to enrich H_2-using members of *Archaeoglobus* and *Pyrolobus* from sites with abundant pyrite deposits by applying FeS/H_2S as a hydrogen source (Drobner, Blöchl and Stetter unpublished).

Surprisingly, in our pyrite formation experiments in the presence of nitrate, large amounts of ammonia were formed abiotically. Subsequent experiments gave evidence for a synergistic effect of H_2S and FeS for the reduction of NO_3^- to NH_3 (Table 15.4). FeS alone causes only minor conversion, and H_2S alone causes none. Because these conditions are geochemically plausible, this kind of nitrate reduction may be assumed to proceed in nature. For the prebiotic Earth, a primitive N-cycle has been postulated (Mancinelli and McKay 1988). NO, produced from N_2 and CO_2 in the early atmosphere by lightning, reacts to form dissolved NO_3^- and NO_2^- in the oceans. This could have been reduced to NH_3 by the reaction found by us, thereby providing NH_3 to the early life forms. Therefore, by this reaction, there may have always been sufficient ammonia supply to life on the early Earth. In addition, our result has a biochemical significance. It suggests a mechanistic communality between the iron–sulfur clusters of nitrite reductases and reduction by FeS/H_2S. The evolution of enzymatic nitrate reduction may now be traced back to abiotic nitrate reduction (Blöchl et al. 1992).

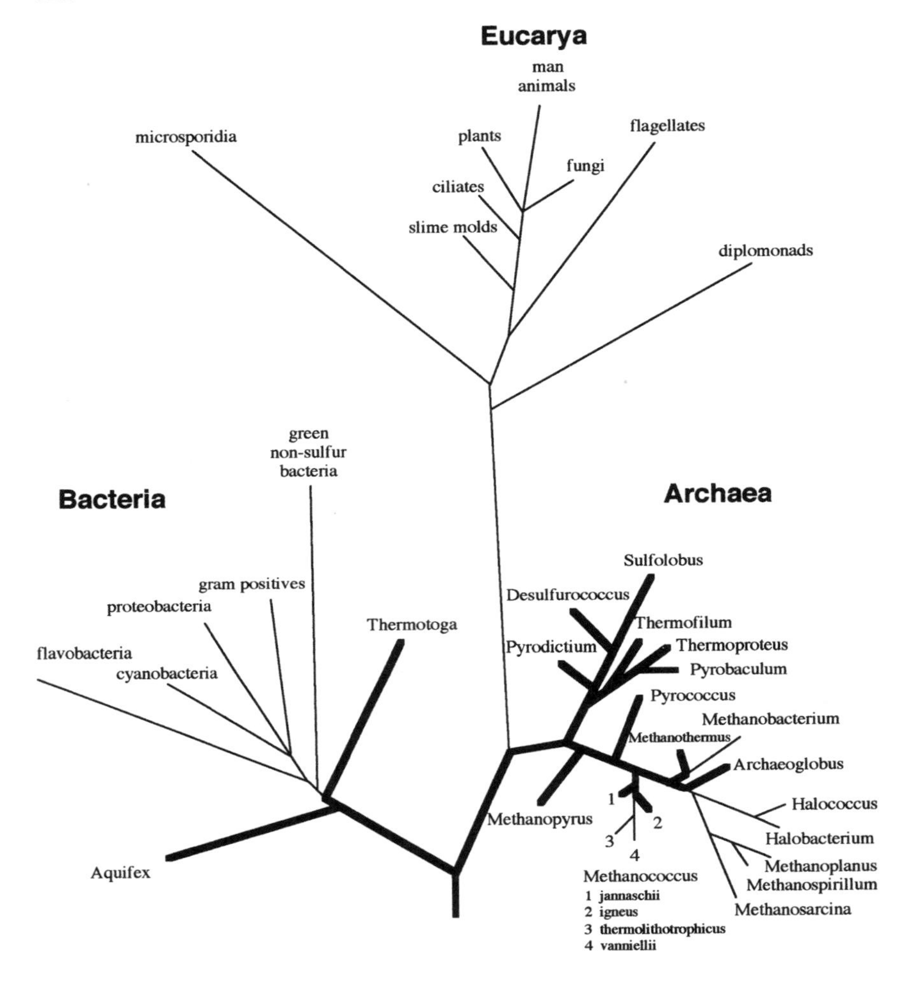

Figure 15.1. Hyperthermophiles within the 16S rRNA–based phylogenetic tree. Redrawn and modified (based on C. R. Woese's model; Woese, Kandler and Wheelis 1990).

4. Molecular universal phylogenetic tree of life

Based on 16/18S rRNA sequence comparisons, a universal phylogenetic tree is now available (Iwabe et al. 1989; Woese, Kandler and Wheelis 1990). It shows a tripartite division of the living world consisting of the domains *Bacteria* (former: eubacteria), *Archaea* (former: archaebacteria), and the *Eucarya* (Figure 15.1). Short phylogenetic lineages indicate a rather slow clock of evolution. Organisms representing them may still be rather similar to their primitive ancestors. Deep branching points are evidence for early separation of two groups. The separation of the *Bacteria* from the *Eucarya–Archaea* lineage is the deepest and earliest branching point known so far (Figure 15.1). Hyperthermophiles are represented among all the deepest and shortest lineages, including *Aquifex* and *Thermotoga* within the *Bacteria;* and *Pyrodictium, Pyrolobus, Pyrobaculum, Desulfurococcus, Sulfolobus, Methanopyrus, Thermococcus, Methanothermus,* and *Archaeoglobus* within the *Archaea* (Figure 15.1, bold lines). Based on these observations, the conclusion can be drawn that hyperthermophiles may still be rather primitive and the last common ancestor may have been a hyperthermophile (Stetter 1994).

5. Physiological and metabolic properties of hyperthermophiles

Hyperthermophiles are well adapted to their biotopes, being able to grow at high temperatures and extremes of pH, redox potential, and salinity (Table 15.2). Within their habitats, hyperthermophiles form complex ecosystems consisting of a variety of primary producers and decomposers of organic matter. Primary producers are chemolithoautotrophs using inorganic electron donors and acceptors in their energy-yielding reactions. In addition, chemolithoautotrophs are able to use CO_2 as their sole carbon source.

Neutrophiles and modest acidophiles

Communities of neutrophilic and slightly acidophilic hyperthermophiles are found in terrestrial solfataric fields, submarine hydrothermal systems, and deep oil reservoirs (Stetter et al. 1993). Most of them are strict anaerobes. Terrestrial solfataric fields harbor members of the genera *Desulfurococcus, Methanothermus, Pyrobaculum, Thermoproteus,* and *Thermofilum.* Cells of *Thermoproteus, Pyrobaculum,* and *Thermofilum* are almost rectangular rods (Figure 15.2a). During the exponential growth phase, spheres are formed at the ends ("golf clubs"). Cells of *Thermofilum* are only about 0.17–0.35 μm in diameter, whereas those of *Pyrobaculum* and *Thermoproteus* are about 0.50 μm. *Pyrobaculum islandicum, Thermoproteus tenax,* and *Thermoproteus neutrophilus* are able to grow chemolithoautotrophically by anaerobic reduction of S° by H_2. *Pyrobaculum aerophilum,* however, is a marine organism able to grow anaerobically by reduction of nitrate by H_2. In addition, under microaerobic conditions this species grows on H_2 and O_2 as its energy source (Völkl et al. 1993). Strains of *Pyrobaculum organotrophum, Thermoproteus uzoniensis,* and *Thermofilum* are obligate heterotrophs growing on organic

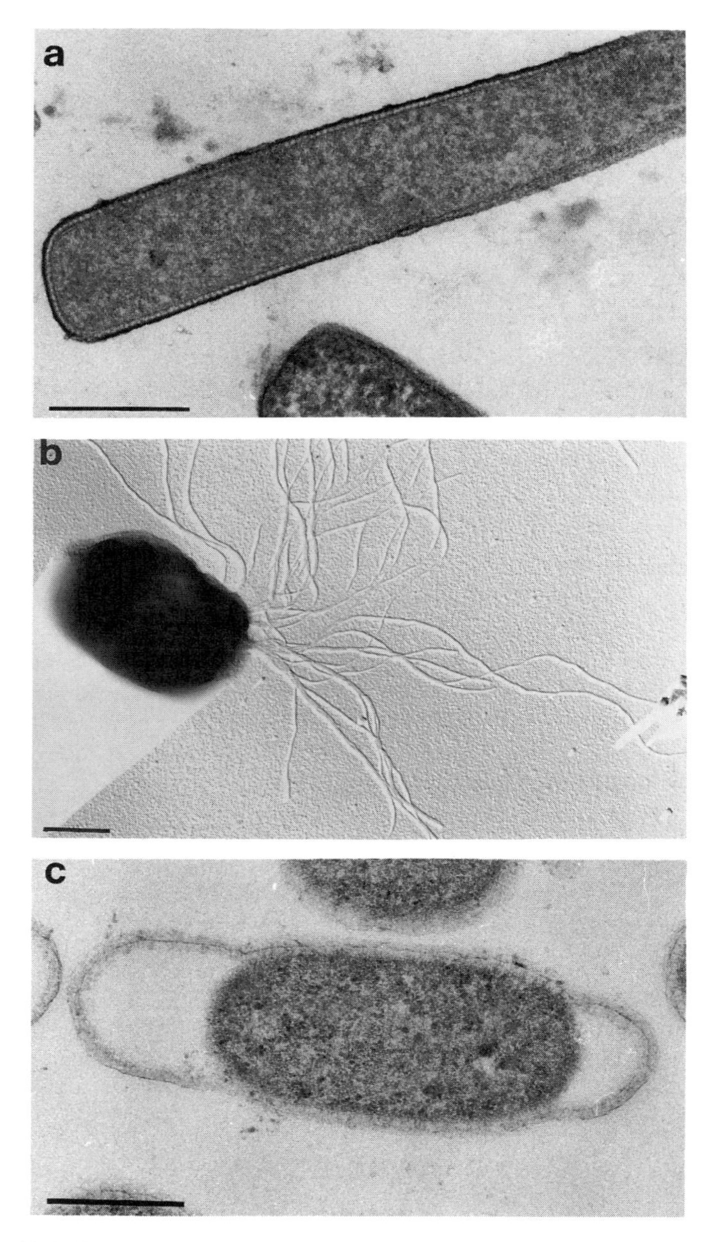

Figure 15.2. Electron micrographs of hyperthermophilic *Archaea* and *Bacteria*. (**a**) *Pyrobaculum aerophilum*, ultrathin section; (**b**) cell of *Pyrococcus furiosus* with flagella, platinum shadowed; (**c**) cell of *Thermotoga maritima*, surrounded by the TTogaU, ultrathin section; (**d**) network of tubules with cells of *Pyrodictium abyssi*, scanning electron micrograph; (**e**) *Methanopyrus kandleri*, ultrathin section; (**f**) cell of *Metallosphaera prunae*, freeze-etched. Bars: 0.55 μm.

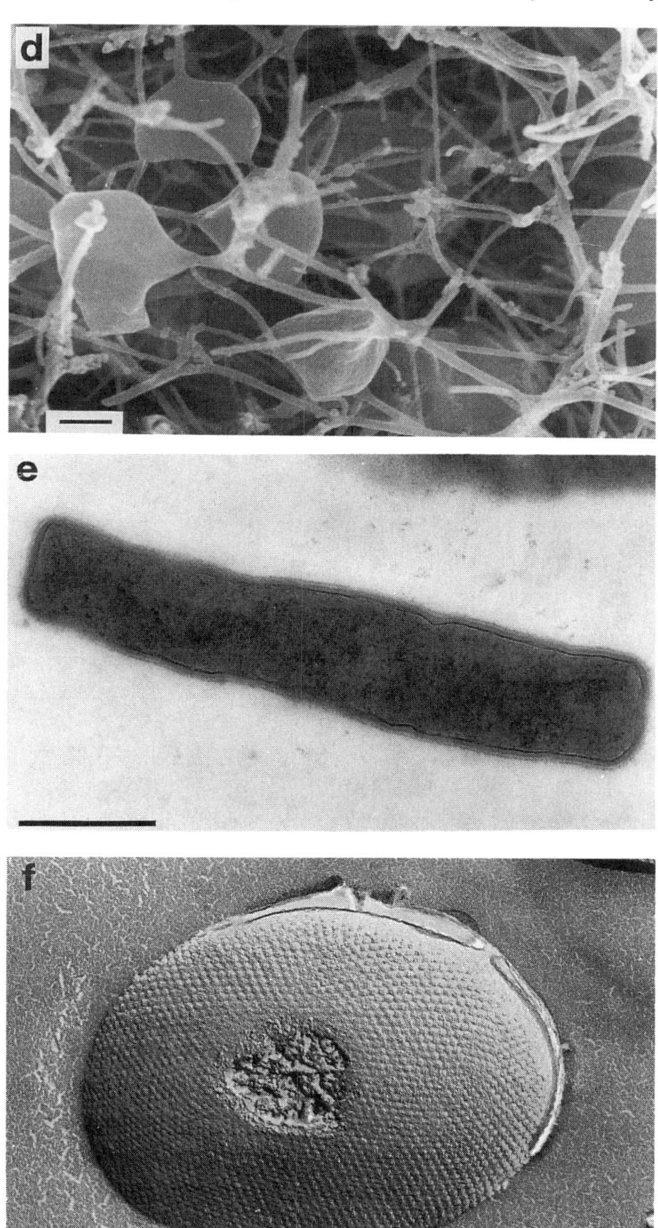
d
e
f

Table 15.5. *Growth conditions and morphological and biochemical features of hyperthermophiles*

Species	Growth conditions					Habitat	DNA	Morphology
	Temp (°C)			pH	Aerobic (ae) or anaerobic (an)	marine (m); terrestrial (t)	G+C (mol%)	
	Minimum	Optimum	Maximum					
Thermotoga maritima	55	80	90	5-9	an	m	46	Rods with a *Toga*
Aquifex pyrophilus	67	85	95	5-7	ae	m	40	Rods
Acidianus infernus	60	88	95	2-5	ae/an	t	31	Lobed cocci
Metallosphaera sedula	50	75	80	1-4	ae	t	45	Cocci
Sulfolobus acidocaldarius	60	75	85	1-5	ae	t	37	Lobed cocci
Stygiolobus azoricus	57	80	89	1-5	an	t	38	Lobed cocci
Thermoproteus tenax	70	88	97	3-6	an	t	56	Regular rods
Pyrobaculum islandicum	74	100	103	5-7	an	t	46	Regular rods
Pyrobaculum aerophilum	75	100	104	6-9	ae/an	m	52	Regular rods
Thermofilum pendens	70	88	95	4-6	an	t	57	Slender regular rods
Desulfurococcus mobilis	70	85	95	5-7	an	t	51	Cocci
Staphylothermus marinus	65	92	98	5-8	an	m	35	Cocci in aggregates
Pyrodictium occultum	82	105	110	5-7	an	m	62	Discs with tubules
Thermodiscus maritimus	75	88	98	5-7	an	m	49	Discs
Pyrococcus furiosus	70	100	105	5-9	an	m	38	Cocci
Archaeoglobus fulgidus	60	83	95	5-7	an	m	46	Irregular cocci
Methanothermus sociabilis	65	88	97	5-7	an	t	33	Rods in clusters
Methanococcus igneus	45	88	91	5-7	an	m	31	Irregular cocci
Methanopyrus kandleri	84	98	110	5-7	an	m	60	Rods in chains

substrates by sulfur respiration. *Pyrobaculum islandicum* and *Thermoproteus tenax* are facultative heterotrophic sulfur respirers. Members of *Desulfurococcus, Staphylothermus,* and *Thermodiscus* are coccoid, strictly heterotrophic, sulfur respirers. Members of *Thermococcus* and *Pyrococcus* are flagellated coccoid cells (Figure 15.2b) that gain energy by the fermentation of peptides, amino acids, and sugars, thereby forming fatty acids, CO_2 and H_2. In the presence of S°, H_2S is formed instead of H_2. *Pyrococcus furiosus* is able to ferment pyruvate, forming acetate, H_2 and CO_2 (Schäfer and Schönheit 1992). *Pyrococcus* and *Thermococcus* species have been found in oil reservoirs (Stetter et al. 1993). *Thermococcus alcaliphilus* is able to grow under conditions up to pH 10.5, and *Thermococcus chitonophagus* represents a hyperthermophilic chitin degrader (R. Huber et al. 1995; Keller et al. 1995). Many terrestrial and submarine hydrothermal fields contain members of the bacterial genus *Thermotoga,* which consists of rod-shaped cells surrounded by a characteristic sheathlike structure (*Toga*) overballooning at the ends (Figure 15.2c; Table 15.5). The *Toga* contains porins and is most likely homologous to the outer membrane of Gram-negative bacteria (Rachel et al. 1990). *Thermotoga* ferments various carbohydrates such as glucose, starch, and xylanes, forming acetate, L-lactate, H_2, and CO_2 as end products (R. Huber et al. 1986). The rod-shaped chemolithoautotrophic *Aquifex pyrophilus* represents the deepest phylogenetic branch within the bacterial domain (Figure 15.1; Burggraf et al. 1992). *Aquifex* gains energy by oxidation of hydrogen or sulfur under microaerobic conditions (R. Huber et al. 1992). Alternatively, *Aquifex* is able to use nitrate as an electron acceptor (Table 15.6). Archaeal coccoid sulfate reducers are members of *Archaeoglobus.* Some species occur within hot oil reservoirs and may be responsible for H_2S formation there ("reservoir souring") (Stetter et al. 1993). *Archaeoglobus fulgidus* and *Archaeoglobus lithotrophicus* are able to gain energy by reduction of SO_4^{2-}. *Archaeoglobus profundus* is an obligate heterotroph. *Archaeoglobus fulgidus* possesses several coenzymes that had been assumed to be unique for methanogens. The organisms with the highest growth temperature are members of *Pyrodictium* and *Methanopyrus,* growing at 110 °C. Cells of *Pyrodictium* are disk shaped and are connected by a network of ultrathin hollow tubules (Figure 15.2d). Strains of *Pyrodictium* are usually chemolithoautotrophs gaining energy through the reduction of S° by H_2. *Pyrodictium abyssi* is a heterotroph growing by peptide fermentation. *Pyrolobus fumarii* is a coccoid, nitrate ammonifier that shows specific phylogenetic relationship to *Pyrodictium* and grows even at temperatures up to 113 °C (Blöchl et al. 1997). *Methanopyrus kandleri* is a rod-shaped methanogen with a pseudomurein cell wall covered by an S-layer (Figure 15.2e). Representing the deepest and

Table 15.6. *Energy-yielding reactions in chemolithoautotrophic hyperthermophiles*

Energy-yielding reaction	Genera
$4H_2 + CO_2 \rightarrow CH_4 + 2H_2O$	*Methanopyrus, Methanothermus, Methanococcus*
$H_2 + S° \rightarrow H_2S$	*Acidianus, Stygiolobus, Pyrobaculum, Thermoproteus, Pyrodictium*
$4H_2 + H_2SO_4 \rightarrow H_2S + 4H_2O$	*Archaeoglobus*
$H_2 + HNO_3 \rightarrow HNO_2 + H_2O$	*Aquifex, Pyrobaculum*
$H_2 + 1/2O_2 \rightarrow H_2O$	*Aquifex, Acidianus, Metallosphaera, Pyrobaculum, Sulfolobus*
$2S° + 3O_2 + 2H_20 \rightarrow 2H_2SO_4$ ($FeS_2 + 7O_2 + 2H_2O \rightarrow 2FeSO_4 + 2H_2SO_4$)	*Sulfolobus, Acidianus, Metallosphaera, Aquifex*

shortest phylogenetic branch within the *Archaea* (Figure 15.1), members of *Methanopyrus* are strict chemolithoautotrophs that grow optimally at 100 °C with a doubling time of 50 minutes (R. Huber et al. 1989; Kurr et al. 1991). In contrast to all other members of *Archaea, Methanopyrus* species contain 2,3-di-O-geranylgeranyl-sn-glycerol as the dominating membrane lipid. It is the nonreduced precursor molecule for 2,3-di-O-phytanyl-sn-glycerol, the common lipid in *Archaea* (Hafenbradl et al. 1993).

Extreme acidophiles

Members of extremely acidophilic hyperthermophiles, including the genera *Sulfolobus, Metallosphaera, Acidianus,* and *Stygiolobus,* are lobed cocci (Figure 15.2f) found in terrestrial and marine solfataric fields and smoldering coal refuse piles (Brock 1978; Brock 1986; Stetter 1992; Fuchs 1994). They grow aerobically, facultatively aerobically, or strictly anaerobically at acidic pH (optimum about pH 3). Members of *Sulfolobus* are strict aerobes growing autotrophically by oxidation of S°, S^{2-}, and H_2, forming sulfuric acid or water as end products (Tables 15.5, 15.6). *Acidianus brierleyi* and *Sulfolobus metallicus* are able to grow by leaching of sulfidic ores (Brierley and Brierley 1973; G. Huber and Stetter 1991). Several *Sulfolobus* isolates are facultative or obligate heterotrophs, growing on sugars, yeast extract, and peptone (Brock

1978). Under microaerobic conditions, *Sulfolobus* isolates are able to reduce ferric iron and molybdate (Brierley and Brierley 1982). Growth of *Sulfolobus* requires low ionic strength; consequently, *Sulfolobus* has not been found in marine solfataric fields. *Metallosphaera sedula,* which differs from *Sulfolobus* species by the much higher GC-content of its DNA (Table 15.5), is a powerful oxidizer of sulfidic ores such as pyrite, chalcopyrite, and sphalerite; it forms sulfuric acid and solubilizes heavy metal ions (Table 15.6) *Acidianus,* similar to *Sulfolobus,* is able to grow by oxidation of S°, sulfides, H_2, and organic matter. In addition, it is able to grow anaerobically by reduction of elemental sulfur with H_2 as the electron donor (Segerer, Stetter and Klink 1985). Members of *Acidianus* are able to grow in the presence of up to 4% salt and have been isolated from a marine hydrothermal system (Segerer et al. 1986). *Stygiolobus* is a strictly anaerobic extreme acidophile, growing obligately chemolithoautotrophically by reduction of S° with H_2 (Segerer et al. 1991).

6. Distribution and spreading

The distribution and modes of spreading of hyperthermophiles are so far unknown. Hyperthermophiles do not form spores and are unable to grow at ambient temperatures. In addition, most of them are strict anaerobes. Usually, there are large, cold, oxygen-rich areas in between the biotopes, which may act as barriers against dissemination. Therefore, there is a good chance for the existence of endemites. This seems to be the case with members of the genus *Methanothermus,* which so far we have found only within the Kerlingarfjöll solfataras southwest of Iceland (Stetter et al. 1981; Lauerer et al. 1986). On the other hand, we were very surprised to find *Pyrodictium occultum, Pyrococcus furiosus, Thermococcus celer,* and *Archaeoglobus fulgidus* in Vulcano, Italy, as well as at Macdonald Seamount, Polynesia (Stetter, König and Stackebrandt 1983; Zillig et al. 1983; Fiala and Stetter 1986; Stetter et al. 1987; R. Huber et al. 1990). Furthermore, we are able to isolate *Methanopyrus kandleri* from the Guaymas deep sea basin within the Pacific as well as from the Kolbeinsey Ridge in the Atlantic, north of Iceland, several thousand miles away (Kurr et al. 1991). There is evidence that at least some strictly anaerobic hyperthermophiles are able to spread over long distances. Although general observations are still lacking, oxygen appears to be toxic to some hyperthermophilic anaerobes only at the high temperatures allowing growth (e.g., *Pyrobaculum*; R. Huber, Kristjansson and Stetter 1987). However, at low temperatures (e.g., 4 °C) in the laboratory, in the presence of oxygen, these anaerobes survive for years, possibly in a "frozen"-like inactive state. Using this capability, hyperthermophiles may be transported by currents within the cold atmosphere and seawater, for example, after volcanic eruptions. Hyperthermophiles might even have successfully infected other planets, such as Mars (e.g., by meteorites), whenever growth conditions were favorable there.

7. Conclusions

High-temperature environments harbor microbial ecosystems with an unexpectedly complex diversity of hyperthermophilic procaryotes. This is evident from 16S rRNA diversity and from unusual phylogenetic properties. Very surprisingly, and in contrast to usual moderate thermophiles, they represent all the deepest and shortest phylogenetic lineages within the phylogenetic tree, where they form a cluster around the root (Figure 15.1). Due to restrictions imposed by the extreme environment, hyperthermophiles may have evolved much more slowly than mesophiles. Therefore, they may still be rather similar to primitive life on the early Earth, testifying to the existence of primitive hyperthermophilic life forms at that time. This conclusion is consistent with our geological view of the Early Archaean, when the Earth's surface was much hotter than today (Ernst 1983). Based on their mode of carbon assimilation, hyperthermophiles can be divided into heterotrophs and chemolithoautotrophs. Assuming a heterotrophic common ancestor (derived from a heterotrophic origin of life), the shortest and deepest branches within the phylogenetic tree would be expected to consist exclusively of heterotrophs, whereas long lineages should harbor mainly chemolithoautotrophs. In striking contrast to this assumption, however, the phylogenetic tree demonstrates that the deepest and shortest branches are represented almost exclusively by strictly chemolithoautrotrophic organisms like *Pyrodictium*, *Methanopyrus,* and *Aquifex*. Therefore, autotrophy appears to be a very ancient feature in the history of life. The energy-yielding reactions in chemolithoautotrophic hyperthermophiles are anaerobic (microaerophilic) types of respiration (Table 15.6). Molecular hydrogen is widely used as an electron donor, whereas CO_2, oxidized sulfur compounds, and oxygen serve as electron acceptors. Since molecular hydrogen, CO_2, and sulfur compounds are present within volcanic environments, anaerobic hyperthermophilic autotrophic organisms using these substances are excellently adapted to this environment. They are completely independent of molecular oxygen and therefore represent life independent of the Sun. They could even exist on other planets that possess hydrothermal activity.

The presence of oxygen-using hydrogen oxidizers among hyperthermophiles is surprising since oxygen appeared globally in the Earth's atmosphere rather late (Schopf, Hayes and Walter 1983). However, the only hyperthermophiles (growing at the lower end of the hot temperature scale) tolerating fully aerobic conditions are members of the thermoacidophilic *Sulfolobales* (G. Huber et al. 1992), which represent a rather fastly evolving phylogenetic lineage (Figure 15.1) Therefore, this metabolic property could be explained by a rather late adaptation to oxygen by these continental organisms. In contrast, all other hyperthermophilic hydrogen oxidizers, such as *Pyrobaculum aerophilum*, *Pyrolobus fumarius*, and *Aquifex phyrophilus,* are much more extreme in their growth temperatures and represent very deep and short lineages within the phylogenetic tree (Figure 15.1). These are all extreme microaerophiles, able to use free oxygen only in traces (as low as 100 to 10 ppm in the atmosphere; Blöchl, Völkl and Stetter unpublished). Alternatively, microaerophiles can be powerful nitrate reducers, which could possibly have been a kind of "precursor metabolism" in these organisms because nitrate could have been formed by lightning from N_2 and CO_2 already at the early Earth (Mancinelli and Mc Kay 1988). In addition, on the primitive Earth, traces of free oxygen (at least locally) most likely could have built up photochemically and could have been present ever since. Therefore, O_2 could have been used by organisms early in the evolution of life (Towe 1988; Towe 1994).

A great variety of hyperthermophiles are able to use sulfur and nitrogen compounds in their energy-yielding reactions. Within hot sulfur-rich terrestrial and submarine environments, communities of hyperthermophiles are capable of accomplishing complete biogenic high-temperature sulfur cycles. Their contribution to the global S-cycle is at present unknown, and mesophilic organisms may play a major part therein. In Early Archaean times, however, hyperthermophiles may have been the dominating life form and could have accomplished biogenic conversions of sulfur. As concluded from the geological record of the isotope composition, sulfates of magmatic origin that contributed to the total flux of oxidized sulfur compounds were already present on the early Earth (Hattori and Cameron 1986). These sulfates may have served as electron acceptors for primitive sulfate reducers of the *Archaeoglobus* type. At present, there is no evidence for a complete nitrogen cycle within hot biotopes. So far, only hyperthermophilic denitrifiers and nitrate ammonifiers (such as *Aquifex pyrophilus*, *Pyrobaculum aerophilum,* and *Pyrolobus fumarius*) have been isolated, and nitrification and nitrogen fixation are unknown among hyperthermophiles. The question of whether primitive (hyperthermophilic) nitrogen-fixing organisms exist at all, cannot be answered now and has to await further cultivation attempts. Even before the advent of nitrate-reducing microorganisms, however, FeS/H_2S-driven geochemical anoxic nitrate ammonification could have operated abiotically (Blöchl et al. 1992), providing the first hyperthermophilic life forms with ammonia. Therefore, in contrast to earlier assumptions (Schidlowsky, Hayes and Kaplan 1983), ammonia supply may not have been a major problem on the early Earth.

Acknowledgments

I wish to thank Reinhard Rachel for the preparation of the electron micrographs. The work presented from my laboratory was supported by grants of the DFG, the BMWF, the EEC, and the Fonds der Chemischen Industrie.

References

Alfredsson, G. A., Kristjansson, J. K., Hjorleifsdottir, S., and Stetter, K. O. 1988. *Rhodothermus marinus*, new genus new species, a thermophilic, halophilic bacterium from submarine hot springs in Iceland. *J. Gen. Microbiol.* 134:299–306.

Barns, S. M., Fundyga, R. E., Jeffries, M. W., and Pace, N. R. 1994. Remarkable archaeal diversity detected in a Yellowstone National Park hot spring environment. *Proc. Natl. Acad. Sci. USA* 91:1609–1613.

Blöchl, E., Keller, M., Wächtershäuser, G., and Stetter, K. O. 1992. Reactions depending on iron sulfide and linking geochemistry with biochemistry. *Proc. Natl. Acad. Sci. USA* 89:8117–8120.

Blöchl, E., Rachel, R., Burggraf, S., Jannasch, H., and Stetter, K. O. 1997. *Pyrolobus fumarii*, gen. and sp. nov. represents a novel group of archaea, extending the upper temperature border of life to 113 °C. *Extremophiles* 1:14–21.

Brierley, C. L., and Brierley, J. A. 1973. A chemolithoautotrophic and thermophilic microorganism isolated from an acidic hot spring. *Canadian J. Microbiol.* 19:193–198.

Brierley, C. L., and Brierley, J. A. 1982. Anaerobic reduction of *Sulfolobus* species. *Zentralblatt f. Bakteriologie und Hygiene, I. Abteilung Originale C3*:289–294.

Brock, T. D. 1978. *Thermophilic Microorganisms and Life at High Temperatures.* New York, Berlin, Heidelberg:Springer-Verlag.

Brock, T. D. 1986. *Thermophiles: General, Molecular and Applied Microbiology.* New York: John Wiley & Sons.

Burggraf, S., Olsen, G. J., Stetter, K. O., and Woese, C. R. 1992. A phylogenetic analysis of *Aquifex pyrophilus*. *System. Appl. Microbiol.* 15:352–356.

Castenholz, R. W. 1979. Evolution and ecology of thermophilic microorganisms. In *Strategies of Microbial Lfe in Extreme Environments*, ed. M. Shilo, pp. 373–392. Weinheim: Verlag Chemie.

Darland, G., Brock, T. D., Samsonoff, W., and Conti, S. F. 1970. A thermophilic, acidophilic mycoplasma isolated from a coal refuse pile. *Science* 170:1416–1418.

Drobner, E., Huber, H., Wächtershäuser, G., Rose, D., and Stetter, K. O. 1990. Pyrite formation linked with hydrogen evolution under anaerobic conditions. *Nature* 346:742–744.

Ernst, W. G. 1983. The early earth and the achean rock record. In *Earth's Earliest Biosphere*, ed. J. W. Schopf, pp. 41–52. Princeton: Princeton University Press.

Fiala, G., and Stetter, K. O. 1986. *Pyrococcus furiosus* sp. nov. represents a novel genus of marine heterotrophic archaebacteria growing optimally at 100 °C. *Arch. Microbiol.* 145:56–61.

Fricke, H., Giere, O., Stetter, K. O., Alfredsson, G. A., Kristjansson, J. K., Stoffers, P. , and Svavarsson, J. 1989. Hydrothermal vent communities at the shallow subpolar Mid-Atlantic Ridge. *Marine Biology* 102:425–429.

Fuchs, T. M. 1994. Physiologische und molekularbiologische Untersuchungen an neu isolierten thermoacidophilen Archaeen. Universität Regensburg, Germany: Diplomarbeit.

Fuchs, T., Huber, H., Teiner, K., Burggraf, S., and Stetter, K. O. 1995. *Metallosphaera prunae*, sp. nov., a novel metal-mobilizing, thermoacidophilic archaeum, isolated from a uranium mine in Germany. *System. Appl. Microbiol.* 18:560–566.

Hafenbradl, D., Keller, M., Thiericke, R., and Stetter, K. O. 1993. A novel unsaturated archaeal ether core lipid from the hyperthermophile *Methanopyrus kandleri*. *System. Appl. Microbiol.* 16:165–169.

Hattori, K., and Cameron, E. M. 1986. Archaean magmatic sulphate. *Nature* 319:45–47.

Huber, G., Drobner, E., Huber, H., and Stetter, K. O. 1992. Growth by aerobic oxidation of molecular hydrogen in Archaea: a metabolic property so far unknown for this domain. *System. Appl. Microbiol.* 15:502–504.

Huber, G., Huber, R., Jones, B. E., Lauerer, G., Neuner, A., Segerer, A., Stetter, K. O., and Degens, E. T. 1991. Hyperthermophilic archaea and bacteria occurring within Indonesian hydrothermal areas. *System. Appl. Microbiol.* 14:397–404.

Huber, G., and Stetter, K. O. 1991. *Sulfolobus metallicus*, new species, a novel strictly chemolithoautotrophic thermophilic archaeal species of metal-mobilizers. *System. Appl. Microbiol.* 14:372–378.

Huber, R., Kristjansson, J. K., and Stetter, K. O. 1987. *Pyrobaculum* gen. nov., a new genus of neutrophilic, rod-shaped archaebacteria from continental solfataras growing optimally at 100 °C. *Arch. Microbiol.* 149:95–101.

Huber, R., Kurr, M., Jannasch, H. W., and Stetter, K. O. 1989. A novel group of abyssal methanogenic archaebacteria (*Methanopyrus* growing at 110 °C. *Nature* 342:833–834.

Huber, R., Langworthy, T. A., König, H., Thomm, M., Woese, C. R., Sleytr, U. B., and Stetter, K. O. 1986. *Thermotoga maritima* sp. nov. represents a new genus of unique extremely thermophilic eubacteria growing up to 90 °C. *Arch. Microbiol.* 144:324–333.

Huber, R., Stoffers, P., Cheminee, J. L., Richnow, H. H., and Stetter, K. O. 1990. Hyperthermophilic archaebacteria within the crater and open-sea plume of erupting Macdonald seamount. *Nature* 345:179–181.

Huber, R., Stöhr, J., Hohenhaus, S., Rachel, R., Burggraf, S., Jannasch, H. W., and Stetter, K. O. 1995. *Thermococcus chitonophagus* sp. nov., a novel, chitin-degrading, hyperthermophilic archaeum from a deep-sea hydrothermal vent environment. *Arch. Microbiol.* 164:255–264.

Huber, R., Rossnagel, P., Woese, C. R., Rachel, R., Langworthy, T. A., and Stetter, K. O. 1996. Formation of ammonium from nitrate during chemolithoautotrophic growth of the extremely thermophilic bacterium *Ammonifex degensii* gen. nov. sp. nov. *System. Appl. Microbiol.* 19:40–49.

Huber, R., Wilharm, T., Huber, D., Trincone, A., Burggraf, S., König, H., Rachel, R., Rockinger, I., Fricke, H., and Stetter, K. O. 1992. *Aquifex pyrophilus*, new genus new species, represents a novel group of marine hyperthermophilic hydrogen-oxidizing bacteria. *System. Appl. Microbiol.* 3:340–351.

Iwabe, N., Kuma, K.-I., Hasegawa, M., Osawa, S., and Miyata, T. 1989. Evolutionary relationship of archaebacteria, eubacteria, and eukaryotes inferred from phylogenetic trees of duplicated genes. *Proc. Natl. Acad. Sci. USA* 86:9355–9359.

Keller, M., Braun, F.-J., Diermeier, R., Hafenbradl, D., Burggraf, S., Rachel, R., and Stetter, K. O. 1995. *Thermococcus alcaliphilus* sp. nov., a new hyperthermophilic archaeum growing on polysulfide at alkaline pH. *Arch. Microbiol.* 164:390–395.

Kristjansson, J. K., and Stetter, K. O. 1992. Thermophilic bacteria. In *Thermophilic Bacteria*, ed. J. K. Kristjansson, pp. 1–18. Boca Raton, Fla.: CRC Press.

Kurr, M., Huber, R., König, H., Jannasch, H. W., Fricke, H., Trincone, A., Kristjansson, J. K., and Stetter, K. O. 1991. *Methanopyrus kandleri,* gen. and sp. nov. represents a novel group of hyperthermophilic methanogens, growing at 110 °C. *Arch. Microbiol.* 156:239–247.

Lauerer, G., Kristjansson, J. K., Langworty, T. A., König, H., and Stetter, K. O. 1986. *Methanothermus sociabilis* sp. nov., a second species within the *Methanothermaceae* growing at 97 °C. *System. Appl. Microbiol.* 8:100–105.

Mancinelli, R. L., and McKay, C. P. 1988. The evolution of nitrogen cycling. *Origins Life Evol. Biosphere* 18:311–328.

Marsh, R. M., and Norris, P. R. 1985. The isolation of some thermophilic, autotrophic iron- and sulfur-oxidizing bacteria. *FEMS Microbiol. Lett.* 17:311–315.

Rachel, R., Engel, A. M., Huber, R., Stetter, K. O., and Baumeister, W. 1990. A porin-type like protein is the major constituent of the cell envelope of the ancestral eubacterium *Thermotoga maritima. FEBS Lett.* 262:64–68.

Schäfer, T., and Schönheit, P. 1992. Maltose fermentation to acetate, CO_2 and H_2 in the anaerobic hyperthermophilic archaeon *Pyrococcus furiosus*: evidence for the operation of a novel sugar fermentation pathway. *Arch. Microbiol.* 158:188–202.

Schidlowski, M., Hayes, J. M., and Kaplan, I. R. 1983. Isotopic inferences of ancient biochemistries: carbon, sulfur, hydrogen, and nitrogen. In *Earth's Earliest Biosphere*, ed. J. W. Schopf, pp. 149–186. Princeton, N.J. : Princeton University Press.

Schopf, J. W., Hayes, J. M., and Walter, M. R. 1983. Evolution of Earth's earliest ecosystems: recent progress and unsolved problems. In *Earth's Earliest Biosphere*, ed. J. W. Schopf, pp. 361–384. Princeton, N. J. : Princeton University Press.

Segerer, A., Neuner, A., Kristjansson, J. K., and Stetter, K. O. 1986. *Acidianus infernus*, new genus new species, and *Acidianus brierleyi*, new combination: Facultatively aerobic, extremely acidophilic thermophilic sulfur-metabolizing archaebacteria. *Int. J. Syst. Bact.* 36:559–564.

Segerer, A., Stetter, K. O., and Klink, F. 1985. Two contrary modes of chemolithotrophy in the same archaebacterium. *Nature* 313(6005):787–789.

Segerer, A. H., Trincone, A., Gahrtz, M., and Stetter, K. O. 1991. *Stygiolobus azoricus*, new genus new species represents a novel genus of anaerobic, extremely thermoacidophilic archaebacteria of the order Sulfolobales. *Int. J. Syst. Bact.* 41:495–501.

Stetter, K. O. 1992. Life at the upper temperature border. In *Frontiers of Life*, ed. J. Tran Than Van, K. Tran Than Van, J. C. Mounolou, J. Schneider, and C. McKay, pp. 195–219. Gif-sur-Yvette: Editions Frontières.

Stetter, K. O. 1994. The lesson of Archaebacteria. In *Nobel Symposium No.84*, ed. S. Bengtson, pp. 143–151. New York: Columbia University Press.

Stetter, K. O. 1996. Hyperthermophilic procaryotes. *FEMS Microbiol. Reviews* 18:149–158.

Stetter, K. O., Huber, R., Blöchl, E., Kurr, M., Eden, R. D., Fiedler, M., Cash, H., and Vance, I. 1993. Hyperthermophilic archaea are thriving in deep North Sea and Alaskan oil reservoirs. *Nature* 365:743–745.

Stetter, K. O., König, H., and Stackebrandt, E. 1983. *Pyrodictium* gen. nov., a new genus of submarine disc-shaped sulphur reducing archaebacteria growing optimally at 105 °C. *System. Appl. Microbiol.* 4:535–551.
Stetter, K. O., Lauerer, G., Thomm, M., and Neuner, A. 1987. Isolation of extremely thermophilic sulfate reducers: Evidence for a novel branch of archaebacteria. *Science* 236:822–824.
Stetter, K. O., Thomm, M., Winter, J., Wildgruber, G., Huber, H., Zillig, W., Janekovic, D., König, H., Palm, P., and Wunderl, S. 1981. *Methanothermus fervidus*, sp. nov., a novel extremely thermophilic methanogen isolated from an Icelandic hot spring. *Zbl. Bakt. Hyg., I. Abt. Orig. C2*:166–178.
Towe, K. M. 1988. Early biochemical innovations, oxygen, and earth history. In *Molecular Evolution and the Fossil Record*, ed. T. W. Broadhead, pp.114–129. Short Course, Washington, D.C.: Paleontological Society.
Towe, K. M. 1994. Earth's early atmosphere: Constraints and opportunities for early evolution. In *Nobel Symposium No.84*, ed. S. Bengtson, pp. 36–47. New York: Columbia University Press.
Völkl, P., Huber, R., Drobner, E., Rachel, R., Burggraf, S., Trincone, A., and Stetter, K. O. 1993. *Pyrobaculum aerophilum* sp. nov., a novel nitrate-reducing hyperthermophilic archaeum. *Appl. Environ. Microbiol.* 59:2918–2926.
Wächtershäuser, G. 1988. Pyrite formation, the first energy source for life: a hypothesis. *System. Appl. Microbiol.* 10:207–210.
Woese, C. R., and Fox, G. E. 1977. Phylogenetic structure of the prokaryotic domain: the primary kingdoms. *Proc. Natl. Acad. Sci. USA* 74:5088–5090.
Woese, C. R., Kandler, O., and Wheelis, M. L. 1990. Towards a natural system of organisms: proposal for the domains *archaea, bacteria and eukarya. Proc. Natl. Acad. Sci. USA* 87:4576–4579.
Zillig, W., Holz, I., Janekovic, D., Schäfer, W., and Reiter, W. D. 1983. The archaebacterium *Thermococcus celer* represents a novel genus within the thermophilic branch of the archaebacteria. *System. Appl. Microbiol.* 4:88–94.

16

Tracing the roots of the Universal Tree of Life

J. WILLIAM SCHOPF
IGPP Center for the Study of Evolution and the Origin of Life
Department of Earth and Space Sciences
and Molecular Biology Institute
University of California

Tracing life's roots: progress and problems

Precambrian paleobiology: a new field of science

In 1859, when Charles Darwin unveiled his monumental monograph, *The Origin of Species*, major chapters in the history of life had already been deciphered. The familiar progressions from seaweeds to land plants, from marine invertebrates to higher mammals, provided Darwin a fossil-based foundation for his grand thesis. This evolutionary sequence from water to land is the history of Phanerozoic life, dating from 543 million years (Ma) ago and the appearance of calcareous algae and shelled invertebrates that marks the beginning of the Cambrian Period of geologic time. But what happened earlier? What were the forerunners of these early algae and primitive invertebrates? In short, how did evolution proceed during the Precambrian Eon, the pre-Phanerozoic seven-eighths of Earth history?

To these questions Darwin had no answers – indeed, until a scant three decades ago, Precambrian biologic history was *terra incognita*. But Darwin did know that if answers were not forthcoming, his theory was in jeopardy:

> If the theory [of evolution] be true it is indisputable that before the lowest Cambrian stratum was deposited ... the world swarmed with living creatures. [Yet] to the question why we do not find rich fossiliferous deposits belonging to these earliest periods ... I can give no satisfactory answer. The case at present must remain inexplicable; and may be truly urged as a valid argument against the views here entertained. (Darwin 1859, Chapter X)

Though Darwin posed the problem, it was not until a century later – in the mid-1960s with the birth of a new field of science, Precambrian paleobiology – that this earliest "missing" fossil record began to be uncovered (Figure 16.1). Progress since the 1960s has been impressive. The documented record of life has been extended steadily and now stretches to nearly 3.5 billion years (Ga) ago, a date in the geologic past approaching the age (4.5 Ga) of the Earth itself.

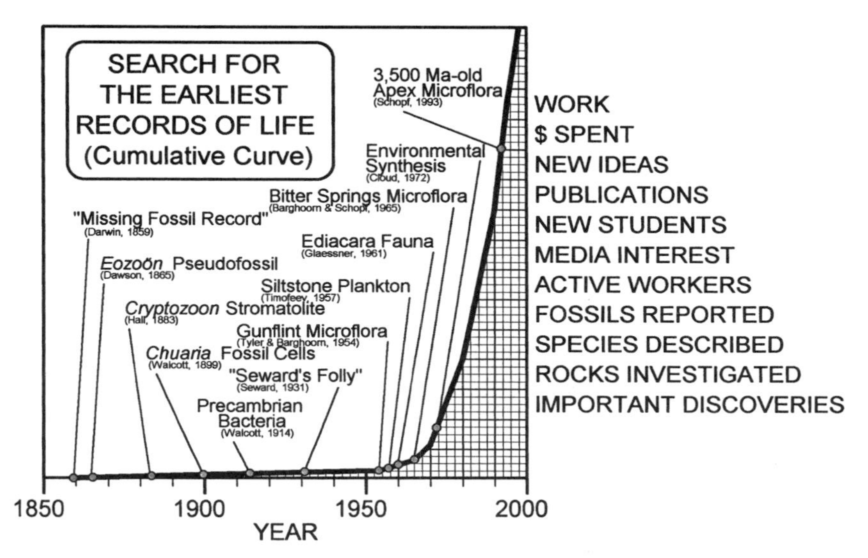

Figure 16.1. Cumulative curve summarizing the historical development of Precambrian paleobiologic studies, 1859 to the present.

The Universal Tree of Life

As the early fossil record came increasingly into focus during the 1970s and 1980s, parallel advances were underway in studies of the molecular biology of living systems. By the late 1980s, these had coalesced into powerful means to determine the evolutionary, phylogenetic relations among *all* organisms alive today (Woese 1987).

The bases of these determinations and for constructing from them a Universal Tree of Life are straightforward:

1. All organisms biosynthesize proteins (such as the enzymes that mediate biochemical reactions), and in all, proteins are synthesized on minute intracellular bodies known as ribosomes.
2. Ribosomes are composed largely of various types of ribonucleic acids, molecules that, like the DNA of chromosomes, contain an aperiodic sequence of four nitrogenous bases (guanine, cytosine, adenine, and uracil, the last in place of the thymine of DNA).
3. The placement of each nitrogenous base in a molecule of ribosomal RNA (rRNA) is determined by information encoded in chromosomal DNA; changes (mutations) in this DNA code are reflected by substitution of bases (changes in the "base sequence") in the resulting rRNA.

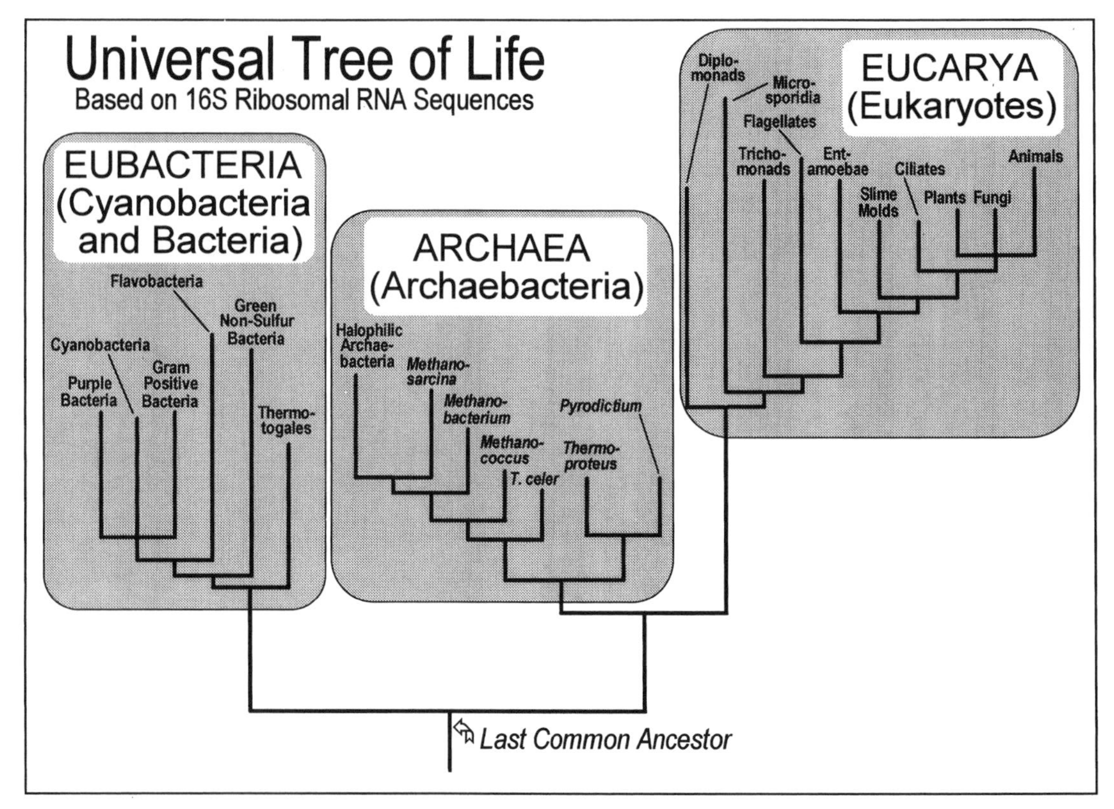

Figure 16.2. Universal Tree of Life, based on 16S ribosomal RNA sequences (data from Olsen and Woese 1993).

4. Because closely related organisms (whether siblings in a family, or members of closely related evolutionary lineages) have closely similar DNAs, their rRNAs are also closely similar, and the more distantly related are any two organisms, the more dissimilar their rRNAs.
5. Thus, comparison of the base sequences of rRNA from diverse organisms provides a simple, logical, universally applicable method to determine how closely or distantly related are all organisms of the living world.

Shown in Figure 16.2 is the Universal Tree of Life (based on 16S rRNA) that results from this type of analysis (Olsen and Woese 1993). In such trees, the vertical length of each branch corresponds quantitatively to the number of base sequence changes that have occurred in that lineage since its divergence from its nearest neighbor.

Four important features stand out. First, despite the seeming dominance in the living world of plants and animals (shown at the upper right in Figure 16.2), they are minor components of the total biota, only 2 of more than 20 principal evolutionary branches. Second, the Tree of Life is overwhelmingly composed of microscopic organisms – of the 23 lineages shown, only 3 (plants, fungi, and animals) contain megascopic members, and each of these includes microscopic forms as well. Third, all organisms alive today are members of but three major groups (formally, "domains"; Woese, Kandler and Whellis 1990): the Eucarya, or eukaryotes (organisms having cells in which chromosomes are encapsulated in a distinct, membrane-bounded nucleus); the Archaea, or the archaebacteria of earlier classifications (nonnucleated microorganisms including methanogenic bacteria and many extremeophiles, microbes that thrive in exceptionally acidic, high-temperature settings); and the Eubacteria, the domain that includes the cyanobacteria and bacteria of traditional classifications. Fourth, the Eucarya is more closely related to the Archaea than to the Eubacteria, and the last common ancestor of all organisms living today – the root stock of the Universal Tree of Life – lies between the Archaea and the Eubacteria (Iwabe et al. 1989).

The Universal Tree of Life shows the order of branching of the major biotic lineages. When in the geologic past did these various branches emerge? The answer, unfortunately, cannot be provided by rRNA trees. Such trees are based wholly on organisms that live today, so if they were precise clocks of evolutionary history, the branch tips of all lineages would extend to the same height (a level corresponding to the present) and the top of the tree would be flat, rather than irregularly undulating as it actually is. Evidently, different lineages have evolved at different rates: Those with relatively long vertical branches have evolved faster (for example, within the Eubacteria, flavobacteria and green nonsulfur bacteria; and within the Eucarya, microsporidia and diplomonads), whereas the rRNA in lineages with relatively short branches (several in the Archaea and eukaryotic ciliates, plants, and fungi) has changed more slowly.

Can fossils time life's history?

Accurate dating of the times of origin of the major biologic lineages has been a long-standing goal in Precambrian paleobiology (Schopf 1970; Schopf, Hayes, and Walter 1983; Schopf 1992a). This is a young field, however, and the early fossil record is as yet too incompletely known to provide precise answers. Moreover, even under the best of circumstances, fossil evidence can record only the first *detected* occurrence of a biologic lineage, not its first *actual* occurrence. Together, fossils and associated sedimentological and geochemical indicators of biologic activity (that is, "paleobiologic" evidence) can establish a minimum age for a lineage, but they cannot reveal how much earlier in the geologic past the lineage actually existed.

The amino acid "clock"

Another way to solve this problem, based on the amino acid composition of proteins, has recently been proposed by Doolittle et al. (1996). Proteins are amino acid polymers, chainlike molecules containing up to 20 different common amino acids linked together head-to-tail. Proteins evolve over time, for much the same reason as do molecules of rRNA: Information encoded in chromosomal DNA determines which of the 20 amino acids is emplaced at each position in a protein polymer, and changes (mutations) in this code can result in substitution of different amino acids in the protein product. There are many different families of proteins (cytochromes, hemoglobins, and the like) and members of the same family occur in diverse organisms, some in lineages of all three biotic domains. Because each biologic lineage has a unique evolutionary history of mutation-driven amino acid substitution, the proteins of any given family vary somewhat from one lineage to another.

On the basis of these considerations, Doolittle et al. (1996) reasoned that a molecular "clock" by which to date times of lineage emergence might be established by comparing the sequences of amino acids in proteins of the same families in diverse groups of organisms. Their study was thorough: It took into account fast and slowly evolving lineages and compared 531 amino acid sequences in 57 families of enzymatic proteins from 15 major groups of organisms. Their analysis yielded an internally consistent amino acid–based tree in which the branches are all more or less the same length and have a branching order in good agreement with the rRNA Universal Tree of Life.

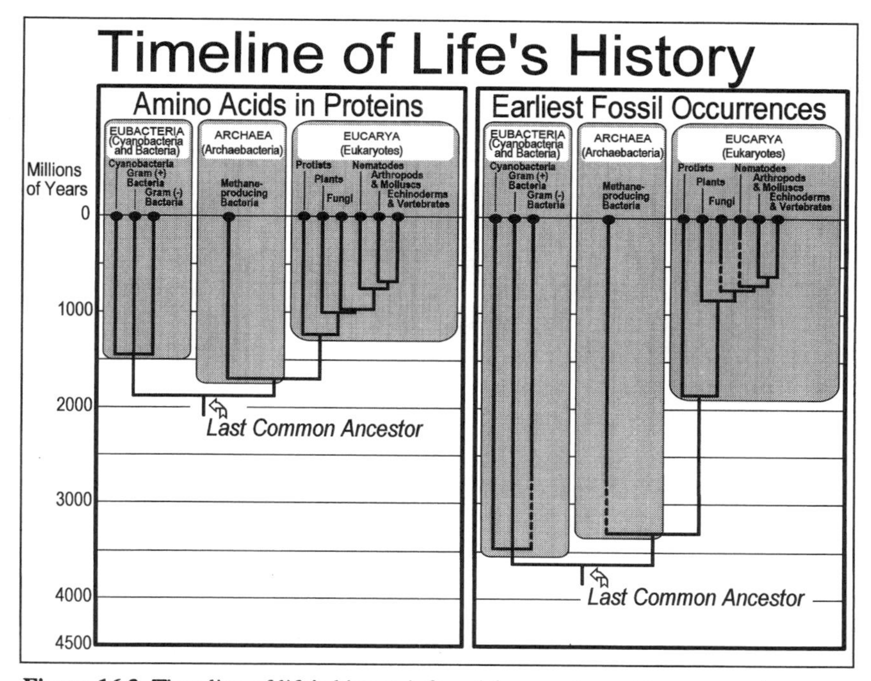

Figure 16.3. Time line of life's history inferred from amino acid sequences in proteins (left; data from Doolittle et al. 1996) compared with time line based on the known Precambrian paleobiologic record (right).

Astonishingly, however, this amino acid clock meshes not at all well with the Precambrian fossil record. As noted previously, paleobiologic evidence can yield only a minimum date for the emergence of a biologic lineage, so the molecular clock would be expected to have yielded *older* ages than those documented by fossils. Yet the Doolittle et al. (1996) study concluded that all early evolving lineages originated at *younger*, not older, ages than those indicated by paleobiology, and the differences are substantial. For example, as shown in Figure 16.3, the amino acid data suggest that cyanobacteria and Gram-positive eubacteria originated about 1,500 Ma ago, two billion years later than inferred from fossil evidence. Similarly, the origins of Gram-negative eubacteria and methane-producing archaebacteria are placed a billion or more years later than indicated by paleobiology, and the last common ancestor – the root of all present-day life – is dated at only ~2,000 Ma ago, whereas cellular fossils of diverse morphologies have been found in rocks nearly 3,500 Ma old (Schopf 1992b, 1993).

Something is amiss! Perhaps the ancient fossils have been misinterpreted and are unrelated to present-day lineages – perhaps life originated, became extinct, then originated a second time about 2,000 Ma ago. Or, perhaps like

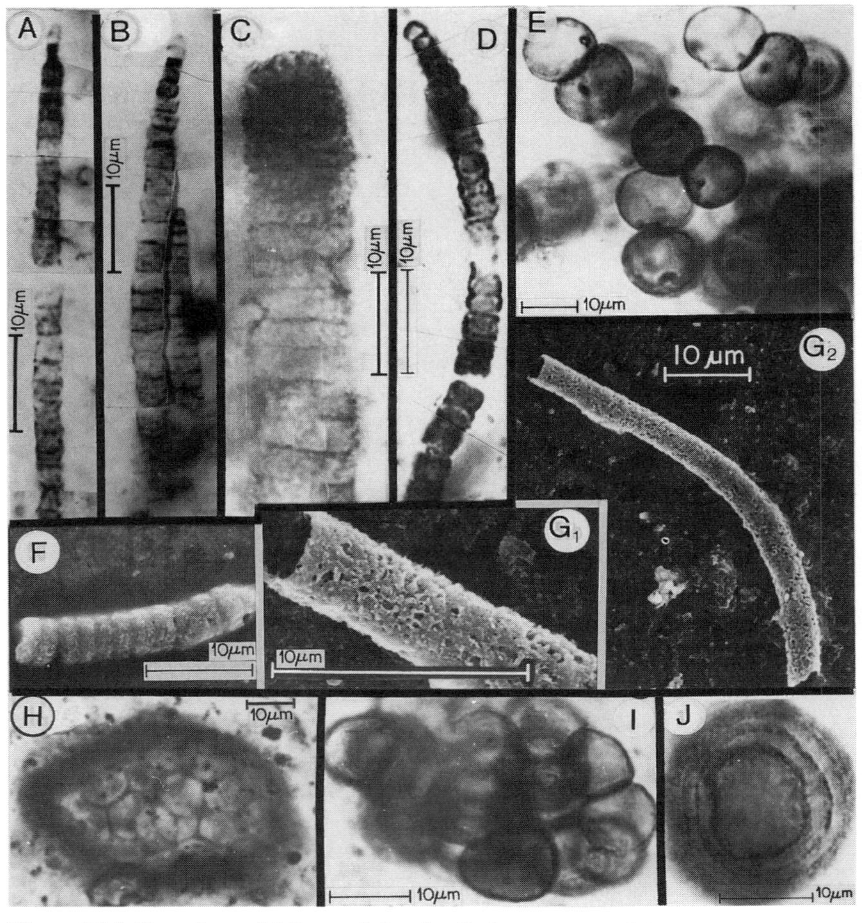

Figure 16.4. Cyanobacterial Precambrian fossils from stromatolitic cherts of the ~850-Ma-old Bitter Springs Formation of central Australia, shown in petrographic thin sections (**A–E**; **H–J**) and acid-resistant residues (**F** and **G**). **A**, **B**, and **D** show composite photomicrographs, necessitated by the three-dimensional preservation of the sinuous filaments; G_1 and G_2 show a single specimen at different magnifications. **A–D**, **F**: Cellular carbonaceous oscillatoriacean filaments. **A** and **B**, *Cephalophytarion grande* Schopf 1968, a taxon exhibiting tapering trichomes and globose terminal cells; **C**, *Palaeolyngbya barghoorniana* Schopf 1968, a *Lyngbya*-like oscillatoriacean having disk-shaped medial cells and rounded terminal cells; **D**, *Cephalophytarion variabile* Schopf & Blacic 1971; **F**, scanning electron micrograph showing the disk-shaped medial cells of an unnamed oscillatoriacean trichome. **G**: Scanning electron micrographs showing a coriaceous tubular oscillatoriacean sheath (*Eomycetopsis robusta* Schopf 1968). **E**, **H–J**: Coccoidal carbonaceous chroococcacean cyanobacteria. **E**, *Glenobotrydion aenigmatis* Schopf 1968, a taxon characterized by unordered colonial spheroidal cells; **H** and **I**, *Myxococcoides minor* Schopf 1968, a chroococcacean exhibiting close-packed colonial cells; **J**, *Gloeodiniopsis lamellosa* Schopf 1968, a *Chroococcus*-like unicellular chroococcacean encompassed by a well-defined multi-lamellated envelope.

the rRNA trees, the amino acid data yield reliable evidence only of the branching order of evolution, not the timing of that branching.

How firmly established *is* the early fossil evidence? What amount of data would have to be overturned for the amino acid clock to prove correct?

Ancient fossil cyanobacteria

Quality and quantity of the known fossil record

The best documented early branch of the Tree of Life is the cyanobacterial lineage, represented in the Precambrian fossil record by ensheathed solitary and colonial unicells (chiefly referred to the living cyanobacterial family Chroococcaceae) and cellular microscopic filaments (for the most part regarded as belonging to the Oscillatoriaceae, taxonomically the most diverse extant cyanobacterial family). Well-preserved examples (Figure 16.4) are indistinguishable in morphology from modern cyanobacteria, an identity first shown nearly three decades ago in detailed studies of the microbial community of the ~850-Ma-old Bitter Springs Formation of central Australia (Schopf 1968; Schopf and Blacic 1971). Since that time, such similarities have been encountered so frequently that it is now common practice among Precambrian paleobiologists to name fossil taxa after their modern morphological counterparts by adding appropriate prefixes (palaeo-, eo-) or suffixes (-opsis, -ites) to the names of present-day cyanobacterial genera (Mendelson and Schopf 1992; Schopf 1994a). More than 40 such namesakes (for example, *Palaeolyngbya*, *Eomicrocoleus*, *Aphanocapsaopsis*, and *Oscillatorites*) are in usage worldwide.

As shown in Figure 16.5, the striking similarity of Precambrian microfossils to modern cyanobacteria is characteristic of a wide range of morphologic types, among which are fossils that are hundreds of millions of years older than the date of ~1,500 Ma suggested by the amino acid clock for the origination of the group. For example, *Eoentophysalis belcherensis* (Figure 16.5H) – compared with species of modern *Entophysalsis* (Figure 16.5G) and, on this basis, referred to the living cyanobacterial family Entophysalidaceae (Golubic and Hofmann 1976) – is preserved in microbially laminated stromatolites of Northwest Territories, Canada, that are ~2,150 Ma old (Hofmann 1976). Though the fossil taxon is decidedly "too old" to be cyanobacterial according to the amino acid clock, the fossil and modern species are morphologically indistinguishable (in cell shape, and in form and arrangement of originally mucilaginous cell-encompassing envelopes); exhibit similar frequency distributions of dividing cells and essentially identical patterns of cellular development (resulting from cell division in three perpendicular planes); form

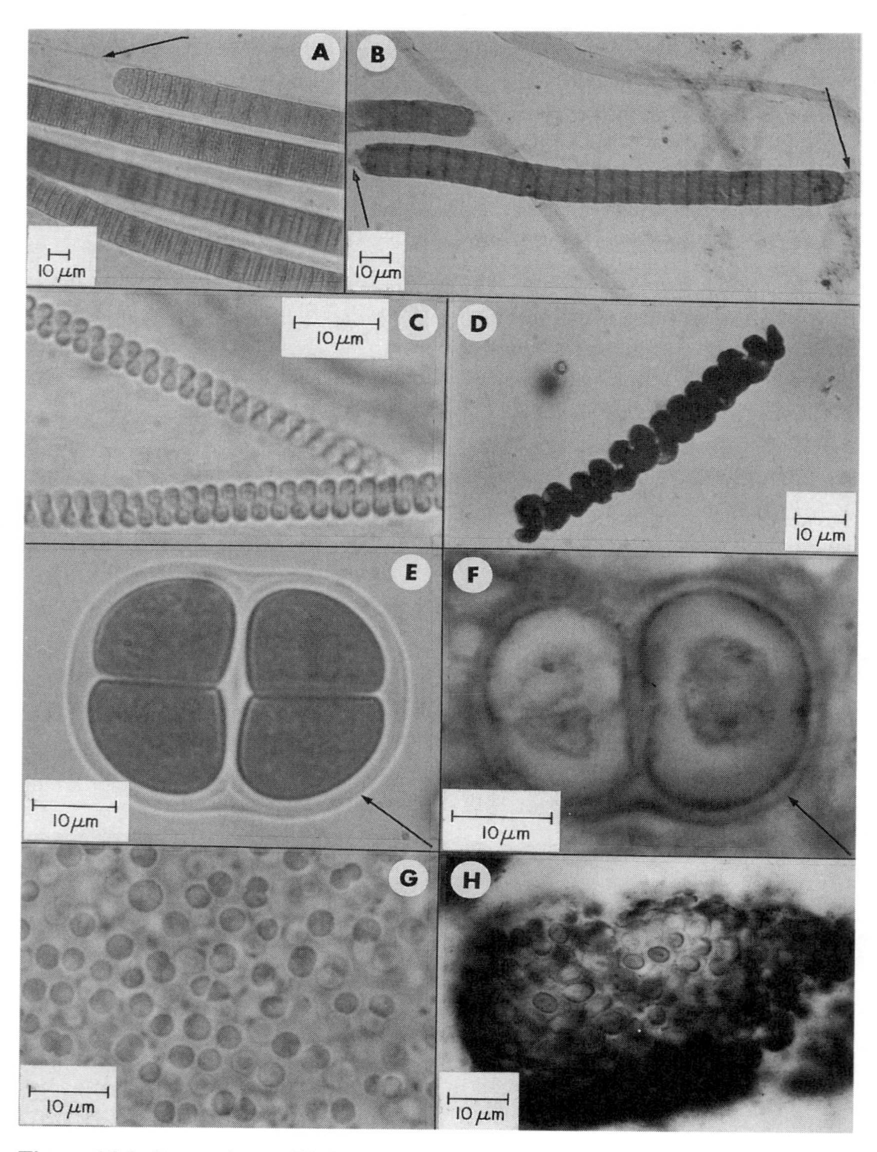

Figure 16.5. Comparison of living and Precambrian cyanobacteria. Living examples (A, C, E, and G) are from mat-building stromatolitic communities of northern Mexico. **A:** *Lyngbya* (Oscillatoriaceae), encompassed by a cylindrical mucilagenous sheath (arrow). **B:** *Palaeolyngbya helva* Hermann 1981, similarly ensheathed (arrows), in an acid-resistant residue of siltstone from the ~950-Ma-old Lakhanda Formation of eastern Siberia. **C:** *Spirulina* (Oscillatoriaceae). **D:** *Heliconema turukhania* Hermann 1981, a *Spirulina*-like cyanobacterium in an acid-resistant residue of siltstone from the ~850-Ma-old Miroedikha Formation of eastern Siberia.

microtexturally similar stromatolitic structures in comparable intertidal to shallow marine environmental settings; undergo essentially identical post-mortem degradation sequences; and occur in microbial communities that are comparable in both species composition and biological diversity (Golubic and Hofmann 1976).

Shown in Figure 16.6 are other examples of cyanobacterial look-alikes, also from intertidal to shallow marine stromatolitic settings like those inhabited by their living oscillatoriacean counterparts, that are "too old" by the amino acid data. The largest of these (Figure 16.6A) is *Oscillatoriopsis majuscula*, a ~2,000-Ma-old filamentous taxon more than 70 μm in diameter and thus very much broader than almost all noncyanobacterial prokaryotic filaments (95% of which are <5μm in diameter; Schopf 1996a). Many other such examples can be cited. Indeed, as summarized in Figure 16.7, fossils interpreted as cyanobacteria and referred to the Chroococcaceae, Entophysalidaceae, and Oscillatoriaceae occur in many geologic units older than permitted by the amino acid clock – if *any* of these thousands of fossils and the more than 120 cyanobacterial species to which they have been referred has been identified correctly, the protein-based amino acid clock is in error.

Testing the "clock": Is there a break in the fossil record ~1,500 Ma ago?

What about the idea that life originated, became extinct, then originated a second time, leading to the advent of cyanobacteria ~1,500 Ma ago? Perhaps Precambrian paleobiologists have so accepted the presumed cyanobacterial affinity of the fossils they have found that they have neglected to consider other possibilities, especially for very ancient forms. Is there a break in the fossil record of "cyanobacteria" ~1,500 Ma ago?

One way to address this problem objectively – free of taxonomic biases and predilections – is through detailed morphometric analyses of the fossil record. Such studies have been carried out on 55 communities of coccoidal or filamentous microfossils ~2,300 to 1,500 Ma in age and the data compared with those for 73 such communities preserved in immediately younger strata (1,500 to 900 Ma in age); numerous morphometric traits (cell size, shape, and range

---

E: *Gloeocapsa* (Chroococcaceae), a four-celled colony encompased by a distinct thick sheath (arrow). **F:** *Gloeodiniopsis uralicus* Krylov & Sergeev 1987, a similarly sheath-enclosed (arrow) *Gloeocapsa*-like colonial cyanobacterium, in a petrographic thin section of stromatolitic chert from the ~1,500-Ma-old Satka Formation of southern Bashkiria. **G:** *Entophysalis* (Entophysalidaceae). **H:** *Entophysalis belcherensis* Hofmann 1976, an *Entophysalis*-like colonial cyanobacterium, in a petrographic thin section of stromatolitic chert from the ~2,150-Ma-old Kasegalik Formation of Northwest Territories, Canada.

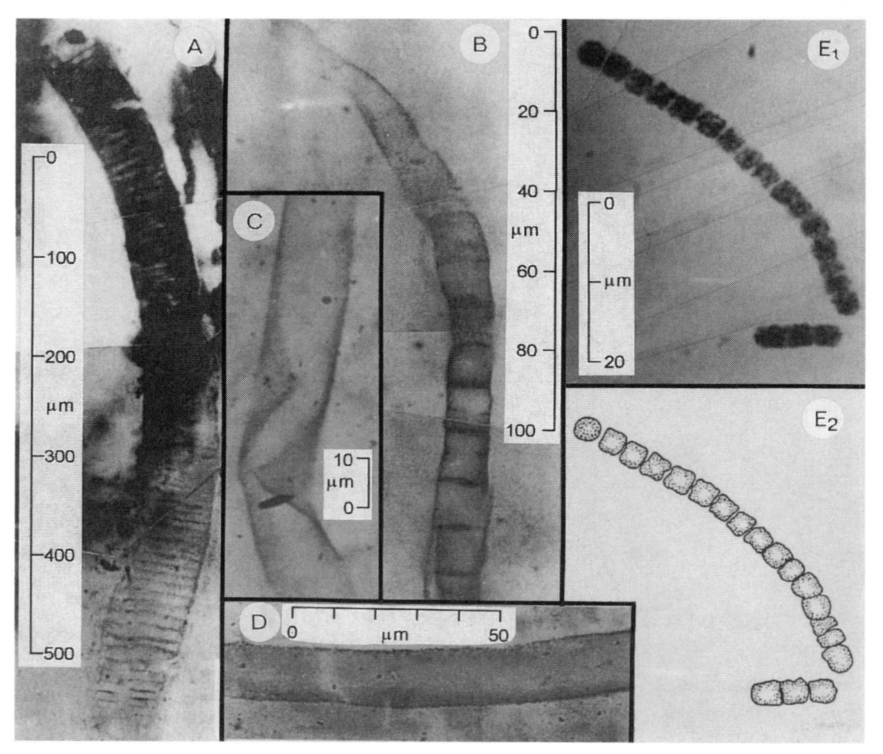

Figure 16.6. Cyanobacterial filamentous carbonaceous fossils in petrographic thin sections of stromatolitic cherts of the ~2,000-Ma-old Duck Creek Dolomite of Western Australia (*A–D*) and the ~2,090 Ma-old Gunflint Formation of Ontario, Canada (E). A, B, and E_1 show composite photomicrographs, necessitated by the three-dimensional preservation of the sinuous filaments; E_2 is an interpretive drawing of the specimen shown in E_1. **A:** *Oscillatoriopsis majuscula* Knoll, Strother & Rossi 1988, an exceptionally broad oscillatoriacean trichome composed of disk-shaped cells up to 78 µm wide and 6–11 µm long. **B:** *Oscillatoriopsis cuboides* Knoll, Strother & Rossi 1988, a trichome composed of distinctive cuboidal cells 11–13 µm wide and long. **C, D:** *Siphonophycus* sp., cylindrical to somewhat flattened (C) tubular sheaths that originally encompassed cellular oscillatoriacean trichomes like those shown in A and B. **E:** An *Oscillatoria*-like trichome, described informally as *"Gunflintia formosa"* (Moore 1993), composed of well-defined quadrate cells 3–4 µm in size. (Parts A–D from Knoll, Strother and Rossi 1988, reprinted with permission.)

of variability; colony form; sheath thickness and structure) were measured for 690 occurrences of coccoidal species ≤60 µm in diameter (if cyanobacteria, members of the Chroococcaceae) and 270 occurrences of cellular, filamentous, oscillatoriacean-like taxa (Schopf 1992c). No discontinuity in community structure, biotic composition, or environments inhabited was detected at 1,500 Ma ago or at any other time horizon of the ~1,400-Ma-long period analyzed.

Any marked biotic discontinuity occurring during this period – like that predicted by the amino acid data – should have been apparent. As a group, cyanobacteria differ distinctly in morphology from both archaebacteria and noncyanobacterial eubacteria, and the diagnostic characters required to draw such distinctions are commonly preserved. Many species of the Archaea are morphologically unlike most other prokaryotes (occurring as lobed cocci, rods in clusters, disks, disks with fibers, and so forth), and virtually all of the 43 species now recognized (Stetter 1996) are readily distinguishable from cyanobacteria. Most other noncyanobacterial prokaryotes can also be distinguished from cyanobacteria. The median diameter of coccoidal noncyanobacterial eubacteria is <1 μm, less than one-fourth that of cyanobacterial analogues, and the median for cellular filamentous taxa is ~1.6 μm, less than one-third that of similar cyanobacteria (Schopf 1996a). If cyanobacteria originated ~1,500 Ma ago, as suggested by the amino acid clock, this event should have left its mark in the fossil record by an easily detectable increase in median cell size, but this demonstrably did not occur (Schopf 1992c).

Two families of cyanobacteria dominate modern intertidal to shallow marine environments (Golubic 1976a, b), the Chroococcaceae and Oscillatoriaceae. Fossils referable to both of these families are abundant in these same settings in Precambrian rock units, some of which are older than (Figures 16.6, 16.7) and some younger (Figures 16.4, 16.5B, D, F) than 1,500 Ma in age. In fact, the great majority of Precambrian cyanobacterium-like fossils, both those older, and those younger, than 1,500 Ma are referable to *living genera* of chroococcaceans or oscillatoriaceans. Moreover, 37% of coccoidal taxa (26 of 70) older than 1,500 Ma and 28% (31 of 109) of those younger are indistinguishable in morphology from *living species* of chroococcaceans; 35% of filamentous taxa (18 of 51) older than 1,500 Ma and 37% (56 of 152) of those younger have *living species–level* morphological counterparts among the Oscillatoriaceae; and virtually all colony forms exhibited by living chroococcaceans occur also both among chroococcacean-like fossils older than 1,500 Ma and among fossils younger than 1,500 Ma (Schopf 1992c, d, e).

When did cyanobacteria originate?

The evidence from cellularly preserved fossils is unambiguous: Cyanobacteria originated well before 1,500 Ma ago and were present earlier than 2,000 Ma ago (the clock date for the divergence of present-day biotic lineages from their last common ancestor). How much farther into the geologic past does this evolutionary continuum extend?

As shown in Figure 16.7, the documented fossil record is abundant, diverse, and continuous back to about 2,200 Ma ago. However, only a few fossiliferous deposits have been discovered in older geologic terrains, and the

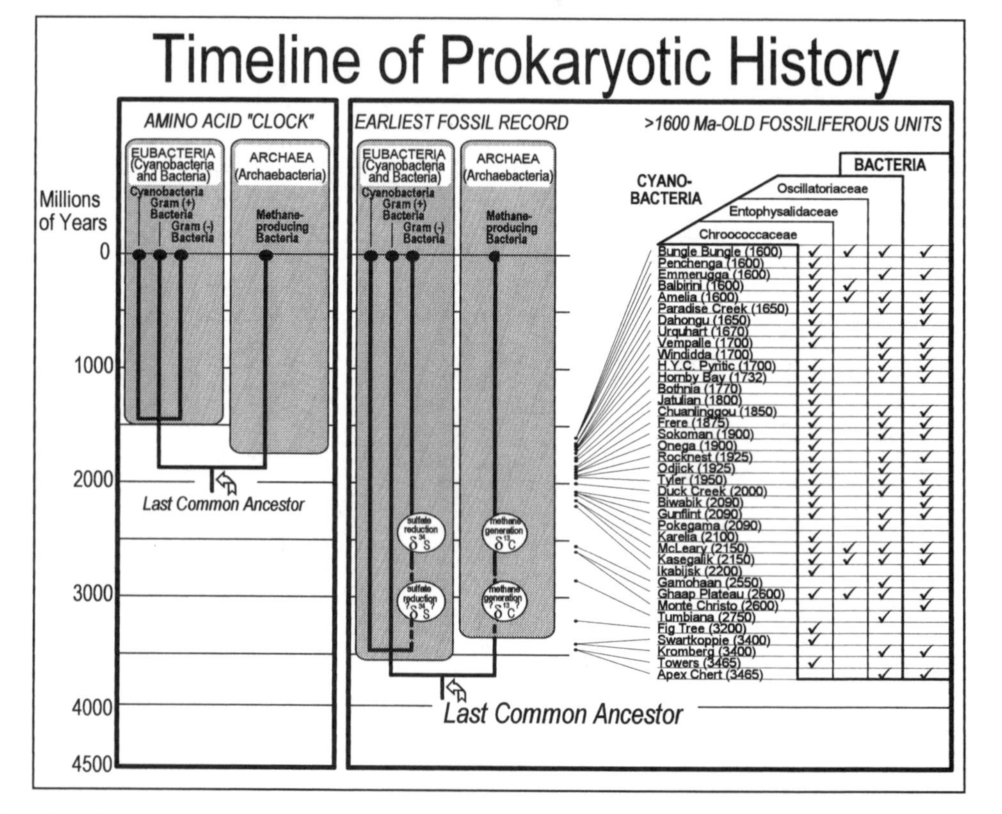

Figure 16.7. Time line of prokaryotic history inferred from amino acid sequences in proteins (left; data from Doolittle et al. 1996) compared with time line based on the known Precambrian paleobiologic record (right). Fossil representatives of three families of cyanobacteria (Chroococcaceae, Entophysalidaceae, and Oscillatoriaceae) and numerous types of noncyanobacterial prokaryotes ("bacteria") have been identified in the listed 38 geologic units older than 1,600 Ma (Hofmann and Schopf 1983; Schopf and Walter 1983; Mendelson and Schopf 1992).

microbial assemblages of these tend to be meager and poorly preserved (see, for example, Altermann and Schopf 1995). Though much of the older rock record is paleontologically unexplored, these deficiencies are largely a result of normal geologic processes. Because of erosion and geological recycling, few rock units from these earlier times have survived to the present and most of these are severely metamorphosed. Significantly, however, the oldest microbial community now known – nearly 3,500 Ma in age – is also the most diverse and among the best preserved of these very ancient assemblages. This community, from the Apex chert of northwestern Western Australia, holds the key to understanding early stages in the history of life and, very likely, of the cyanobacterial lineage.

Evidence from the ~3.5-Ga-old Apex chert

Though there is reason to question some reports of exceedingly ancient "microfossils" (Schopf and Walter 1983), such uncertainty does not apply to the microbial assemblage recently described from the ~3,500-Ma-old conglomeratic Apex chert (Schopf 1992b, 1993):

1. The geographic and stratigraphic source of the fossiliferous horizon is known with certainty (fossil-bearing samples having been collected from outcrop at the locality by numerous workers on multiple occasions).
2. The early Precambrian age (specifically, >3,458 ± 1.9 Ma; <3,471 ± 5 Ma) of the chert is well established, based on U-Pb zircon ages of immediately overlying and stratigraphically underlying units of the rock sequence (Blake and McNaughton 1984; Thorpe et al. 1992).
3. The fossils are demonstrably indigenous to the chert, as shown by their occurrence in petrographic thin sections (Figures 16.8, 16.9).
4. The fossils are preserved in transported and redeposited rounded pebbles (Figures 16.8A, B) that are assuredly syngenetic with deposition of the silicified sedimentary conglomerate in which they occur (and the fossils themselves therefore predate deposition of this bedded chert unit).
5. Carbon isotopic data (Schopf 1993, 1994b) and the morphological complexity and carbonaceous composition of the fossils establish that they are unquestionably biogenic – 11 filamentous species have been identified, ranging from 0.5 to 19.5 μm in diameter and exhibiting rounded to conical terminal cells; quadrate, disk-shaped, or barrel-shaped medial cells; taxon-specific degrees of filament attenuation; and evidence of cell division similar to that occurring in living filamentous prokaryotes.

The Apex assemblage stands out not only because of its great age but also because of its diversity. Indeed, the next oldest known, comparably diverse

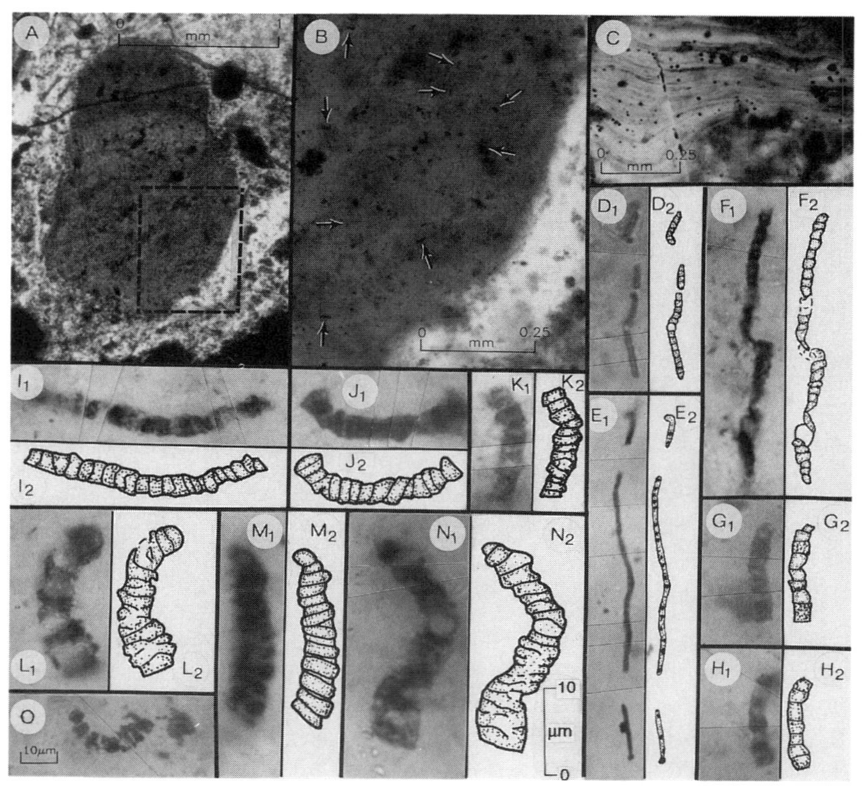

Figure 16.8. Microfossiliferous (A and B) and stromatolitic (C) conglomeratic clasts, and iron-stained (L) and carbonaceous cellular filamentous fossils (with interpretive drawings) shown in petrographic thin sections of the ~3,465-Ma-old Apex chert of northwestern Western Australia. Except as otherwise indicated (in A–C and O), magnification of all parts is denoted by the scale in N. D–K, N, and O show composite photomicrographs, necessitated by the three-dimensional preservation of the sinuous filaments. Narrow taxa, <1.5 μm wide (D–F), are interpreted as probably noncyanobacterial eubacteria; broad taxa, >3.5 μm wide (L–O), as probably cyanobacteria; and intermediate-diameter taxa (G–K) as undifferentiated prokaryotes (Schopf 1993). **A**: Microfossiliferous clast; area denoted by dashed lines is shown in B. **B**: Arrows point to minute filamentous microfossils, randomly oriented in the clast. **C**: Portion of a clast showing stromatolitic laminae. **D, E**: *Archaeotrichion septatum* Schopf 1993. **F**: *Eoleptonema apex* Schopf 1993. **G, H**: *Primaevifilum minutum* Schopf 1993. **I–K**: *Primaevifilum delicatulum* Schopf 1992b. **L–O**: *Archaeoscillatoriopsis disciformis* Schopf 1993.

microbiotas (Figure 16.7; Barghoorn and Tyler 1965; Hofmann 1976; Altermann and Schopf 1995) are a billion or more years younger. Moreover, the Apex fossils provide unequalled insight into the evolutionary status of the early biota. In particular, of the 11 species identified in the assemblage, seven (comprising ~63% of measured specimens) are notably similar in cellular

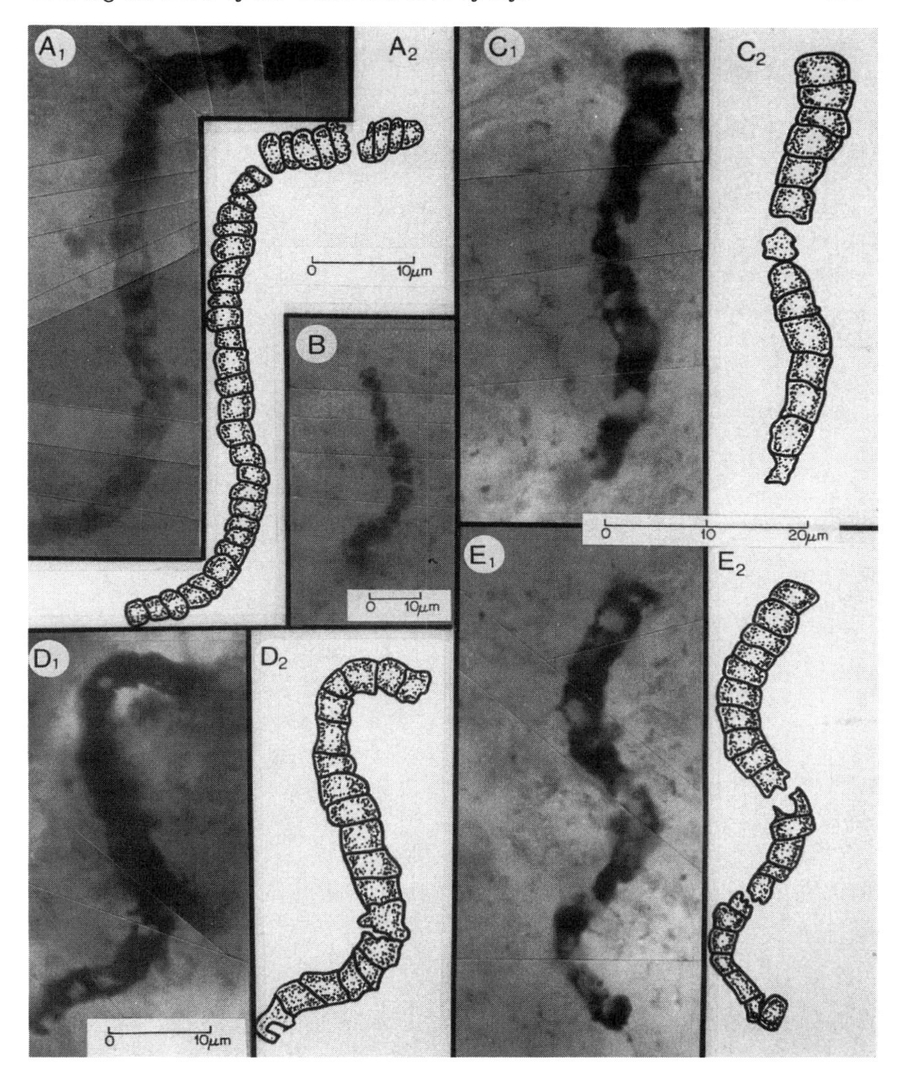

Figure 16.9. Cyanobacterium-like filamentous carbonaceous fossils (*Primaevifilum amoenum* Schopf 1992b), and interpretive drawings, shown in petrographic thin sections of the ~3,465-Ma-old Apex chert of northwestern Western Australia. Because of their sinuous three-dimensional preservation, all fossils are shown in composite photomicrographs.

organization to oscillatoriacean cyanobacteria and, in size range, median dimensions, and pattern of size distribution, they are much more like oscillatoriaceans than other filamentous prokaryotes (Schopf 1992b, 1993). Several of these taxa are essentially indistinguishable at the species level from common especially well known oscillatoriaceans, both fossil (*Oscillatoriopsis* spp.) and modern (*Oscillatoria* spp.).

The presence of oscillatoriacean cyanobacteria in the Apex assemblage is consistent also with four other lines of evidence. First, models of the early global ecosystem and trace element (cerium and europium) concentrations in Apex-age sediments show that aerobic respiration and O_2-producing photosynthesis – processes characteristic of cyanobacteria, photosynthesis not being exhibited by any other prokaryotes – had probably evolved by this early stage in Earth history (Towe 1990, 1991). Second, the isotopic compositions of organic and carbonate carbon in the Apex chert and associated sedimentary units evidence the occurrence of photosynthetic CO_2 fixation, evidently like that occurring in extant cyanobacterial populations grown in CO_2-rich environments (Schidlowski, Hayes and Kaplan 1983; Schopf 1994b). Third, the occurrence within the Apex filaments of bifurcated cells and cell pairs very likely reflects the original presence of partial septations and, thus, cell division like that occurring in extant oscillatoriaceans (Schopf 1992a, 1993). Fourth, as shown by analyses of rRNA, the Oscillatoriaceae is among the earliest evolved of extant cyanobacterial families (Giavannoni et al. 1988).

Taken together, these various lines of evidence seem persuasive: There is little doubt that if the Apex filaments had been discovered in younger Precambrian sediments – where fossil oscillatoriaceans are well known and widespread – or if they had been detected in a modern microbial community, and morphology were the only criterion by which to infer biologic relations, most would be interpreted as oscillatoriacean cyanobacteria.

Paleobiologic evidence of early evolution

Traditionally, paleontology has focused almost entirely on fossils – their morphology, anatomy, communities, affinities, evolution, paleoecology, paleogeography, biostratigraphy. Other paleobiologic issues have drawn less attention, among them biotic interactions with long-term environmental change and the physiology and biochemistry of ancient organisms. For example, though the Phanerozoic is punctuated by repeated ice ages and fluctuations in sea level, aridity, volcanism, and the like, all these are episodic, relatively short-term events. Truly major changes in the global environment – in the composition of the Earth's atmosphere, for example – are too long-term to fall within the scope of Phanerozoic studies. Similarly, because virtually all Phanerozoic fossils are readily relatable to living biologic groups, questions regarding their basic metabolic characteristics do not arise. Uncertainty may exist whether one or another group of dinosaurs was more or less warm-blooded, but it is safe to assume all were oxygen-requiring obligate aerobes, like their reptilian and avian descendants; and early-evolving land plants assuredly were oxygen-producing photosynthesizers, even though this lineage, like the dinosaurs, is long extinct.

The Precambrian Eon, however, is nearly an order of magnitude longer than the Phanerozoic, and events that are minor or even imperceptible over Phanerozoic time can have cumulative impact if they span appreciable portions of earlier Earth history. For example, the dramatic change from an early, anoxic global environment to the current oxygen-rich (~21%) atmosphere, if spanning the time from the Apex fossils to the present, would have required sustained addition of only 0.006% O_2 per million years. Yet were this same rate of increase to date from the beginning of the Phanerozoic, the total oxygen rise would be little more than 3%, roughly half that needed to support aerobic higher animals. Moreover, most major events in early evolution centered around the development of metabolic processes (Schopf 1996b), but because prokaryotes are highly diverse physiologically – and those of similar morphology can differ markedly in metabolism – the sorting out of such capabilities requires more than traditional fossil-focused paleontology. Data from geology and geochemistry are also crucial to unraveling the early history of life.

Evidence from geology

The capability to carry out oxygen-producing photosynthesis is a universal characteristic of cyanobacteria that distinguishes them from all other prokaryotes. If the inference is correct that cyanobacteria are represented among the Apex fossils, the Apex sequence and coeval geologic units should contain evidence of the reactants required for oxygenic photosynthesis (H_2O and CO_2) as well as the products produced (reduced organic carbon and O_2).

Both of the reactants are well evidenced in the early rock record. The fossiliferous Apex chert is a bedded, water-laid, conglomeratic deposit, part of an interbedded sequence of volcanic and sedimentary rock that contains abundant evidence of subaqueous deposition (Groves, Dunlop and Buick 1981). The scene was dominated by extensive shallow seas, with scattered volcanic islands fringed by sedimentary debris (river gravels and sands), mud flats, and interspersed evaporitic lagoons (Barley et al. 1979). Further evidence of liquid water is provided by ripple marks and pillow lavas, present both in the Apex sequence and in units about the same age in South Africa (Figures 16.10A, B). Similarly, in both the Apex and South African sequences there is ample evidence of the other reactant of photosynthesis, carbon dioxide, in the form of $CaCO_3$-rich limestones deposited as a result of aqueous reaction between Ca^{++}, derived from weathering of the land surface, and bicarbonate (HCO_3^-), produced by dissolution of atmospheric CO_2.

Figure 16.10. Geologic evidence of the nature of the early environment. **A:** ~3,300-Ma-old ripple marks, evidencing the early Precambrian presence of liquid water, preserved in the upper Swaziland Supergroup (Moodies Group, Joe's Luck Formation) at Hollebrand's Pass on Bonanza Gold Mine Road near Barberton, eastern Transvaal, South Africa. **B:** ~3,500-Ma-old mound-shaped pillow lava, formed as a molten volcanic lava-flow cooled rapidly on contact with early Precambrian oceanic waters, from the lower Swaziland Supergroup (Onverwacht Group, Komati Formation) exposed in the Komati River Valley 30 km south of Barberton, eastern Trasvaal, South Africa. **C:** ~2,250-Ma-old iron oxide–rich banded iron formation, evidencing the presence of trace amounts of dissolved oxygen in the early Precambrian ocean, shown in a 2.5-cm-diameter subsurface core sample obtained from the Transvaal

The two products of cyanobacterial photosynthesis, organic matter and O_2, are also evidenced in the early rock record. Not only is particulate degraded organic carbon, (kerogen) ubiquitous and abundant (ranging from ~0.5% to ~0.8% by weight; Strauss and Moore 1992) in sediments of the Apex sequence but it also makes up the carbonaceous cell walls of the petrified microscopic fossils (Figures 16.8, 16.9). And at least small amounts of free oxygen were present in the environment, reflected by the occurrence of iron oxide–rich sedimentary units known as banded iron formations (BIFs), the world's major source of iron ore. Units of this type are widespread in geologic terrains older, but not younger, than about 2,000 Ma and, together with uranium-rich pyritic conglomerates, provide evidence of the early history of the atmosphere.

A particularly well banded specimen of BIF is shown in Figure 16.10C. The banding in such units is produced by an alternation of iron-rich and iron-poor layers, and because the ferruginous beds are composed of fine rustlike particles of hematite (Fe_2O_3) and, in some deposits, magnetite (Fe_3O_4), BIFs have a characteristic dull to bright red color. The iron evidently owes its origin to the hydrothermal circulation of seawater through oceanic crust, primarily at deep submarine ridge systems. In a dissolved (ferrous) state, it then circulated upward into shallower reaches of the water column – often seasonally, giving rise to the distinctive millimetric banding – where it was oxidized to ferric iron, chiefly by reaction with dissolved molecular oxygen, and precipitated as a fine rusty rain of minute insoluble iron oxide particles.

The Earth's atmosphere is chiefly an accumulated product of volcanic outgassing of the planetary interior over geologic time, but unlike the other principal components of the atmosphere (N_2, H_2O, CO_2), molecular oxygen (O_2) is not released from rocks when they are heated. Other inorganic sources of molecular oxygen (for example, thermal dissociation or UV-induced photodissociation of water vapor) similarly are wanting, leaving oxygenic photosynthesis as the only quantitatively plausible source for the enormous amount of oxygen sequestered in Precambrian BIFs. The presence of these deposits, however, does not indicate the environment was oxygen rich. On the contrary, BIF-containing basins are large, typically several hundred kilometers in length and breadth, and dissolved ferrous iron could not have been distributed over such broad areas unless at least the lower portions of the water column

---

Supergroup (Postmasburg Group, Hotazel Formation) about 50 km northwest of Kuruman, northern Cape Province, South Africa. **D, E:** ~2,700-Ma-old uraninite-bearing quartz–pyrite conglomerate, shown in a hand specimen (D) and a longitudinally sliced core sample (E), containing well-rounded pebbles of detrital pyrite (E, arrows) that would have been oxidized and dissolved if the early Precambrian atmosphere had contained large amounts of free oxygen, from the Ventersdorp Supergroup (Venterspost Carbon Reef "Member," Venterspost Formation, Platberg Group) at Libanon Gold Mine, 25 km west of Johannesburg, Transvaal, South Africa.

were anoxic. Thus, two cardinal inferences can be drawn from the continuous presence of abundant, widespread BIFs 3,500 to ~2,000 Ma ago (Klein and Beukes 1992): (1) Copious amounts of oxygen were being pumped into the environment by oxygenic (cyanobacterial) photosynthesis, but (2) the oceanic water column was not fully oxygenated (nor was the atmosphere) because the molecular oxygen was removed from the system by its reaction with iron to form sedimented iron oxide minerals.

The existence of a low-O_2 environment up to ~2,000 Ma ago is supported also by other geologic evidence. Resistance of some minerals to weathering is strongly affected by the presence of O_2. Uraninite (UO_2) and pyrite (FeS_2), both minerals oxidized and dissolved in the presence of molecular oxygen, are good examples. Extensive detrital accumulations of these minerals do not occur in sediments today because the minerals are easily destroyed during weathering in the present oxygen-rich atmosphere. Nevertheless, large sedimentary ore deposits that contain detrital pebbles of uraninite and pyrite (Figures 16.10D, E) occur in geologic units older, but not younger, than ~2,300 Ma. The persistence of pebbles of uraninite and pyrite during weathering, fluvial transport, and deposition in conglomeratic deposits more than 2,300 Ma ago is consistent with a low-O_2 atmosphere up to this time, as are other geologic data, most notably the mineralogy of paleosols (ancient soil horizons). Taken together, these and related lines of evidence support the conclusion that the O_2 content of the atmosphere increased dramatically between 2,200 and 1,900 Ma ago, evidently from $\leq$1% to $\geq$15% of the present atmospheric level (Holland 1994).

Geologic and paleontologic evidence thus agree with the inferred existence of oxygen-producing cyanobacteria as early as ~3,500 Ma ago, about two billion years earlier than suggested by the amino acid clock. Data from isotopic geochemistry are similarly consistent with this interpretation and, like the record of cellularly preserved noncyanobacterial prokaryotes (Figures 16.7, 16.8D–F), evidence the presence of other early-evolved microbial lineages as well.

Evidence from geochemistry

The process of photosynthesis involves an enzyme-mediated fractionation of the two stable isotopes of carbon such that the organic compounds biosynthesized are enriched in the lighter isotope, ^{12}C, relative to the $^{12}C/^{13}C$ ratio of the CO_2 carbon source (Schidlowski, Hayes and Kaplan 1983). As a result, photosynthetically produced organic matter sequestered in sedimentary rocks as kerogen is ^{12}C-enriched relative to the carbon isotopic ratio of concurrently deposited carbonate minerals (which reflect the isotopic composition of atmos-

Table 16.1. *Geochemical evidence of early-evolving metabolic processes*

O_2 PHOTOSYNTHESIS & RESPIRATION (Cyanobacteria & Gram-positive Eubacteria) $CO_2 + H_2O$ ———> CH_2O + O_2 ⇩ ⇩ ^{12}C-ENRICHED KEROGEN BANDED IRON FM.
BACTERIAL (NON-O_2) PHOTOSYNTHESIS (Purple & Green Eubacteria) $CO_2 + H_2S$ ———> CH_2O ⇩ ^{12}C-ENRICHED KEROGEN
SULFATE REDUCTION (Gram-negative Eubacteria) SO_4 ———> H_2S ⇩ ^{32}S-ENRICHED PYRITE
METHANE GENERATION (Methanogenic Archaebacteria) $CO_2 + H_2$ ———> CH_4 ⇩ *HIGHLY* ^{12}C-ENRICHED KEROGEN

pheric and dissolved CO_2), a difference typically ranging from 15‰ to 35‰ (expressed as a $\delta^{13}C$ value) depending partly on ambient CO_2 concentrations (Schidlowski, Hayes and Kaplan 1983; Hayes 1994; Schopf 1994b). As summarized in Table 16.1, the occurrence of such ^{12}C-enriched kerogen therefore reflects the existence of photosynthesis (whether cyanobacterial and oxygenic, or bacterial and nonoxygenic) and, if coupled with the evidence of oxidation provided by BIFs, the presence both of oxygenic photosynthesis and of aerobic respiration (Towe 1990, 1991; Schopf 1996b). Geochemical analyses of thousands of samples establish that the carbon isotopic signature of photosynthetic activity is continuous from the present back to 3,500 Ma ago (Schidlowski, Hayes and Kaplan 1983; Strauss et al. 1992; Hayes 1994; Schopf 1994b). These data, together with the geologic and paleontologic evidence discussed earlier, support the conclusion that the cyanobacterial, photosynthetic bacterial, and Gram-positive eubacterial lineages have been extant since at least 3,500 Ma ago, about two billion years earlier than suggested by the amino acid clock.

Isotopic evidence also records the presence of Gram-negative eubacteria at a time much "too old" by the amino acid clock. This lineage includes sulfate-reducing bacteria, such as *Desulfovibrio desulfuricans*, that derive energy from hydrogenation of oceanic sulfate, SO_4^{2-}, to sulfide, H_2S (that is, reduction of S^{6+} to S^{2-}). As in photosynthesis, this process involves an enzymatic fractionation of two stable isotopes, but of sulfur rather than carbon, such that the sulfide generated is enriched in the lighter isotope, ^{32}S, relative to the $^{32}S/^{34}S$ ratio of the SO_4^{2-} sulfur source (Schidlowski, Hayes and Kaplan 1983).

The bacterially generated H_2S reacts with iron-bearing minerals in sediments to form pyrite, FeS_2, which consequently is also enriched in the lighter sulfur isotope (Table 16.1). Biogenic pyrites in modern sediments typically have ratios of sulfur isotopes that are highly variable, encompassing a ~60‰ range of $\delta^{34}S$ values, and are thus distinguishable from the sulfide minerals in igneous rocks, which fall generally within a narrow range of about 5‰. The isotopic signature of microbial sulfate reduction and, thus, of Gram-negative eubacteria, can be traced back to at least 2,700 Ma ago (Schidlowski, Hayes and Kaplan 1983) and the lineage may well have been extant as early as 3,400 Ma ago (Ohmoto, Kakegawa and Lowe 1993; Kakegawa, Kawai, and Ohmoto 1994), nearly two billion years "too early" by the amino acid clock.

In addition to the eubacterial lineages previously discussed, geochemical evidence records the presence of the Archaea much earlier than permitted by the amino acid data. Methanogenic archaebacteria derive energy by hydrogenating carbon dioxide or carbon monoxide to produce methane by the overall reactions:

$$CO_2 + 4H_2 \longrightarrow CH_4 + 2H_2O + 32 \text{ kcal}$$
$$4CO + 2H_2O \longrightarrow CH_4 + 3CO_2 + 50 \text{ kcal.}$$

The methane generated tends to be extremely ^{12}C-enriched relative to the $^{12}C/^{13}C$ ratio of the inorganic carbon source, commonly by 60‰ or more (Games, Hayes and Gunsalus 1978), and is thus readily distinguishable from ^{12}C-enriched products of photosynthesis. Incorporation of this biogenic methane into sedimented organic matter leaves an unmistakable signature in the preserved kerogen (Kaplan and Nissenbaum 1966) that can be traced to at least 2,900 Ma ago (Hayes 1983, 1994) and perhaps even earlier (Mojzsis, Nutman and Arrhenius 1994; Schopf 1994b).

Paleobiology: The court of last resort

Taken together, paleontology, geology, and geochemistry provide convincing evidence of the antiquity and evolutionary continuity of the eubacterial and archaebacterial domains. The lines of paleobiologic evidence are distinct yet mutually reinforcing; the data, voluminous. Stromatolitic microbial ecosystems, evidently including cyanobacteria and photosynthetic and other non-cyanobacterial eubacteria, were extant ~3,500 Ma ago; methanogenic archaebacteria, by ~2,900 Ma ago; and Gram-negative sulfate-reducing bacteria, at least as early as ~2,700 Ma ago. The discrepancies between these dates and the times suggested by the amino acid clock (Table 16.2) are too great and too consistent to be ignored.

Table 16.2. *Earliest occurrences of prokaryotes based on paleobiologic (paleontological, geological, and geochemical) evidence compared with those inferred from the amino acid "clock"*

Microorganisms	Geologic Evidence: Paleontological	Geologic Evidence: Geochemical	Amino Acid "Clock"	**Discrepancy**
MICROBIAL ECOSYSTEMS	~3.5 Ga Stromatolites	~3.5 Ga $\delta^{13}C$	~1.9 Ga	*1.6 Ga*
OXYGEN-PRODUCING CYANOBACTERIA	~3.5 Ga Microfossils	~3.5 Ga $\delta^{13}C$ + BIF*	~1.5 Ga	*2.0 Ga*
PHOTOSYNTHETIC BACTERIA		~3.5 Ga $\delta^{13}C$	~1.5 Ga	*2.0 Ga*
NONCYANOBACTERIAL EUBACTERIA	~3.5 Ga Microfossils		~1.5 Ga	*2.0 Ga*
GRAM-NEGATIVE EUBACTERIA		~2.7 Ga $\delta^{34}S$	~1.5 Ga	*1.2 Ga*
METHANOGENIC ARCHAEBACTERIA		~2.9 Ga $\delta^{13}C$	~1.9 Ga	*1.0 Ga*

*BIF = Banded Iron Formation

The amino acid clock is evidently in error, perhaps for one or more of the following reasons. Doolittle et al. (1996) were forced to assume that amino acids are substituted in different parts of a given protein at the same constant rate, but site-to-site variation and multiple substitution at the same site no doubt occur, both of which lead to underestimation of lineage divergence times. Similarly, if detectable amino acid substitution was slower in the Precambrian and more rapid in the Phanerozoic (due, for example, to prokaryotic–eukaryotic differences in the efficacy of protein repair), the clock dates would be too young. And, of course, the amino acid clock might be dating episodes of horizontal gene transfer and the endosymbiotic origin of eukaryotes, rather than times of lineage divergence and the age of the last common ancestor.

As Darwin well recognized, the sole source of direct evidence of the nature, products, and timing of the evolutionary process is the geologic record. Fossils, geology, and geochemistry, together, must serve as the final arbiter of competing theories about the history of life – the science of paleobiology is the court of last resort.

References

Altermann, W., and Schopf, J. W. 1995. Microfossils from the Neoarchean Campbell Group, Griqualand West Sequence of the Transvaal Supergroup, and their paleoenvironmental and evolutionary implications. *Precambian Res.* 75:65–90.

Barghoorn, E. S., and Tyler, S. A. 1965. Microorganisms from the Gunflint chert. *Science* 147:563–577.

Barley, M. E., Dunlop, J. S. R., Glover, J. E., and Groves, D. I. 1979. Sedimentary evidence for an Archean shallow-water volcanic-sedimentary facies, eastern Pilbara Block, Western Australia. *Earth Planet. Sci. Lett.* 43:74–84.

Blake, T. S., and McNaughton, N. J. 1984. A geochronological framework for the Pilbara region. In *Archean & Proterozoic Basins of the Pilbara, Western Australia: Solution and Mineralization Potential (Publication 9),* ed. J. R. Muhling, D. K. Groves, and T. S. Blake, pp. 1–22. Perth, Australia: University of Western Australia. Geology Department and University Extension.

Darwin, C. R. 1859. *The Origin of Species by Means of Natural Selection.* London: John Murray.

Doolittle, R. F., Feng, D-F., Tsang, S., Cho, G., and Little, E. 1996. Determining divergence times of the major kingdoms of living organisms with a protein clock. *Science* 271:470–477.

Games, L. M., Hayes, J. M., and Gunsalus, R. P. 1978. Methane-producing bacteria: Natural fractionation of the stable carbon isotopes. *Geochim. Cosmochim. Acta* 42:1295–1297.

Giavannoni, S. J., Turner, S., Olsen, G. J., Barns, S., Lane, D. J., and Pace, N. R. 1988. Evolutionary relationships among cyanobacteria and green chloroplasts. *J. Bacteriol.* 170:3584–3592.

Golubic, S. 1976a. Organisms that build stromatolites. In *Stromatolites, Developments in Sedimentology 20,* ed. M. R. Walter, pp. 113–126. Amsterdam: Elsevier.

Golubic, S. 1976b. Taxonomy of extant stromatolite-building cyanophytes. In *Stromatolites, Developments in Sedimentology 20,* ed. M. R. Walter, pp. 127–140. Amsterdam: Elsevier.

Golubic, S., and Hofmann, H. J. 1976. Comparison of Holocene and mid-Precambrian Entophysalidaceae (Cyanophyta) in stromatolitic mats: Cell division and degradation. *J. Paleontol.* 50:1074–1082.

Groves, D. I., Dunlop, J. S. R., and Buick, R. 1981. An early habitat of life. *Scient. Am.* 245:64–73.

Hayes, J. M. 1983. Geochemical evidence bearing on the origin of aerobiosis, a speculative hypothesis. In *Earth's Earliest Biosphere,* ed. J. W. Schopf, pp. 291–301. Princeton, N.J.: Princeton University Press.

Hayes, J. M. 1994. Global methanotrophy at the Archean–Proterozoic transition. In *Early Life on Earth,* ed. S. Bengtson, pp. 220–236. New York: Columbia University Press.

Hermann, T. N. 1981. Nitchatye mikroorganizmy lakhandinskoj svity reki Mai [Filamentous microorganisms from the Lakhanda Formation on the Maya River]. *Palaeontol. Zhur.* 1981(2):126–131 (in Russian).

Hofmann, H. J. 1976. Precambrian microflora, Belcher Islands, Canada: Significance and systematics. *J. Paleontol.* 50:1040–1073.

Hofmann, H. J., and Schopf, J. W. 1983. Early Proterozoic microfossils. In *Earth's Earliest Biosphere,* ed. J. W. Schopf, pp. 2321–360. Princeton, N.J.: Princeton University Press.

Holland, H. D. 1994. Early Proterozoic atmospheric change. In *Early Life on Earth,* ed. S. Bengtson, pp. 237–244. New York: Columbia University Press.

Iwabe, N., Kuma, K., Hasegawa, M., Osawa, S., and Miyata, T. 1989. Evolutionary relationships of archaebacteria, eubacteria and eukaryotes inferred from phylogenetic trees of duplicated genes. *Proc. Nat. Acad. Sci. USA* 86:9355–9359.

Kakegawa, T., Kawai, H., and Ohmoto, H. 1994. Biological activities and hydrothermal activity recorded in the ~2.5 Ga Mount McRae Shale, Hamersley district, western Australia. II. Sulfur isotopic composition of pyrite. *Resource Geol.* 44:284–285.

Kaplan, I. R., and Nissenbaum, A. 1966. Anomalous carbon isotope ratios in non-volatile organic material. *Science* 153:744–745.

Klein, C., and Beukes, N. J. 1992. Time distribution, stratigraphy, sedimentologic setting, and geochemistry of Precambrian iron-formations. In *The Proterozoic Biosphere,* ed. J. W. Schopf and C. Klein, pp. 139–146. New York: Cambridge University Press.
Knoll, A. H., Strother, P. K., and Rossi, S. 1988. Distribution and diagenesis of microfossils from the Lower Proterozoic Duck Creek Dolomite, Western Australia. *Precambrian Res.* 38:257–279.
Krylov, I. N., and Sergeev, V. N. 1987. Rifeiskie mikrofossilii uzhnogo Urala v raione goroda Kusa [Riphean microfossils of the southern Urals near the town of Kusa]. *Palaeontol. Zhur.* 1987(2):107–116 (in Russian).
Mendelson, C. V., and Schopf, J. W. 1992. Proterozoic and selected Early Cambrian microfossils and microfossil-like objects. In *The Proterozoic Biosphere,* ed. J. W. Schopf and C. Klein, pp. 865–951. New York: Cambridge University Press.
Mojzsis, S., Nutman, A. P., and Arrhenius, G. 1994. Evidence for a marine sedimentary system at >3.87 Ga in southern West Greenland. *EOS* 75(44):690.
Moore, T. B. 1993. Micropaleontology of the Early Proterozoic Gunflint Formation. Ph.D. Dissertation, Department of Earth and Space Sciences, University of Calififornia, Los Angeles.
Ohmoto, H., Kakegawa, T., and Lowe, D. R. 1993. 3.4-billion-year-old biogenic pyrites from Barberton, South Africa: Sulfur isotope evidence. *Nature* 262:555–557.
Olsen, G. J., and Woese, C. R. 1993. Ribosomal RNA: A key to phylogeny. *FASEB J.* 7:113–123.
Schidlowski, M., Hayes, J. M., and Kaplan, I. R. 1983. Isotopic inferences of ancient biochemistries: Carbon, sulfur, hydrogen, and nitrogen. In *Earth's Earliest Biosphere,* ed. J. W. Schopf, pp. 149–186. Princeton, N.J.: Princeton University Press.
Schopf, J. W. 1968. Microflora of the Bitter Springs Formation, Late Precambrian, central Australia. *J. Paleontol.* 42:651–688.
Schopf, J. W. 1970. Precambrian micro-organisms and evolutionary events prior to the origin of vascular plants. *Biol. Rev. Cambridge Phil. Soc.* 45:319–352.
Schopf, J. W. 1992a. Times of origin and earliest evidence of major biologic groups. In *The Proterozoic Biosphere,* ed. J. W. Schopf and C. Klein, pp. 587–593. New York: Cambridge University Press.
Schopf, J. W. 1992b. Paleobiology of the Archean. In The Proterozoic Biosphere, ed. J. W. Schopf and C. Klein, pp. 25–39. New York: Cambridge University Press.
Schopf, J. W. 1992c. Proterozoic prokaryotes: Affinities, geologic distribution, and evolutionary trends. In *The Proterozoic Biosphere,* ed. J. W. Schopf and C. Klein, pp. 195–218. New York: Cambridge University Press.
Schopf, J. W. 1992d. Informal revised classification of Proterozoic microfossils. In *The Proterozoic Biosphere,* ed. J. W. Schopf and C. Klein, pp. 1119–1168. New York: Cambridge University Press.
Schopf, J. W. 1992e. Atlas of representative Proterozoic microfossils. In *The Proterozoic Biosphere,* ed. J. W. Schopf and C. Klein, pp. 1055–1117. New York: Cambridge University Press.
Schopf, J. W. 1993. Microfossils of the Early Archean Apex chert: New evidence of the antiquity of life. *Science* 260:640–646.
Schopf, J. W. 1994a. Disparate rates, differing fates: Tempo and mode of evolution changed from the Precambrian to the Phanerozoic. *Proc. Nat. Acad. Sci. USA* 91:6735–6742.

Schopf, J. W. 1994b. The oldest known records of life: Stromatolites, microfossils, and organic matter from the Early Archean of South Africa and Western Australia. In *Early Life on Earth*, ed. S. Bengtson, pp. 193–206. New York: Columbia University Press.

Schopf, J. W. 1996a. Cyanobacteria: Pioneers of the early Earth. *Nova Hedwigia* 112:13–32.

Schopf, J. W. 1996b. Metabolic memories of Earth's earliest biosphere. In *Evolution and the Molecular Revolution*, ed. C. R. Marshall and J. W. Schopf, pp. 73–107. Boston: Jones & Bartlett.

Schopf, J. W., and Blacic, J. M. 1971. New microorganisms from the Bitter Springs Formation (Late Precambrian) of the north-central Amadeus Basin, Australia. *J. Paleontol.* 45:925–961.

Schopf, J. W., Hayes, J. M., and Walter, M. R. 1983. Evolution of Earth's earliest ecosystems: Recent progress and unsolved problems. In *Earth's Earliest Biosphere*, ed. J. W. Schopf, pp. 361–384. Princeton, N.J.: Princeton University Press.

Schopf, J. W., and Walter, M. R. 1983. Archean microfossils: New evidence of ancient microbes. In *Earth's Earliest Biosphere*, ed. J. W. Schopf, pp. 214–239. Princeton, N.J.: Princeton University Press.

Stetter, K. O. 1996. Hyperthermophilic procaryotes. *FEMS Microbiol. Rev.* 18:149–158.

Strauss, H., DesMarais, D. J., Hayes, J. M., and Summons, R. E. 1992. The carbon-isotopic record. In *The Proterozoic Biosphere,* ed. J. W. Schopf and C. Klein, pp. 117–127. New York: Cambridge University Press.

Strauss, H., and Moore, T. B. 1992. Abundance and isotopic composition of carbon and sulfur species in whole rock and kerogen samples. In *The Proterozoic Biosphere,* ed. J. W. Schopf and C. Klein, pp. 709–798. New York: Cambridge University Press.

Thorpe, R. I., Hickman, A. H., Davis, D. W., Mortensen, J. K., and Trendall, A. F. 1992. U-Pb zircon geochronology of Archaean felsic units in the Marble Bar region, Pilbara Craton, Western Australia. *Precambrian Res.* 56:169–189.

Towe, K. M. 1990. Aerobic respiration in the Archaean? *Nature* 348:54–56.

Towe, K. M. 1991. Aerobic carbon cycling and cerium oxidation: Significance for Archean oxygen levels and banded iron-formation deposition. *Palaeogeog. Palaeoclimatol. Palaeoecol.* 97:113–123.

Woese, C. R. 1987. Bacterial evolution. *Microbiol. Rev.* 51:221–271.

Woese, C. R., Kandler, O., and Whellis, M. L. 1990. Towards a natural system of organisms: Proposal for the domains Archaea, Bacteria, and Eucarya. *Proc. Nat. Acad. Sci. USA* 87:4576–4579.

Part V

Clues from other planets

17

Titan

FRANÇOIS RAULIN
Laboratoire Interuniversitaire des Systèmes Atmosphériques
Universités Paris 7 et Paris 12 and CNRS
Créteil, France

It is increasingly obvious that to study the origins of life on Earth we cannot limit our investigations to our own planet, but we have to look at the various planetary objects of the Solar System. All traces of the history of the first half-billion years on Earth have been erased. By looking in detail at Venus and Mars now, in particular with the help of spacecrafts' atmospheric probes and landers, we can perform a systematic comparison of the present status of the three terrestrial planets. From such a comparison we can deduce the main physical–chemical processes that governed these planets' evolution from their formation up to now. We can thus model for each planet, including the Earth, not only its evolution but also the initial conditions, which are currently unreachable by direct observation.

By looking in detail at comets and meteorites, we can study the organic matter currently carried by these extraterrestrial objects. We can thus approach the chemical nature of the organics that may have been imported onto the primitive Earth from these carriers.

By looking in detail at the outer Solar System, we can study the organic physical and chemical processes that are currently involved in all the atmospheres of the outer planets and, first of all, as this chapter will try to demonstrate, in the atmosphere of Titan, the largest satellite of Saturn. We can thus constrain our models of the organic physical and chemical evolution on Earth, before the emergence of life.

1. A need for a control prebiotic laboratory on a planetary scale

In most scientific fields there are three main approaches: direct observation, theoretical modeling, and experimental works (including simulations) (Bruston et al. 1994). Studying the origins of life on Earth in a satisfactory manner should involve following these three complementary approaches. As clearly evidenced in the previous chapters, mainly the two first have been used. In particular, since the first publication of Stanley Miller's experiment

(1953), experimental simulations in the laboratory have been a very fruitful means to study the main pathways of prebiotic chemistry. However, there is a giant gap between the conditions used in such experimental works and those that are involved in a true planetary environment. In the laboratory, it is very difficult to simulate the complexity of a planetary atmosphere, including the absence of walls; the variation with altitude of pressure, temperature, and UV and other energy fluxes; the possible influence of clouds and hazes, eddy diffusion, and so forth. In fact, no experiment simulating all atmospheric characteristics has been carried out thus far.

Such limitations in the simulation can yield strong differences between the laboratory results and the true behavior of the atmospheric environment supposedly being simulated. For instance, wall effects in the laboratory reactor frequently induce adsorption phenomena and third body reactions that are absent in the real atmosphere. In addition to this space limitation, there may also be a strong time limitation: It is far from obvious that we can accurately simulate within a few days (or even a few years) a few million or tens of millions of years (if not more) of chemical evolution of the terrestrial environment.

Consequently, a secure extrapolation of laboratory simulation experiments to a planetary scale requires real planetary data. The primitive Earth is not directly available now, but planetary exploration has discovered another place that can constrain our models of the chemical evolution on the early Earth. With its environment very rich in organics, and many of its physical conditions similar to that of the Earth, Titan can play the role of the needed control prebiotic laboratory.

2. Titan before and after Voyager

Titan was discovered by the Dutch astronomer Huygens in 1655. Such an early discovery is certainly related to the large size of Titan: With a mean diameter of 5,150 km (as determined much later), it is the largest satellite of Saturn and the second in size of the Solar System (Table 17.1). But its more surprising peculiarity is the presence of an atmosphere. Suggested in 1908 by the Spanish astronomer Comas-Sola, who observed limb darkening, it has been confirmed by the American astronomer Kuiper in 1944, who discovered the infrared signature of gaseous methane on Titan. Later on, ethane, ethylene, and acetylene were detected from Earth observations and calculations by the American planetologist Hunten (1978), who predicted molecular nitrogen as main constituent of a thick atmosphere. Pre-Voyager observations also indicated the presence of hazes obscuring the low atmosphere.

Table 17.1. *Main characteristics of Titan*

Surface radius	2,575 km
Surface gravity	1.35 m s^{-2} (0.14 Earth's value)
Mean volumic mass	1.88 kg dm^{-3} (0.34 Earth's value)
Distance from Saturn	20 Rs (about 1.2×10^6 km)
Orbit period around Saturn	15.95 days
Orbit period around Sun	29.5 years
Temperatures	
Surface	96 ± 4K
Tropopause (42 km)	71.4 K
Stratosphere (200 km)	about 170 K
Pressures	
Surface	1,496 mbar
Tropopause	135 mbar

However, most of our current knowledge of Titan's atmosphere comes from the Voyager data, after Voyager 1 and Voyager 2 flybys of Titan in 1980 and 1981 respectively. Both were unable to image Titan's surface, masked by very thick layers of aerosols and clouds. However, several instruments carried out synergistic measurements, the retrieval of which provided a tremendous amount of information on Titan's atmosphere and even on its surface. The Radio Science Subsystem (RSS), through an Earth occultation experiment coupled with the data from the Infrared Radiometer and Interferometric Spectrometer (IRIS), determined the temperature and pressure profiles of the atmosphere. IRIS, helped with the data from the UltraViolet Spectrometer (UVS), was able to determine the main molecular composition of the atmosphere, and also to detect many trace species and to determine their abundance, with some information on their vertical and latitudinal distribution.

The data confirmed that the atmosphere is dominated by N_2, with a noticeable fraction (up a few percent) of CH_4, and a low abundance of H_2 (Table 17.2). Although its has not been detected, argon (essentially nonradiogenic Argon, i.e., ^{36}Ar and ^{38}Ar) was expected to be present for cosmogonical reasons. However, profound reexamination of the far-infrared spectrum of Titan has recently allowed Courtin, Gautier and McKay (1995) to derive an upper limit of abundance for this element of only 6% (at the 3 sigma level). This compound may also not be present at all in Titan's atmosphere. Water is nondetectable in the atmosphere (its abundance must be extremely low because of the cold environment). The only O-compounds detected are CO (a few ten ppm) and CO_2 (around 10 ppb). It is a very dense atmosphere: The combined

Table 17.2. *Chemical composition of Titan's stratosphere: observational data and mean concentrations. (Argon has not been detected but, based on cosmogonical considerations, may be present. It may also be totally absent. The possible prebiotic role of the detected organics is also indicated.)*

Compounds	Stratosphere mixing ratio (E = Equ.; N = North Pole)	Plausible prebiotic source of:
Main constituents		
Nitrogen N_2	0.90–0.99	
Methane CH_4	0.017–0.045	
Argon Ar	0–0.06	
Hydrogen H_2	0.0006–0.0014	
Hydrocarbons		
Ethane C_2H_6	1.3×10^{-5} E	
Acetylene C_2H_2	2.2×10^{-6} E	-> CH oligomers
Propane C_3H_8	7.0×10^{-7} E	
Ethylene C_2H_4	9.0×10^{-8} E	-> CH oligomers
Propyne C_3H_4	1.7×10^{-8} N	-> CH oligomers
Diacetylene C_4H_2	2.2×10^{-8} N	-> CH oligomers
N-Organics		
Hydrogen cyanide HCN	6.0×10^{-7} N	-> CHN oligomers, amino acids, purines, pyrimidines
Cyanoacetylene HC_3N	7.0×10^{-8} N	-> CHN oligomers pyrimidines
Cyanogen C_2N_2	4.5×10^{-9} N	-> Condensing agent
Acetonitrile CH_3CN	few 10^{-9}	
Dicyanoacetylene C_4N_2	Solid Pha. N	-> CN (H) oligomers
O-compounds		
Carbon dioxide CO_2	1.4×10^{-8} E	) source
	$<7.0 \times 10^{-9}$ N	) of
Carbon monoxide CO	2.0×10^{-5}	) O-compounds

factors of a higher surface pressure (1.5 bar) and a three times lower temperature (95 K), compared to the Earth, result in a 4.5 times denser atmosphere. However, there are many analogies between Titan and the Earth. First, although Titan is much colder than Earth, the vertical temperature profile (Lindal et al. 1983) is quite similar to the terrestrial one (Figure 17.1). It has

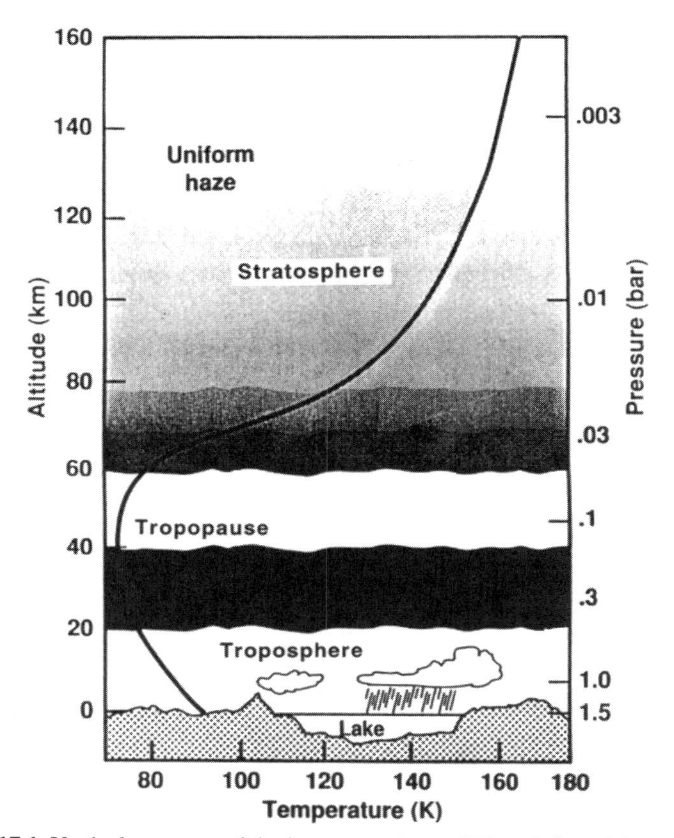

Figure 17.1. Vertical structure of the low atmosphere of Titan (adapted from Owen et al. 1992). The variation of temperature and pressure with altitude, from the Voyager observational data, shows a (relative) behavior very similar to that of the Earth, with the same kind of atmospheric zones, including a troposphere and a stratosphere.

a troposphere where temperature decreases with altitude, a tropopause where it goes through a minimum (70 K), and a stratosphere where it increases with altitude (up to about 170 K). Such a profile on Earth is caused mainly by the presence of two greenhouse gases: CO_2, which cannot condense in the Earth's atmosphere, and H_2O, which can condense. On Titan, the situation is similar, with the presence of two other greenhouse gases: noncondensable H_2 and condensable CH_4. In addition, both atmospheres also include anti-greenhouse species: clouds and hazes. Moreover, because of the size of both planets, in spite of the temperature differences, hydrogen escape must be efficient in both cases, maintaining a low H_2 mole fraction in their atmospheres.

Detailed analysis and modeling of the Voyager data clearly show that Titan's atmosphere is rich in organic compounds, including those involved in terrestrial prebiotic chemistry, and strongly suggest that they are present not only in the gas phase, but also in the aerosols. Such analyses and modeling also suggest the presence of liquid bodies at Titan's surface, where, again, organic chemistry is deeply involved.

Thus, after the Voyager encounter, the largest satellite of Saturn is suddenly appearing as an Earth-like planet – or, more precisely, as a primitive Earth–like planet – because of the many organic compounds (in particular, compounds of prebiotic interest) and processes present on Titan (Clark and Ferris 1997). In fact, as we will see now, such prebiotic-like chemistry must be involved in the three parts of what could be named (by comparison with the Earth) "Titan's geofluid": gas, aerosols, and "oceans" (Raulin et al. 1992).

3. An organic atmosphere in the gas phase . . .

Since the first, now classical, experiment by Miller (1953), many laboratory works have been carried out to simulate the chemical evolution of planetary atmospheres by submitting various mixtures of simple gases, as models of given atmospheres, to energy fluxes, in particular to electrons or photons irradiation. A very wide range of molecular compositions of the starting gas mixture has been explored, from highly reducing (CH_4–H_2–NH_3–H_2O) to oxidized (CO_2–N_2–H_2O) media. Comparison of the different results obtained from these experiments shows that the gas mixture the most favorable for organic syntheses of prebiotic interest is a mid-reducing mixture of N_2–CH_4–H_2O, with a ratio of N/C greater than 1 (Toupance, Raulin and Buvet 1975; Bossard et al. 1982; Raulin 1992) and with an efficient H_2 escape maintaining the H_2 mole fraction at a low value (Raulin, Mourey and Toupance 1982; Bossard, Mourey and Raulin 1983). Such conditions provide the greatest variety of organic products: linear and ramified alkanes, alkenes and alkynes, and various nitriles and dinitriles with saturated and unsaturated carbon chains, including HCN, HC_3N, and C_2N_2, which play an important role in prebiotic chemistry.

With more than 90% N_2, several percent of CH_4, and a mole fraction of H_2 smaller than about 0.5%, the main composition of Titan's atmosphere, with the exception of water (which, at least in the gas phase, is not present in noticeable amounts in this too cold environment) corresponds to this optimal gas mixture. Indeed, many organic compounds have already been detected in the atmosphere of Titan. As shown on Table 17.2, the list includes hydrocarbons – the three C_2-hydrocarbons, propane, propyne, and diacetylene – as well as various nitriles: alkane-nitriles, one alkynenitrile, and two dinitriles. There is quite good agreement between this list and the list of organics detected in Titan simulation experiments. In particular, experiments performed by Thompson et al. (1991) using a continuous flow plasma discharge reactor and low-energy dose at two different total pressures (17 and 0.24 mbar) show the important effect of the total pressure parameter on the yields of synthesis. However, even at low pressure, in spite of a decrease in the importance of third body reactions, Thompson and co-workers observed the formation of organic molecules with high degrees of multiple bonding. Moreover, the nature and relative yields of synthesis of organics were very similar to those already detected in Titan.

In fact, until recently, only one organic compound has been detected in Titan's atmosphere but not in any related simulation experiment: dicyanoacetylene (butynedinitrile), C_4N_2. This was probably due to the very low stability (at room temperature) of this acetylenic dinitrile, and to the difficulty of its chemical analysis by GC–MS techniques (Aflalaye et al. 1995). A new series of simulation experiments is currently in development using low-temperature reactors (more appropriate to simulate Titan's conditions) and analytical procedures compatible with the quantitative analysis of compounds thermally unstable at room temperature, but stable at the low temperatures of Titan (Coll et al. 1995, 1998; de Vanssay et al. 1995). The preliminary results show the formation of polyynes (C_6H_2) and cyanopolyyne (HC_5N), in addition to the similar products (C_2H_2, C_4H_2, HC_3N) already reported from previous experiments. These two compounds, C_6H_2 and HC_5N, have not been detected (yet) in Titan's atmosphere (in particular, not from the available Titan infrared spectra), but their upper limit of abundance in Titan, which can be calculated from this nondetection (a fraction of ppb) is fully compatible with their relative abundance in the simulation experiment. Very recent results obtained in the frame of this program, using a flow reactor at low temperature, also show, at last, the formation of C_4N_2 (Coll 1997; Coll et al. in preparation).

The relatively good agreement between the detection of minor organic constituents in Titan and the data from simulation experiments (Table 17.3) strongly suggests that one can extrapolate the results of these experiments to the case of Titan, at least to get information on the nature and relative abundances of organics that may be present in Titan's atmosphere. Such extrapolation indicates the possible presence of many organics (additional hydrocarbons and nitriles) at the ppb or sub-ppb level, with a concentration that systematically decreases, for a given chemical family, when the number of C + N atoms increases in the molecule (Sagan and Thompson 1984).

The chemistry of Titan's atmosphere has also been studied by several authors through kinetic modeling. These models give important information on the main chemical processes involved in the atmosphere and on the vertical distribution of the atmospheric constituents, although they do not fully

Table 17.3. *Organic compounds already detected in Titan's atmosphere compared to previsions from experimental and theoretical modelings*

	TITAN	PREVISIONS	
PRODUCTS	**Equator or (n) North Pole (ppb)**	**Simulation Experiments**	**Photochemical Modeling**
C_2H_6	13 000	×	×
C_2H_4	150	×	×
C_2H_2	3 000	×	×
C_3H_8	500	×	×
C_3H_6		×	×
CH_3C_2H	37 n	×	×
CH_2CCH_2		×	×
C_4H_{10}		×	×
C_4H_2	27 n	×	×
C_6H_6		×	×
C_6H_2, C_8H_2		×	×
Other hydrocarbons		×	
HCN	1 500 n	×	×
C_2N_2	22 n	×	×
CH_3CN	Detected	×	×
C_2H_3CN		×	×
CH_2CHCN		×	×
CHCCN	70 n	×	×
C_3H_3CN		×	
CH_3C_2CN		×	
C_4N_2	solid n		×
Other nitriles		×	
CH_3N_3		×	
CH_2N_2			×
HC_5N		×	

explain yet the cause of the differences between the polar and equatorial abundances observed for several species. The detailed model developed by Yung, Allen and Pinto (1984) and updated by Toublanc et al. (1995) indicates two main zones of synthesis. HCN, C_2H_2, CO, and CO_2 are formed mainly at very high altitudes. HCN would involve the dissociation of N_2 by energetic electrons (from Saturn's magnetosphere) and far-UV solar photons. In the same atmospheric zone, the dissociation of CH_4 would produce C_2H_2, C_2H_4, and C_3H_4. These products are transported by eddy diffusion down to the stratosphere, where only less energetic UV photons are available. These products, being much easier to photolyze than CH_4 and N_2, can be photodissociated in the stratosphere. The resulting photolysis of C_2H_2 catalyzes the dissociation of CH_4 in CH_3:

$$\begin{array}{lll} & C_2H_2 + \text{photon} & \longrightarrow C_2H + H \\ & C_2H + CH_4 & \longrightarrow CH_3 + C_2H_2 \\ \hline \text{Net}: & CH_4 + \text{photon} & \xrightarrow{-C_2H_2} CH_3 + H. \end{array}$$

The CH_3 radical can yield C_2H_6 and many other hydrocarbon species. Similarly, HCN is easily photolyzed in the stratosphere :

$$HCN + \text{photon} \longrightarrow CN + H.$$

The resulting CN radical can yield higher nitriles by reaction with hydrocarbons.

These photochemical processes would also produce many species not yet detected in Titan, but detected in simulation experiments, such as other hydrocarbons, allene, butanes, polyynes (including C_6H_2 and C_8H_2), and additional nitriles, such as acrylonitrile and cyanopolyynes (including HC_5N). The presence of other compounds of low stability is also suggested by theoretical modeling, such as the very unstable and reactive diazomethane (CH_2N_2) and azidomethane (CH_3N_3).

The photochemical models also provide explanations for the presence of CO and CO_2 in Titan's atmosphere. They involve a meteoritic flux of water ice producing a very low mole fraction of gaseous water in the atmosphere. The photodissociation of water produces OH. This radical can react with CH_3, producing CO:

$$CH_3 + OH \longrightarrow CO + CH_4,$$

and with CO, producing CO_2:

$$CO + OH \longrightarrow CO_2 + H.$$

But the radical OH can also induce the formation of O-organics, in particular CH_3OH and HCHO. Even if these two products are not expected to be present at a mole ratio higher than a fraction of ppb, their possible presence greatly increases the variety of organic compounds and of organic processes that are involved in Titan's atmosphere. This is particularly the case with formaldehyde, the chemical reactivity and the prebiotic importance of which must be pointed out.

Thus, the organic chemistry of Titan's atmosphere involves several of the organic compounds considered to be key molecules in terrestrial prebiotic chemistry. Consequently, Titan's chemistry is of major interest to test our tools (experimental as well as theoretical). But even more interesting, such a chemistry is not restricted to the gas phase.

4. . . . and in the aerosols

Simulation experiments produce not only volatile organics, but also solid particles. Those, often called tholins, can be considered as good laboratory analogues of the submicron particles contributing to Titan's haze, which is observed in the 300–500-km altitude range of the atmosphere (Sagan and Thompson 1984). Until recently, the experimental procedure followed to prepare and to recover these analogues did not prevent their contamination by the air in the laboratory. Consequently, the resulting solid products, because of their high reactivity with several of the (terrestrial) atmosphere constituents (mainly O_2 and H_2O), included a noticeable fraction of O atoms (up to about 10%). In a new series of experiments (Coll et al. 1995, 1998; Coll 1997), a specific protocol has been developed to recover these reactive solid materials under an inert atmosphere (N_2) in a glove box. The preliminary results obtained on the elemental composition of the now uncontaminated material show that it is made of H, C, and N atoms, with a gross formula of about $H_{11}C_{11}N$. The most recent data using low-pressure cold plasma and low-temperature cold plasma give a C/N ratio of 2.8 (Coll et al. 1997). Such a composition clearly indicates that N molecules must be involved in the building up of those solid products, although at a relatively low level. Because of the nature and chemical reactivity of many unsaturated species detected in the gas phase during the simulation experiment, such as polyynes $C_{2n}H_2$ and cyanopolyynes $C_{2n}HCN$, it seems likely that theses compounds must be directly involved in the formation of the tholins.

At this point, it is very interesting to note that contamination problems, which have been raised during these studies, can also concern experiments simulating the primitive Earth's environment. Thus, the laboratory studies that have prompted the development of techniques and procedures designed to study Titan's atmosphere can be applied to other environments, including the primitive Earth.

Because of the atmospheric temperature gradient, all the organic compounds

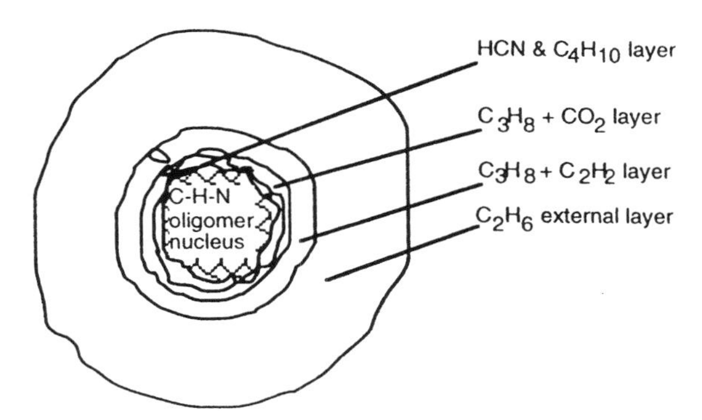

Figure 17.2. Chemical composition and structure of aerosols in Titan's low stratosphere, from microphysical modeling. The condensation phenomena give rise to an onion structure, with layers of increasingly volatile compounds from the center to the outside of the particle.

(with the exception of CH_4) must condense in the low stratosphere of Titan. Such condensation is probably induced by the submicron particles of the high atmospheric hazes. After aggregation and sedimentation down to lower altitudes, those particles may act as condensation nuclei in the stratosphere. After successive icing with increasingly volatile compounds, the particle would increase in size from a few tenths of a micron to several microns at the tropopause. There, its structure would consist of a nucleus made of tholins, covered by roughly concentric layers of low-molecular-weight organics, mainly HCN, C_3H_8, C_2H_2 and an external layer of C_2H_6 (Figure 17.2). Below the tropopause, the condensation of methane, which may also form clouds in the troposphere, would increase the mean diameter of the particle up to several hundred microns.

The entire process results in the irreversible transportation of the organic compounds, including tholins and volatiles, from the atmosphere where they have been formed, to the surface.

5. Oceans on Titan?

If we take into account the mean rate of methane photolysis, all gaseous methane currently present in the atmosphere of Titan should be depleted within about 50 million years. This is quite a short time compared to the age of the satellite (4.5 billion years, the same as for the rest of the Solar System). The probability that we are observing Titan at a very particular moment of its history, when methane would be (almost) suddenly degassing, is very low. Thus it seems very likely that there is a methane reservoir on Titan.

The temperature and partial pressures at Titan's surface are not compatible with a reservoir of pure liquid methane (plus dissolved nitrogen). However, they are fully compatible with a mixture of liquid methane plus ethane (Lunine, Stevenson and Yung 1983). Moreover, they are also fully compatible with the evolution of a reservoir of liquid hydrocarbon at Titan's surface, initially made only of methane. The photolysis of methane in the high atmosphere decreases the methane content of the liquid reservoir but produces C_2H_6 and other organics, which are irreversibly transported down to the surface, through the aerosols, continuously increasing the ethane fraction in the reservoir. The presence of methane in liquid ethane decreases the atmospheric methane abundance, providing conditions that are consistent with the observational data.

Thus, there could be oceans on Titan's surface. Voyager did not provide any direct, clear evidence on the nature of Titan's surface. Several attempts were made later to sound this surface, in particular by sending a radar signal from the Earth to Titan and by looking at the nature of the echo signal received on Earth after reflection from Titan's surface (Muhleman et al. 1990, 1992). The resulting data totally exclude the possibility of a global ocean. Recent observations of Titan by Smith et al. (1996) using the Hubble Space Telescope indicate the presence of several different structures on Titan's surface that also exclude the presence of a global ocean. However, those data do not exclude the presence of oceans partially covering Titan, frothy oceans, or ones containing a significant amount of aerosol constituents dispersed on the surface (Lunine 1993). A subsurface ocean, covered by a partially porous surface, is also possible (Lunine 1993).

Knowing the surface temperature and the adjacent atmospheric composition, one can calculate the bulk composition of these oceans. Such thermodynamical modeling, assuming quasi equilibrium between the near-surface atmosphere and the ocean, has been carried out to predict the behavior of the aerosols once they reach such a surface and the consequences for the chemical composition of the hypothetical ocean (Raulin 1987; Dubouloz et al. 1989). The resulting models indicate that, due to the uncertainties of surface temperature and atmospheric methane abundance near the surface, the composition of the ocean could possibly be dominated by methane, or by ethane, including a noticeable fraction (up to 10% for the warm, methane-rich ocean) of dissolved nitrogen.

When the aerosol reaches the ocean surface, part or all of it, depending on its solubility, will dissolve, the rest will float or sink, depending on its density relative to the ocean. Knowing the flux of ethane and other organics down to the surface, and the solubility of these compounds in a CH_4–N_2–C_2H_6–Ar liquid mixture, it is possible to calculate their concentration in the oceans. Such a modeling has been systematically developed using thermodynamical calculations for estimating the solubility and density of most of the potential solutes in the hypothetical oceans of Titan. The results of such modeling are the following (Dubouloz et al. 1989):

1. Almost all solutes are denser than the oceans, whatever the main component (CH_4/C_2H_6) of the ocean is.
2. Several solutes, including nitriles, in spite of their high polarity are noticeably soluble in the liquid methane–ethane mixture of the oceans.
3. In contrast, the macromolecular constituents of the aerosols are very unsoluble and should not be present as solutes in the liquid phase.
4. The surface of Titan's oceans should then be free of any iceberg of solid organics.
5. However, the bottom of these oceans should be covered by a deep layer of ices, made mainly of acetylene and hydrogen cyanide, and their oligomers.

Thus, Titan's oceans appear as a liquid medium very rich in complex organics. Such a cold environment seems very unfavorable to any chemical evolution, since reaction rates at 90–100 K are generally very low. However, Titan's surface is continuously bombarded by very high energy cosmic rays. These particles, although not abundant, may induce new reactions in the ocean involving not only the main constituents, but also the dissolved minor species, in particular CO, and eventually NH_3, and allowing the formation of many other organics, from the simple and very reactive diazomethane, to additional $HCN–C_2H_2$ heterooligomers. Thus, Titan's oceans are not a static system. They continuously evolve, but their evolution should be very different from that of the terrestrial oceans, since this is an apolar system evolving at a very slow rate and at a very low temperature, in the absence of any liquid water. From this point, chemical evolution on Titan may have followed a very different path compared to that on the Earth, and hence chemical evolution on Titan may differ markedly from that on the Earth.

6. Chemical and physical couplings in Titan's geofluid

In spite of the strong differences between Titan and the (early and present) Earth, which are essentially due to the huge temperature differences and the resulting absence of liquid water on Titan, there are many similarities between both planetary objects. Thus, studies of Titan should give access to important new ideas and crucial information for our understanding of the prebiotic early Earth.

Titan provides a real-scale example of a planetary body having a size of the same order of magnitude as that of the Earth; a dense atmosphere presenting a terrestrial-like, vertical thermal structure; the likely presence of liquid bodies on its surface; and organics, including organic molecules of prebiotic interest, widely distributed in the different parts (gas, aerosols, solid and liquid surface) of its environment.

There must be strong couplings between the high atmosphere–low pressure and the lower atmosphere–higher pressure processes. Eddy diffusion should play an important role in these couplings, as should the atmospheric aerosols. There must be strong couplings between the atmospheric processes occurring in the high atmosphere and the chemical evolution of Titan's surface. The organic compounds formed by UV and magnetospheric electrons in Titan's mesosphere and thermosphere – simple organics or heterooligomers – diffuse down to the stratosphere, act as condensation nuclei, and induce the condensation of organic compounds of smaller molecular weight. The resulting particles precipitate down to the troposphere, with dimensions increasing as altitude decreases. Their layered structure is formed of compounds increasingly volatile from its core to its outside. Haze particles in the stratosphere become hail in the high troposphere and rain in the low troposphere. These particles irreversibly carry most of the organics of very low vapor pressure from the high atmosphere down to the surface. Such processes remove the organic materials from the atmospheric zones where they were synthesized and carry them to less chemically active zones, where they are protected from UV and electron impact. There, they can accumulate and be available for slower-rate chemical processes.

Because of such couplings, Titan's organic chemistry must be considered as a whole (Figure 17.3). Depending on planetary environment, chemical evolution may evolve differently. On the primitive Earth, in the presence of liquid water, prebiotic organic chemistry starting from simple reactive molecules allowed the emergence of life. On Titan, with very low temperature conditions, chemical evolution is still going on, but in the absence of liquid water. Studies of the prebiotic organic chemistry on this planetary body may provide information on the atmospheric processes that occurred on the primitive Earth. But such studies could also indirectly furnish important information on the role of liquid water in chemical evolution, since prebiotic chemistry on Titan has evolved in the absence of this universal solvent, which seems to be a requirement for the emergence of life. However, Titan's oceans may have evolved in the presence of noticeable traces of dissolved ammonia, which played the role of water. If this were the case, there may be a pseudobiochemistry still going on in Titan's ocean, where ammonia substitutes for water.

Is Titan's chemistry even more complex than we expect? What is the cause of the differences between the polar and the near-equatorial chemistry? Are

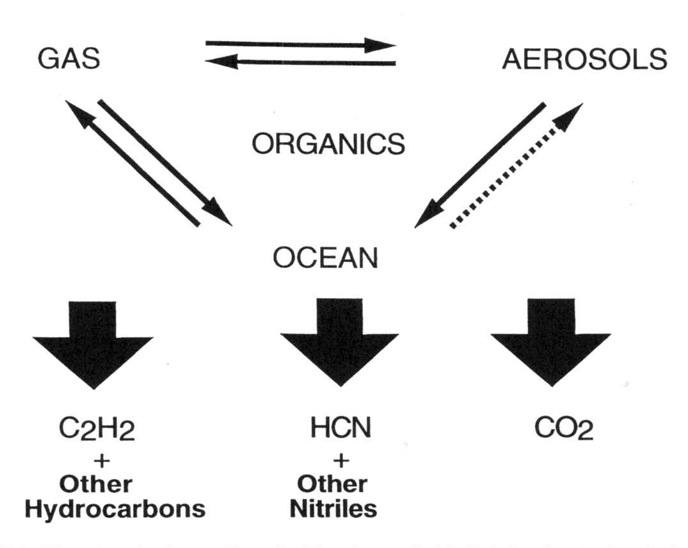

Figure 17.3. The chemical couplings in Titan's geofluid. Subtle physical and chemical exchanges between the gas, the aerosols, and the (liquid?) surface provide a complex chemistry in Titan's environment, including sources, transport, and sinks of complex organic compounds.

additional chemical processes involved at Titan's surface, induced by galactic cosmic rays or by internal heating? Are O atoms involved in the surface organic chemistry, because of dissolved CO? Are there purine and pyrimidine bases present on Titan? Are amino acids or analogues present? Are pseudopolypeptides included in Titan's organic oligomers? To check such hypotheses and to try to answer the many associated questions, it is necessary to study in detail. Titan's prebiotic geofluid. This is precisely one of the main objectives of the Cassini–Huygens mission to Saturn and Titan.

7. The Cassini–Huygens mission

In October 1997, a Titan IV rocket launched a NASA spacecraft named Cassini, which carries several scientific instruments, and a European atmospheric probe, named Huygens. The spacecraft, after flybying Venus twice, then the Earth, to gain energy from the gravity assistance of these planets, will flyby Jupiter in 2000 and reach the Saturn system in 2004. The spacecraft trajectory will then be modified, allowing Cassini to become an artificial satel-

Figure 17.4. The Cassini–Huygens mission. In 2004, after a 7-year space journey, the Cassini orbiter will become an artificial satellite around Saturn and Titan. Simultaneously, the Huygens probe, which will have been released from the spacecraft a few months earlier, will penetrate Titan's atmosphere. After a forceful braking by Titan's high atmosphere, the Huygens probe velocity will be reduced by a factor of 13, to only 1,400 km/h at 270 km altitude. Then the probe will slowly descend through the atmosphere, allowing the six scientific instruments to perform in situ measurements of the physical and chemical characteristics of Titan's stratosphere, troposphere, and surface.

lite orbiting around Saturn and Titan. Meanwhile, the probe will be released and will follow the spacecraft for two weeks before entering Titan's atmosphere. There, the front shield of the Huygens probe will absorb the main heat flow resulting from its impact in Titan's atmosphere and will protect the six experiments located inside the probe (Lebreton 1997). A pilot parachute, then a main parachute will be released, allowing the probe to slowly descend through the atmosphere, down to the surface (Figure 17.4). During the 120–150 minutes of descent, the six instruments (Table 17.4) will carry out physical and chemical measurements of the atmosphere and, after landing, if they survive, of the surface.

Among the instruments of the scientific payload, the most important for studying Titan's organic chemistry are the GC–MS and the ACP. GC–MS is a combined Gas Chromatograph–Mass Spectrometer; it is, in fact, the second instrument of its kind ever used in a space mission of planetary exploration. The GC subsystem allows the chromatographic separation of most of the molecular constituents from gaseous samples of Titan's atmosphere; then the MS subsystem detects and identifies those constituents. The whole instrument can provide vertical concentration profiles of the main atmospheric constituents and of minor species, which have already been detected, as well as of new organics, in particular those involved in the general evolution of Titan's geofluid. The ACP (Aerosol-Collector Pyrolyzer) instrument collects the aerosol particles on a filter during the descent of the probe, then the filter is automatically retracted into an oven. The oven is isolated and heated at different step-temperatures, up to a pyrolyzing temperature of 650 °C. For each temperature, the resulting gases are transferred to the GC–MS instrument for chemical analysis. With these operations, ACP can provide data on the chemical composition and relative abundances of the condensed volatiles and organic core constituting the aerosols.

The rest of the payload is mainly devoted to physical and physical–chemical measurements. HASI, the Huygens Atmospheric Structure Instrument, can determine the temperature and pressure vertical profiles essential for modeling the gas and aerosol phases. DISR, the Descent Imager Spectral Radiometer, gives information on the cloud structure and surface physical state, and also on the photon fluxes versus altitude, crucial data to constrain photochemical modeling of the atmosphere. DWE, the Doppler Wind Experiment, measures the zonal wind and atmospheric turbulences. SSP, the Surface Science Package, provides direct data on the surface physical state and on its chemical composition, essential to model surface organic chemistry.

The nominal duration of the mission will be 4 years, allowing more than 30 orbits of the Cassini satellite around Saturn and Titan. Several of the 12

Table 17.4. *The scientific payload of the Huygens probe*

Acronym	Instrument	Main goal	Principal investigator
GC–MS	Gas Chromatograph/Mass Spectrometer	Atmosphere (gas & aerosols) chemical composition, and (gas) vertical profile	H. Niemann, U.S.A.
ACP	Aerosol Collector & Pyrolyzer	Aerosol chemical composition profile	G. Israël, France
HASI	Huygens Atmosphere Structure Instrument	Temperature/pressure profiles, winds and turbulence, lightning	M. Fulchignoni, Italy
DI/SR	Descent Imager/Spectral Radiometer	Atmosphere composition, cloud structure, radiative measurements & surface imaging	M. Tomasko, U.S.A.
DWE	Doppler Wind Experiment	Zonal wind profile	M. Bird, Germany
SSP	Surface Science Package	Surface state and composition	J. Zarnecki, U.K.

instruments on board the Cassini orbiter will also provide information of crucial importance for understanding Titan's chemistry, and strong interactions are expected between the data from the probe and those from the orbiter. For instance, the infrared (CIRS instrument) and ultraviolet (UVIS instrument) spectrometers have capabilities that complement the experiments on board the probe related to chemical analysis. The UVIS spectrometer explores a high atmospheric region, permitting the study of the far-UV and high-energy-electron processes. On the other hand, CIRS appears to be a unique and essential instrument for chemical analysis of Titan's atmosphere and a global mapping of the atmospheric composition of the planet. By coupling the atmospheric vertical profiles with the latitudinal variations, given in particular by CIRS, it should be possible to develop a two-dimensional model of Titan's organic chemistry, including atmospheric circulation and variation of sunlight with latitude.

The Cassini–Huygens mission will provide a fantastic campaign to study the origin, nature, and possible evolution of the organic compounds, as well as their distribution in the different parts of Titan's geofluid. These parts include the gas phase, liquid phase (ocean), solid phase (sedimentary deposits), and condensed atmospheric phases (aerosols). By coupling the observational data provided by the Cassini–Huygens mission with experimental (laboratory simulation) and theoretical (photochemical, microphysical, and thermodynamical modeling) data, it will be possible to study and model organic chemistry in the different parts of Titan's geofluid and to get a better understanding of the importance of the interactions among these parts. This should open new avenues in the studies of the origin of life on the Earth and, more generally, in exobiology. As a matter of fact, the Cassini–Huygens mission, through its development, has already induced the opening of such new approaches.

References

Aflalaye, A., Andrieux, D., Bénilan, Y., Bruston, P., Coll, P., Coscia, D., Gazeau, M.-C., Khlifi, M., Paillous, P., Sternberg, R., de Vanssay, E., Guillemin, J.-C., and Raulin, F. 1995. Thermally instable polyynes and N-organics of planetological interest: new laboratory data and implications for their detection by in situ and remote sensing techniques. *Adv. Space Res.* 15 (10):5–11.

Bossard, A., Mourey, D., and Raulin, F. 1983. The escape of molecular hydrogen and the synthesis of organic nitriles in planetary atmospheres. *Adv. Space Res.* 3 (9):39–43.

Bossard, A., Raulin, F., Mourey, D., and Toupance, G. 1982. Organic synthesis from reducing models of the atmosphere of the primitive earth with UV light and electric discharges. *J. Mol. Evol.* 18:173–178.

Bruston, P., Khlifi, M., Benilan, Y., and Raulin, F. 1994. Laboratory studies of organic chemistry in planetary atmospheres: from simulation experiments to

spectroscopic determinations. *J. Geophys. Res.* 99 (E9):19047–19061.
Clarke, D. W., and Ferris, J. P. 1997. Chemical evolution on Titan: comparisons to the prebiotic Earth. *Origins Life Evol. Biosphere* 27: 225–248.
Coll, P. 1997. Modélisation expérimentale de l'atmosphère de Titan: production et caractérisations physico-chimiques d'analogues des aérosols et de la phase gazeuse enfin représentatifs. Thèse de Doctorat de l'Université Paris 12.
Coll, P., Coscia, D., Gazeau, M. C., de Vanssay, E., Guillemin, J. C., and Raulin, F. 1995. Organic chemistry in Titan's atmosphere: new data from laboratory simulations at low temperature. *Adv. Space Res.* 16 (2):93–104.
Coll, P., Coscia, D., Gazeau, M. C., and Raulin, F. 1998. Review and latest results of laboratory investigations of Titan's aerosols. *Origins Life Evol. Biosphere* 28:195–213.
Comas-Sola, J. 1908. Observations des satellites principaux de Jupiter et de Titan. *Astron. Nachr.* 179 (4290):289–290.
Courtin, R., Gautier, D., and McKay, C. P. 1995. Titan's thermal emission spectrum: reanalysis of the Voyager infrared measurements. *Icarus* 114:144–162.
Dubouloz, N., Raulin, F., Lellouch, E., and Gautier, D. 1989. Titan's hypothesized ocean properties: the influence of surface temperature and atmospheric composition uncertainties. *Icarus* 82:81–96.
Hunten, D. M. 1978. A Titan atmopshere with a surface temperature of 200 K. In *The Saturn System*, ed. D. M. Hunten and D. Morrison, pp. 671–759. *NASA Conf. Publ.* CP–2068.
Kuiper, G. P. 1944. A satellite with an atmosphere. *Astrophys. J.* 100:378–383.
Lebreton J. P. (ed.). 1997. Huygens: Science Payload and Mission. ESA-SP1177, ESA Pub. Noordwijk, The Netherlands.
Lindal, G. F., Wood, G. E., Hotz, H. B., Sweetnam, D. N., Eshleman, V. R., and Tyler, G. L. 1983. An analysis of the Voyager radio occultation measurements. *Icarus* 53:348–363.
Lunine, J. I. 1993. Does Titan have an ocean? A review of current understanding of Titan's surface. *Rev. Geophys.* 31:133–149.
Lunine, J. I., Stevenson, D. J., and Yung, Y. L. 1983. Ethane ocean on Titan. *Science* 222:1229–1230.
Miller, S. L. 1953. A production of aminoacids under possible primitive earth conditions. *Science* 117:528–529.
Muhleman, D. O., Grossman, A. W., Butler, B. J., and Slade, M. A. 1990. Radar reflectivity of Titan. *Science* 248:975–80.
Muhleman, D. O., Grossman, A. W., Slade, M. A., and Butler, B. J. 1992. The surface of Titan and Titan's rotation: what is radar telling us? *Bull. Am. Astron. Soc.* 24:954–55.
Owen, T., Gautier, D., Raulin, F., and Scattergood, T. 1992. Titan. In *Exobiology in Solar System Exploration*, ed. G. Carle, D. Schwartz, and J. Huntington, pp. 127–143. NASA SP 512.
Raulin, F. 1987. Organic chemistry in the oceans of Titan. *Adv. Space Res.* 7(5):71–81.
Raulin, F. 1992. Prebiotic chemistry in planetary environments. *Lecture Notes in Physics* 390:141–148.
Raulin, F., Frere, C., Do, L., Khlifi, M., Paillous, P., and de Vanssay, E. 1992. Organic chemistry on Titan versus terrestrial prebiotic chemistry: exobiological implications. *ESA Special Publications* SP 338:149–160.
Raulin, F., Mourey, D., and Toupance, G. 1982. Organic syntheses from CH_4–N_2 atmospheres: implications for Titan. *Origins of Life* 12:267–279.
Sagan, C., and Thompson, W. R. 1984. Production and condensation of organic gases in the atmosphere of Titan. *Icarus* 59:133–161.

Smith, P. H., Lemmon, M. T., Lorenz, R. D., Sromovsky, L. A., Caldwell, J. J., and Allison, M. D. 1996. Titan's surface revealed by HST imaging. *Icarus* 119:336–349.

Thompson, W., Todd, H., Schwartz, J., Khare, B., and Sagan, C. 1991. Plasma discharge in N_2 + CH_4 at low pressures: experimental results and applications to Titan. *Icarus* 90:57–73.

Toublanc, D., Parisot, J. P., Brillet, J., Gautier, D., Raulin, F., and McKay, C. P. 1995. Photochemical modeling of Titan's atmosphere. *Icarus* 13:2–26.

Toupance, G., Raulin, F., and Buvet, R. 1975. Formation of prebiological compounds in models of the primitive Earth's atmosphere. I. CH_4–NH_3 and CH_4–N_2 atmospheres. *Origins of Life* 6:83–90.

de Vanssay, E., Gazeau, M. C., Guillemin, J. C., and Raulin, F. 1995. Experimental simulation of Titan's organic chemistry at low temperature. *Planet. Space Sci.* 43:25–31.

Yung, Y. L., Allen, M., and Pinto, J. P. 1984. Photochemistry of the atmosphere of Titan: comparison between model and observations. *Astrophys. J. Suppl. Ser.* 55:465–506.

18

Life on Mars

CHRISTOPHER P. McKAY
Space Science Division
NASA Ames Research Center
Moffett Field, California

1. Introduction

Biochemically, life on Earth is essentially a single phenomenon. All life forms on this planet share the same basic biochemical and genetic pattern and appear to have descended from a common ancestor. We don't know where, when, or how this form of life first originated, nor do we know what features of its biochemistry are general and would be found in any type of life and what features are the result of the particular history of Earth life. One way that may provide a better understanding and generalization of life would be the discovery of a second type of life. Three research programs are currently directed toward discovering a second type of life: (1) laboratory synthesis of life, (2) the search for extraterrestrial intelligence, and (3) the search for life on other planets. This chapter deals with the search for life on Mars, since this is the planet most likely to have, or have had, life.

Table 18.1, compiled from Kieffer et al. (1992), compares the surface environments on Earth and Mars. The atmosphere of Mars contains the basic elements necessary to support life – including H_2O. However, it lacks sufficient atmospheric pressure and temperature for that water to exist as a liquid. The absence of water in the liquid form at any location or season on Mars is the primary reason that the surface is believed to be lifeless at the present time. However, there is direct evidence that early in its history Mars did have stable liquid water on its surface. The main epoch of liquid water on Mars corresponds to the time period of the oldest fossil of life on Earth, 3.8 to 3.5 Gyr ago. Water is the quintessential ecological requirement for life, and its presence on early Mars has led to considerable speculation that there could have been life on Mars during this early, possibly Earth-like, period.

If life did arise on Mars, it could represent an independent origin from life on Earth – a second genesis. Alternatively, life on Mars could be phylogenetically related to life on Earth due to the exchange of biological materials via

Table 18.1. *Mars and Earth*

Property	Mars	Earth
Surface gravity	0.38 g	1 g
Surface pressure	6 hPa	1,013.25 hPa
Average surface temperature	- 60 °C	+15 °C
Atmosphere	CO_2 95% N_2 2.7% Ar 1.6%	N_2 78% O_2 21% Ar 1%
Biogenic elements	C, H, N, O in atmosphere S in soil P likely in soil	All present
Liquid water	~ 3.5 Gyr ago; None today	Continuously

meteors during the heavy late bombardment, 3.8 Gyr ago. Definitive evidence for life on Mars must await fossils and organic remains that may be found in sediments on the Martian surface. Currently, the only samples we have from Mars are meteorites, and they have provided intriguing, but inconclusive, evidence regarding the question of life. In the near future, more carefully selected samples may be collected on sample return missions and could provide a level of evidence for life comparable with the fossil record of life on the early Earth.

2. Viking results

From the beginning of the Space Age, it was clear that Mars was the planet most likely to harbor life. For this reason the first in-depth biology missions to another planet were the Viking missions to Mars in 1976. Their primary objective was the search for life in the soil of Mars.

The two Viking landers carried several experiments that were key to the study of life on Mars; there were three biology incubation experiments, an X-ray instrument for elemental analysis, and a combination gas chromatograph/mass spectrometer (GC–MS) for organic analysis.

The three Viking biology experiments are diagramed in Figure 18.1. The pyrolytic release experiment (PR) was based on the assumption that Martian life would have the capability to photosynthesize, that is, to incorporate

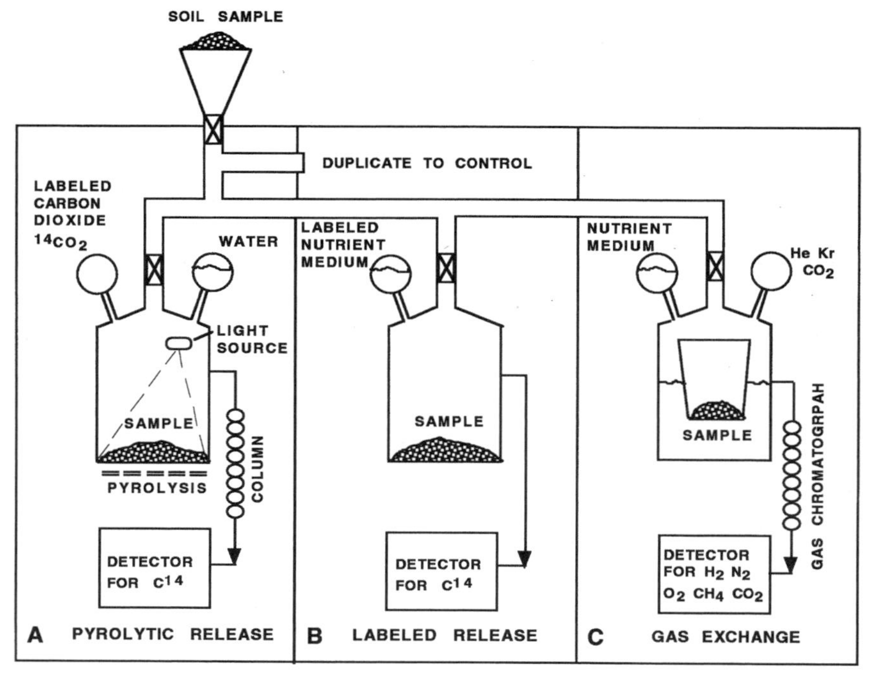

Figure 18.1. Diagram of the three Viking lander biology experiments.

radioactively labeled carbon dioxide in the presence of sunlight (Horowitz and Hobby 1977). The labeled release experiment (LR) sought to detect life by the release of radioactively labeled carbon initially incorporated into organic compounds in a nutrient solution (Levin and Straat 1977). The gas exchange experiment (GEx) was designed to determine if Martian life could metabolize and exchange gaseous products in the presence of water vapor and in a nutrient solution (Oyama and Berdahl 1977).

All three Viking biology instruments returned signs of activity when Martian soil was added. In the GEx, the soil released O_2 upon humidification in amounts ranging from 70 to 770 nmoles cm^{-3} (Oyama and Berdahl 1977; 1979). Heating the sample to 145 °C for 3.5 hours reduced the amount of O_2 released by about 50%. There was a slow evolution of CO_2 when nutrient solution was added to the soil. The LR experiment indicated the rapid release of CO_2, followed by a prolonged slow release of CO_2, from radioactively labeled C in a nutrient solution. The effect was completely removed by heating to 160 °C for 3 hours, partially destroyed at 40–60 °C, relatively stable for short periods at 18 °C, but lost after long-term storage at 18 °C.

The most surprising results of the Viking soil analyses were those of the

GC–MS. This instrument was unable to detect organics at ppb levels in surface samples or in samples from below the surface (maximum depth sampled was about 10 cm) (Biemann et al. 1977; Biemann 1979). Since the infall of meteorites and interplanetary dust should be carrying organics to Mars at a rate of over 10^5 kg yr^{-1} (Flynn 1996), the absence of organics suggests that they are being actively destroyed. The destruction rate need be about only 7×10^{-11} g cm^{-2} yr^{-1} to balance the infall rate and could be due solely to destruction by solar UV (Oró and Holzer 1979).

The results of the labeled release experiment were particularly controversial because they met all the preflight criteria for a biological reaction. If considered alone, the LR results would be indicative of life on Mars. However, when LR and GEx biology experiments are considered together with the lack of organics as determined by the GC–MS, the most likely explanation for the observed reactivity is chemical, not biological, activity (for a review see Klein 1978, 1979; Horowitz 1986; for an opposite view, see Levin and Straat 1981). The likely cause of this chemical activity is the presence in the Martian soil of one or more photochemically produced oxidants such as H_2O_2 (for a recent review, see Zent and McKay 1994).

The Viking landers also carried an X-ray fluorescence experiment that analyzed the elemental composition of the soil at the Viking lander sites but could detect only elements with an atomic number greater than Mg (Clark et al. 1982). There was no direct measurement of the important biogenic elements C, H, N, or O in the soil material – although all are present in the Martian atmosphere. Phosphorus, although not directly measured because its signal is masked by S and Si, is nevertheless thought to be present (Toulmin et al. 1976), especially since it is found at concentrations of about 0.3% by weight in meteorites thought to have come from Mars (Stoker et al. 1993). Thus, all the biogenic elements (C, H, N, O, P, and S) are present on the surface of Mars. Most of the trace elements used by Earth life, such as Fe, Mg, and Al, have been directly detected as well, either by the Viking landers or in Martian meteorites.

From a biological perspective, the most interesting results of the Viking missions to Mars (and the Mariner 9 mission before that) were not from the search for life on the landers but from the orbital images. These images showed clear evidence of valley features that are generally accepted to have been cut by water (possibly under a relatively thin layer of ice). Other possible fluids – wind, ice, low-viscosity lavas, liquid CO_2, SO_2 and hydrocarbons – have been suggested, but are not considered as likely as water, to explain the morphology and distribution of the fluvial features.

Some of the fluvial features on Mars, those termed valley networks (Carr

1981; 1996), exhibit a morphology that implies the stable and extended flow of liquid water. Figure 18.2 shows one such feature. The evidence of the former existence of liquid water on the surface of Mars is the basis for the current interest in the search for ancient life on Mars.

The placement of the network channels on the heavily cratered southern terrain of Mars indicates that they formed contemporaneously with the end of the late bombardment some 3.8 Gyr ago (Carr 1981, 1996; Carr and Clow 1981). However, there are network channels found in locations on Mars that are considerably younger. For example, Gulick and Baker (1989) have reported channels on the slopes of Alba Patera with an expected age of far less than 3 Gyr. The outflow and flood features also appear to have formed over extended periods of time. Masursky et al. (1977) presented evidence of interlayering of flood channels and lava flows at Kasei Valles, suggesting that the flood flows were episodic in Martian history.

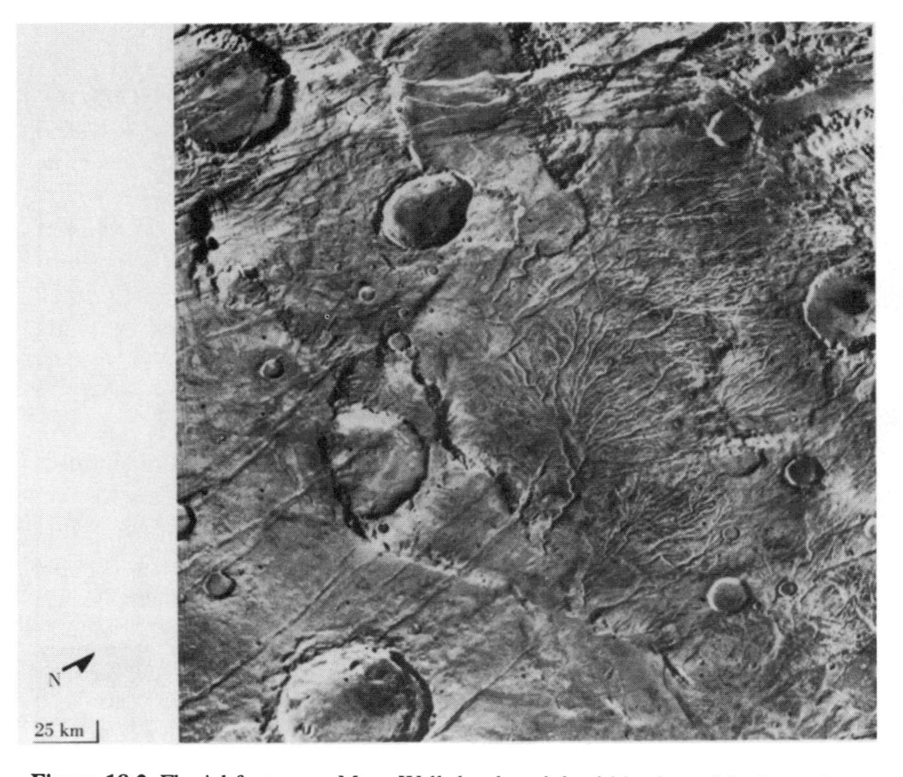

Figure 18.2. Fluvial feature on Mars. Well-developed dendritic channel in the ancient cratered terrain in Warrego Valles (48°S, 98°W; Viking frame 63A09, 250 km across).

3. How to search for life?

A standard piece of hardware in science fiction is a device that can remotely detect life forms. This hypothetical device is capable of sensing some essential property of life that is not possessed by inanimate matter even of similar elemental composition. Unfortunately, current technology does not provide us with such a device and the search for life on other planets is the more difficult for this. To search for life on Mars requires an operational definition of life and its correlates (McKay 1991).

This definition must be based on Earth life because the science of biology is based entirely on the study of life on Earth. There is essentially only one type of life form on Earth descendant from a common ancestor. All life is constructed from the same small set of building blocks: proteins composed of 20 amino acids; information-bearing molecules constructed from 8 nucleotides, in turn made up of 5 nucleotide bases; and polysaccharides made up mostly from a few simple sugars (e.g., Lehninger 1975). For all life forms, Darwinian evolution – the defining process of living systems – operates on a genetic program based on DNA. From *E. coli* to the blue whale, life on Earth is the same stuff and follows the same patterns. Analysis of the RNA of living forms has allowed the construction of universal family trees showing that all organisms are related in an evolutionary context and suggesting a common ancestor for all of Earth's life (Woese 1987). Despite the unity of biochemistry and the genetic relationships between organisms, there are various and peculiar forms – for example, viruses – that have frustrated all attempts to provide a rigorous definition of life. Life on Earth is singular and thus does not easily yield generalizations and, at the same time, life is too broad in phenomenon to allow for a complete characterization. Thus, the task of extending biology to life beyond Earth and of addressing questions such as the origin of life – exobiology (Lederberg 1960) – is met with difficulties.

McKay (1991) has suggested, not a definition of life, but a listing of its requirements. These are shown in Table 18.2. The most significant requirement for life – and the one with the most implications in the planetary context – is the requirement for liquid water (Mazur 1980; Kushner 1981). Water not only acts as a liquid medium for reaction among biomolecules, but also actively influences the shape and behavior of many biomolecules by, for example, hydrogen bonding with hydrophilic groups and repelling hydrophobic groups.

Because of the singular biochemistry of terrestrial biology, we are not in a position to determine which of the biochemical details of Earth's life are fundamental and which are not. On the other hand, suggestions of drastically different life forms have been based upon other liquids (e.g., NH_3) or upon Si in place of C. Although these speculations cannot be proven false, they fail to

Table 18.2. *Ecological requirements for life (McKay 1991)*

Requirement	Occurrence in the Solar System
Energy (usually sunlight)	Common
Carbon	Common as CO_2 and CH_4
Liquid water	Rare
N, P, S, and possibly other elements	Common

connect to the only example of life available and have become relegated to the domain of interesting models that have not made significant contributions to either biology or exobiology.

Table 18.2 also compares the requirements for life with the environments available within our own Solar System; liquid water emerges as the limiting factor. For this reason the search for life in planetary environments is, operationally, a search for liquid water habitats – past and present.

4. An ecological comparison of Earth and Mars

The primary basis for the search for life on Mars is the similarity of the early Earth and early Mars. Figure 18.3 shows the history of Earth and Mars from 4.5 Gyr ago to the present. The two major events in the history of life on Earth were the origin of life before either 3.5 or 3.8 Gyr ago and the development of an O_2-rich atmosphere about 1 Gyr later, which prompted the rise of complex life forms. For Mars, the main period of valley formation is thought to be coincident with the time of the origin of life on Earth, between 3.9 and 3.5 Gyr ago. During this time period, it is probable that Earth and Mars were ecologically similar in that all the potential habitats for the origin and sustenance of life that existed on the Earth would also have existed on Mars. Both planets are thought to have had a thick CO_2 atmosphere, extensive volcanic activity, and liquid water present on the surface.

Before the end of the formation of the planets, 3.8 Gyr ago, the rate of impacts on the surface of both Earth and Mars is thought to have been considerably higher than at present (e.g., see Chyba et al. 1990). It is known that large rocks can be ejected from Mars by such impacts and eventually land intact on the surface of the Earth. Presumably the reverse transport is possible as well, although at a reduced rate due to the asymmetry in mass between the two planets. It has been speculated that microorganisms can be carried from one planet to another by this ejection processes (Melosh 1988). Thus life that was present on Mars, especially before 3.8 Gyr ago, could have been carried to the Earth and, likewise, life from Earth could have been carried to Mars. Thus it is possible that life on the two planets share a common origin, which could have been on either Earth or Mars. To test this possibility it will be necessary to compare the RNA sequencing of Martian life to the RNA phylogeny of life on Earth.

The fluvial features, such as the one shown in Figure 18.2, are direct evidence that early Mars had liquid water on its surface. Presumably, warmer surface temperatures resulting from a thicker CO_2 atmosphere allowed for the stability of liquid water even though the Sun was about 30% dimmer at this early epoch (e.g., Pollack et al. 1987). However, the formation of CO_2 condensate limits the greenhouse efficacy of CO_2 on early Mars, and climate models are currently not able to produce average temperatures above freezing on Mars (Kasting 1991). These models are simplistic, and more detailed modeling may relieve this dilemma. Nonetheless, it is clear that early Mars was quite cold even if liquid water was present on its surface. It is important to note here that, from a biological point of view, the presence of liquid water, *by itself*, is the critical requirement for life. The surface temperature and atmospheric composition are immaterial if liquid water is present. An interesting and relevant example of the utility of liquid water in cold environments is that of the Antarctic dry valleys.

In the dry valleys precipitation is low (less than a few cm of snow) and mean annual temperatures are about -20 °C (Clow et al. 1988). The dry, cold conditions make the valleys virtually lifeless. However, there are two major microbial ecosystems in the dry valleys: one found beneath the ice of the perennially ice-covered lakes (Parker et al. 1982) and the other within the porous subsurface of sandstone rocks (Friedmann 1982).

The ice covers of the Antarctic dry valley lakes are about 3–6 m deep. Beneath the ice cover, liquid water (10–50 m deep) is present and the bottom is rich in microbial life (Parker et al. 1982; Wharton et al. 1982). Summer temperatures above freezing are the key to maintaining liquid water under the relatively thin ice cover. Thermodynamic models can explain the thickness of the ice and the conditions under which ice-covered liquid water would persist (McKay et al. 1985). McKay and Davis (1991) have applied this analysis of the thermal balance of ice-covered lakes in Antarctica to the stability and ice thickness of lakes on early Mars. They concluded that if there were a source of ice to provide meltwater, liquid water habitats could have been maintained under relatively thin ice covers for several hundred million years ($7 \pm 3 \times 10^8$ years) *after* mean global temperatures fell below the freezing point. At that point, the summertime peak temperatures never rose above freezing, and there was no further production of meltwater. Thus, the time scale for liquid water

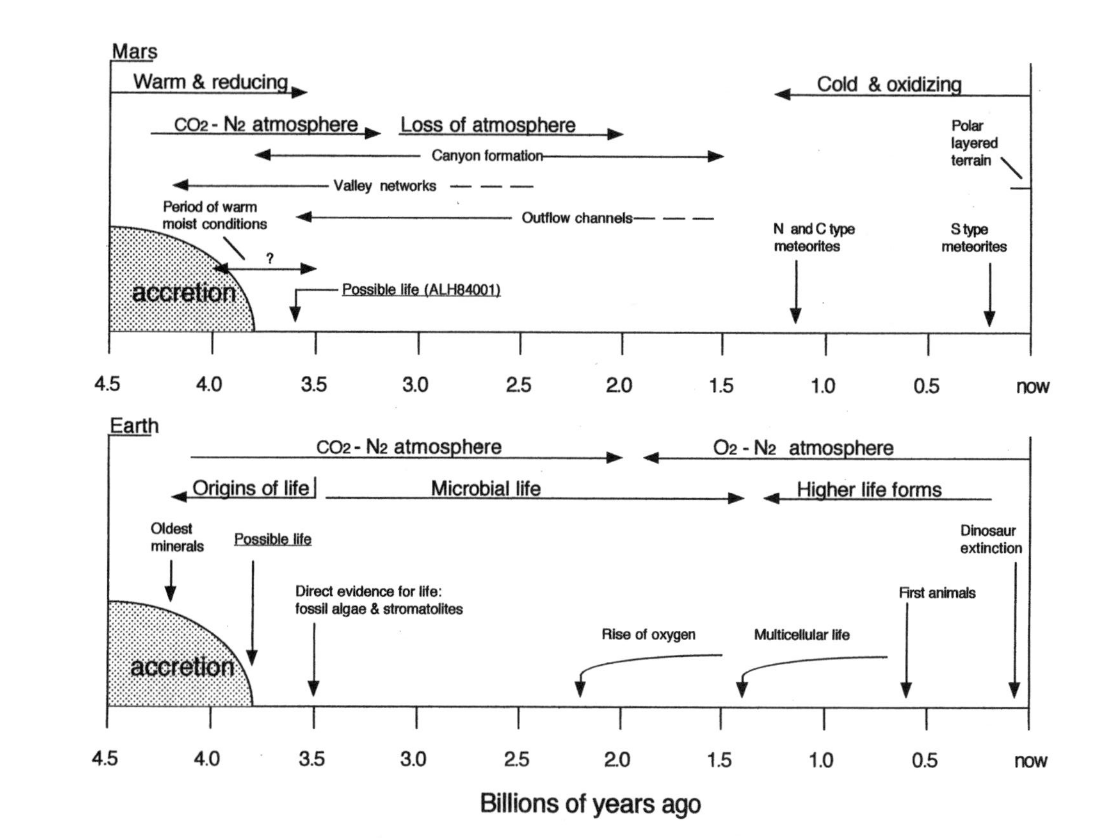

Figure 18.3. History of Earth and Mars. (McKay 1997; with kind permission from Kluwer Academic Publishers.)

on early Mars seems to be comparable to the time scale for the origination of life on Earth. If groundwater or a geothermal heat source were present, the longevity of ice-covered lakes would be extended further.

The second microbial ecosystem in the dry valleys, the cryptoendolithic microbial ecosystem, provides another example of a possible habitat on Mars (Friedmann 1982; Friedmann, McKay and Nienow 1987). The cryptoendolithic microbial ecosystem is found just below the surface of sandstone rocks at elevations typically 1,500 m and more above the dry valley floors (and the lakes). The mean summer air temperatures at this elevation are about 10 °C colder than on the valley floor, and even peak summer air temperatures are virtually always below freezing (Friedmann, McKay and Nienow 1987). As a result, there is no summer melt, and ice-covered lakes cannot persist.

Even with air temperatures below freezing, solar heating on dark surfaces of rocks can result in rock temperatures above 15 °C (Friedmann, McKay and Nienow 1987). The translucence of the sandstone allows sufficient light to penetrate so that photosynthesis is possible to depths of about a centimeter (Nienow, McKay and Friedman 1988). The solar heating of the rock causes the occasional snowfall to melt and then percolate into the rock, where it is retained as capillary water in the pore spaces even at reduced ambient humidity (Friedmann, McKay and Nienow 1987). As a result, the interior of the rock provides a liquid water habitat suitable for life, whereas locations outside the rock, and even on the rock surface, are not favorable sites for life due to the lack of consistent supplies of liquid water. The cryptoendolithic microbial community provides an analogue of a possible ecosystem that may have existed on Mars after the ice-covered lake habitats vanished.

The early, thick CO_2 atmosphere of Mars was not long lived. Reaction of the CO_2 with silicate rock in the presence of water would have formed carbonate minerals, which would have precipitated in the sediments. The time scale for decreasing atmospheric CO_2 from 1 atm to its present value by carbonate formation is estimated to be a few times 10^7 yr (Fanale 1976; Pollack et al. 1987), although more complex models (Schaefer 1993) give longer time scales. However, it is difficult to completely remove the atmosphere because the weathering rates become exceedingly slow at low temperatures. McKay and Davis (1991) found that 0.5 atm remained after weathering slowed to negligible levels.

Thus, in the absence of recycling, the lifetime of a warm early atmosphere would have been very short indeed. On an active planet like the Earth, subduction of ocean sediments at plate boundaries, followed by decomposition of carbonates in the mantle, is the primary mechanism for completing the long-term geochemical CO_2 cycle. Mars does not have sufficient heat flow at present to cause the global-scale recycling of volatiles incorporated into crustal rocks, nor is there any sign that Mars has, or ever had, crustal dynamics similar to plate tectonics – rather, its features are consistent with a one-plate planet (Solomon 1978; Head and Solomon 1981; but compare Sleep 1994). Without these processes, there appears to be no long-term geological mechanism on Mars to recycle CO_2-sink materials back into the atmosphere.

Active volcanism (Pollack et al. 1987) and impact volatilization (Carr 1989) could have recycled carbonates before the decline in the impact rates 3.8 Gyr ago, maintaining a thick CO_2 atmosphere on early Mars. Impact-related recycling of carbonates would tie the existence of the dense CO_2 atmosphere to the impact history and would be consistent with the previously mentioned decline in erosion rate (and atmospheric pressure) after the end of the early bombardment.

No direct detection of carbonates on Mars has been achieved despite ground-based spectroscopic searches (see, e.g., McCord, Clark and Singer 1982; Singer 1985) and searches of past spacecraft data (Roush et al. 1986; McKay and Nedell 1988). This may be simply due to the presence of a thin layer of aeolian dust covering the surface at the scales of the observations (typically hundreds of kilometers or more). Carbonates are thought to be stable under the present Martian environment (Booth and Kieffer 1978; Gooding 1978; but see also Clark and Van Hart 1981), although there are recent suggestions that ultraviolet light could decompose any carbonate exposed on the surface (Mukhin et al. 1996). However, carbonates have been detected in the meteorites that are thought to have originated on Mars (Gooding, Wentworth and Zalensky 1988; McKay et al. 1996).

McKay and Nedell (1988) have suggested that the sedimentary material in the Valles Marineris canyons could be formed by carbonate deposition under persistent ice covers, as in the Antarctic dry valley lakes (Wharton et al. 1982). The amount of carbonates required is equivalent to ~30 hPa of atmospheric CO_2. This result illustrates that carbonate precipitation could provide enough material to account for the layered deposits in the Valles Marineris (Nedell, Squyers and Andersen 1987). It also suggests that many Martian canyons and playa basins must contain carbonates if more than 1 bar of atmospheric CO_2 is sequestered as carbonates on Mars.

There is direct exobiological importance to the detection of carbonates. In lakes and shallow water environments on Earth, the presence of microorganisms often causes the precipitation of carbonate by the removal of CO_2 from the local environment (see, e.g., Golubic 1973; Wharton et al. 1982). For this reason, microbial mats and other benthic microflora are often encrusted in carbonate deposits. Indeed, carbonate layers in lake sediments are often good sites for microfossils.

Although there is evidence that there were sporadic events of fluvial formation throughout Martian history, it is unlikely that Mars had Earth-like conditions for any geologically significant period of time after 3.5 Gyr ago (Gulick and Baker 1989). The primary argument against an Earth-like climate is the extremely low erosion rates on surfaces of this age and younger (Carr 1996).

5. Origins of life on Mars

A key issue in the search for life on Mars is the possible origin of life. Davis and McKay (1996) considered this by comparing two theories for the origin of life on Earth. There is no consensus theory for the origins of life on Earth, although the standard theory would certainly be that based on the abiotic synthesis of organics first demonstrated by Miller (1953, 1992). For comparing the origin of life on Earth to its possible duplication on Mars, Davis and McKay systematically considered all theories for the origins of life on Earth (Figure 18.4).

At the most fundamental level, theories for the origins of life on Earth can be divided into those that posit that life arose on the early Earth and those that invoke an extraterrestrial origin (the top branch in Figure 18.4). These extraterrestrial theories – known as panspermia – do not remove the origins of

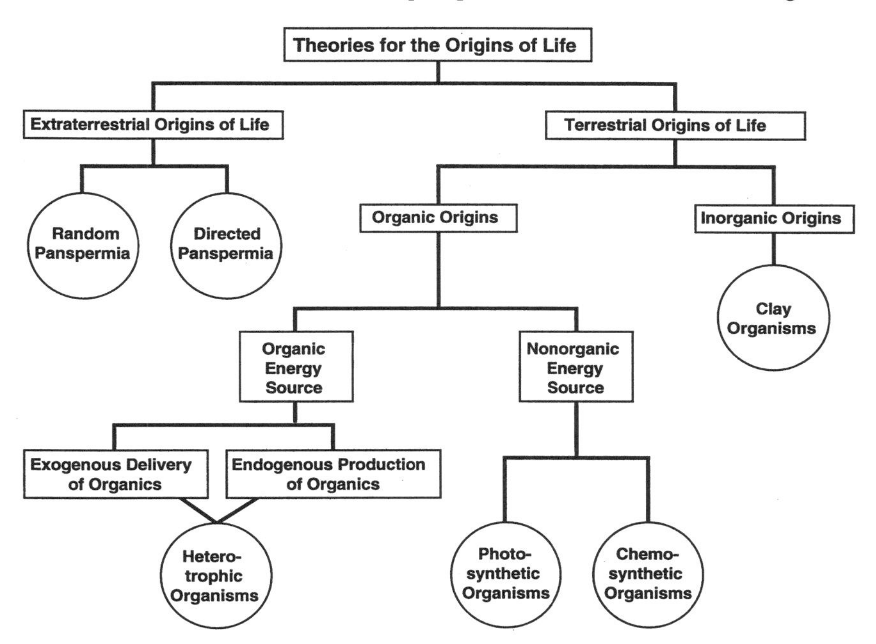

Figure 18.4. Diagram of the theories for the origins of life on Earth. Any of these theories would work on Mars as well as on Earth. (Davis and McKay 1996; with kind permission from Kluwer Academic Publishers.)

life from scientific study but merely suggest that planets are not isolated biologically; this would be consistent with the idea that life on Mars was seeded from the early Earth.

Most theories for the origins of life consider a terrestrial origin. These can be broadly divided into those theories that assume the first life-forms were composed of organic material similar to present life-forms, and those theories that suggest organisms were first composed of clay minerals and that the transition to organic substrates was a subsequent evolutionary development (see discussion in Davis and McKay 1996). Within the area of theories for the organic origins of life, there are differences in the energy source for the first life-forms. Systems based on light, chemical, and organic energy have been postulated. For life composed of organics, and initially consuming these abiotic organics as an energy source, there are two possible sources for the organic matter. It can be produced on the Earth itself or carried to the Earth from elsewhere, presumably by comets, asteroids, and interplanetary dust.

Considering the various theories for the origins of life, Davis and McKay (1996) observed that the one requirement common to all theories is liquid water. The direct evidence of liquid water on early Mars is consistent, therefore, with all theories for the origins of life. Thus there are no conceptual obstacles to postulating that whatever led to the origins of life on Earth also led to the origins of life on Mars – provided that life-sustaining conditions persisted on Mars for an adequate duration of time.

Besides the general requirement for liquid water, the various theories diagramed in Figure 18.4 invoke different specific requirements for life. Both early Earth and early Mars had active volcanism and associated sulfurous hydrothermal regions, subsurface hydrological systems, small ephemeral ponds, large stable bodies of water, meteorite and cometary impacts, anoxic conditions, and airborne particles and bubbles. Indeed, all the major habitats and microenvironments that would have existed on the early Earth would be expected on early Mars as well. The one exception may have been tidal pools, although the solar tide on Mars would still be about 10% of the lunar tide on the contemporary Earth.

6. Meteorites from Mars

There are 12 meteorites on Earth that are thought to have come from Mars. That these meteorites came from a single parent body is evidenced by common ratios of the oxygen isotopes – values distinct from terrestrial, lunar, or asteroidal ratios. The Martian meteorites can be grouped into four classes. Three of these classes contain 11 of the 12 samples and are known by the name of the type specimen; the S (Shergotty), N (Nakhla), and C (Chassigny) class meteorites. The S, N, and C meteorites are relatively young, having formed on Mars between 200 and 1,300 million years ago (e.g., Gooding 1992). Gas inclusions in two of the S type meteorites contain gases similar to the present Martian atmosphere as measured by the Viking landers (e.g., Bogard and Johnson 1983; Pepin 1991; McSween 1994), thus indicating that all of these meteorites came from Mars and that the Martian atmosphere has not changed much during the past 200 Myr.

The fourth class of Martian meteorite is represented by the single specimen known as ALH84001 (Mittlefehldt, 1994). Studies of this meteorite indicate that it formed on Mars about 4.5 Gyr ago (Jagoutz et al. 1994; Treiman 1995) under warm, reducing conditions (Romanek et al. 1994). There are even indications that it contains Martian organic material (as do some SNC meteorites; Wright, Grady and Pillinger 1989) and appears to have experienced aqueous alteration 3.6 billion years ago (McKay et al. 1996). This rock formed during the time period when Mars is thought to have had a warm, wet climate capable of supporting life.

McKay et al. (1996) suggest that ALH84001 contains plausible evidence for past life on Mars. They base their conclusion on four observations. (1) Complex organic material (polycyclic aromatic hydrocarbons) consistent with a biogenic source are present inside ALH84001. (2) Carbonate globules indigenous to the meteorite are enriched in ^{13}C, possibly a result of preferential biological removal of ^{12}C. The isotopic shift is in the range that, on Earth, can result only from biogenic activity. (3) Magnetite and iron sulfide particles are present, and the distribution and shape of these particles are most naturally explained by microbial activity. (4) Microfossils that could be actual traces of microbial life are seen. Subsequent work on ALH84001 has provided alternative explanations for the origin of the PAHs (Becker, Glavin and Bada 1997) and the origin of the magnetite phases (Bradley, Harvey and McSween 1996).

Even allowing for the McKay et al. (1996) interpretation, ALH84001 provides no direct proof of past life on Mars, but it does support the suggestion that conditions suitable for life were present on Mars early in its history. When compared to the SNC meteorites, ALH84001 indicates that Mars experienced a transition from a warm, reducing environment with organic material present to a cold, oxidizing environment in which organic material was unstable.

It is interesting to note that the mineral and fossil evidence for life in the Martian meteorites is very much like the evidence for life on Earth at 3.8 Gyr ago – it is consistent with life but it is not yet compelling. This is unlike the situation for life 3.5 Gyr ago on Earth. In this case, direct stromatolites and fossil algae are found that are clearly biological structures and can be compared directly to modern organisms (Schopf and Packer 1987; Schopf 1993). There is no consensus that we have found evidence for life on Mars. We need better data.

There is a small chance that better data are to be found in ALH84001 and that further analysis will yield a compelling case for life on Mars. It is also possible that we will find more Martian meteorites that date back to the early time of Martian history – perhaps other pieces ejected from the same impact that kicked ALH84001 into space. Even given these possible developments, it is likely that to really answer the question of life on Mars we will have to return to Earth a carefully selected set of rocks for study. If a randomly selected rock, like ALH84001, can contain such interesting evidence of life on Mars, a well-selected rock should be much more informative.

7. Sites for sample return from Mars

More than 50% of the surface of Mars consists of ancient, heavily cratered terrain older than 3.5 Gyr old. Thus, even though there may be no life on Mars today, it may hold the best record of the events 4.0 to 3.5 Gyr ago, events that led to life on Earth and possibly on Mars. Within this terrain, sites of standing water are considered the best locations in the search for Martian fossils.

As discussed previously, the Antarctic dry valleys show that ice-covered lakes could have provided a long-lived habitat for life on Mars as mean temperatures fell below freezing. Of particular interest in this regard are the canyons of the Valles Marineris. The floors of many of the canyons contain deposits of horizontally layered material where individual layers are laterally continuous over tens of kilometers (Figure 18.5). It has been suggested that these materials were deposited in lakes that existed in the canyons early in Martian history (e.g., McCauley 1978; Lucchitta 1981; Lucchitta and Ferguson 1983; Nedell, Squyres and Andersen 1987; Carr 1996). McKay and Nedell (1988) have suggested that a considerable fraction of the possible Martian paleolake sediment could be carbonate material that was precipitated in standing water under conditions of high atmospheric pressures of CO_2.

In addition to the Valles Marineris valleys, there are numerous craters and basins in the ancient terrain that may have been lakes at one time (Masursky, Strobell and Dial 1979; Goldspiel and Squyres 1991; Scott, Rice and Dohm 1991; De Hon 1992; Chapman 1994; Cabrol and Grin 1995; Wharton et al. 1995). Cabrol and Grin (1996) have shown that Gusev crater (15° S, 185° W) is probably filled with delta and bottom sediments from the lake that once filled this crater. This crater is probably the most promising, and practical, landing site for exploration of lake sediments on near term missions.

Aqueous mineral deposits on Mars associated with thermal springs may also offer opportunities for detecting evidence of life, either in mineralized remains (Walter and DesMarais 1993) or in salt crusts (Rothschild 1990).

Figure 18.5. Viking orbiter image of Hebes Chasma (0° S, 75° W). This canyon is a box canyon about 280 km long. The mesa in the center of the canyon shows layered sediments that are believed to have been deposited in standing bodies of water and may be composed of carbonates. (Nedell, Squyres and Andersen 1987; McKay and Nedell 1988.)

Studies have previously been conducted in springs of temperate environments with ambient temperature conditions that probably never existed on Mars. Recently, however, work has begun on springs in permafrost regions providing a better analogue to Mars.

To understand Martian life and to compare it to terrestrial biochemistry will require more than fossilized forms; it will require the analysis of Martian biomolecules in well-preserved specimens. Searching for remnants of an early Martian biota preserved in permafrost for over 3 Gyr provides this possibility. On Earth, there are no continuously frozen locations older than 65 million years. The oldest permafrost that has been sampled for microorganisms is the 3-million-year-old sediments in Siberia (Gilichinsky et al. 1992). Viable bacteria have been recovered from these sediments. Furthermore, it has been reported (but not yet independently confirmed) that bacteria can be preserved in amber for 25 million years (Cano and Borucki 1995). There is even speculation, as yet unpublished, that halophilic bacteria have survived in salt for over 200 million years.

On Mars, it is possible that in the polar regions of the ancient cratered terrain in the Southern Hemisphere there are sediments deep enough to have remained frozen since the end of the late bombardment, 3.8 Gyr ago. However, it is unlikely that organisms held in permafrost for over a billion years are viable, due to the accumulated radiation dose naturally resulting from Martian material, primarily generated by potassium, uranium, and thorium (McKay 1997). Even though organisms in the Martian permafrost would not be viable, they are likely to be organically preserved (Kanavarioti and Mancinelli 1990; Bada and McDonald 1995). The material from these organisms – amino acids, for example – could then be analyzed directly and compared to terrestrial biochemistry.

8. Missions

The sequence of missions to Mars that can comprise a search for life would begin with orbital and airborne coverage to determine geological and mineralogical units that may have been deposited in lakes and hot springs. Such reconnaissance would be followed by surface missions, probably rovers capable of investigating in detail the sites and confirming their identification as past habitats for life. From these surface missions, a selection of rocks and soil samples would be identified for return to Earth in a sample return mission. Well-collected samples from biologically significant sites on Mars would then be available for study in terrestrial laboratories. It is at this stage that the conclusive evidence for life might be found.

9. Conclusions

The search for life on Mars is a search for understanding the generality of the phenomenon of life. Our knowledge of Mars indicates that it is the place in our Solar System most likely to have had, or still have, life outside of Earth. Because of the similarities of the environments on early Earth and early Mars, it is reasonable to expect life on Mars to be comparable to life as we know it. Although life on Mars may represent a separate, independent origin of life, it is possible that life on Mars is phylogenetically related to life on Earth due to the exchange of biotic material in the early history of the Solar System.

If fossils are found, they will confirm that life existed on Mars. If direct organic remains are found, a biochemical comparison of Martian life and terrestrial life can be made. Whether the answer is yes or no, the question of life on Mars will be a crucial first step in extrapolating from life on Earth to life in the universe.

References

Bada, J. L., and McDonald, G. D. 1995. Amino acid racemization on Mars: implications for the preservation of biomolecules from an extinct martian biota. *Icarus* 144:139–143.

Becker, L., Glavin, D. P., and Bada, J. L. 1997. Polycyclic aromatic hydrocarbons (PAHs) in Antarctic Martian meteorites, carbonaceous chondrites, and polar ice. *Geochim. Cosmochim. Acta* 61:475–481.

Biemann, K. 1979. The implications and limitations of the findings of the Viking Organic Analysis Experiment. *J. Mol. Evol.* 14:65–70.

Biemann, K., Oró, J., Toulmin, P., III, Orgel, L. E., Nier, A. O., Anderson, D. M., Simmonds, P. G., Flory, D., Diaz, A. V., Rushneck, D. R., Biller, J. E., and Fleur, A. L. 1977. The search for organic substances and inorganic volatile compounds in the surface of Mars. *J. Geophys. Res.* 82:4651–4658.

Bogard, D. D., and Johnson, P. 1983. Martian gases in an Antarctic meteorite? *Science* 221:651–654.

Booth, M. C., and Kieffer, H. H. 1978. Carbonate formation in Marslike environments. *J. Geophys. Res.* 83:1809–1815.

Bradley, J. P., Harvey, R. P., and McSween, H. Y., Jr. 1996. Magnetite whiskers and platelets in the ALH84001 Martian meteorite: Evidence of vapor phase growth. *Geochim. Cosmochim. Acta* 60:5149–5155.

Cabrol, N. A., and Grin, E. A. 1995. A morphological view on potential niches for exobiology on Mars. *Planet. Space Sci.* 43:179–185.

Cabrol, N. A., and Grin, E. 1996. Ma'adim Vallis revisited through new topographic data: Evidence for an ancient intravalley lake. *Icarus* 123:269–283.

Cano R. J., and Borucki, M. K. 1995. Revival and identification of bacterial spores in 25- to 40-million-year-old Dominican amber. *Science* 268:1060–1064.

Carr, M. H. 1981. *The Surface of Mars.* New Haven: Yale University Press.

Carr, M. H. 1989. Recharge of the early atmosphere of Mars by impact induced release of CO_2. *Icarus* 79:311–327.

Carr, M. H. 1996. *Water on Mars.* New York: Cambridge University Press.

Carr, M. H., and Clow, G. D. 1981. Martian channels and valleys: Their characteristics, distribution, and age. *Icarus* 48:91–117.

Chapman, M. G. 1994. Evidence, age, and thickness of a frozen paleolake in Utopia Planitia, Mars. *Icarus* 109:393–406.

Chyba, C. F., Thomas, P. J., Brookshaw, L., and Sagan, C. 1990. Cometary delivery of organic molecules to the early Earth. *Science* 249:366–373.

Clark, B. C., Baird, A. K., Weldon, R. J., Tsusaki, D. M., Schnabel, L., and Candelaria, M. P. 1982. Chemical composition of martian fines. *J. Geophys. Res.* 87:10059–10067.

Clark, B. C., and Van Hart, D. C. 1981. The salts of Mars. *Icarus* 45:370–387.

Clow, G. D., McKay, C. P., Simmons, G. M., Jr., and Wharton, R. A., Jr. 1988. Climatological observations and predicted sublimation rates at Lake Hoare, Antarctica. *J. Climate* 1:715–728.

Davis, W. L., and McKay, C. P. 1996. Origins of life: a comparison of theories and application to Mars. *Origins Life Evol. Biosphere* 26:61–73.

De Hon, R. A. 1992. Martian lake basins and lacustrine plains. *Earth, Moon, and Planets* 56:95–122.

Fanale, F. P. 1976. Martian volatiles: Their degassing history and geochemical fate. *Icarus* 28:179–202.

Flynn, G. J. 1996. The delivery of organic matter from asteroids and comets to the early surface of Mars. *Earth, Moon, and Planets* 72:469–474.

Friedmann, E. I. 1982. Endolithic microorganisms in the Antarctic cold desert. *Science* 215:1045–1053.

Friedmann, E. I., McKay, C. P., and Nienow, J. A. 1987. The cryptoendolithic microbial environment in the Ross Desert of Antarctica: Nanoclimate data, 1984 to 1986. *Polar Biology* 7:273–287.

Gilichinsky, D. A., Vorbyova, E. A., Erokhina, D. G., Fyordorov-Dayvdov, D. G., and Chaikovskaya, N. R. 1992. Long-term preservation of microbial ecosystems in permafrost. *Adv. Space Res.* 12(4):255–263.

Goldspiel, J. M., and Squyres, S. W. 1991. Ancient aqueous sedimentation on Mars. *Icarus* 89:392–410.

Golubic, S. 1973. The relationship between blue-green algae and carbonate deposits. In *The Biology of Bluegreen Algae*, ed. N. C. Carr and B. A. Whitton, pp. 434–472. Berkeley: University of California Press.

Gooding, J. L. 1978. Chemical weathering on Mars: thermodynamic stabilities of primary minerals (and their alteration products) from mafic igneous rocks. *Icarus* 33:483–518.

Gooding, J. L. 1992. Soil mineralogy and chemistry on Mars: Possible clues from salts and clays in SNC meteorites. *Icarus* 99:28–41.

Gooding, J. L., Wentworth, S. J., and Zolensky, M. E. 1988. Calcium carbonate and sulfate of possible extraterrestrial origin in the EETA 79001 meteorite. *Geochim. Cosmochim. Acta* 52:909–915.

Gulick, V. C., and Baker, V. R. 1989. Fluvial valleys and martian paleoclimates. *Nature* 341:514–516.

Head, J. W., and Solomon, S. C. 1981. Tectonic evolution of the terrestrial planets. *Science* 213:62–76.

Horowitz, N. H. 1986. *To Utopia and Back: The Search for Life in the Solar System.* New York: W. H. Freeman and Co.

Horowitz, N. H., and Hobby, G. L. 1977. Viking on Mars: The carbon assimilation experiments. *J. Geophys. Res.* 82:4659–4662.

Jagoutz, E., Sorowka, A, Vogel, J. D., and Wanke, H. 1994. ALH84001: Alien or progenitor of the SNC family? *Meteoritics* 29:478–479.

Kanavarioti, A., and Mancinelli, R. L. 1990. Could organic matter have been preserved on Mars for 3.5 billion years? *Icarus* 84:196–202.

Kasting, J. F. 1991. CO_2 condensation and the climate of early Mars. *Icarus* 94:1–13.

Kieffer, H. H., Jakosky, B. M., Snyder, C. W., and Matthews, M. S. 1992. *Mars.* Tucson: University of Arizona Press.

Klein, H. P. 1978. The Viking biological experiments on Mars. *Icarus* 34:666–674.

Klein, H. P. 1979. The Viking mission and the search for life on Mars. *Rev. Geophys. and Space Phys.* 17:1655–1662.

Kushner, D. 1981. Extreme environments: Are there any limits to life? In *Comets and the Origin of Life*, ed. C. Ponnamperuma, pp. 241–248. Dordrecht: D. Reidel.

Lederberg, J. 1960. Exobiology: Approaches to life beyond the Earth. *Science* 132:393–400.

Lehninger, A. L. 1975. *Biochemistry.* New York: Worth.

Levin, G. V., and Straat, P. A. 1977. Recent results from the Viking Labeled Release Experiment on Mars. *J. Geophys. Res.* 82:4663–4667.

Levin, G. V., and Straat, P. A. 1981. A search for nonbiological explanation of the Viking labeled release life detection experiment. *Icarus* 45:494–516.

Lucchitta, B. K. 1981. Mars and Earth: Comparison of cold-climate features. *Icarus* 45:264–303.

Lucchitta, B. K., and Ferguson, H. M. 1983. Chryse basin channels: low gradients and ponded flows. *J. Geophys. Res. Suppl.* 88:A553–A586.

McCauley, J. F. 1978. Geologic map of the Coprates Quadrangle of Mars. USGS Miscellaneous Investigation Series Map I–897.

McCord, T. M., Clark, R. N., and Singer, R. B. 1982. Mars: Near-infrared spectral reflectance of surface regions and compositional implications. *J. Geophys. Res.* 87:3021–3032.

McKay, C. P. 1991. Urey prize lecture: Planetary evolution and the origin of life. *Icarus* 91:92–100.

McKay, C. P. 1997. The search for life on Mars. *Origins Life Evol. Biosphere* 27:263–289.

McKay, C. P., Clow, G. A., Wharton, R. A., Jr., and Squyres, S. W. 1985. Thickness of ice on perennially frozen lakes. *Nature* 313:561–562.

McKay, C. P., and Davis, W. L. 1991. Duration of liquid water habitats on early Mars. *Icarus* 90:214–221.

McKay, D. S., Gibson, E. K., Thomas-Keprta, K. L., Vail, H., Romanek, C. S., Clement, S. J., Chillier, X. D. F., Maechling, C. R., and Zare, R. N. 1996. Search for past life on Mars: possible relic biogenic activity in Martian meteorite ALH84001. *Science* 273:924–930.

McKay, C. P., and Nedell, S. S. 1988. Are there carbonate deposits in Valles Marineris, Mars? *Icarus* 73:142–148.

McSween, H. Y. 1994. What have we learned about Mars from SNC meteorites? *Meteoritics* 29:757–779.

Masursky, H., Boyce, J. M., Dial, A. L., Schaber G. G., and Strobel, M. E. 1977. Classification and time of formation of martian channels based on Viking data. *J. Geophys. Res.* 82:4016–4038.

Masursky, H., Strobell, M. E., and Dial, A. L. 1979. Martian channels and the search for extraterrestrial life. *J. Mol. Evol.* 14:39–55.

Mazur, P. 1980. Limits to life at low temperatures and at reduced water contents and water activities. *Origins of Life* 10:137–159.

Melosh, H. J. 1988. The rocky road to panspermia. *Nature* 332:687–688.

Miller, S. L. 1953. A production of amino acids under possible primitive Earth conditions. *Science* 117:528–529.

Miller, S. L. 1992. The prebiotic synthesis of organic compounds as a step toward the origin of life. In *Major Events in the History of Life,* ed. J. W. Schopf, pp. 1–28. Boston: Jones and Bartlett Publishers.

Mittlefehldt, D. W. 1994. ALH 84001, a cumulate orthopyroxenite member of the martian meteorite clan. *Meteoritics* 29:2114–2121.

Mukhin, A. P., Koscheev, Yu., Dikov, P., Huth, J., and Wanke, H. 1996. Experimental simulations of the decomposition of carbonates and sulphates on Mars. *Nature* 379:141–143.

Nedell, S. S., Squyres, S. W., and Andersen, D. W. 1987. Origin and evolution of the layered deposits in the Valles Marineris, Mars. *Icarus* 70:409–441.

Nienow, J. A., McKay, C. P., and Friedmann, E. I. 1988. The cryptoendolithic microbial environment in the Ross Desert of Antarctica: Light in the photosynthetically active region. *Microb. Ecol.* 16:271–289.

Oró, J., and Holzer, G. 1979. The photolytic degradation and oxidation of organic compounds under simulated martian conditions. *J. Mol. Evol.* 14:153–160.

Oyama, V. I., and Berdahl, B. J. 1977. The Viking gas exchange experiment results from Chryse and Utopia surface samples. *J. Geophys. Res* 82:4669–4676.

Oyama, V. I., and Berdahl, B. J. 1979. A model of Martian surface chemistry. *J. Mol. Evol.* 14:199–210.

Parker, B. C., Simmons, G. M., Jr., Wharton, R. A., Jr., Seaburg, K. G., and Love, F. G. 1982. Removal of organic and inorganic matter from Antarctic lakes

by aerial escape of blue-green algal mats. *J. Phycol.* 18:72–78.
Pepin, R. O. 1991. On the origin and early evolution of terrestrial planet atmospheres and meteoritic volatiles. *Icarus* 92:2–79.
Pollack, J. B., Kasting, J. F., Richardson, S. M., and Poliakoff, K. 1987. The case for a wet, warm climate on early Mars. *Icarus* 71:203–224.
Romanek, C. S., Grady, M. M., Wright, I. P., Mittlefehldt, D. W., Socki, R. A., Pillinger, C. T., and Gibson, E. K. 1994. Record of fluid–rock interactions on Mars from the meteorite ALH84001. *Nature* 372:655–657.
Rothschild, L. J. 1990. Earth analogs for martian life. Microbes in evaporites, a new model system for life on Mars. *Icarus* 88:246–260.
Roush, T. L., Blaney, D., McCord, T. B., and Singer, R. B. 1986. Carbonates on Mars: Searching the Mariner 6 and 7 IRS measurements. *Lunar Planet. Sci.* XVII:732–733.
Schaefer, M. W. 1993. Volcanic recycling of carbonates on Mars. *Geophys. Res. Lett.* 20:827–830.
Schopf, J. W. 1993. Microfossils of the early Archean apex chert: New evidence for the antiquity of life. *Science* 260:640–646.
Schopf, J. W., and Packer, B. M. 1987. Early Archean (3.3-billion to 3.5-billion-year-old) microfossils from Warrawoona Group, Australia. *Science* 237:70–73.
Scott, D. H., Rice, J. W., Jr., and Dohm, J. M. 1991. Martian paleolakes and waterways: Exobiological implications. *Origins Life Evol. Biosphere* 21:189–198.
Singer, R. B. 1985. Spectroscopic observations of Mars. *Adv. Space Res.* 5:59–68.
Sleep, N. H. 1994. Martian plate tectonics. *J. Geophys. Res.* 99:5639–5655.
Solomon, S. C. 1978. On volcanism and thermal tectonics on one-plate planets. *Geophys. Res. Lett.* 5:461–464.
Stoker, C. R., Gooding, J. L., Roush, T., Banin, A., Burt, D., Clark, B. C., Flynn, G., and Gwynne, O. 1993. The physical and chemical properties of and resource potential of Martian surface soils. In *Resources of Near-Earth Space*, ed. J. Lewis, M. S. Matthews, and M. L. Guerrieri. Tucson: University of Arizona Press.
Toulmin, P., III, Baird, A. K., Clark, B. C., Keil, K., and Rose, H. J., Jr. 1976. Preliminary results from the Viking X-ray fluorescence experiment: The first sample from Chryse Planitia, Mars. *Science* 194:81–84.
Treiman, A. H. 1995. S ≠ NC: Multiple source areas for Martian meteorites. *J. Geophys. Res.* 100:5329–5340.
Walter, M. R., and DesMarais, D. J. 1993. Preservation of biological information in thermal springs deposits: Developing a strategy for the search for fossil life on Mars. *Icarus* 101:129–143.
Wharton, R. A., Jr., Crosby, J. M., McKay, C. P., and Rice, J. W., Jr. 1995. Paleolakes on Mars. *J. Paleolimnology* 13:267–283.
Wharton, R. A., Jr., Parker, B. C., Simmons, G. M., Jr., Seaburg, K. G., and Love, F. G. 1982. Biogenic calcite structures forming in Lake Fryxell, Antarctica. *Nature* 295:403–405.
Woese, C. R. 1987. Bacterial evolution. *Microbiol. Rev.* 51:221–271.
Wright, I. P., Grady, M. M., and Pillinger, C. T. 1989. Organic materials in a martian meteorite. *Nature* 340:220–222.
Zent, A. P., and McKay, C. P. 1994. The chemical reactivity of the martian soil and implications for future missions. *Icarus* 108:146–157.

Conclusion

ANDRÉ BRACK

The period of time during which life emerged on the primitive Earth has been narrowed thanks to the geological record and to a better understanding of the primitive terrestrial atmosphere and hydrosphere. Terrestrial life probably appeared between 4.0 and 3.8 billion years ago. The inventory of organic molecules of prebiotic interest that were likely to be present at that time is satisfactory except for the RNA sugar component, ribose. Delivery of extraterrestrial organic molecules is of special interest since it is a process that still occurs today. The existence of such an import process on the early Earth required only an atmosphere to decelerate the particles. Such an atmosphere existed 3.8 billion years ago, as illustrated by, for instance, the Greenland sediments.

Among these organic molecules processed by liquid water, some began to transfer their molecular information and to evolve by making a few accidental transfer errors. For the sake of simplicity – for chemists, prebiotic chemistry was simple – one is tempted to assume that the chemical information and the transfer machinery were provided by the same molecules. Self-replicating RNA molecules fulfill these requirements. However, RNA molecules are not really simple, and whether they started life on Earth is still questionable. Self-replicating RNAs are, by definition, autocatalytic molecules that transfer their linear sequence information by an accurate residue-by-residue copying process, thanks to the template chemistry of complementary strands. Other examples of short autocatalytic molecules capable of making more of themselves by themselves are known in chemistry. We can designate such a molecule with the sequence A–B. Although the level of linear sequence information is low, it may potentially lead to longer A–B–A–B ... informative template molecules capable of selectively aligning A and B monomers and catalyzing the formation of new A–B–A–B ... molecules at the expense of other sequences. The autocatalytic molecules will then become self-replicating, that is, adding information transfer to the function of chemical catalysis.

They may also be able to evolve by making accidental errors in the copying process. This field of research is now open to organic chemists and new results are eagerly anticipated.

Homochirality was probably a prerequisite for life. The enantiomeric excess of certain amino acids found in the Murchison meteorite may help to understand the emergence of a primitive homochiral life. A meteoritic preference for one enantiomer would push the problem of the origin of biological chirality out into the cosmos. As already mentioned, autocatalytic systems may lead to homochirality under certain conditions. It is likely that primitive living systems developed optical activity as a result of molecular selection processes that formed structures of higher order and complexity with a concomitant gain of stability. The increased stability helped the systems to resist spontaneous racemization, the half-time of which lies at about 10^5–10^6 years at ambient temperature for amino acids. Racemization was not the only danger. Other calamities were also occurring on the primitive Earth, such as meteoritic and cometary impacts. Perhaps two mirror-image populations of primitive living systems developed and competed at the beginning, until one of them won the competition. Most probably, there was never any racemic life using both L and D monomeric units in the chains. In the context of life's origin, the most relevant aspect of chirality is homochirality. The choice of the sign L or D may be just due to a chance event.

And importing life itself? The idea that life arrived on Earth from another star system is known as panspermia and was originally put forward by Richter and Arrhenius. Recent discoveries have given new support to the idea of panspermia. These are (1) the identification of meteorites of lunar and probably also of Martian origin; (2) the probability that small particles reached escape velocities by the impact of large meteorites on a planet; (3) the ability of bacterial spores to survive the shock waves of a simulated meteorite impact; (4) the high UV resistance of microorganisms at the low temperature of deep space; and (5) the high survival of bacterial spores over extended periods in space provided they are shielded against the intense solar ultraviolet radiation or are coated by a mantle of absorbing material that attenuates the solar UV radiation. Even if panspermia is possible, it will not tell us how life started in another star system.

Chemists are quite confident that they will be able to reconstruct in a test tube living systems fulfilling the requirements of self-replication and mutation. Such chemical living systems will probably be simpler than a cell. As a consequence, one must accept the reductionistic idea that it is possible to make life by a combination of simple chemical reactions. Life has always been associated with complexity. Reducing the complexity of life to the subcellular level, down to single molecules, is a disturbing concept, clashing with the philosophical and religious background attached to the concept of life. Using the term "living systems" instead of "life" may be a way of escaping from the philosophical background attached to life. In the realm of science, that is, of fact-based logical arguments, the assumption of a divine creation of life is not relevant. Such an assumption is the concern of personal belief and faith, which, in essence, do not have to be proven by facts. Accordingly, the proponents of a divine creation of life should not seek to extend their range of influence to science.

The clues that may help chemists to understand the emergence of life on Earth about 4 billion years ago have been erased by plate tectonics, the permanent presence of liquid water, the unshielded solar ultraviolet radiation, and life itself. However, the early histories of Earth and Mars show similarities. Liquid water was probably once stable on the surface of Mars, attesting to the presence of an atmosphere capable of decelerating micrometeorites. Therefore, primitive life might also have developed on Mars. Liquid water seems to have disappeared very early from the Martian surface, about 3.8 billion years ago, before life, if any, became very active. In addition, Mars probably never experienced plate tectonics. Mars may store some well-preserved clues of a still-hypothetical primitive bacterial life. The recent analysis of the Martian meteorite ALH84001 yielded some evidence, but no real proof, of ancient microbial life on Mars. On July 6, 1997, the small rover Sojourner tread upon the Martian surface. Two days before, the Pathfinder probe landed in Ares Vallis, a valley identified after Viking 1 as a mouth of an ancient river. This nice technological achievement, followed in real time by hundreds of millions of people on the Internet, ended a dramatic series of unsuccessful missions to Mars, including the two Phobos missions, Mars Observer and Mars 96. The pictures taken by Pathfinder confirm that early Mars was once covered by large amounts of water. The 11.5-kg six-wheel rover Sojourner moved at 1cm/s. It took pictures of the surrounding soils and rocks and analyzed their elemental composition. It was equipped with an alpha–proton–X-ray spectrometer for the analysis of the light elements (up to nickel; with hydrogen and helium excepted) present in selected soil and rock samples. The samples were bombarded with alpha particles from a radioisotope source. The instrument measured energy and intensity of elastically backscattered alpha particles, protons from nuclear reactions, and characteristic X-rays. The percentages of C, O, Na, Mg, Al, Si, P, S, Cl, P, Ca, Ti, Cr, Mn, Fe, and Ni are of special interest for mineral characterization. The analyzed rocks are similar in composition to terrestrial andesites and close to the mean composition of Earth's crust (Rieder et al. 1997).

American, Japanese, and European space agencies have planned a very intensive exploration of Mars. The NASA program includes Mars Global

Surveyor, which arrived at the Mars vicinity in September 1997. In addition to the high-resolution imaging of the surface, the scientific objectives include the study of both the surface and the atmosphere. Mars Surveyor-98 Orbiter will monitor the daily weather and atmospheric conditions (wind, temperature profiles, water vapor, dust), and the lander will record local meteorological conditions near the Martian South Pole and analyze samples of the polar deposits for volatiles (water, carbon dioxide). As part of the Mars Surveyor Program, NASA plans to send one or two spacecraft to Mars during both 2001 and 2003 launch windows to support the return of samples using the 2005 opportunity and to bring them to Earth in 2008. Mars Surveyor 2001 Orbiter will conduct mineralogical mapping of the entire planet. The corresponding lander will deliver a small rover to collect rock, soil, and gas samples for later return to Earth. Exobiology interests are included, especially in the rover mission objective to characterize in situ sites in the ancient highlands where the environmental conditions may have been favorable for the preservation of evidence of possible prebiotic or biotic processes. The NASA Mars 2003 Mission is expected to be a reconnaissance mission to prepare for the Mars sample return mission in 2005. Japanese Planet-B Mission is planned to be launched in early August 1998. Its primary goal is to study the Martian aeronomy, with particular emphasis on the interaction of the upper atmosphere with the solar wind. The European Space Agency (ESA) has initiated an intensive feasibility study of a Mars Express Mission with the objective of a launch in June 2003. It is anticipated that the orbiter will include instruments to study the surface mineralogy and the atmosphere. If the launch capability permits, landers will be designed to run exobiology and seismology studies of the subsurface. All these missions will contribute to a more precise picture of Mars. Some of them include strategies and potential techniques for detecting extinct indigenous life, either fossils or biomarkers, with experiments devoted to the search for a past atmosphere, past and present liquid water, carbonates, organics, and fossilized microorganisms. Finding extinct life on Mars would be a major discovery. It would help us to understand the origin of terrestrial life but would also demonstrate that life is not restricted to the planet Earth.

Within the Solar System, Europa (one of the four large satellites of Jupiter) may have an ocean of liquid water beneath its icy crust, as suggested by data and theory. Europa undergoes tidal heating similar to that driving the volcanism on the Jovian moon Io, and this heating may have been sufficient to melt the ice and maintain a liquid ocean. If submarine volcanism exists on Europa, the question arises of whether such activity could support life as on volcano–hydrothermal sites on the Earth's sea floor. New planets have been recently discovered beyond the Solar System. On October 6, 1995, Michel

Mayor and Didier Queloz of Geneva Observatory announced the discovery of an extrasolar planet orbiting around an 8-billion-year old-star called 51 Pegasus, 42 light years away within the Milky Way. The suspected planet takes just four days to orbit 51 Pegasus. It has a surface temperature around 1,000 °C and a mass about one-half the mass of Jupiter. One year later, seven other extrasolar planets were identified. Among them, 47 Ursa Major has a surface temperature estimated to be around that of Mars (-90 to -20 °C) and 70 Virginis has a surface temperature estimated around 70–160 °C. The latter is the first known extrasolar planet whose temperature might allow the presence of liquid water and perhaps some form of life. How to detect such hypothetical life will probably be a formidable challenge for astronomers and radioastronomers in the next century. The detection of water and oxygen in the atmosphere will be a good indication but will not constitute a proof. The detection of an electromagnetic signal would be more convincing but probably more problematic. The 28 papers dedicated to SETI (Search for Extraterrestrial Intelligence) presented at the 5th International Conference on Bioastronomy in 1996 at Capri testify that the scientific community is very active in the field despite severe credit restrictions. Such discoveries would overjoy chemists, who suspect that life is universal because organic molecules and water are universal, but who need experimental proofs to claim it.

References

Reider, R., Economou, T., Wänke, H., Turkevich, A., Crisp, J., Brückner, J., Dreibus, G., and McSween, H. Y., Jr. 1997. The chemical composition of Martian soil and rocks returned by the mobile alpha proton X-ray spectrometer: preliminary results from the X-ray mode. *Science* 278:1771–1774.

Cosmovici, C. B., Bowyer, S., and Werthimer, D. (eds.). 1997. *Astronomical and Biochemical Origin and the Search for Life in the Universe.* Bologna: Editrice Compositori.

Index

818